Protein Phosphatases and Stress Management in Plants

Girdhar K. Pandey
Editor

Protein Phosphatases and Stress Management in Plants

Functional Genomic Perspective

Springer

Editor
Girdhar K. Pandey
Department of Plant Molecular Biology
University of Delhi South Campus
New Delhi, India

ISBN 978-3-030-48732-4 ISBN 978-3-030-48733-1 (eBook)
https://doi.org/10.1007/978-3-030-48733-1

This Springer imprint is published by the registered company Springer Nature Switzerland AG
The registered company address is: Gewerbestrasse 11, 6330 Cham, Switzerland

Preface

The famous Sun temple of Konark in India represents the chariot of the Sun God (Surya) that is pulled by seven horses. This beautiful structure carved out of stone is a good example of management where the seven horses must be controlled for the proper movement of the chariot. The charioteer controls the movement of the chariot through a "leash" that is tied to each of the horses. Nature too has its own set of controls that act as leashes to modulate the nuances of cellular functions. This book is dedicated to the understanding of one such group of controls in plants, which function as highly effective components of the signal transduction mechanism.

In today's scenario, human health, food security, natural resources like water, air, and soil (pollution as well as conservation) are the pertinent issues affecting mankind. Though food security is a major challenge at all levels, it is particularly important for the plant biologists, especially in the context of climate change and scarcity of arable land. Environmental stresses (both abiotic and biotic) and extensive anthropogenic activities have contributed significantly to the drastic decline in crop productivity over the years. Researchers have made significant agronomic advances by crop improvement through extensive breeding and to some extent by genetic manipulation. The genomic and post-genomic eras have witnessed the use of advanced tools of gene manipulation and genetic engineering, to target a large number of genes for imparting stress tolerance in plants. Plant biologists are trying to understand and explore the structural and functional relationship of gene(s) and gene families, in the context of different physiological and developmental aspects of the plant life cycle. However, the efforts to develop stress-tolerant crop varieties have not been much successful. In lieu of this, we still require a detailed understanding of the mechanisms underlying stress perception, transduction as molecular signals, and finally translation into defense or adaptive responses by plants.

Signaling pathways act as "nodes and hubs," regulating myriad stimuli including stress signals. These nodes and hubs also regulate the cross talk and disparate channeling of stress signals and hence fine-tune the stimulus–response-coupling process with the generation of adaptive responses. These mechanisms maintain a homeostatic balance in the living systems in concomitance with stress perception and the ensuing responses. There are several controls or molecular switches which turn these biological processes and signaling pathways "on and off." Posttranslational modifications of a protein can act as one of the key molecular switches in an organism. Protein phosphorylation is one such covalent modification that regulates signaling cascades through the activation or deactivation of the components involved in maintenance of a homeostatic state. This process of reversible regulation is carried out through two groups of enzymes, kinases, and phosphatases. Kinases phosphorylate target proteins by adding a phosphate to the hydroxyl group on amino acid residues, while phosphatases dephosphorylate a phosphorylated protein, thereby forming a cellular switch to initiate or terminate diverse cellular processes. Protein phosphosites mainly include nine amino acids: tyrosine (Tyr), serine (Ser), threonine (Thr), cysteine (Cys), arginine (Arg), lysine (Lys), aspartate (Asp), glutamate (Glu), and histidine (His). Ser-, Thr-, and Tyr- are the most commonly phosphorylated residues with profound implications in the regulatory pathways in eukaryotic cells.

Protein kinases (PK) are known to be activated by primary stress response pathways (such as in Ca^{2+} signaling) which subsequently elicit either a short-term quick response like closing and opening of the stomata or a long-term response like activation of transcription factors. However, if this response is not switched off and persists, the plant would be allocating its resources toward adaptive responses even in the absence of stress. Thus, the resource allocation of the plant will tip more toward protecting itself rather than a holistic growth response even after the stress has been mitigated. This aspect of the regulation of stress response pathways dictates the need for a discussion on protein phosphatases (PPs). These PPs are important components which control many regulatory circuits in living organisms by modulating the conformation, activity, localization, and stability of substrate proteins. PPs are categorized depending on their biochemical properties. The three main families of PP are Ser-/Thr-, Asp-, and Tyr-based. The Ser/Thr-based PPs are divided into the phosphoprotein phosphatase (PPP) family and the metallo-dependent PPs (PPM/PP2C) family. The Asp-based PPs are divided into the FCP-like/CPL and HAD families. The Tyr-based PPs are further divided into the protein Tyr phosphatases (PTPs) and dual specificity phosphatases (DsPTPs).

The phosphorylation–dephosphorylation-regulated "cellular switch" that monitors plant physiology, growth, and development has immense potential in crop systems. Much of the information pertaining to this regulatory mechanism in plants is still in the nascent stages, coming largely from model plants, Arabidopsis and rice. The use of genetic and biochemical approaches aided by "omic" approaches are currently enabling the unraveling of key components involved in the regulation of stress tolerance. These key components of phosphorylation–dephosphorylation such as kinases or phosphatases could be exploited to develop crop varieties better equipped to handle adverse environmental conditions and hence lead to enhancement of agricultural productivity.

This book entitled *Protein phosphatases and stress management in plants: Functional genomic perspective* comprises of 17 chapters contributed by several well-known plant biologists working in the field of Protein phosphatases and stress management with a special emphasis on Functional Genomic aspect. This book elaborates on the state-of-art scientific advances in the field of "signaling under stress conditions," which will formulate a holistic understanding on the subject.

The first chapter describes the role of ancient chloroplast and mitochondrial PPs, the Shewanella-like PPs (SLP1 and SLP2), of bacterial origin. They are remarkably conserved in plants, suggesting that they play fundamental roles in chloroplast and mitochondrial biology. The detailed functional role of SLP1 and SLP2 is being investigated in several plant species, especially in the physiological and functional context.

Chapters 2 and 3 elaborate the role of the purple acid phosphatases (PAPs). The PAPs are involved in phosphate (Pi) homoeostasis and several other diverse functions such as regulation of seed traits, root development, osmotic, oxidative and salt stress tolerance in plants. Chapter 4 discusses the PP2A class of protein phosphatases that are composed of three subunits (catalytic "C"; scaffolding "A" and regulatory "R"). The Arabidopsis genome encodes multiple isoforms of these subunits (3-As, 5-Cs, and 17-Bs subunits), and different combinations of these subunits are expected to give rise to almost 255 different PP2A holoenzymes. Though PP2A are mostly implicated in the regulation of developmental pathways, but in this chapter, authors discuss their possible role in responses to salinity stress.

PP2Cs belonging to PPM family are the largest class of PPs in plants. Arabidopsis and rice encode more than 76 and 90 PP2Cs, respectively, that are classified into 10 or more subgroups (A–K) with diverse functions. The best studied PPs include the A-subclade of PP2C, which negatively regulate the ABA signaling pathway. Chapters 5 and 6 discuss the role of PP2Cs in regulating diverse abiotic stresses and ABA signaling. Chapter 7 presents a detailed insight into the role of PP2A and PP2C families in sugar as well as hormone signaling and consequently in the maintenance of balance between stress and growth in plants.

The opening and closing of the stomata are governed by guard cell dynamics which control their turgid state. Guard cell signaling is one of the most well-studied physiological processes in plants, wherein an intricate interplay of PKs and PPs is at work. Chapter 8 gives a detailed account of several PPs such as PP1s, PP2As, and PP2Cs in the regulation of the stomatal movements. Chapter 9 presents an insight into the involvement of several PPs in the regulation of plant responses under salt stress in different species. Chapter 10 discusses the role of phosphatases and different phosphatase gene families involved in stress signaling pathways, involved in the regulation of stress tolerance.

Because crop productivity is directly dependent on the soil fertility and nutrient content, mineral nutrient deficiency in plants is an important area which demands greater attention from plant biologists. Deficiency of the major- and micro-nutrients in the soil leads to a drastic penalty in growth and development, thus affecting the crop yield and productivity. A large number of fertilizers are added to different crops to enhance the yield and productivity. Among the fertilizers, NPK (nitrogen, phosphorus, and potassium) is the most commonly preferred combination. Chapter 11 presents the role of various PPs in the regulation of responses to the K^+ deficiency and the signaling therein. Moreover, this chapter also discusses the importance of Ca^{2+}-mediated CBL-CIPK (a homologue of the animal PP2B class phosphatase calcineurin) and PP2C modules in the regulation of K^+ transport, ABA and abiotic stress signaling pathways in the model plant Arabidopsis. Chapter 12 elaborates the role of PPs in nitrogen response and nitrogen-use efficiency (NUE) in different crops. Several PPs such as PP2Cs, PP2As, and others, identified in different N uptake, assimilation, and remobilization regulatory pathways, are emerging as important candidate genes for genetic manipulation. Chapter 13 elaborates on the genome-wide identification of PPs from major cereals and small grain crops, their structural organization as well as their involvement in diverse stress regulatory pathways. This chapter lays emphasis on PPs from the perspective of crop plants.

Besides regulating the cytoskeletal network comprising of microtubule and actin filaments, PPs act as important determinants of cell cycle progression and thus regulators of cell division. Chapter 14 presents the roles of various PPs in mitotic processes and cytoskeleton regulation. In addition to abiotic stresses, biotic challenges posed by pests and pathogens affect crop productivity drastically. To cope with biotic stresses, plants have different layers of defense systems such as pattern-triggered immunity (PTI) and effector triggered immunity (ETI). However, parallelly, forces of natural selection aid pathogens in evolving a more effective arsenal of defense mechanisms. Successful invasion by pathogens and the consequent defense responses in plants solely depends on the host–pathogen interactions, which comprise of a large number of components that trigger several signaling pathways. Chapter 15 provides an extensive account of involvement of PPs in host–pathogen interactions in both host and pathogen systems.

In animals, both Ser/Thr and Tyr phosphorylation–dephosphorylation regulate a large number of physiological and developmental processes. However, till date, no receptor tyrosine kinase (RTK) has been identified in plants, though several reports suggest Tyr phosphorylation by non-canonical Tyr kinases. Tyr dephosphorylation by Tyr-specific phosphatases is also not much explored. Based on the genome sequence analysis of several plant species, not many PTPs have been identified. Chapters 16 and 17 present an account of dual specificity phosphatases (DSPs; which act on both phosphorylated Ser/Thr and Tyr) and Tyr-specific phosphatases (PTP) in different plants. Their involvement in the regulation of different metabolic (starch degradation), physiological (biotic and abiotic stresses), and various developmental processes is discussed.

Plants need to acquire a large number of reprograming in their biological processes that enable them to withstand the changing nature of their environment. Based on the extensive work done in the field of stress perception and signal transduction, it is evident that research in the area of signal transduction is a key determinant in the implementation of enhanced stress tolerance in plants. My best efforts were rendered toward the inclusion of all aspects of PPs and their role in stress management in this book. However, some aspects still await elaboration due to space constraint and other limitations. Regardless of this, I firmly believe that this book will be able to serve its purpose for students, researchers, and academicians seeking an understanding of stress-mediated signaling in the context of PPs.

I express my gratitude to all the authors whose contributions have made it possible to bring vast information on one platform. I also express my sincere thanks to

Dr. Malathi Bheri, Dr. Sibaji K. Sanyal, and Dr. Deepti M. Nambiar (Department of Plant Molecular Biology, University of Delhi South Campus, New Delhi, India) for their critical and constructive suggestions. I am also thankful to Delhi University, University Grant Commission, Department of Biotechnology, Science and Engineering Research Board, and Council of Scientific and Industrial Research, India, for supporting the research in my laboratory.

New Delhi, India Girdhar K. Pandey

Contents

1 **SLP1 and SLP2: Ancient Chloroplast and Mitochondrial Protein Phosphatases** 1
Jayde J. Johnson, Chris White-Gloria, Ryan Toth, Anne-Marie Labandera, R. Glen Uhrig, and Greg B. Moorhead

2 **Phosphoprotein Phosphatase Function of Secreted Purple Acid Phosphatases** 11
Mina Ghahremani and William C. Plaxton

3 **Purple Acid Phosphatases (PAPs): Molecular Regulation and Diverse Physiological Roles in Plants** 29
Poonam Mehra and Jitender Giri

4 **Role of Serine/Threonine Phosphatase PP2A Class and Its Regulators in Salinity Stress Tolerance in Plants** 53
Srishti Chawla, Deeksha Marothia, and Pratap Kumar Pati

5 **Type 2C Protein Phosphatases in Plant Signaling Pathways under Abiotic Stress** 67
Nguyen Nguyen Chuong, Duong Hoang Trong Nghia, Van-Anh Le Thi, Lam-Son Phan Tran, Xuan Lan Thi Hoang, and Nguyen Phuong Thao

6 **Plant Protein Phosphatase 2C: Critical Negative Regulator of ABA Signaling** 83
Lokesh K. Saini, Nidhi Singh, and Girdhar K. Pandey

7 **Protein Phosphatases at the Interface of Sugar and Hormone Signaling Pathways to Balance Growth and Stress Responses in Plants** 103
Harshita B. Saksena, Dhriti Singh, Manvi Sharma, Muhammed Jamsheer K., Sunita Jindal, Mohan Sharma, Archna Tiwari, Prakhar, Sanjay Singh Rawat, and Ashverya Laxmi

8 Protein Phosphatases in Guard Cells: Key Role in Stomatal Closure and Opening 125
Shashibhushan Gahir, Vaidya Sunitha, Pulimamidi Bharath, and Agepati S. Raghavendra

9 Deciphering the Roles of Protein Phosphatases in the Regulation of Salt-Induced Signaling Responses in Plants 149
Aditya Banerjee and Aryadeep Roychoudhury

10 Phosphatases: The Critical Regulator of Abiotic Stress Tolerance in Plants 163
Sripati Abhiram Sahoo, Rishiraj Raghuvanshi, Ashish Kumar Srivastava, and Penna Suprasanna

11 Role of Protein Phosphatases in Signaling, Potassium Transport, and Abiotic Stress Responses 203
Sibaji K. Sanyal, G. Rajasheker, P. B. Kavi Kishor, S. Anil Kumar, P. Hima Kumari, K. V. Saritha, P. Rathnagiri, and Girdhar K. Pandey

12 Protein Phosphatases in N Response and NUE in Crops 233
Supriya Kumari and Nandula Raghuram

13 Protein Phosphatases of Cereals and Millets: Identification, Structural Organization, and Their Involvement in the Regulation of Abiotic Stresses 245
Roshan Kumar Singh, Ashish Prasad, and Manoj Prasad

14 Interplay of Protein Phosphatases with Cytoskeleton Signaling in Response to Stress Factors in Plants 261
Dariia O. Samofalova, Pavlo A. Karpov, Alexey V. Raevsky, and Yaroslav B. Blume

15 Protein Phosphatase Mediated Responses in Plant Host-Pathogen Interactions 289
Anjan Barman and Suvendra Kumar Ray

16 Role of Dual Specificity Phosphatase in Stress and Starch Metabolism 331
Kanwaljeet Kaur, Manas Kumar Tripathy, and Girdhar K. Pandey

17 Protein Tyrosine Phosphatases: Implications in the Regulation of Stress Responses in Plants 353
Malathi Bheri and Girdhar K. Pandey

Index 377

Author's Biography

Girdhar K. Pandey received his B.Sc. (Hon.) in Biochemistry from Delhi University in 1992 and M.Sc. in Biotechnology in the year 1994 from Banaras Hindu University (BHU). Subsequently, he joined PhD in the School of Life Sciences, Jawaharlal Nehru University (JNU), and worked in the field of calcium signal transduction under abiotic stresses in plants. He was awarded the PhD degree in the year 1999 and then pursued postdoctoral career at Department of Plant and Microbial Biology, University of California at Berkeley in the year 2000. There, he extended his work in the field of calcium-mediated signaling in Arabidopsis by studying CBL-CIPKs, phosphatases, channels/transporters, and transcription factors involved in abiotic stresses. Currently, he is working as Professor in the Department of Plant Molecular Biology, Delhi University South Campus.

Dr. Pandey's research interests involve detail mechanistic interplay of signal transduction networks in plants under mineral nutrient deficiency (mostly potassium, calcium, and nitrate) and abiotic stresses such as drought, salinity, and oxidative stresses induced by heavy metals. His laboratory is working on the coding and decoding of mineral nutrient deficiency and abiotic stress signals by studying several signaling components such as calcium sensors such as calcineurin B-like (CBL) and CBL-interacting protein kinases (CIPK), phosphatases (mainly PP2C and DSP), transcription factors (AP2-domain containing or ERF, WRKY), transporters and channels proteins (potassium and calcium channels/transporters) in both Arabidopsis and rice. The long-term goal of his research group is to establish the mechanistic interplay and cross talk of mineral nutrient-deficient conditions and different abiotic stress signaling cascades in Arabidopsis and rice model system by using the advance tools of bioinformatics, genetics, cell biology, biochemistry, and physiology with greater emphasis on functional genomics approaches.

See Dr. Pandey's web page for further information about his lab and research work: https://sites.google.com/site/gkplab/home; http://www.dpmb.ac.in/index.php?page=girdhar-pandey

Chapter 1
SLP1 and SLP2: Ancient Chloroplast and Mitochondrial Protein Phosphatases

Jayde J. Johnson, Chris White-Gloria, Ryan Toth, Anne-Marie Labandera, R. Glen Uhrig, and Greg B. Moorhead

1.1 Introduction

The covalent modification of proteins is now regarded as a common post-translational mechanism to regulate protein function in all organisms. Phosphorylation was the first protein covalent modification to be discovered and has its origins in the history of glycogen metabolism and signal transduction research (Brautigan and Shenolikar 2018). Up to ten different amino acids that occur in proteins can be phosphorylated, with serine, threonine, and tyrosine being the most common. Although varying slightly across organisms and cellular conditions, a typical phospho-proteome is about 86% phospho-serine, 12% phospho-threonine, and 2% phospho-tyrosine (Sharma et al. 2014; van Wijk et al. 2014; White-Gloria et al. 2018). The recent development of monoclonal antibodies that specifically recognize phospho-histidine has uncovered roles for this modification in eukaryotes (Adam and Hunter 2018). The development of mass spectrometry technologies related to phosphoproteomics, especially quantitative mass spectrometry, has established protein phosphorylation as the most common covalent modification in all organisms explored, including a variety of plant species (Sharma et al. 2014; van Wijk et al. 2014; White-Gloria et al. 2018).

Protein phosphorylation is not just a cytosolic and nuclear phenomenon; new mass spectrometry data have also established protein phosphorylation as a common event in chloroplasts (White-Gloria et al. 2018; Baginsky and Gruissem 2009;

J. J. Johnson · C. White-Gloria · R. Toth · G. B. Moorhead (✉)
Department of Biological Sciences, University of Calgary, Calgary, AB, Canada
e-mail: moorhead@ucalgary.ca

A.-M. Labandera
Plant Proteolysis and Signaling Laboratory, School of Biosciences, University of Birmingham, Birmingham, UK

R. G. Uhrig
Department of Biological Sciences, University of Alberta, Edmonton, AB, Canada

G. K. Pandey (ed.), *Protein Phosphatases and Stress Management in Plants*,
https://doi.org/10.1007/978-3-030-48733-1_1

Reiland et al. 2009; Richter et al. 2016; Pagliarini and Dixon 2006), mitochondria (Baginsky and Gruissem 2009; Pagliarini and Dixon 2006; Bykova et al. 2003; Grimsrud et al. 2012), and peroxisomes (Oeljeklaus et al. 2016; Kataya et al. 2019). Uncovering the abundant cache of protein phosphorylation events in eukaryotes is no surprise given the enormous size of the protein kinase and phosphatase gene families. For instance, human and Arabidopsis genomes encode ~518 and ~942 protein kinases and ~189 and ~150 protein phosphatase catalytic subunits, respectively (Chen et al. 2017; Kerk et al. 2008; Uhrig et al. 2013a). Unlike kinases, the number of protein phosphatases is in fact less than the above-quoted figures as several of these do not act on protein substrates but are included in this group based on sequence. Many protein phosphatases, predominantly the PPP family enzymes, have additional regulatory subunits that dictate their function. The association of a variety of unrelated regulatory subunits with a common catalytic subunit balances the apparent disparity in protein kinase and phosphatase numbers. This multitude of regulatory subunits also brings specificity to what are regarded as somewhat promiscuous catalytic subunits (Brautigan and Shenolikar 2018; Moorhead et al. 2007, 2008, 2009; Bollen et al. 2010; Heroes et al. 2013; Nasa et al. 2018).

1.2 Protein Phosphatases in Eukaryotes

The protein phosphatases in eukaryotes belong to four separate families known as PPP (phosphoprotein phosphatases), PPM/PP2C (Mg^{2+}-dependent protein phosphatases), Asp-based protein phosphatases, and PTP (phospho-tyrosine phosphatases). The majority of phospho-serine and phospho-threonine dephosphorylation is catalyzed by PPP and PPM family members. PPP members include PP1, PP2 (PP2A), PP3 (PP2B), and PP4–7. It is notable that plants do not possess PP3 (PP2B) class of phosphatases but instead are endowed with additional novel members such as ALPH, RLPH, SLP1, and SLP2 (Pagliarini and Dixon 2006; Kerk et al. 2008; Uhrig et al. 2013b). Here, we provide an update on the SLP1 and SLP2 enzymes that function in chloroplasts and mitochondria, respectively. As mentioned above, it is the variety of additional subunits that bind the PPP catalytic subunits and bring specificity to the enzymes, and we predict this is also true for the SLPs.

1.3 Chloroplast and Mitochondrial Protein Phosphorylation

It is now well accepted that protein phosphorylation is the most common covalent modification of proteins in eukaryotes, with a majority of phosphoproteins residing in the cytosol and nucleus. Less is known about the phospho-proteome of mitochondria and chloroplasts, although phosphoproteins were identified in these organelles in 1969 (in animals) (Linn et al. 1969; Miernyk and Randall 1987) and 1977 (Miernyk and Randall 1987; Bennett 1977), respectively. In fact, the first *in vitro* demonstration of protein kinase activity in 1954 was the phosphorylation of casein by a mitochondrial extract (Pagliarini and Dixon 2006; Burnett and Kennedy 1954). The pyruvate dehydrogenase complex (PDC) is composed of three components with the E1 subunit being the first mitochondrial phosphoprotein to be discovered in animals, followed soon by the plant E1 subunit (Miernyk and Randall 1987). PDC catalyzes the reaction that yields acetyl-CoA and NADH from pyruvate and NAD^+. PDC activity is regulated by the inactivating phosphorylation by pyruvate dehydrogenase kinase (PDK) and the activating dephosphorylation by the PP2C-like enzyme, a phospho-pyruvate dehydrogenase phosphatase (PDP). Multiple proteins in the mitochondrial matrix and intermembrane space have now been identified as phosphoproteins. In 2013, 64 phosphorylated proteins and 10 protein kinases were identified in plant mitochondria (Havelund et al. 2013). More recently, Law et al. (2018) found that out of 802 mitochondrial proteins, 103 were found to have experimentally determined phosphorylation sites in just the first 60 N-terminal amino acids with implications in mitochondrial targeting (Law et al. 2018).

Recent chloroplast specific and general phospho-proteomic studies have illustrated widespread protein phosphorylation in the chloroplast (Baginsky and Gruissem 2009; Reiland et al. 2009; Richter et al. 2016). This is consistent with studies identifying multiple protein kinases and phosphatases in this organelle (Baginsky and Gruissem 2009; Richter et al. 2016; Andreeva and Kutuzov 2004), including casein kinase 2α4 (CK2α4), STN7, STN8, three thylakoid-associated kinases (TAKs), chloroplast sensor kinase (CSK), a family of atypical protein kinases (Activity of BC1 Complex Kinase or ABC1K), several Plastid Protein Kinases With Unknown Function (PKUs), seven type-2C phosphatases (PP2C), TAP38, PBCP, and SLP1 (White-Gloria et al. 2018; Uhrig and Moorhead 2011). We recently reviewed phosphorylation of the chloroplast starch metabolic machinery and cataloged phosphorylation of most of these enzymes (White-Gloria et al. 2018). Note that the plastid enzymes SEX4, LSF1, and LSF2 are designated phosphatases based on sequence, yet they are not protein phosphatases, acting as either scaffolds or starch phosphatases (Silver et al. 2014).

Clearly, protein phosphorylation is a regulatory mechanism that has been well established in mitochondria and chloroplasts. The explosion of phospho-proteomic data has revealed greater phosphorylation in these organelles than originally anticipated. This observation is simply consistent with a flood of phospho-proteomic data revealing abundant protein phosphorylation in bacteria, the ancient origin of these organelles.

1.4 Discovery and Bioinformatics of SLP1 and SLP2

Shewanella-like protein phosphatases 1 and 2, or SLP1 and SLP2, were identified in a bioinformatic study and given the name *Shewanella*-like protein phosphatase based on their sequence relationship to a PPP-like serine/threonine phosphatase of this bacterium (Andreeva and Kutuzov 2004). This bacterial origin is consistent with the mitochondrial and chloroplast localization in eukaryotes as these two organelles have been hypothesized to have had their origins in symbiotic bacteria. All key residues that define the PPP family phosphatases (GDxHG, GDxVDRG, GNHE, and HGG (Shi 2009)) are present in the SLPs, suggesting they are true protein phosphatases (Fig. 1.1) (Andreeva and Kutuzov 2004; Uhrig and Moorhead 2011). Orthologues of SLP1 and SLP2 are present in organisms across four of the five major eukaryotic supergroups (plantae, opisthokonts, chromalveolates, and excavates). The SLPs predominate in photosynthetic organisms with no SLP genes in animals (Uhrig et al. 2013b). All plant species examined contain SLP1 and SLP2 genes. In addition, we uncovered a third, more ancient group of SLP phosphatases (SLP3 phosphatases) in green algae (Uhrig et al. 2013b). Arabidopsis *thaliana* SLP2 is an intronless protein phosphatase, and exploring SLP2 across higher plants shows an almost complete absence of introns in the SLP2 gene (Uhrig and Moorhead 2017). The lack of introns is consistent with many mitochondrial destined proteins (Uhrig and Moorhead 2017). Key features, including the chloroplast transit peptide (cTP) on SLP1 orthologues, are shown in the alignment displayed in Fig. 1.1.

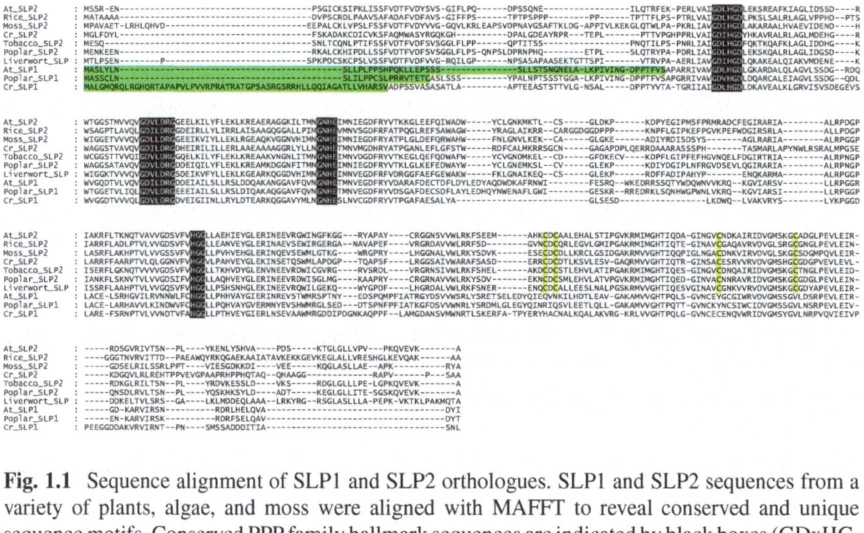

Fig. 1.1 Sequence alignment of SLP1 and SLP2 orthologues. SLP1 and SLP2 sequences from a variety of plants, algae, and moss were aligned with MAFFT to reveal conserved and unique sequence motifs. Conserved PPP family hallmark sequences are indicated by black boxes (GDxHG, GDxVDRG, GNHE, and HGG). SLP1 orthologues contain a chloroplast transit peptide (cTP, green), which is lacking in mitochondrially destined SLP2. The cysteine pairs in SLP2 orthologues (yellow) form disulfide bonds via the action of Mia40. Arabidopsis *thaliana* SLP1 (AtSLP1) and SLP2 (AtSLP2) sequences were input into BLASTp and used to retrieve sequences for poplar (*Populus trichocarpa*), tobacco (*Nicotiana tabacum*), moss (*Physcomitrella patens*), liverwort (*Marchantia polymorpha*), rice (*Oryza sativa*), and *Chlamydomonas* or Cr (*Chlamydomonas reinhardtii*). ChloroP 1.1 and Uhrig et al. (2013b) were used to predict chloroplast transit peptides in SLP1 sequences

1.5 SLP1 Is a Chloroplast-Localized Serine/Threonine Protein Phosphatase

Using fluorescent protein tagged version of AtSLP1 and multiple markers for cellular compartments, AtSLP1 was demonstrated to be chloroplast localized, consistent with bioinformatics that predicted a chloroplast transit peptide in AtSLP1 and most other SLP1 orthologues (Uhrig et al. 2013b; Uhrig and Moorhead 2011) (Fig. 1.1). This is consistent with western blotting of various tissues revealing that AtSLP1 is only expressed in photosynthetic tissues. Biochemically, AtSLP1 is insensitive to the classic PPP protein phosphatase inhibitors okadaic acid and microcystin but is remarkably sensitive to inorganic phosphate (Pi) and pyrophosphate (PPi). Although AtSLP1 has the hallmarks of a PPP family serine/threonine phosphatase, it displays activity against serine, threonine, and tyrosine phosphorylated peptides (Uhrig and Moorhead 2011). Current evidence suggests that no tyrosine phosphorylation occurs in the chloroplast (White-Gloria et al. 2018). AtSLP1 activity against tyrosine phosphorylated peptides may reflect the fact that it resides in the chloroplast and may not need to maintain stringent or specific serine/threonine phosphatase activity. This activity against phospho-tyrosine was displayed by both the bacterial expressed protein and TAP-AtSLP1 produced *in planta* (Uhrig et al. 2016). To date, no SLP1 substrates or regulatory subunits have been identified.

1.6 SLP2 Is a Mitochondrial Intermembrane Space Serine/Threonine Protein Phosphatase

Although the closest relative of SLP1 is SLP2, sequence differences are readily apparent. Arabidopsis *thaliana* SLP2 (AtSLP2) does not have a chloroplast transit peptide (cTP) but does have a series of cysteines critical to AtSLP2 function (Fig. 1.1). We have biochemically characterized AtSLP2 and used TAP (tandem affinity purification)-tag coupled to mass spectrometry to identify binding partners. Using this approach, a single clear binding partner for SLP2 has emerged: the oxidoreductase Mia40. Reverse TAP (TAP-Mia40) and co-immunoprecipitation confirmed Mia40 binding to SLP2 (Uhrig et al. 2017). Mia40 has been characterized in yeast and human cells as a mitochondrial intermembrane space protein that mediates the formation of disulfide bonds on target proteins. It has been demonstrated that both Mia40 and SLP2 reside in the mitochondrial intermembrane space, while some population of Mia40 also localizes to peroxisomes (Uhrig et al. 2017). Although a direct interactor of AtSLP2, Mia40 only modifies the enzyme and is not regarded as a regulatory subunit. Like AtSLP1, no regulatory subunits have been identified for AtSLP2, and it is also insensitive to the inhibitors okadaic acid and microcystin. Also, similar to SLP1, bacterially produced AtSLP2 displays activity against serine, threonine, and tyrosine phosphorylated peptides, but the *in planta* made TAP-AtSLP2 showed a remarkable preference for phospho-threonine over phospho-tyrosine using the same peptide substrates (Uhrig et al. 2017).

An insertional knockout of AtSLP2 (atslp2-2) was screened for growth phenotypes, and an accelerated germination phenotype was uncovered (Uhrig et al. 2017). This could be reversed when the knockout line was complemented with AtSLP2 driven by the endogenous promoter. Overexpression of AtSLP2 delays germination. In a knockout of Mia40 (*atmia40*), seeds exhibited a moderate accelerated germination phenotype consistent with SLP2 being the driver of the phenotype and likely displaying partial activity *in vivo* in the absence of Mia40 (Uhrig and Moorhead 2011).

1.7 Mia40 as a Redox Regulator

Specific mechanisms for the translocation of proteins through the outer mitochondrial membrane using the translocase of outer membrane (TOM) complex and then into specific mitochondrial sub-compartments have been known for some time (Fig. 1.2). However, it was not until 2004 that a protein was discovered in *Saccharomyces cerevisiae* which specifically targeted proteins to the mitochondrial intermembrane space (IMS), such as the small TIM proteins (Chacinska et al. 2004). They termed this protein as mitochondrial intermembrane space import and assembly protein 40, or Mia40. Experimental evidence showed that small TIM proteins

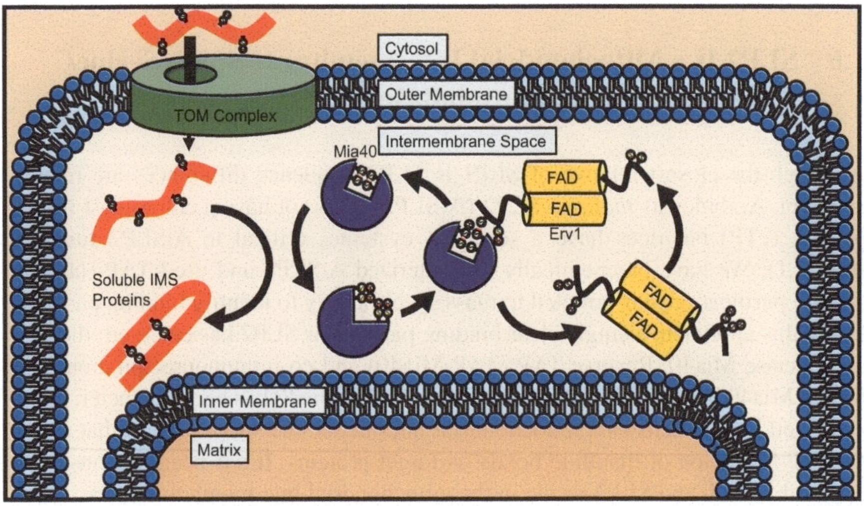

Fig. 1.2 Oxidoreductase Mia40 activates target proteins, including SLP2, in the mitochondrial intermembrane space. Proteins destined to reside in the mitochondrial intermembrane space (IMS) enter through the TOM complex and upon association with oxidized Mia40 form intramolecular disulfide bonds via pairs of conserved cysteines (see Fig. 1.1). IMS destined proteins (red) enter with reduced cysteines that are targeted by Mia40. By accepting electrons from target proteins, Mia40 is reduced and must be reoxidized by Erv1. In yeast, it has been demonstrated that Erv1 is reoxidized by transferring electrons to cytochrome c; this step has yet to be formally shown in plants

are imported into the IMS by TOM in a partly folded conformation and Mia40 then aids in the proper folding of proteins in the IMS. The mechanism of Mia40 was elucidated in 2005 (Mesecke et al. 2005) when it was found that proteins imported through TOM contained conserved cysteine motifs necessary for their import. A disulfide relay system was proposed in which Mia40 and another protein, Erv1, a sulfhydryl oxidase, compose the disulfide relay (Fig. 1.2). In this relay, Erv1 oxidizes cysteine residues in Mia40 allowing it to modify cysteines of imported proteins. Through disulfide bond isomerization, the proteins are then folded in the IMS to their fully functional, native structures. Yeast cells which lack Erv1 and therefore harbor Mia40 protein with reduced cysteines in their mitochondria lack viability due to the inability to successfully import and fold mitochondrial IMS proteins.

Mia40 activates the phosphatase activity of recombinant AtSLP2 ~4-fold when the artificial substrate pNPP is used, and this is dependent upon reductant, consistent with Mia40 generating disulfides on the enzyme to activate it. Although Mia40 does not alter the substrate specificity of AtSLP2 (i.e., pSer, pThr versus pTyr), it increases activity against substrate peptides up to 35-fold. Importantly, we demonstrated that Mia40 has no effect on AtSLP1 (Uhrig et al. 2017). To date, no AtSLP2 substrates have been identified.

1.8 MS-Based Substrate Discovery: The Future of Protein Phosphatases?

The discovery of protein phosphatase substrates has been technically challenging and always lags behind advances in protein kinase substrate discovery. Advances in mass spectrometry methods, in particular quantitative mass spectrometry, have changed the scenario now (Nasa et al. 2018; Rusin et al. 2015). It is now possible to knock out specific protein phosphatases and through quantitative analysis identify phosphopeptides that increase in the absence of the phosphatase, making the proteins these phospho-peptides are derived from as putative direct substrates. Although there are limitations in this approach, it is expected to revolutionize protein phosphatase substrate elucidation, and we expect this will be a common approach for protein phosphatase studies in the near future.

1.9 Conclusions

Bioinformatic, cell biological, and biochemical studies have established SLP1 and SLP2 as protein phosphatases that reside in the chloroplast and mitochondrial intermembrane space, respectively. Sequence analysis of SLP1 and SLP2 shows they are "bare" catalytic subunits with no accessory domains to regulate their activity. PPP family phosphatases typically associate with other proteins that regulate their function in the cell. To date, no regulatory subunits for either SLP1 or SLP2 have been identified. Understanding the functions and roles for each enzyme will require identifying these regulatory subunits and finding their substrates. Only then can we assign clear biological functions for these proteins. We speculate that this will be aided by quantitative mass spectrometry.

References

Adam K, Hunter T (2018) Histidine kinases and the missing phosphoproteome from prokaryotes to eukaryotes. Lab Investig 98:233–247

Andreeva AV, Kutuzov MA (2004) Widespread presence of "bacterial-like" PPP phosphatases in eukaryotes. BMC Evol Biol 4:47

Baginsky S, Gruissem W (2009) The chloroplast kinase network: new insights from large-scale phosphoproteome profiling. Mol Plant 2:1141–1153

Bennett J (1977) Phosphorylation of chloroplast membrane polypeptides. Nature 269:344–346

Bollen M, Peti W, Ragusa MJ, Beullens M (2010) The extended PP1 toolkit: designed to create specificity. Trends Biochem Sci 35:450–458

Brautigan DL, Shenolikar S (2018) Protein serine/threonine phosphatases: keys to unlocking regulators and substrates. Annu Rev Biochem 87:921–964

Burnett G, Kennedy EP (1954) The enzymatic phosphorylation of proteins. J Biol Chem 211:969–980

Bykova NV, Egsgaard H, Moller IM (2003) Identification of 14 new phosphoproteins involved in important plant mitochondrial processes. FEBS Lett 540:141–146

Chacinska A, Pfannschmidt S, Wiedemann N, Kozjak V, Sanjuan Szklarz LK, Schulze-Specking A, Truscott KN, Guiard B, Meisinger C, Pfanner N (2004) Essential role of Mia40 in import and assembly of mitochondrial intermembrane space proteins. EMBO J 23:3735–3746

Chen MJ, Dixon JE, Manning G (2017) Genomics and evolution of protein phosphatases. Sci Signal 10:eaag1796

Grimsrud PA, Carson JJ, Hebert AS, Hubler SL, Niemi NM, Bailey DJ, Jochem A, Stapleton DS, Keller MP, Westphall MS, Yandell BS, Attie AD, Coon JJ, Pagliarini DJ (2012) A quantitative map of the liver mitochondrial phosphoproteome reveals posttranslational control of ketogenesis. Cell Metab 16:672–683

Havelund JF, Thelen JJ, Moller IM (2013) Biochemistry, proteomics, and phosphoproteomics of plant mitochondria from non-photosynthetic cells. Front Plant Sci 4:51

Heroes E, Lesage B, Gornemann J, Beullens M, Van Meervelt L, Bollen M (2013) The PP1 binding code: a molecular-lego strategy that governs specificity. FEBS J 280:584–595

Kataya AR, Muench DG, Moorhead G (2019) A framework to investigate peroxisomal protein phosphorylation in arabidopsis. Trends Plant Sci 24:366–381

Kerk D, Templeton G, Moorhead GB (2008) Evolutionary radiation pattern of novel protein phosphatases revealed by analysis of protein data from the completely sequenced genomes of humans, green algae, and higher plants. Plant Physiol 146:351–367

Law YS, Ngan L, Yan J, Kwok LY, Sun Y, Cheng S, Schwenkert S, Lim BL (2018) Multiple kinases can phosphorylate the N-terminal sequences of mitochondrial proteins in Arabidopsis thaliana. Front Plant Sci 9:982

Linn TC, Pettit FH, Reed LJ (1969) Alpha-keto acid dehydrogenase complexes. X. Regulation of the activity of the pyruvate dehydrogenase complex from beef kidney mitochondria by phosphorylation and dephosphorylation. Proc Natl Acad Sci U S A 62:234–241

Mesecke N, Terziyska N, Kozany C, Baumann F, Neupert W, Hell K, Herrmann JM (2005) A disulfide relay system in the intermembrane space of mitochondria that mediates protein import. Cell 121:1059–1069

Miernyk JA, Randall DD (1987) Some properties of pea mitochondrial phospho-pyruvate dehydrogenase-phosphatase. Plant Physiol 83:311–315

Moorhead GB, Trinkle-Mulcahy L, Ulke-Lemee A (2007) Emerging roles of nuclear protein phosphatases. Nat Rev Mol Cell Biol 8:234–244

Moorhead GB, Trinkle-Mulcahy L, Nimick M, De Wever V, Campbell DG, Gourlay R, Lam YW, Lamond AI (2008) Displacement affinity chromatography of protein phosphatase one (PP1) complexes. BMC Biochem 9:28

Moorhead GB, De Wever V, Templeton G, Kerk D (2009) Evolution of protein phosphatases in plants and animals. Biochem J 417:401–409

Nasa I, Rusin SF, Kettenbach AN, Moorhead GB (2018) Aurora B opposes PP1 function in mitosis by phosphorylating the conserved PP1-binding RVxF motif in PP1 regulatory proteins. Sci Signal 11:eaai8669

Oeljeklaus S, Schummer A, Mastalski T, Platta HW, Warscheid B (2016) Regulation of peroxisome dynamics by phosphorylation. Biochim Biophys Acta 1863:1027–1037

Pagliarini DJ, Dixon JE (2006) Mitochondrial modulation: reversible phosphorylation takes center stage? Trends Biochem Sci 31:26–34

Reiland S, Messerli G, Baerenfaller K, Gerrits B, Endler A, Grossmann J, Gruissem W, Baginsky S (2009) Large-scale Arabidopsis phosphoproteome profiling reveals novel chloroplast kinase substrates and phosphorylation networks. Plant Physiol 150:889–903

Richter AS, Gartmann H, Fechler M, Rodiger A, Baginsky S, Grimm B (2016) Identification of four plastid-localized protein kinases. FEBS Lett 590:1749–1756

Rusin SF, Schlosser KA, Adamo ME, Kettenbach AN (2015) Quantitative phosphoproteomics reveals new roles for the protein phosphatase PP6 in mitotic cells. Sci Signal 8:rs12

Sharma K, D'Souza RC, Tyanova S, Schaab C, Wisniewski JR, Cox J, Mann M (2014) Ultradeep human phosphoproteome reveals a distinct regulatory nature of Tyr and Ser/Thr-based signaling. Cell Rep 8:1583–1594

Shi Y (2009) Serine/threonine phosphatases: mechanism through structure. Cell 139:468–484

Silver DM, Kotting O, Moorhead GB (2014) Phosphoglucan phosphatase function sheds light on starch degradation. Trends Plant Sci 19:471–478

Uhrig RG, Moorhead GB (2011) Two ancient bacterial-like PPP family phosphatases from Arabidopsis are highly conserved plant proteins that possess unique properties. Plant Physiol 157:1778–1792

Uhrig RG, Moorhead G (2017) AtSLP2 is an intronless protein phosphatase that co-expresses with intronless mitochondrial pentatricopeptide repeat (PPR) and tetratricopeptide (TPR) protein encoding genes. Plant Signal Behav 12:e1307493

Uhrig RG, Labandera AM, Moorhead GB (2013a) Arabidopsis PPP family of serine/threonine protein phosphatases: many targets but few engines. Trends Plant Sci 18:505–513

Uhrig RG, Kerk D, Moorhead GB (2013b) Evolution of bacteria-like phosphoprotein phosphatases in photosynthetic eukaryotes features ancestral mitochondrial or archaeal origin and possible lateral gene transfer. Plant Physiol 163:1829–1843

Uhrig RG, Labandera AM, Muhammad J, Samuel M, Moorhead GB (2016) Rhizobiales-like phosphatase 2 from Arabidopsis thaliana is a novel phospho-tyrosine-specific phospho-protein phosphatase (PPP) family protein phosphatase. J Biol Chem 291:5926–5934

Uhrig RG, Labandera AM, Tang LY, Sieben NA, Goudreault M, Yeung E, Gingras AC, Samuel MA, Moorhead GB (2017) Activation of mitochondrial protein phosphatase SLP2 by MIA40 regulates seed germination. Plant Physiol 173:956–969

van Wijk KJ, Friso G, Walther D, Schulze WX (2014) Meta-analysis of Arabidopsis thaliana phospho-proteomics data reveals compartmentalization of phosphorylation motifs. Plant Cell 26:2367–2389

White-Gloria C, Johnson JJ, Marritt K, Kataya A, Vahab A, Moorhead GB (2018) Protein kinases and phosphatases of the plastid and their potential role in starch metabolism. Front Plant Sci 9:1032

Chapter 2
Phosphoprotein Phosphatase Function of Secreted Purple Acid Phosphatases

Mina Ghahremani and William C. Plaxton

Abbreviations

APase	Acid phosphatase
ECM	Extracellular matrix
ER	Endoplasmic reticulum
HAD	Haloacid dehalogenase
LC-MS/MS	Liquid chromatography-tandem mass spectrometry
MS	Mass spectrometry
PAP	Purple acid phosphatase
Pi	Orthophosphate
P-Ser	Phosphoserine
PSI	Pi starvation-inducible
P-Thr	Phosphothreonine
PTM	Posttranslational modification
P-Tyr	Phosphotyrosine
VLK	Vertebrate lonesome kinase

2.1 Introduction

Acid phosphatases (APases; E.C. 3.1.3.2) catalyze the hydrolysis of orthophosphate (Pi, HPO_4^{2-}) from Pi monoesters and anhydrides with acidic pH optima. Purple APases (PAPs) represent the largest class of plant APases and exist as a diverse family of metallohydrolases involved in a multitude of biological processes. These include Pi-ester hydrolysis to facilitate plant Pi acquisition and the generation of

M. Ghahremani
Department of Biology, University of Ottawa, Ottawa, ON, Canada

Department of Biology, Queen's University, Kingston, ON, Canada

W. C. Plaxton (✉)
Department of Biology, Queen's University, Kingston, ON, Canada
e-mail: plaxton@queensu.ca

G. K. Pandey (ed.), *Protein Phosphatases and Stress Management in Plants*,
https://doi.org/10.1007/978-3-030-48733-1_2

reactive oxygen species as an immune response in mammals (Schenk et al. 2013; Tran et al. 2010a; Wang and Liu 2018). In contrast to other APases, PAPs are not inhibited by L(+)-tartrate. Hence, PAPs are also known as tartrate-resistant APases. Eukaryotic PAPs are glycoproteins that display highly variable amino acid sequences and sizes. However, their active sites are highly conserved, with seven invariant amino acid side chains coordinating Fe^{3+} and M^{2+} metal cations (M = Fe^{2+} in animals; Zn^{2+} or Mn^{2+} in plants) that participate in the catalytic mechanism for Pi-ester hydrolysis (Schenk et al. 2013). A charge-transfer transition from a conserved metal-coordinating tyrosine residue to the Fe^{3+} metal ligand leads to a typical absorption peak around 510–560 nm and a distinctive pink or purple color of PAPs in solution. Plant PAPs belong to a relatively large gene family encoding a diverse suite of isozymes. For example, the genome of the model plant Arabidopsis *thaliana* contains 29 *PAP* genes whose transcription is dependent upon various developmental and environmental factors (Li et al. 2002) (Fig. 2.1). These PAPs have been classified into two major groups according to their size, i.e., low molecular weight "mammalian-like" PAPs of approximately 35 to 45-kDa and higher molecular weight PAPs that range from about 50 to 70 kDa (Fig. 2.1). By contrast, only a single PAP isozyme of about 35-kDa (ACP5) has been characterized from animals. As discussed below, human ACP5 (HsACP5) is associated with microbial killing and bone resorption through its peroxidase and protein phosphatase activities, respectively (Schenk et al. 2013). Bioinformatics has identified a second *PAP-like* gene termed *ACP7* in mammals and other animal phyla that encodes a 55-kDa polypeptide that is more closely related to high molecular weight plant PAPs than it is to the low molecular weight ACP5 (Fig. 2.1) (Flanagan et al. 2006). Subsequent transcriptome profiling via RNA-seq indicated that human *ACP7* (*HsACP7*) is transcribed in various tissues, particularly skin and brain (https://www.ncbi.nlm.nih.gov/gene/390928). However, there have been no follow-up studies of the protein expression levels, biochemical properties, or function of the putative plant-like, high molecular weight ACP7 of humans or other animals (G. Schenk, personal communication).

Most PAPs that have been biochemically characterized are classified as nonspecific APases that catalyze Pi hydrolysis from a broad spectrum of Pi monoesters (Schenk et al. 2013; Tran et al. 2010a). This is consistent with their central role in cellular Pi metabolism, particularly scavenging and recycling Pi from intra- and extracellular Pi esters during nutritional Pi deprivation or senescence of vascular plants (Stigter and Plaxton 2015; Tran et al. 2010a; Wang and Liu 2018). However, HsACP5 expressed in macrophages also plays a role in immunity via the generation of reactive oxygen species (via a Fenton reaction involving the "redox-active" Fe^{2+} of their catalytic site) (Schenk et al. 2013). Similarly, several plant PAPs that function as APases also exhibit peroxidase activity that may contribute to the metabolism of reactive oxygen species during biotic or abiotic stress (Li et al. 2008; Tran et al. 2010a). The aim of this chapter is to briefly review the central role of PAPs in mediating plant Pi acquisition and use, followed by a discussion (1) of protein phosphorylation networks in the extracellular matrix (ECM) of animal and plant tissues

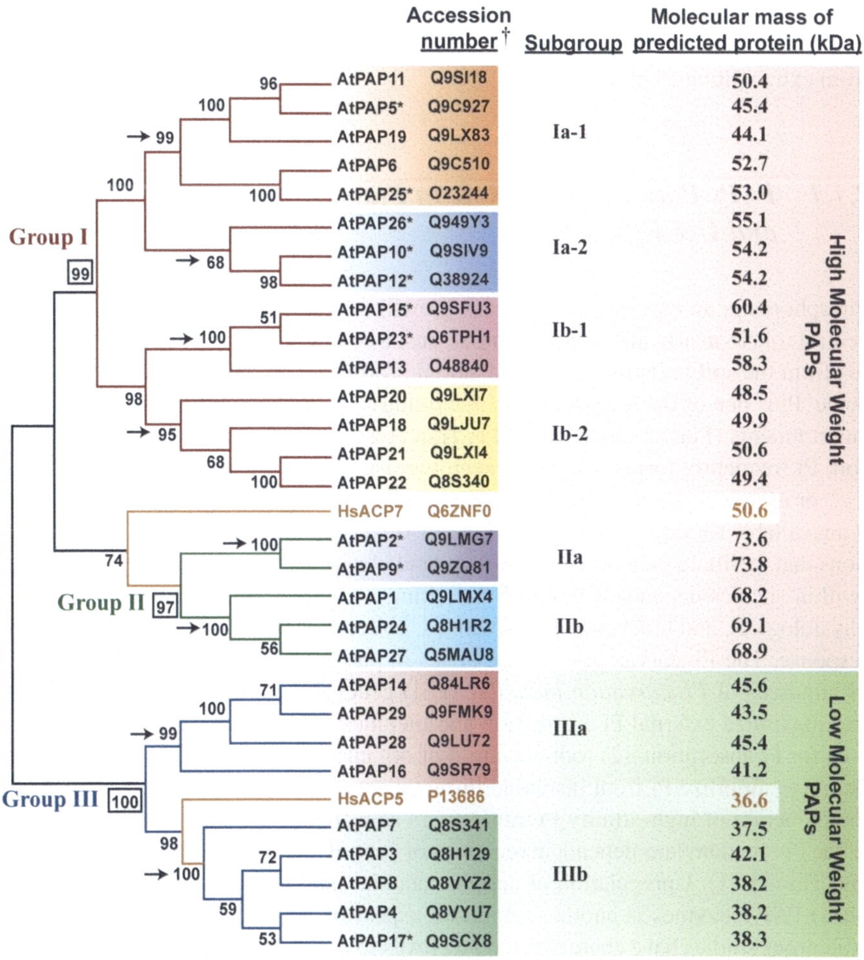

†Taken from UniProt.
*Functionally and/or biochemically characterized *Arabidopsis* PAPs.

Fig. 2.1 Classification of Arabidopsis *thaliana* PAPs (AtPAPs) and *Homo sapiens* PAPs (HsACP5 and HsACP7) based on clustering analysis of amino acid sequence. For AtPAPs, the clustering analysis used amino acid sequences of 19 predicted PAPs and those of ten PAPs (AtPAP3, AtPAP7–AtPAP13, AtPAP17, and AtPAP18) derived from cDNA analysis. AtPAPs possess three main groups (groups I, II, and III), which are further divided into subgroups. The deduced amino acid sequences of two human PAPs, HsACP7 and HsACP5, were obtained from UniProt (https://www.uniprot.org) and aligned with AtPAPs in MUSCLE 3.8 using ClustalW. A maximum likelihood tree was constructed in MEGA 7.0 using WAG model with the gamma distributed with invariant sites (G + I) and the partial deletion options. The bootstrap values for the three main groups are boxed, and the bootstrap values for the subgroups are indicated by arrows. The predicted molecular masses of the deduced polypeptides are listed in the last column. (Figure modified from Li et al. 2002)

and (2) that certain PAP isozymes secreted by animal and plant cells appear to function as phosphoprotein phosphatases rather than as nonspecific scavengers of Pi from extracellular Pi monoesters.

2.1.1 PAPs Play a Central Role in Plant Pi Acquisition and Use Efficiency

Phosphorus is an essential element for growth and metabolism because it plays a central role in nearly all-metabolic processes. Roots preferentially absorb phosphorus from the soil in its fully oxidized anionic form, Pi (Fig. 2.2). Despite its importance, Pi is one of the least available macronutrients in many terrestrial and aquatic environments (Plaxton and Tran 2011; Tran et al. 2010a; Veneklaas et al. 2012). In soil, Pi frequently forms insoluble precipitates with metal cations such as Al^{3+} and Ca^{2+} or is converted into organic P molecules by soil microbes that therefore render it unavailable for direct root uptake (Fig. 2.2). Thus, plants needed to evolve adaptations that facilitate their acclimation to extended periods of nutritional Pi deficiency (within species-dependent limits) by eliciting a complex array of morphological, physiological, and biochemical adaptations, collectively known as the Pi-starvation response. The Pi-starvation response arises in part from the coordinated induction of hundreds of *Pi-starvation-inducible* (PSI) genes that reprioritize internal Pi use and maximize external Pi acquisition and includes (1) extending the root's surface area for Pi absorption, (2) root excretion of organic acid anions such as malate and citrate to mobilize Pi from insoluble Pi-metal cation complexes in the soil, and (3) the induction of high-affinity Pi transporters as well as alternative bypass enzymes to the Pi- or adenylate-dependent reactions of central metabolism (Fig. 2.3) (Plaxton and Tran 2011). Upregulation of certain vacuolar and secreted (cell wall and apoplast) PAP isozymes is another important aspect of plant Pi-starvation responses. Numerous studies have characterized PSI PAPs to define the molecular mechanisms underlying this archetypal response of Pi-deprived plants, as well as to identify potential targets for the biotechnological improvement of crop Pi acquisition and use efficiency (Tran et al. 2010a; Wang and Liu 2018). Transgenic PAP expression offers a promising approach for sustainable crop Pi nutrition since organic P typically constitutes at least 50% of the total P in soils and is the predominant form of P found in soil solutions (Fig. 2.2).

Integrated biochemical and functional genomic studies have identified the closely related AtPAP10, AtPAP12, and particularly AtPAP26 (Fig. 2.1) as the predominant PAP isozymes that are upregulated and secreted into the ECM by Pi-deprived Arabidopsis suspension cells and seedlings (Hurley et al. 2010; Robinson et al. 2012b; Tran et al. 2010b; Veljanovski et al. 2006; Wang et al. 2014). Their widespread and reversible upregulation in roots and shoots of Pi-starved Arabidopsis, overlapping but nonidentical substrate selectivities and pH-activity profiles, and high specific APase activities support the hypothesis that they collectively mediate

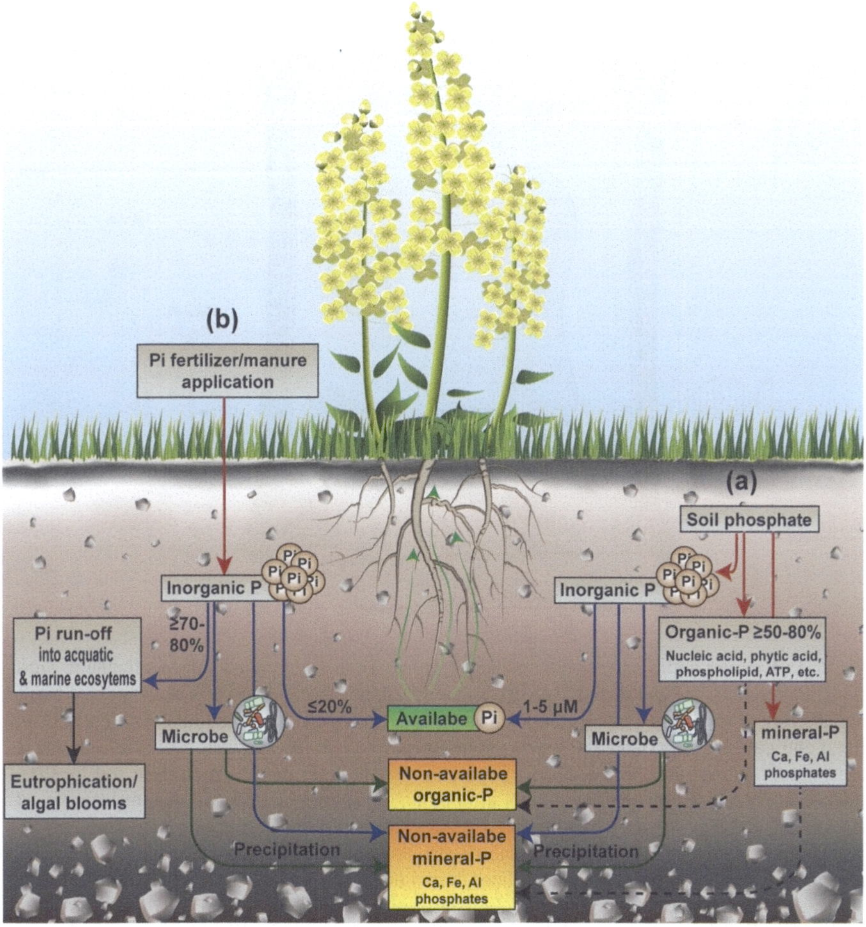

Fig. 2.2 Model of soil phosphorus (P) cycle. Most soils across the world are highly Pi limited since their soluble Pi concentration typically ranges between 1 and 5 μM, which is far below the intracellular Pi concentration (5–20 mM) required for optimal plant growth. (**a**) The lack of available Pi is due to several factors: (1) Pi is leached out of the soil due to the negative charge of clay, (2) Pi is converted into organic P forms by microorganisms, and (3) Pi can precipitate as insoluble calcium salts and iron or aluminum oxides in alkaline and acidic soils, respectively. (**b**) Less than 20% of Pi fertilizer applied worldwide per year is typically assimilated by crops. The remainder is either (1) converted into organic P by soil microbes; (2) bound by metal cations such as Al^{3+}, forming insoluble complexes; or (3) lost as runoff into nearby surface waters, resulting in the nutrient enrichment of aquatic and marine ecosystems and consequent "blooms" of green algae and toxic cyanobacteria; this stubborn environmental problem has caused extensive eutrophication

efficient Pi scavenging and recycling from a broad range of extracellular Pi esters (Fig. 2.3). Indeed, growth of wild-type Arabidopsis seedlings on media containing glycerol-3-phosphate (an effective *in vitro* substrate of AtPAP10, AtPAP12, and AtPAP26) as their sole source of exogenous P was indistinguishable from that of

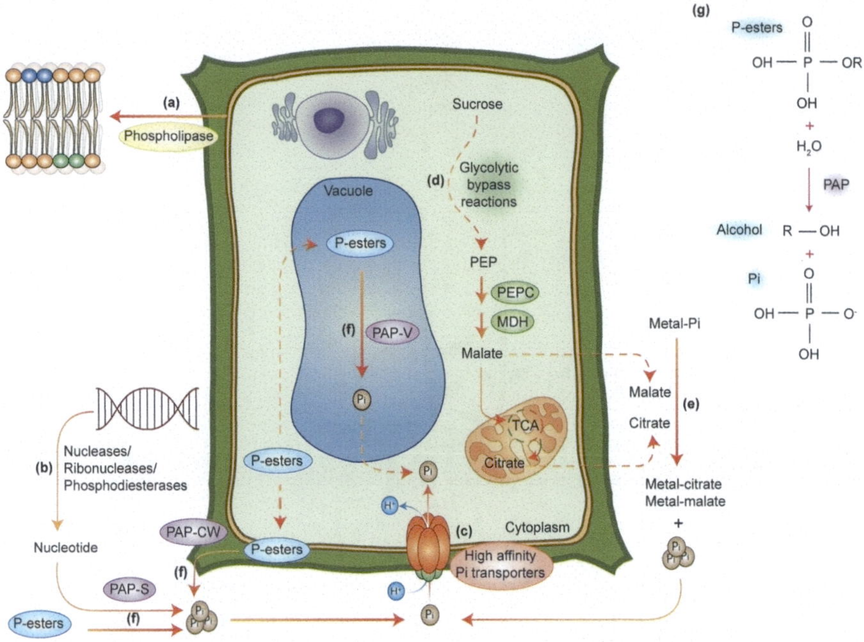

Fig. 2.3 A model outlining adaptive metabolic processes that help plant cells acclimate to nutritional Pi deficiency (Plaxton and Tran 2011). (**a**) Phospholipase induction is accompanied by the replacement of membrane phospholipids (orange sphere) with non-P containing amphipathic sulfonyl and galactolipids (blue and green sphere, respectively). (**b**) Secreted nucleases, ribonucleases, phosphodiesterases, and PAPs participate in systematic Pi mobilization from soil-localized nucleic acids. (**c**) Upregulation of high-affinity Pi transporters of the plasma membrane. (**d**) Alternative pathways of cytosolic glycolysis, mitochondrial electron transport, and tonoplast H^+-pumping facilitate respiration and vacuolar pH maintenance by Pi-deprived plant cells. (**e**) Organic acid anion excretion solubilizes mineralized forms of Pi and organic P as well as increases the ability of secreted PAPs to scavenge Pi from soil-localized organic Pi monoesters. (**f**, **g**) Upregulation of intracellular, cell wall (CW), and apoplast/rhizosphere targeted PAPs enhances the Pi acquisition and use efficiency of Pi-deprived plants (Tran et al. 2010b; Wang et al. 2014). PAPs catalyze Pi hydrolysis from a broad and overlapping range of Pi monoesters with an acidic pH optimum and function in the production, transport, and recycling of Pi

Pi-replete seedlings. AtPAP12 and AtPAP26 function likely includes scavenging Pi from 3′-(d)NMPs derived from nuclease-mediated nucleic acid hydrolysis (Fig. 2.3). This was supported by the impaired development of *atpap12/atpap26* T-DNA double insertion mutant seedlings during growth on media containing salmon sperm DNA as their sole source of exogenous P (Robinson et al. 2012b). Vacuolar and cell wall-targeted AtPAP26 were also strongly upregulated by senescing leaves of Pi-replete plants to remobilize Pi from endogenous Pi-ester pools (Robinson et al. 2012a; Shane et al. 2014). Senescing leaves of an *atpap26* T-DNA mutant exhibited a >90% decrease in APase activity, impaired Pi remobilization, and delayed senescence (Robinson et al. 2012a). The collective results have defined AtPAP26 as a principal contributor to intra- and extracellular APase activity, and that AtPAP26 loss of function elicits dramatic effects on Arabidopsis Pi metabolism that cannot be compensated for by any other AtPAP isozyme. As outlined below, however, several animal and plant PAP isozymes that are secreted into the ECM effectively hydrolyze Pi from phosphoamino acid and phosphoprotein substrates, suggesting that they might function *in planta* as a phosphoprotein phosphatase rather than as nonspecific scavengers of Pi from organic P molecules. However, this discussion first warrants a summary of recent and compelling evidence for extensive and dynamic extracellular protein phosphorylation networks in the animal and plant kingdoms.

2.2 Extracellular Protein Phosphorylation Networks of Animals and Plants: The Neglected PTM

Reversible protein phosphorylation is the most important posttranslational modification (PTM) of eukaryotic proteins since it participates in the control of virtually all aspects of cell physiology and development including signal transduction, cell differentiation, cytoskeleton organization, active transport (ion pumping), gene expression, disease and stress responses, and metabolic fluxes (Moorhead and Tran 2006). Phosphoproteomic studies indicate that phosphorylation occurs in at least 70% of all eukaryotic proteins, with the majority having multiple phosphorylation sites. Protein kinases and phosphatases catalyze the covalent incorporation or hydrolysis, respectively, of Pi groups on target proteins.

2.2.1 Animals

The occurrence of extracellular protein phosphorylation was debated for many years, despite the fact that casein, a secreted storage protein of a mother's milk, was the first phosphoprotein to be discovered over 130 years ago. Numerous secreted proteins[1] have since been shown to be phosphorylated in vertebrate and invertebrate animals (Yalak et al. 2014). For example, phosphoproteomic screens have identified 25 to 85 different phosphoproteins in different human body fluids including cerebrospinal fluid, blood plasma, and saliva. Further studies detected over 500 phosphoproteins in human serum (Yalak et al. 2014), whereas over 1000 animal phosphoproteins listed in the PhosphoSitePlus database (https://www.phosphosite.org) have been annotated as being extracellular or transmembrane proteins (Klement and Medzihradszky 2017). A remarkable feature of the mammalian phosphoproteome (and possibly plant; see below) is that a substantial proportion of the ECM

[1] Proteins occurring in the luminal side of the ER or the Golgi, as well as interior of digestive vacuole (i.e., lysosome in animals, cell vacuole of plants), are also considered to be part of the secreted proteome (i.e., the "secretome").

proteome of cultured or primary tissue samples contain phosphotyrosine (P-Tyr), in addition to phosphoserine (P-Ser) and phosphothreonine (P-Thr) (Bordoli et al. 2014).

In the classical secretion pathway, a short transit peptide targets the protein to the endoplasmic reticulum (ER) and then the Golgi network after which it is packaged into secretory vesicles that are released into the ECM (Fig. 2.4) or targeted to the lysosome. During this transport, the protein may undergo various PTMs apart from glycosylation (a typical PTM of secreted eukaryotic proteins), including dithiol-disulfide interconversion and phosphorylation (Canut et al. 2016; Ghahremani et al. 2016). Proteins of the secretory pathway that are not fully secreted may also be phosphorylated (e.g., the extracellular domain of plasma membrane-spanning receptors). Proteins lacking a transit peptide and thus not entering the classical ER/Golgi secretory pathway can also be exported to the ECM by unconventional means, i.e., via exocytosis or direct translocation across the plasma membrane. Phosphorylation of secreted proteins may precede their export or occur post-secretion via extracellular protein kinases (Fig. 2.4) (Klement and Medzihradszky 2017; Yalak et al. 2014).

Secreted protein kinases, evolutionarily and structurally distinct from cytoplasmic kinases, have been identified in the ECM of mammals, as well as several invertebrate animals (Gerson-Gurwitz et al. 2018; Sreelatha et al. 2015; Tagliabracci et al. 2013; Yalak et al. 2014). For example, Fam20C is a ubiquitous mammalian serine kinase dedicated to phosphorylating a wide range of secreted and highly acidic milk-, salivary-, enamel-, dentin-, and bone-specific proteins (which typically contain an S-x-E-pS consensus motif) involved in diverse processes such as biomineralization (i.e., bone and tooth formation), lipid homeostasis, wound healing, cell adhesion, and cell migration (Sreelatha et al. 2015; Tagliabracci et al. 2013, 2015; Yalak and Vogel 2012). Fam20C resides inside the Golgi but also occurs as an N-terminally truncated, fully secreted form. FAMK-1, a secreted Fam20C ortholog, contributes to fertility, embryogenesis, and development in the nematode worm *Caenorhabditis elegans* (Gerson-Gurwitz et al. 2018; Tagliabracci et al. 2015). In addition, "vertebrate lonesome kinase" (VLK) is a novel secreted protein kinase of mammals that phosphorylates a broad range of ECM proteins on tyrosine residues and is vital for embryonic development (Bordoli et al. 2014; Tagliabracci et al. 2015). High VLK expression occurs in platelets, where it is rapidly and quantitatively secreted in response to specific stimuli. Besides phosphorylating substrate proteins within the Golgi, secreted VLK also phosphorylates tyrosine residues in various protein targets in the ECM using endogenous secreted ATP[2] sources (e.g.,

[2] The protein kinase co-substrate ATP cannot passively diffuse across the plasma membrane owing to its high charge. Thus, extracellular ATP originates from cytosolic ATP via its regulated secretion in the absence of cell lysis; that is, cytoplasmic vesicles laden with ATP secrete their cargo into the ECM of animal and plant cells via exocytosis (Bordoli et al. 2014; Chivasa and Slabas 2012; Yalak and Vogel 2012). It is notable that extracellular ATP is an important stimulus for cell signaling that functions in many aspects of animal and plant physiology, including growth, development, and stress responses (Cao et al. 2014; Chivasa and Slabas 2012; Yalak and Vogel 2012).

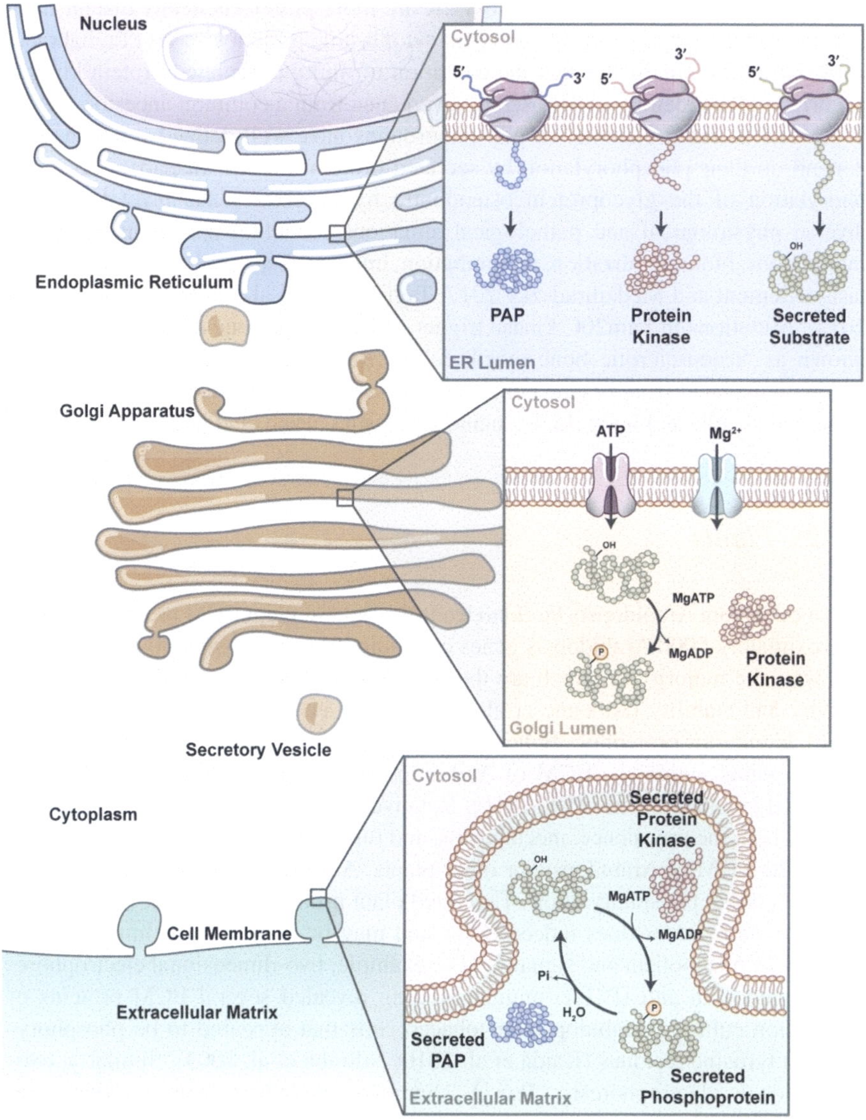

Fig. 2.4 Phosphorylation and dephosphorylation of ECM proteins by secreted protein kinases and PAPs, respectively. Signal-peptide-containing proteins that are directed to the ECM are transported through the secretory pathway. The newly synthesized proteins enter the endoplasmic reticulum as nascent proteins containing a short sequence of hydrophobic amino acids at the N-terminus known as the signal peptide. The signal peptide is recognized and cleaved by signal peptidase, localized in the ER. The proteins are shuttled to the Golgi apparatus for further PTMs including glycosylation. Specific type of Golgi membrane proteins transports ATP and divalent metal cations (e.g., Mg^{2+}) into the Golgi lumen to provide the protein kinase with phosphate-donating molecules. The secreted proteins can be packaged into vesicles that are transported and released into the ECM via exocytosis. Proteins of the secretory pathway become phosphorylated either in Golgi lumen or after secretion to the ECM. Secreted PAP catalyzes the dephosphorylation of the substrate in the ECM

see model of Fig. 2.4). Fam20C and VLK are more phylogenetically distant from one another than they are from canonical cytoplasmic protein kinases (Tagliabracci et al. 2013). This indicates that the evolution of multiple secreted protein kinases occurred independently, as opposed to divergence from a common ancestor.

Several secreted mammalian phosphoproteins have well-defined functions that depend on their phosphorylation by secreted protein kinases. In particular, phosphorylation of the glycoprotein osteopontin by Fam20C plays pivotal roles in diverse physiological and pathological situations including osteopontin-receptor interactions, biomineralization, inflammation, immune responses, and tumor metastasis (Klement and Medzihradszky 2017; Tagliabracci et al. 2015; Yalak and Vogel 2012). Mutations in Fam20C kinase trigger a devastating bone disorder in humans known as "osteosclerotic bone dysplasia" or Raine syndrome. Osteopontin was originally isolated from bovine bones but was subsequently detected in a variety of mammalian cells and body fluids, including plasma, urine, and milk.

2.2.2 Plants

The model plant Arabidopsis *thaliana* leads the field of plant ECM proteomics, with approximately 5000 Arabidopsis genes encoding proteins targeted to the secretory pathway, the majority of which are thought to require at least one PTM to confer activity and stability (Albenne et al. 2013; Canut et al. 2016; Ghahremani et al. 2016). Numerous proteomic studies have documented a wide array of enzymes and other proteins within the ECM of Arabidopsis and other plant species (Albenne et al. 2013; Ghahremani et al. 2016). Relative to mammals, however, far less is known about the prevalence, mechanisms, and functions of protein phosphorylation within the ECM of Arabidopsis or other plants. A limited number of studies have established that phosphorylation of secreted plant proteins on tyrosine, serine, and/or threonine residues does indeed occur and may be of functional importance to plant ECM metabolism and signaling. For example, two-dimensional electrophoresis coupled with anti (P-Tyr)-immunoblotting revealed several ECM proteins of suspension-cultured Arabidopsis or tobacco cells that appeared to be phosphorylated on tyrosine residues (Kaida et al. 2010; Ndimba et al. 2003). Similarly, two-dimensional electrophoresis, Pro-Q Diamond phosphoprotein staining, and matrix-assisted laser desorption/ionization time-of-flight mass spectrometry (MS) indicated that up to 31 different cell wall-targeted glycoproteins were phosphorylated during cell wall regeneration by protoplasts of Arabidopsis suspension-cultured cells (Kwon et al. 2005). The list of Arabidopsis cell wall phosphoproteins included expansins, α-xylosidase, β-xylosidase, β-galactosidase, α-mannosidase, phosphatases, phosphoesterases, isomerases, lectins, dehydrogenases, and proteases (Kwon et al. 2005). Likewise, two-dimensional electrophoresis coupled with phosphoprotein staining and high-resolution liquid chromatography-tandem MS

(LC-MS/MS) identified 35 phosphoproteins involved in a wide range of biological processes and molecular functions (including cell wall and reactive oxygen species metabolism) in the ECM of cultured potato cells (Elagamey et al. 2016). Orbitrap LC-MS/MS and immunoblotting recently revealed that cell wall-localized AtGAL1, a glycosylated 55-kDa mannose-binding and apple domain lectin that is upregulated and secreted into ECM and cell vacuole of Pi-starved Arabidopsis, is bisphosphorylated at Tyr^{38} and Thr^{39} (Ghahremani et al. 2019b). Conservation of AtGAL1's $pTyr^{38}$ and $pThr^{39}$ phosphosites in its paralogs AtGAL2 and AtGAL3, as well as several of its closest orthologs from other plant species, indicated that this unusual PTM might be of widespread importance for members of this mannose-binding lectin subgroup. Determining the mechanisms and functions of AtGAL1 (bis)phosphorylation will be an important avenue for future studies. Interestingly, parallel research demonstrated that AtGAL1 specifically interacts with a high-mannose glycoform of AtPAP26 and that this association enhances AtPAP26's APase activity and physical stability (Ghahremani et al. 2019a). Biochemical, proteomic, and loss-of-function mutant studies have provided definitive evidence that AtPAP26 is the predominant vacuolar as well as a major secreted PAP isozyme involved in Pi scavenging and recycling during Pi deprivation or leaf senescence (Veljanovski et al. 2006; Hurley et al. 2010; Tran et al. 2010b; Robinson et al. 2012a, b; Shane et al. 2014; Wang et al. 2014).

It is notable that developmental- or stress-induced dynamic changes in the phosphorylation status of plant ECM proteins have been reported, implying that reversible protein phosphorylation is not restricted to the intracellular phosphoproteome of plant cells. For example, when Arabidopsis suspension cells were treated with fungal elicitors or subjected to nutritional Pi deprivation, marked changes in the profile of the extracellular proteome occurred, including differential phosphorylation of certain ECM proteins (Ndimba et al. 2003; H. Tran, K. Ellis, W. Plaxton, unpublished research). These changes were hypothesized to be part of a signal transduction cascade mediating cellular responses to fungal infection or Pi deprivation. Alterations in the phosphorylation status of many ECM proteins also occurred during early protoplast regeneration of Arabidopsis cell cultures (Kwon et al. 2005). However, it remains unknown whether any secreted plant phosphoprotein is phosphorylated in the ER or Golgi lumen during the sorting pathway and/or phosphorylated post-secretion by extracellular protein kinases (Fig. 2.4). Pioneering studies of Citovsky and coworkers (Citovsky et al. 1993) demonstrated that P30, a tobacco mosaic virus (TMV) cell-to-cell movement protein, was phosphorylated by a protein kinase contained within the cell wall of tobacco leaves. The cell wall-associated protein kinase was proposed to reduce TMV virulence by phosphorylating P30 at Ser258, Thr261, and Ser265 and thereby sequestering it within the cell walls. Although high-throughput proteomic studies have listed mitogen-activated and Ser/Thr protein kinases as members of the plant ECM proteome, further research is needed to validate their occurrence and protein kinase activities, along with potential ECM protein targets that they phosphorylate *in planta*.

2.3 Phosphoprotein Phosphatase Function of Secreted PAPs

Eukaryotic protein phosphatase type I and calcineurin (protein phosphatase 2B) contain binuclear metal centers that possess striking similarity to PAPs in the coordination environment of the active site (Ljusberg et al. 1999; Schenk et al. 2013; Vincent and Averill 1990). This implies that PAPs might also function to reverse protein kinase action by dephosphorylating phosphoproteins. Indeed, numerous PAPs purified from the ECM of animal or plant cells effectively hydrolyze Pi from phosphoproteins, phosphopeptides, and/or phosphoamino acids *in vitro* indicating that they might also function as phosphoprotein phosphatases *in vivo* (Del Vecchio et al. 2014; Halleen et al. 1998; Kaida et al. 2008; Ljusberg et al. 1999; Tran et al. 2010a). This is supported by the aforementioned discoveries of dynamic protein phosphorylation events within the ECM of animal and plant cells.

2.3.1 Animals

One of the best documented examples for a phosphoprotein phosphatase function of a secreted PAP concerns the involvement of HsACP5, the 35-kDa human PAP (Fig. 2.1), in bone resorption by osteoclasts (Oddie et al. 2000; Schenk et al. 2013). Expression of HsACP5 is mainly restricted to differentiated cells of nuclear phagocytic lineage, primarily osteoclasts and macrophages. In osteoclasts involved in active bone resorption, HsACP5 is secreted into the bone ECM where it dephosphorylates phosphoprotein substrates, particularly osteopontin and osteonectin (Halleen et al. 1998; Ljusberg et al. 1999; Marshall et al. 1997; Oddie et al. 2000). Studies of transgenic mice over- and underexpressing their endogenous HsACP5 ortholog provided definitive evidence that secretion of this PAP by osteoclasts is an essential prerequisite for proper mineralization of cartilage in developing bone and bone matrix resorption by adult bones (Oddie et al. 2000; Schenk et al. 2013). Thus, the major function of HsACP5 in bone resorption appears to be the catabolic degradation of bone matrix phosphoproteins, thereby facilitating access by specific proteases. This is analogous to the role proposed for HsACP5 in the dephosphorylation of red blood cell membrane and cytoskeletal phosphoproteins during erythrophagocytosis by macrophages. Although HsACP5 exhibits optimal phosphoprotein phosphatase activity with acidic phosphopeptides containing P-Tyr, it also dephosphorylates protein-bound P-Ser and P-Thr as well (Halleen et al. 1998; Ljusberg et al. 1999; Marshall et al. 1997; Oddie et al. 2000). One of its most important physiological substrates, osteopontin, was *in vivo* phosphorylated on multiple serine and threonine residues, in addition to tyrosine, when overexpressed in human embryonic kidney cell cultures (Li et al. 2015).

2.3.2 Plants

Although purified plant PAPs inevitably appear to show phosphatase activity with phosphoamino acids or phosphopeptides as substrates (Tran et al. 2010a), several exhibit novel kinetic features that would make them particularly well geared for a phosphoprotein phosphatase function *in planta*. For example, the major PAP purified from potato tubers or cell walls of tobacco suspension cell cultures exhibited maximal activity with P-Tyr relative to any other substrate that was tested (Gellatly et al. 1994). The purified potato PAP exhibited a P-Tyr hydrolyzing specific activity of >1900 μmol Pi produced min^{-1} mg^{-1}, which to the best of our knowledge remains the highest APase activity reported for any PAP studied to date. Immunoblotting using antibodies raised against P-Tyr demonstrated that several endogenous phosphotyrosylated tuber polypeptides could serve as *in vitro* substrates for the purified potato PAP (Gellatly et al. 1994). Although the physiological significance of the potato PAP's substantial *in vitro* activity with P-Tyr and endogenous phosphotyrosylated proteins remains obscure, the possibility that this PAP may function to dephosphorylate certain protein-located P-Tyr residues *in vivo* was suggested (Gellatly et al. 1994; Kaida et al. 2008). This might include patatin, the major storage protein of potatoes which accounts for up to 45% of the total soluble protein in the tubers, and that becomes heavily phosphorylated during tuber development (Bernal et al. 2017; Elagamey et al. 2016). As with other vegetative storage proteins (and PAPs), patatin is also targeted to the cell vacuole. Similarly, NtPAP12 is a cell wall-localized tobacco PAP that is highly active against phosphotyrosylated peptides (Kaida et al. 2008). Interestingly, transgenic NtPAP12 expression resulted in altered cell wall composition and enhanced β-glucan synthase activity, implying that this PAP isozyme might function in the ECM as a phosphoprotein phosphatase involved in the control of cell wall biosynthesis (Kaida et al. 2009, 2010). Follow-up studies demonstrated that NtPAP12 efficiently dephosphorylated three phosphoproteins of the tobacco ECM, including α-xylosidase and β-glucosidase (Kaida et al. 2010). Moreover, dephosphorylation inhibited α-xylosidase activity, whereas overexpression of NtPAP12 in tobacco cell cultures not only decreased β-glycosidase activity but also increased levels of xyloglucan oligosaccharides and cello-oligosaccharides within the ECM. This research provided strong evidence that NtPAP12 helps control the phosphorylation status and thus activity of α-xylosidase and β-glucosidase, which are responsible for the degradation of xyloglucan oligosaccharides and cello-oligosaccharides within the cell wall (Kaida et al. 2010).

AtPAP25 is a member of the high molecular weight Arabidopsis PAP family (Fig. 2.1) that appears to be exclusively expressed and targeted to cell walls of Pi-starved plants, while exhibiting kinetic features consistent with its potential function as a phosphoprotein phosphatase (Del Vecchio et al. 2014). Transcript profiling and immunoblotting with anti-AtPAP25 immune serum indicated that AtPAP25 is strictly expressed and secreted during Pi deficiency. Coupled with AtPAP25's potent mixed-type (allosteric) inhibition by Pi (I_{50} = 50 μM), this indicates a very tight feedback control by Pi that would prevent AtPAP25 from being synthesized or functioning as a phosphatase except when Pi levels are extremely low. Promoter:β-glucuronidase reporter assays revealed specific *AtPAP25* expression in shoot vascular tissue of Pi-starved plants. Development of an *atpap25* T-DNA insertion mutant was arrested during cultivation on soluble Pi-deficient soils but rescued upon Pi fertilization or complementation with *AtPAP25*. Quantification of transcripts encoding five well-documented *PSI* genes (i.e., *At4*, *AtPPCK1*, *AtRNS1*, *AtPAP12*, and *AtPAP17*) indicated that Pi starvation signaling was attenuated in shoots of the soil-cultivated *atpap25* mutant. Since AtPAP25 exhibited near optimal APase activity with several phosphoamino acids and phosphoproteins as substrates, it was hypothesized to play a key signaling role during Pi deprivation by functioning as a phosphoprotein phosphatase, rather than as a nonspecific scavenger of Pi from extracellular Pi esters (Del Vecchio et al. 2014). Additional studies are required to confirm AtPAP25's putative phosphoprotein phosphatase role in the ECM of shoot vascular tissue of Pi-deprived plants, as well as how it might participate in the signaling pathways by which Arabidopsis responds to nutritional Pi deprivation at the molecular level. In this regard, transcriptomic and (phospho)proteomic profiling to assess global gene and (phospho)protein expression changes in the *atpap25* mutant could help to further establish the degree to which Pi starvation signaling was attenuated in the mutant.

2.4 Plant Haloacid Dehalogenase-Like APases May Function as Cytoplasmic Phosphoprotein Phosphatases

Apart from PAPs, APases belonging to the haloacid dehalogenase (HAD) superfamily are also upregulated by Pi-deficient plants (Baldwin et al. 2001, 2008; Hur et al. 2007; Pandey et al. 2017). The HAD superfamily is represented by a diverse assortment of enzymes including ATPases, epoxide hydrolases, dehalogenases, phosphomutases, and phosphoserine phosphatases. *LePS2* was the first PSI *HAD superfamily* gene characterized in tomato (Baldwin et al. 2001, 2008). The phosphoprotein phosphatase activity of overexpressed LePS2 was hypothesized to trigger a signaling cascade that resulted in increased APase activity, anthocyanin accumulation, and delayed flowering in tomato (Baldwin et al. 2008). Another PSI HAD, OsHAD1, is a cytosolic APase of rice that exhibits a broad pH-activity profile and substrate specificity including phosphoamino acids and phytic acid (Pandey et al. 2017). Overexpression of OsHAD1 in rice led to increased APase activity, Pi accumulation, and improved growth under restricted Pi supply. OsHAD1 overexpression was suggested to reduce the phosphorylation status of several target phosphoproteins, leading to the induction of Pi transporters, phytases, phosphatases, and genes involved in organic acid production. Moreover, pull-down assays indicated that OsHAD1 associates with protein kinases known to play important roles in several signal transduction pathways involved in abiotic and biotic stress acclimation. Thus, OsHAD1 in coordination with protein kinases was hypothesized to mediate the phosphorylation status of downstream targets and thereby contribute to the control of rice Pi metabolism (Pandey et al. 2017).

2.5 Concluding Remarks

The biochemical and molecular properties, expression patterns, and subcellular location of plant PAPs indicate that this family of metalloenzymes was significantly modified during the evolution of vascular plants, generating an assortment of high and low molecular weight isozymes (Fig. 2.1) suited for a broad variety of metabolic, developmental, and environmental situations. Plant PAP studies have traditionally focused on the roles that they play in Pi scavenging and recycling from Pi monoesters and anhydrides during Pi starvation or senescence (Stigter and Plaxton 2015; Tran et al. 2010a; Wang and Liu 2018). However, the possibility that certain secreted PAP isozymes have a specific phosphoprotein phosphatase function within the ECM of plant cells is supported by (1) their ability to efficiently dephosphorylate various phosphoamino acids, phosphopeptides, and phosphoproteins *in vitro*; (2) the existence of extensive and dynamic protein phosphorylation events within the plant ECM; and (3) the well-established precedent for an extracellular phosphoprotein phosphatase role of the secreted low molecular weight human PAP, HsACP5. Nevertheless, studies of the prevalence, functions, and mechanisms of reversible protein phosphorylation within the plant ECM are still in their infancy. Although enhanced proteomic and MS techniques are continually expanding the list of plant ECM proteins (Albenne et al. 2013), and the majority of plant ECM proteins appear to undergo multiple PTMs (Canut et al. 2016; Ghahremani et al. 2016), few studies to date have focused on detecting or understanding site-specific PTMs such as phosphorylation that may be pivotal to ECM protein function and metabolism. A challenging yet important goal for future research will therefore be to document the mechanisms and functional consequences of extracellular protein phosphorylation in the signaling and metabolic pathways involved in plant development and acclimation to biotic and abiotic stresses. Whether biotic or abiotic, the imposed stress can induce dramatic alterations within the ECM proteome that facilitate plant acclimation to unavoidable stressful conditions. Signal transduction cascades initiated within the apoplast can trigger changes in cell wall composition and intracellular metabolism that mediate stress acclimation via the action of a diverse suite of enzymes. Understanding how, when, and why plant ECM proteins are phosphorylated could provide crucial insights into the dynamic nature of the cell wall and how plant cells respond to external stressors and stimuli. Gaining a full understanding of the molecular pathways involved in these PTMs will likely contribute toward future biotechnological efforts of engineering stress-tolerant crops and trees. Continued identification and characterization of Golgi and ECM phosphoproteins under various physiological situations, along with pinpointing the occurrence and targets of secreted protein kinases, and phosphoprotein phosphatase function of secreted PAPs will be required before our understanding of reversible protein phosphorylation within the plant ECM is complete. These key questions are not simple issues; however, they are critical to continued pursuit of the mechanisms and functions of protein phosphorylation in plant cell biology.

Acknowledgments WCP is indebted to past and present members of his laboratory who have characterized the molecular, biochemical, and functional properties of plant PAPs and/or the role of enzyme phosphorylation in the control of plant metabolism. Our research has been generously supported by grants from the Natural Sciences and Engineering Research Council of Canada (NSERC) and the Queen's Research Chairs program. We are also grateful to Prof. Gary Schenk (Univ. of Queensland) for helpful discussions about the biochemistry and functions of animal and plant PAPs.

References

Albenne C, Canut H, Jamet E (2013) Plant cell wall proteomics: the leadership of Arabidopsis *thaliana*. Front Plant Sci 4:111

Baldwin JC, Karthikeyan AS, Raghothama KG (2001) LePS2, a phosphorus starvation-induced novel acid phosphatase from tomato. Plant Physiol 125:728–737

Baldwin JC, Karthikeyan AS, Cao A, Raghothama KG (2008) Biochemical and molecular analysis of LePS2: a phosphate starvation induced protein phosphatase gene from tomato. Planta 228:273–280

Bernal J, López-Pedrouso M, Franco D, Bravo S, García L, Zapata C (2017) Identification and mapping of phosphorylated isoforms of the major storage protein of potato based on two-dimensional electrophoresis. In: Jimenez-Lopez JC (ed) Advances in seed biology. InTech, Rijeka, pp 65–82

Bordoli MR, Yum J, Breitkopf SB, Thon JN, Italiano JE, Xiao J et al (2014) A secreted tyrosine kinase acts in the extracellular environment. Cell 158:1033–1044

Canut H, Albenne C, Jamet E (2016) Post-translational modifications of plant cell wall proteins and peptides: A survey from a proteomics point of view. Biochim Biophys Acta 1864:983–990

Cao Y, Tanaka K, Nguyen CT, Stacey G (2014) Extracellular ATP is a central signaling molecule in plant stress responses. Curr Opin Plant Biol 20:82–87

Chivasa S, Slabas AR (2012) Plant extracellular ATP signaling: new insight from proteomics. Mol BioSyst 8:445–452

Citovsky V, McLean BG, Zupan JR, Zambryski P (1993) Phosphorylation of tobacco mosaic virus cell-to-cell movement protein by a developmentally regulated plant cell wall-associated protein kinase. Genes Dev 7:904–910

Del Vecchio HA, Ying S, Park J, Knowles VL, Kanno S, Tanoi K et al (2014) The cell wall-targeted purple acid phosphatase AtPAP25 is critical for acclimation of Arabidopsis *thaliana* to nutritional phosphorus deprivation. Plant J 80:569–581

Elagamey E, Narula K, Sinha A, Aggarwal P, Ghosh S, Chakraborty N et al (2016) Extracellular matrix proteome and phosphoproteome of potato reveals functionally distinct and diverse canonical and non-canonical proteoforms. Proteomes 4:20

Flanagan JU, Cassady AI, Schenk G, Guddat LW, Hume DA (2006) Identification and molecular modeling of a novel, plant-like, human purple acid phosphatase. Gene 377:12–20

Gellatly KS, Moorhead GBG, Duff SMG, Lefebvre DD, Plaxton WC (1994) Purification and characterization of a potato tuber acid phosphatase having significant phosphotyrosine phosphatase activity. Plant Physiol 106:223–232

Gerson-Gurwitz A, Worby CA, Lee K-Y, Khaliullin R, Bouffard J, Cheerambathur D et al (2018) Ancestral roles of the Fam20C family of secreted protein kinases revealed by functional analysis in *C. elegans*. bioRxiv. https://doi.org/10.1101/363440

Ghahremani M, Stigter KA, Plaxton WC (2016) Extraction and characterization of extracellular proteins and their post-translational modifications from Arabidopsis *thaliana* suspension cell cultures and seedlings: a critical review. Proteomes 4:25

Ghahremani M, Park J, Anderson EM, Marty-Howard NJ, Mullen RT, Plaxton WC (2019a) Lectin AtGAL1 interacts with high-mannose glycoform of the purple acid phosphatase AtPAP26 secreted by phosphate-starved Arabidopsis. Plant Cell Environ 42:1158–1166

Ghahremani M, Tran H, Biglou SG, O'Gallagher B, She Y-M, Plaxton WC (2019b) A glycoform of the purple acid phosphatase AtPAP26 co-purifies with a mannose-binding lectin (AtGAL1) secreted by phosphate-starved Arabidopsis. Plant Cell Environ 42:1139–1157

Halleen JM, Kaija H, Stepan JJ, Vihko P, Väänänen HK (1998) Studies on the protein tyrosine phosphatase activity of tartrate-resistant acid phosphatase. Arch Biochem Biophys 352:97–102

Hur YJ, Lee HG, Jeon EJ, Lee YY, Nam MH, Yi G et al (2007) A phosphate starvation-induced acid phosphatase from *Oryza sativa*: phosphate regulation and transgenic expression. Biotechnol Lett 29:829–835

Hurley BA, Tran HT, Marty NJ, Park J, Snedden WA, Mullen RT et al (2010) The dual-targeted purple acid phosphatase AtPAP26 is essential for efficient acclimation of Arabidopsis *thaliana* to nutritional phosphate deprivation. Plant Physiol 153:1112–1122

Kaida R, Hayashi T, Kaneko TS (2008) Purple acid phosphatase in the walls of tobacco cells. Phytochemistry 69:2546–2551

Kaida R, Satoh Y, Bulone V, Yamada Y, Kaku T, Hayashi T et al (2009) Activation of β-glucan synthases by wall-bound purple acid phosphatase in tobacco cells. Plant Physiol 150:1822–1830

Kaida R, Serada S, Norioka N, Norioka S, Neumetzler L, Pauly M et al (2010) Potential role for purple acid phosphatase in the dephosphorylation of wall proteins in tobacco cells. Plant Physiol 153:603–610

Klement E, Medzihradszky KF (2017) Extracellular protein phosphorylation, the neglected side of the modification. Mol Cell Proteomics 16:1–7

Kwon H-K, Yokoyama R, Nishitani K (2005) A proteomic approach to apoplastic proteins involved in cell wall regeneration in protoplasts of Arabidopsis suspension-cultured cells. Plant Cell Physiol 46:843–857

Li D, Zhu H, Liu K, Liu X, Leggewie G, Udvardi M et al (2002) Purple acid phosphatases of Arabidopsis *thaliana*. Comparative analysis and differential regulation by phosphate deprivation. J Biol Chem 277:27772–27781

Li W-YF, Shao G, Lam H-M (2008) Ectopic expression of *GmPAP3* alleviates oxidative damage caused by salinity and osmotic stresses. New Phytol 178:80–91

Li H, Shen H, Yan G, Zhang Y, Liu M, Fang P et al (2015) Site-specific structural characterization of O-glycosylation and identification of phosphorylation sites of recombinant osteopontin. Biochim Biophys Acta 1854:581–591

Ljusberg J, Ek-Rylander B, Andersson G (1999) Tartrate-resistant purple acid phosphatase is synthesized as a latent proenzyme and activated by cysteine proteinases. Biochem J 343:63–69

Marshall K, Nash K, Haussman G, Cassady I, Hume D, de Jersey J et al (1997) Recombinant human and mouse purple acid phosphatases: expression and characterization. Arch Biochem Biophys 345:230–236

Moorhead GBG, Tran HT (2006) Role of protein kinases, phosphatases and 14-3-3 proteins in the control of primary plant metabolism. In: Plaxton W, McManus M (eds) Annual plant reviews. Blackwell Publishing Ltd., New Delhi, pp 121–144

Ndimba BK, Chivasa S, Hamilton JM, Simon WJ, Slabas AR (2003) Proteomic analysis of changes in the extracellular matrix of Arabidopsis cell suspension cultures induced by fungal elicitors. Proteomics 3:1047–1059

Oddie GW, Schenk G, Angel NZ, Walsh N, Guddat LW, de Jersey J et al (2000) Structure, function, and regulation of tartrate-resistant acid phosphatase. Bone 27:575–584

Pandey BK, Mehra P, Verma L, Bhadouria J, Giri J (2017) OsHAD1, a haloacid dehalogenase-like APase, enhances phosphate accumulation. Plant Physiol 174:2316–2332

Plaxton WC, Tran HT (2011) Metabolic adaptations of phosphate-starved plants. Plant Physiol 156:1006–1015

Robinson WD, Carson I, Ying S, Ellis K, Plaxton WC (2012a) Eliminating the purple acid phosphatase AtPAP26 in Arabidopsis *thaliana* delays leaf senescence and impairs phosphorus remobilization. New Phytol 196:1024–1029

Robinson WD, Park J, Tran HT, Vecchio D, A H, Ying S et al (2012b) The secreted purple acid phosphatase isozymes AtPAP12 and AtPAP26 play a pivotal role in extracellular phosphate-scavenging by Arabidopsis *thaliana*. J Exp Bot 63:6531–6542

Schenk G, Mitić N, Hanson GR, Comba P (2013) Purple acid phosphatase: a journey into the function and mechanism of a colorful enzyme. Coord Chem Rev 257:473–482

Shane MW, Stigter K, Fedosejevs ET, Plaxton WC (2014) Senescence-inducible cell wall and intracellular purple acid phosphatases: implications for phosphorus remobilization in *Hakea prostrata* (Proteaceae) and Arabidopsis *thaliana* (Brassicaceae). J Exp Bot 65:6097–6106

Sreelatha A, Kinch LN, Tagliabracci VS (2015) The secretory pathway kinases. Biochim Biophys Acta 1854:1687–1693

Stigter KA, Plaxton WC (2015) Molecular mechanisms of phosphorus metabolism and transport during leaf senescence. Plants 4:773–798

Tagliabracci VS, Pinna LA, Dixon JE (2013) Secreted protein kinases. Trends Biochem Sci 38:121–130

Tagliabracci VS, Wiley SE, Guo X, Kinch LN, Durrant E, Wen J et al (2015) A single kinase generates the majority of the secreted phosphoproteome. Cell 161:1619–1632

Tran HT, Hurley BA, Plaxton WC (2010a) Feeding hungry plants: the role of purple acid phosphatases in phosphate nutrition. Plant Sci 179:14–27

Tran HT, Qian W, Hurley BA, She Y-M, Wang D, Plaxton WC (2010b) Biochemical and molecular characterization of AtPAP12 and AtPAP26: the predominant purple acid phosphatase isozymes secreted by phosphate-starved Arabidopsis *thaliana*. Plant Cell Environ 33:1789–1803

Veljanovski V, Vanderbeld B, Knowles VL, Snedden WA, Plaxton WC (2006) Biochemical and molecular characterization of AtPAP26, a vacuolar purple acid phosphatase up-regulated in phosphate-deprived Arabidopsis suspension cells and seedlings. Plant Physiol 142:1282–1293

Veneklaas EJ, Lambers H, Bragg J, Finnegan PM, Lovelock CE, Plaxton WC et al (2012) Opportunities for improving phosphorus-use efficiency in crop plants. New Phytol 195:306–320

Vincent JB, Averill BA (1990) Sequence homology between purple acid phosphatases and phosphoprotein phosphatases. Are phosphoprotein phosphatases metalloproteins containing oxide-bridged dinuclear metal centers? FEBS Lett 263:265–268

Wang L, Liu D (2018) Functions and regulation of phosphate starvation-induced secreted acid phosphatases in higher plants. Plant Sci 271:108–116

Wang L, Lu S, Zhang Y, Li Z, Du X, Liu D (2014) Comparative genetic analysis of Arabidopsis purple acid phosphatases AtPAP10, AtPAP12, and AtPAP26 provides new insights into their roles in plant adaptation to phosphate deprivation. J Integr Plant Biol 56:299–314

Yalak G, Vogel V (2012) Extracellular phosphorylation and phosphorylated proteins: not just curiosities but physiologically important. Sci Signal 5:re7

Yalak G, Ehrlich YH, Olsen BR (2014) Ecto-protein kinases and phosphatases: an emerging field for translational medicine. J Transl Med 12:165

Chapter 3
Purple Acid Phosphatases (PAPs): Molecular Regulation and Diverse Physiological Roles in Plants

Poonam Mehra and Jitender Giri

3.1 Introduction

Purple acid phosphatases (PAPs) represent the largest group of acid phosphatases (APases; E.C. 3.1.3.2) that are characterized by their pink or purple color in water solution. This color of PAPs is derived from a charge transfer transition from Tyr residue to chromopheric Fe (III) in the dinuclear metal center (Oddie et al. 2000; Wang et al. 2015). PAPs possess metallophos domain and belong to metallophosphoesterase superfamily, which also includes exonucleases and protein phosphatases (Tran et al. 2010a). PAPs catalyze the hydrolysis of several P-containing compounds at acidic pH (pH 4.0–7.0). In mammals, PAPs have been studied for many biological functions (*reviewed in* Olczak et al. 2003). Mammalian PAPs are involved in ROS generation, iron transport, and bone sorption. Some of these roles such as iron transport are not associated with their enzymatic activity. Due to their role in bone metabolism, mammalian PAPs have been identified as potential targets for curing osteoporosis and other bone ailments (Oddie et al. 2000; Mitić et al. 2006). Bifunctional mammalian PAPs catalyze peroxidation as well as hydrolytic reactions. However, plant PAPs are widely studied for their roles under Pi deficiency (Tran et al. 2010a). The transcript induction of PAPs is considered as one of the explicit molecular signatures for Pi deficiency. Under Pi deficiency, PAPs catalyze the hydrolysis of several P containing organic compounds present intracellularly or in the rhizosphere. Thus, PAPs are key players involved in Pi acquisition and redistribution (Kuang et al. 2009). With progressively more investigations into their biological functions, some PAPs have now also been characterized for their roles in signal transduction, oxidative stress, nodule formation, mycorrhizal symbiosis, and senescence (Kaida et al. 2008; Li et al. 2008). This chapter describes the diverse

P. Mehra · J. Giri (✉)
National Institute of Plant Genome Research, New Delhi, India
e-mail: jitender@nipgr.ac.in

G. K. Pandey (ed.), *Protein Phosphatases and Stress Management in Plants*,
https://doi.org/10.1007/978-3-030-48733-1_3

roles of PAPs in plants and contemplates new lines of investigations to discover their novel functions and mechanisms.

3.2 Structure and Classification of PAPs

In plants, the protein sequences of PAPs are reported to have an N-terminal non-active domain and a C-terminal active domain (Schenk et al. 2005, 2013). All PAPs are characterized with five conserved blocks of amino acid residues (**D**X**G**/G**D**XX**Y**/GNH(D/E)/VXX**H**/G**H**X**H**) with seven metal-ligating residues (in bold letters) that are involved in metal coordination at dinuclear metal center (Li et al. 2002). These metal-ligating residues constitute the active site of these enzymes and are highly conserved from bacterial to mammalian PAPs (Li et al. 2002; Olczak et al. 2003). However, some recent studies in Arabidopsis, rice, and soybean have reported some alterations in the composition of the seven invariant residues (Li et al. 2002, 2012a; Zhang et al. 2011). PAPs are present in a diverse class of organisms including bacteria, fungi, plants, and mammals (Kuang et al. 2009). The bimetallic active site of mammalian PAPs possesses Fe^{3+}–Fe^{3+} center, whereas the active site of plant PAPs contains Fe^{3+}–Mn^{2+} or Fe^{3+}–Zn^{2+} center (Schenk et al. 1999, 2005; Flanagan et al. 2006). Replacement of ferric ion with Mn^{2+} in the dinuclear center led to complete loss of PAP activity in sweet potato (Mitić et al. 2009). This indicates the absolute requirement of ferric ion for the phosphatase activity of PAPs. All characterized mammalian PAPs are either monomeric proteins of ~35 kDa (LMW; low molecular weight) or homodimeric proteins of ~55 kDa (HMW; high molecular weight) (Schenk et al. 2000). Similar LMWs and HMWs also exist in plants; however, the HMW PAPs in plants are of ~45–75 kDa molecular weight (Tran et al. 2010a). Structural analysis of plant PAPs has revealed dimerization of HMW PAPs by disulfide bridges formed through cysteine residues (Olczak et al. 2003). Depending on their location of activity preferences, PAPs are also classified as intracellular and extracellular. Intracellular PAPs can remobilize phosphorus (P) from cellular P reserves, whereas secreted PAPs can hydrolyze P from bound organic P sources. Noticeably, a significant fraction of soil P, nearly 50–80%, occurs as organic P and constitutes the bound P pool, not readily available for plant use (Wang et al. 2009). Secretory PAPs can hydrolyze these bound organic P sources to make them available for plant use. Thus, PAPs are considered important candidates for improving low Pi tolerance of plants.

3.3 PAPs Exist as a Multigene Family

In plants, PAPs are present as a multigene family. PAP family members have been identified in Arabidopsis (29 AtPAPs) (Li et al. 2002), rice (26 OsPAPs) (Zhang et al. 2011), soybean (35 GmPAPs) (Li et al. 2012a), maize (33 ZmPAPs)

(González-Muñoz et al. 2015), chickpea (Bhadouria et al. 2017), and *Jatropha curcas* (25 JcrPAPs) (Venkidasamy et al. 2019). In a recent study, PAP genes were identified in ten vegetable species belonging to Brassicaceae, Solanaceae, and Cucurbitaceae (Xie and Shang 2018).

According to the phylogenetic classification, all rice and Arabidopsis PAPs are classified into three major groups (I, II, and III) and seven subgroups (Ia, Ib, Ic, IIa, IIb, IIIa, and IIIb) (Zhang et al. 2011). In a recent classification, PAPs from Arabidopsis and different crops (*Phaseolus vulgaris*, *Zea mays*, *Vigna radiata*, *Hordeum vulgare*, *Nicotiana tabacum*, *Triticum aestivum*, *Oryza sativa*, and *Medicago truncatula*) were divided into four major groups: I, II, III, and IV (Tian and Liao 2015). The largest group, group I, was further divided into subgroups: I-1 and I-2 (Tian and Liao 2015). Noticeably, many PAPs of the subgroup I-2 have been reported to possess phytase activity. These include AtPAP15, OsPAP23 (OsPHY1), MtPHY1, GmPAP19, and CaPAP7 (Hegeman and Grabau 2001; Xiao et al. 2005; Kuang et al. 2009; Li et al. 2012b; Bhadouria et al. 2017). On the other hand, almost all of the PAPs of group I except for AtPAP2 are not yet characterized. Interestingly, Pi starvation does not influence the expression of AtPAP2, which is localized in plastids and mitochondria (Sun et al. 2012). Further, AtPAP2 was found to be involved in carbon metabolism (Sun et al. 2012). Most of the PAPs of group II (AtPAP10, AtPAP12, AtPAP25, AtPAP26, OsPAP10a, OsPAP10c) are characterized for their roles in improving low Pi tolerance of plants (Tran et al. 2010b; Wang et al. 2011; Tian et al. 2012; Del Vecchio et al. 2014; Lu et al. 2016). Group III, comprising only 11 PAPs, constitutes the smallest group. No PAP of this group has been characterized so far. Similarly, the functional role of very few PAPs from group IV has been elucidated. This includes AtPAP17 (del Pozo et al. 1999), PvPAP3, PvPAP4, and PvPAP5 (Liang et al. 2012). Therefore, given the functional diversity of plant PAPs, a large number of PAPs still need to be investigated for realizing their potential in multiple stresses.

3.4 Plant PAPs Are Nonspecific Phosphatases

Depending on their pH requirements for optimum activity, plant phosphatases belong to two classes: alkaline phosphatases and acid phosphatases. While alkaline phosphatase exhibits absolute substrate specificity, acid phosphatases are relatively nonspecific (Duff et al. 1994). APases (acid phosphatases) are further classified as specific and nonspecific APases (Duff et al. 1994). Specific APases are specialized in nature and possess substrate specificities, which are not absolute but specific to some extent. These include 3-phosphoglycerate phosphatases and phosphoenolpyruvate phosphatase (Duff et al. 1989). On the other hand, nonspecific APases act upon a broad range of substrates. Majority of the PAPs belong to the class of nonspecific APases. However, some PAPs such as OsPAP23 and AtPAP15 are specific APases due to their high preference for phytate as substrate (Kuang et al. 2009; Li et al. 2012b). Similarly, cell wall-localized AtPAP25 specifically functions as phosphoprotein phosphatase (Del Vecchio et al. 2014). Nonspecific PAPs such as AtPAP10 and PvPAP3 showed the highest activity with an organic substrate, ATP (Liang et al. 2010; Wang et al. 2011), whereas secretory Arabidopsis PAPs (AtPAP12 and AtPAP26) and tomato PAPs (LeSAP1 and LeSAP2) showed maximum activity with PEP (phosphoenolpyruvate) (Bozzo et al. 2002; Tran et al. 2010b). In rice, OsPAP21b was shown to possess the highest activity against di- and tri-nucleotides as well as sugar phosphates and phosphoproteins (Mehra et al. 2017). In the absence of detailed biochemical characterization of several known and unknown PAPs, it is difficult to develop a concluding remark about overall substrate specificities of PAPs. However, there is growing evidence that phosphatase activity of many PAPs is inhibited by the higher concentration of Pi suggesting a feedback loop regulating PAPs (Bozzo et al. 2002; Veljanovski et al. 2006; Liang et al. 2010; Mehra et al. 2017).

3.5 Regulation of Purple Acid Phosphatases in Plants

PAPs are reported to be regulated at transcriptional, translational, and/or posttranslational levels in plants (Fig. 3.1). Several PAPs have been reported to be induced spatiotemporally under Pi deficiency in Arabidopsis and crop plants (Wang et al. 2011; Zhang et al. 2011; González-Muñoz et al. 2015; Mehra et al. 2016). Induction of PAPs under Pi deficiency is regulated by transcription factors such as PHR1, WRKY75, and ZAT6 (Rubio et al. 2001; Devaiah et al. 2007a, b; Zhang et al. 2011). Recently, *OsPAP21b* was shown to be directly regulated by OsPHR2 through *in vitro* studies (Mehra et al. 2017). Overexpression of *ZAT6* increased the activity of APases in roots of Arabidopsis as compared to WT (Devaiah et al. 2007b). Similarly, overexpression lines of rice *PHR2* showed increased expression of *OsPAPs* (Zhang et al. 2011). Further, rice mutant, *ospho2*, and RNAi lines of *OsSPX1* also showed increased activity of APases in rice roots as compared to WT under Pi deficiency indicating *PHO2* and *SPX* to be negative regulators of *OsPAPs* (Zhang et al. 2011). However, any direct evidence of such regulation is not revealed yet. Few studies report the posttranscriptional/posttranslational regulation of PAPs. Recently, transcripts and protein of *OsPAP21b* in rice overexpression lines were shown to undergo downregulation in Pi-replete conditions (Mehra et al. 2017). This study also showed that expression of *OsPAP21b* was systemically regulated by plant Pi status. Also, some PAPs (e.g., AtPAP26) undergo posttranslational modifications such as glycosylation, which is supposed to influence localization, kinetic properties, and stability of enzymes (Tran et al. 2010b). It has been found that transcripts of *AtPAP26* are constitutively expressed irrespective of plant Pi status (Veljanovski et al. 2006). However, AtPAP26 protein showed significant accumulation under Pi deprivation (Veljanovski et al. 2006). Furthermore, under Pi deficiency, AtPAP26 is secreted as two distinct glycoforms; AtPAP26-S1 and AtPAP26-S2. The AtPAP26-S2 form was found to interact with secretory bisphosphorylated, Pi-inducible lectin, AtGAL1 (Ghahremani et al. 2018). Interaction of AtGAL1 with AtPAP26-S2 glycosylated

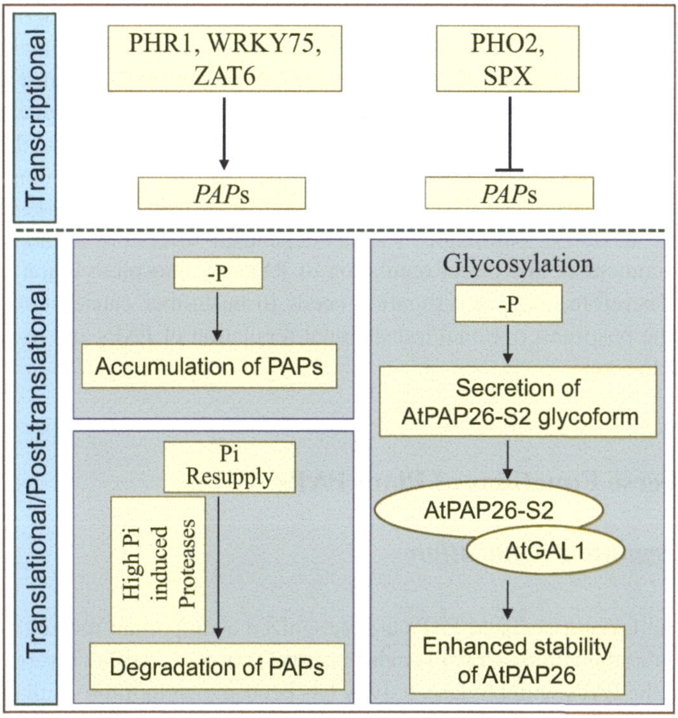

Fig. 3.1 Regulation of purple acid phosphatase (PAP) activity in plants. Schematic representation of known molecular regulatory mechanisms of PAPs. Expression of PAPs is positively regulated at transcriptional levels by transcription factors such as PHR1, WRKY75, and ZAT6, whereas PHO2 and SPX protein negatively regulate PAPs. Phosphorus deficiency (–P) induces accumulation of most of the PAP proteins. Repletion of phosphorus degrades PAPs by high Pi-inducible proteases. Many PAPs are known to undergo posttranslational modification such as glycosylation. AtPAP26-S2, a glycoform of AtPAP26, is produced during Pi deficiency. Interaction of AtPAP26-S2 with Pi-inducible lectin AtGAL1 is known to stabilize AtPAP26 under Pi deficiency

form enhances stability and APase activity of AtPAP26. This suggests that specific posttranslational modifications of PAPs may have a significant impact on their activities.

Some other studies also indicate the existence of such regulations. For instance, the proteolytic turnover of PAPs dramatically increased after Pi resupply in Pi-starved tomato suspension cells (Bozzo et al. 2004). This was also accompanied by the induction of some proteases that possibly bring about the degradation of PAPs (Bozzo et al. 2004). Similarly, the protein level of AtPAP10 was significantly increased in overexpression lines of AtPAP10 under Pi deficiency as compared to Pi-replete conditions (Zhang et al. 2014). Few posttranscriptional and posttranslational regulatory modules have been revealed under Pi deficiency. Among these, regulation by AtPHO2-*miR399*-AtPHR1-*At4* module constitutes one of the highly characterized regulations under Pi deficiency (Chiou et al. 2006; Doerner 2008). However, no reports are suggesting any role of this module in regulating PAPs.

Moreover, reversible phosphorylation and dephosphorylation are also known to regulate some enzymes under Pi deficiency. In Arabidopsis, low Pi-inducible phosphoenolpyruvate carboxylase (PEPC) is activated by phosphorylation under Pi deficiency (Gregory et al. 2009). Interestingly, transcripts of genes encoding for PEP carboxylase kinases (AtPPCK1/2) are also significantly induced upon Pi starvation (Gregory et al. 2009). Although Pi status-dependent reversible accumulation of PAPs is documented, any direct regulation of PAPs by phosphorylation is not yet explored. Therefore, more investigation needs to be further carried out to reveal details of the posttranscriptional/translational regulation of PAPs and its relevance under Pi deficiency.

3.6 Diverse Functions of Plant PAPs

3.6.1 Organic P Utilization

Secretory and root-associated PAPs are recognized as important molecular players for the plant's adaptation to low Pi environments (Tran et al. 2010a). Overexpression of some of the PAPs with phytase activity has been shown to improve plant growth on organic P supplemented growth medium. Since phytate is the most abundant (30–65%) organic P form in soil (Harrison 1987), targeting PAPs with phytase activity is an active research area to increase organic P solubilization by crop plants. Overexpression of *MtPAP1* derived from *Medicago truncatula*, *GmPAP4* from soybean, and *AtPAP15* from Arabidopsis led to the increased Pi acquisition and plant growth on phytate as a sole P source (Xiao et al. 2006; Wang et al. 2009; Kong et al. 2014). AtPAP15 is a nonsecretory PAP. Therefore, AtPAP15 was fused with the signal peptide of carrot to enhance extracellular phytate utilization (Wang et al. 2009). In the model plant Arabidopsis, 29 putative PAPs have been reported, of which 11 are low Pi inducible (Wang et al. 2014). Some of the Arabidopsis PAPs, such as AtPAP17, AtPAP26, AtPAP12, AtPAP25, AtPAP15, and AtPAP10, have been well characterized for their important roles under Pi deficiency (Kuang et al. 2009; Tran et al. 2010b; Wang et al. 2011; Del Vecchio et al. 2014). AtPAP12 and AtPAP26 are the major PAPs secreted by Arabidopsis roots under Pi deficiency (Tran et al. 2010b). Secreted AtPAP12 and AtPAP26 were shown to hydrolyze exogenously supplied herring sperm DNA for providing phosphate nutrition to Pi-starved Arabidopsis seedlings (Robinson et al. 2012), whereas *atpap12atpap26* double mutant was unable to hydrolyze extracellular DNA leading to poor growth and development as compared to WT (Robinson et al. 2012). AtPAP12 and AtPAP26 were also found to hydrolyze phosphate monoesters such as glycerol-3-phosphate when supplied as sole P source to Arabidopsis seedlings (Robinson et al. 2012).

Similarly, few PAPs have been biochemically and/or functionally characterized in crop plants. These include KbPAP and PvPAP3 from common bean (Cashikar et al. 1997; Liang et al. 2012), NtPAP from tobacco (Lung et al. 2008) and LeSAP1 and LeSAP2 from tomato (Bozzo et al. 2002). Overexpression lines of low molecular weight *PAP*, *PvPAP3*, increased utilization of extracellular dNTPs in common bean leading to enhanced biomass as compared to WT (Liang et al. 2010). PvPAP3 was reported to be localized in the apoplast, and plasma membrane is indicating that PvPAP3 can also mobilize phosphate monoesters in apoplast (Liang et al. 2010). Other most characterized PAPs, LeSAP1 and LeSAP2, were identified from culture media of Pi-starved tomato cell suspension and were found to be the major secretory tomato PAPs with broad substrate specificity (Bozzo et al. 2002). In soybean, GmPAP4 and GmPAP14 have been shown to enhance organic P utilization (Kong et al. 2014, 2018). In a recent study, plasma membrane-localized SgPAP23 from *Stylosanthes guianensis* was reported to be a primary enzyme involved in extracellular phytate utilization in forage crop, stylo (*Stylosanthes* spp.) (Liu et al. 2018). Overexpression of *SgPAP23* enhanced phytase activity in both bean hairy root and Arabidopsis. Likewise, overexpression of other *SgPAPs* (*SgPAP7*, *SgPAP12*, and *SgPAP26*) also increased root-associated APase activities in transgenic bean hairy root (Liu et al. 2016).

In rice, PAPs exist as a large family of 26 members (Zhang et al. 2011). Most of these PAPs are low Pi inducible and are principally expressed in roots (Zhang et al. 2011). However, there are no reports revealing localization of rice PAPs at the tissue or organellar levels. Rice PAP OsPAP23 has been shown to possess phytase activity and is recognized as OsPHY1 (Li et al. 2012b). Overexpression of *OsPHY1* with potato signal peptide increased accumulation of total and soluble P in transgenic tobacco as compared to WT (Li et al. 2012b). *OsPHY1* was found to be highly expressed in seeds indicating their role in solubilization of seed phytate during seed germination. Notably, 70% of the seed P reserves are stored as phytate, which is hydrolyzed by phytases in germinating seeds (Dionisio et al. 2011). Another rice PAP, OsPAP10a, was reported to increase utilization of exogenously supplied ATP. Rice overexpression transgenics of *OsPAP10a* showed significantly enhanced secretory APase activity as compared to WT (Tian et al. 2012). Another closer homolog of OsPAP10a, i.e., OsPAP10c, has also been reported to possess similar functions as OsPAP10a (Lu et al. 2016). Notably, both rice PAPs, OsPAP10a and OsPAP10c, are close homologs of well-characterized Arabidopsis PAP, AtPAP10. Another rice PAP, OsPAP21b, is a secretory protein that hydrolyzes soil organophosphates making them available for plant use (Mehra et al. 2017). Despite the very low phosphorus use efficiency (PUE) and high economic importance of rice, the majority of the rice PAPs are still uncharacterized.

3.6.2 Seed Germination and Abiotic/Biotic Stress Tolerance

Though most of the PAPs are low Pi inducible, Pi deficiency was not found to influence the expression of some of the PAPs (Veljanovski et al. 2006; Sun et al. 2012). Functional analysis of many PAPs revealed their diverse functional roles (summarized in Table 3.1). For example, AtPAP15 is known to possess phytase activity and is involved in the hydrolysis of phytate during seed and pollen germination (Kuang

Table 3.1 Functional details of PAPs characterized in different plant species

PAP	Gene ID	Monomer/dimer	Low Pi inducibility (transcript/protein)	Preferred substrate	Subcellular localization	Phenotype/function	References
Arabidopsis *thaliana*							
AtPAP2	At1g13900	ND	–/ND	ND	Plastid and mitochondrion	Overexpression of *AtPAP2* increased plant growth and seed yield in Arabidopsis by regulating carbon metabolism	Sun et al. (2012)
						Overexpression of *AtPAP2* in *Camelina sativa* led to faster plant growth, photosynthetic rate, seed size, and yield	Zhang et al. (2012)
AtPAP10	At2g16430	Oligomer	+/+	ATP	Cell wall	*atpap10* mutant showed decreased fresh weight, primary root length, and lateral root length and density under Pi deficiency	Wang et al. (2011)
						AtPAP10 overexpression lines showed enhanced biomass, root length, and lateral root length and density	Wang et al. (2011)
AtPAP12	At2g27190	Homodimer	+/+	PEP	Secretory	P solubilization and acquisition from extracellular environment under Pi deficiency	Tran et al. (2010b), Robinson et al. (2012)

AtPAP15	At3g07130	Monomer	−/−	PEP and Na-phytate	ND	*atpap15* mutant displayed lower phytase and phosphatase activity in germinating seedling and pollen and lower seed and pollen germination	Kuang et al. (2009)
						Overexpression of *AtPAP15* with carrot signal peptide in soybean increased dry weight and total P, pod number and seeds on phytate as a sole P source	Wang et al. (2009)
AtPAP17	At3g17790	Monomer	+/ND	ND	ND	Proposed to play role in phosphate mobilization	del Pozo et al. (1999)
						Also possesses peroxidase activity and induced by salt and ABA; therefore, plays a role in ROS scavenging under abiotic stresses	del Pozo et al., (1999)
AtPAP23	At4g13700	ND	ND	ATP, dATP, phosphoserine	ND	Overexpression lines of AtPAP23 showed enhanced Fe and Mn content	Zhu et al. (2005)
AtPAP25	At4g36350	Monomer	+/+	PEP and phosphoamino acids	Cell wall	*atpap25* mutant showed arrested growth, low P accumulation, and high anthocyanin content under Pi deficiency	Del Vecchio et al. (2014)

(continued)

Table 3.1 (continued)

PAP	Gene ID	Monomer/ dimer	Low Pi inducibility (transcript/ protein)	Preferred substrate	Subcellular localization	Phenotype/function	References
AtPAP26	At5g34850	Monomer	−/+	PEP	Dual-targeted; vacuolar and secretory	*atpap26* mutant showed lower intracellular and secretory APase activity, P concentrations, and impaired plant growth under Pi deficiency	Tran et al. (2010b), Hurley et al. (2010)
						Senescing leaves of *atpap26* mutant depicted ~90% decrease in phosphatase activity, seed P content, and yield	Robinson et al. (2012)
Oryza sativa							
OsPHY1 (OsPAP23)	Os08g17784	ND	+/NA	Phytate	ND	Involved in the mobilization of seed phytate during germination. Overexpression of *OsPHY1* fused with potato signal peptide increased phytase activity, P content, and dry weight of transgenic tobacco	Li et al. (2012b)
OsPAP10a	Os01g56880	ND	+/+	ATP	ND	Overexpression of *OsPAP10a* increased secretory APase activity and P content of transgenic rice under Pi deficiency. Overexpression lines also showed an increased number of tillers	Tian et al. (2012)

OsPAP10c	Os12g44020	ND	+/+	ATP	ND	Overexpression of *OsPAP10c* in rice led to higher intracellular and secretory APase activity and P content when ATP was supplied as a sole P source. OE lines also possessed a greater number of tillers	Lu et al. (2016)
OsPAP21b	Os11g05400	ND	+/+	ADP	ND	Overexpression of *OsPAP21b* in rice improved hydrolysis of exogenous organophosphates in the soil as well as hydroponics leading to enhanced growth	Mehra et al. (2017)
OsPAP26	Os06g43640	ND	−/+	ND	Apoplast	Overexpression of *OsPAP26* increased remobilization of Pi from senescing to non-senescing leaves under Pi deficiency which led to improved plant growth as compared to WT	Gao et al. (2017)
Glycine max							
GmPAP3	Glyma03g35190	ND	−/−	ND	Mitochondrion	Induced by salt, osmotic, and oxidative stress. Overexpression of *GmPAP3* increased root length under NaCl, PEG, and paraquat treatment in Arabidopsis	Liao et al. (2003), Li et al. (2008)

(continued)

Table 3.1 (continued)

PAP	Gene ID	Monomer/ dimer	Low Pi inducibility (transcript/ protein)	Preferred substrate	Subcellular localization	Phenotype/function	References
GmPAP4	Glyma03g35200	ND	+/ND	Phytate	Cytoplasm or plasma membrane	Arabidopsis overexpression lines of *GmPAP4* accumulated higher biomass and P content when grown on phytate as a sole P source	Kong et al. (2014)
GmPAP14	Glyma08g09880	ND	+/ND	Phytate	Secretory	Overexpression of *GmPAP14* increased phosphate content and phytate utilization in Arabidopsis	Kong et al. (2018)
GmPHY (GmPAP19)	Glyma08g46550	ND	ND	Phytate	ND	Overexpression of *GmPHY* increased phytase activity in transformed soybean cells	Hegeman and Grabau (2001)
GmPAP21	Glyma10g08300	ND	+/ND	ND	Intracellular	Overexpression of *GmPAP21* inhibited nodule formation in soybean upon infection with rhizobia	Li et al. (2017)
GmPAP33	Glyma19g37860	ND	+/ND	Phospholipids	Plasma membrane	Overexpression of *GmPAP33* led to the formation of large arbuscules and increased yield in soybean as compared to control plant	Li et al. (2019)
Phaseolus vulgaris							
PvPAP3	FJ464333	Monomer	+/+	ATP	Plasma membrane and apoplast	Overexpression of *PvPAP3* increased growth and Pi content of soybean transgenics when ATP was supplied as a sole P source	Liang et al. (2010)
KbPAP (PvPAP2)	AJ001270	Dimer	+/ND	ATP	Cytoplasm	Overexpression of *PvPAP2* was found toxic to transgenic bean hairy root	Cashikar et al. (1997), Lianget al. (2012)

Lycopersicon esculentum							
LeSAP1	ND	Monomer	+/+	PEP	Secretory	Proposed to be involved in the hydrolysis of extracellular organic P compounds	Bozzo et al. (2002, 2006)
LeSAP2	ND	Monomer	+/+	PEP	Secretory	Proposed to be involved in the hydrolysis of extracellular organic P compounds	Bozzo et al. (2002, 2006)
SlIAP	ND	Heterodimer	+/+	Phosphotyrosine	ND	Proposed to hydrolyze intracellular phosphate esters under Pi deficiency in tomato	Bozzo et al. (2002, 2006)
						Also shows peroxidase activity, therefore involved in ROS metabolism	Bozzo et al. (2002, 2006)
Nicotiana tabacum							
NtPAP	EF397753	Monomer	ND/+	dNTPs and phytate	Secretory	Proposed to mobilize organic P in the rhizosphere	Lung et al. (2008)
NtPAP12	AB017967	Tetramer	ND	phosphotyrosine	Cell wall	Overexpression of *NtPAP12* increased deposition of cellulose and β-glucan synthesis in the cell wall of transformed tobacco cells	Kaida et al. (2008)
Medicago truncatula							
MtPAP1	AY804257	ND	+/ND	Phytate	Apoplast	Overexpression of *MtPAP1* in Arabidopsis increased biomass, phosphatase activity, and P accumulation when grown on phytate as a sole P source	Xiao et al. (2006)

ND not determined, "+" upregulation, "–" downregulation

et al. 2009). Overexpression of *AtPAP15* was shown to alleviate salt and osmotic stress besides reducing phytate content in seed as compared to WT (Zhang et al. 2008). Overexpression of another Arabidopsis PAP, *AtPAP2*, enhances growth and yield through enhanced sucrose phosphate synthase (SPS) activity in Arabidopsis transgenics as compared to WT (Sun et al. 2012). Similarly, mitochondrion localized soybean PAP, GmPAP3, was found to be induced by salt and oxidative stress instead of phosphate deficiency (Liao et al. 2003). Overexpression of *GmPAP3* in Arabidopsis increased tolerance to osmotic, salt, and oxidative stresses as compared to WT (Li et al. 2008). This study suggested the role of *GmPAP3* in ROS scavenging for providing stress tolerance. Similarly, *PgPAP18* from *Pennisetum glaucum* was also reported to be involved in multiple stress tolerance (Reddy et al. 2017). Another PAP, AtPAP17, was also reported to exhibit peroxidase activity and is induced by salt and ABA (del Pozo et al. 1999). However, *AtPAP17* is also induced by Pi starvation. This upholds the possibility that low Pi-inducible PAPs can also play multiple roles under different abiotic stresses like other Pi-independent PAPs.

Very few studies have also unearthed the role of PAPs in disease resistance. Expression of Arabidopsis *Purple Acid Phosphatase 5 (AtPAP5)* is induced by infection of the bacterial pathogen, *Pseudomonas syringae pv. tomato DC3000 (Pst DC3000)* (Ravichandran et al. 2013). *atpap5* mutants showed enhanced susceptibility to *Pst DC3000* and displayed reduced expression of defense-related genes such as *PR1* (Pathogenesis-Related Gene 1), *ICS1* (Isochorismate Synthase 1), and *PDF1.2* (Plant Defensin 1.2). Exogenous application of SA analog, BTH, restored expression of *PR1* in *atpap5* indicating the role of *AtPAP5* upstream to SA pathway. This study provided evidence indicating novel roles of PAPs in plant defense mechanisms.

3.6.3 Root Architecture Modulation

Phosphatase activity of PAPs can also modulate root system architecture (RSA). Overexpression lines of root surface-associated Arabidopsis PAP, *AtPAP10*, developed larger roots as compared to WT under Pi-deficient and ADP supplemented conditions (Wang et al. 2011). *AtPAP10* overexpression lines possessed increased lateral root length, lateral root density, and primary root length as compared to WT under Pi deficiency. On the other hand, *atpap10* mutants formed weaker roots as compared to WT (Wang et al. 2011). AtPAP10-mediated phosphorylation of associated root proteins involved in growth and development was suggested as a possible mechanism for increased root growth in *AtPAP10* overexpression lines. Similarly, overexpression of a tobacco PAP, NtPAP12, increased deposition of cellulose and β-glucan synthesis in the cell wall of transformed tobacco cells (Kaida et al. 2009). NtPAP12 is believed to dephosphorylate cell wall-associated enzymes, α-xylosidase and β-glucosidase for regulating cell wall biosynthesis (Kaida et al. 2009). In rice also, overexpression of *OsPAP21b* enhanced primary as well as lateral root length under Pi deficiency as compared to WT (Mehra et al. 2017). Opposite to this, silencing of *OsPAP21b* reduced root biomass in rice transgenics as compared to WT. These findings indicate some plausible direct or indirect roles of PAPs in RSA modulation under Pi deficiency. Further, some phosphate compounds, which act as signaling molecules, could be potential targets of PAPs to bring about physiological and morphological alterations in plants under Pi deficiency. However, this hypothesis needs further investigations.

3.6.4 Nodule Formation and Arbuscular Mycorrhizal (AM) Symbiosis

Leguminous crops have high phosphate requirement for driving energy-demanding processes such as nitrogen fixation (Mehra et al. 2018). Few PAPs have been studied for their roles in nitrogen fixation and nodule formation. Recently, soybean PAP *GmPAP21* was reported to be involved in energy charge control in nodules of soybean (Li et al. 2017). *GmPAP21* showed high expression in nodules apart from roots, leaves, stems, and pods. However, overexpression of *GmPAP21* inhibited nodule formation in soybean. Similarly, overexpression of another PAP, *AsPPD1*, also inhibited nodulation in transgenic *Astragalus sinicus* hairy roots (Wang et al. 2015). Moreover, no nodule formation was observed in half of the overexpressed roots. On the other hand, silencing of *AsPPD1* increased the number of nodules, albeit nodules were smaller and possessed very low nitrogenase activity. Furthermore, silencing of *AsPPD1* led to early nodule senescence consequently impairing nitrogen fixation.

Besides legume-rhizobium symbiosis, PAPs participate in AM symbiosis. Soybean PAP, GmPAP33, was recently found to be involved in arbuscule degeneration (Li et al. 2019). Investigations revealed that GmPAP33 hydrolyzes phospholipids (phosphatidylcholine) and phosphatidic acid present in arbuscular membranes. Expression of *GmPAP33* is induced by AM inoculation irrespective of Pi availability indicating crucial roles of *GmPAP33* in AM symbiosis rather Pi availability. Overexpression of *GmPAP33* led to the formation of large arbuscules and increased yield in soybean as compared to control plants. On the contrary, RNAi lines showed smaller arbuscules and high content of phospholipids. These results point toward the roles of PAPs in arbuscule turnover and AM symbiosis.

3.6.5 Regulation of Flowering

Expression studies of PAP family members in different tissue have revealed their prominent expression in reproductive tissues. Although their accurate biological functions in reproductive tissues are still unclear, accumulating evidence constantly indicates their role in flowering and seed development. Interestingly, all 28 of the expressed *AtPAPs* showed expression in flower. Moreover, nine of the *AtPAPs* showed expression in flower only (Zhu et al. 2005). Histochemical GUS assays showed expression of *AtPAP23* in the floral apical meristem, immature flower, petals, and anthers in Arabidopsis (Zhu et al. 2005). However, overexpression, knockout, or silencing of *AtPAP23* could not produce any visible differences in flower development between WT and transgenics. This may be attributed to functional redundancy among different members of Arabidopsis PAPs. Similar to Arabidopsis, most of the chickpea PAPs were predominantly expressed in flower bud (Bhadouria et al. 2017). In rice also, overexpression of PAPs such as *OsPAP21b* and *OsPAP10a* led to early flowering (Mehra et al. 2017; Tian et al. 2012). These findings suggest that PAPs may influence flowering in plants, probably by regulating Pi homeostasis. Several reports indicate the role of phosphorus in flowering and fruit set in different plant species (Menary and Staden 1976; Anuradha et al. 1990; Erel et al. 2008; Petraglia et al. 2014). Interestingly, Pi deficiency is also reported to influence the expression of several genes involved in flowering such as *FLC* (flowering locus C), *FT* (flowering locus T), and *LFY* (leafy) (Kant et al. 2011). However, a direct mechanism between the activity of PAPs and flowering is yet missing and needs further investigations.

3.6.6 Seed Development

Phosphorus nutrition is known to play important roles in seed development (Millar and Turk 1943). Phytate content in seeds is directly associated with seed weight (Dwivedi et al. 2017). In a recent study, one of the chickpea PAPs, CaPAP7, was found to possess phytase activity (Bhadouria et al. 2017). Interestingly, expression of *CaPAP7* was significantly higher in chickpea genotypes with lower seed weight and seed phytate content. This study revealed the role of PAPs as "functional phytases" in regulating seed phytate and consequently seed weight.

Functions of some of the PAPs have also been implicated in grain filling in crops. Out of 26 rice PAPs, five PAPs were found upregulated at 15 DAA (days after anthesis) (Jeong et al. 2017). Increased expression of PAPs was suggested to be involved in phosphate remobilization from senescing flag leaves to developing grains. Notably, expression of *OsPAP26* was found to be induced in senescing leaves, whereas Pi deficiency had no impact on the expression of *OsPAP26* (Gao et al. 2017). This suggests an explicit role of some PAPs in leaf senescence and Pi remobilization. Similarly, *AtPAP26* (Arabidopsis orthologue of rice *OsPAP26*) also plays a role in leaf senescence and phosphate remobilization to seeds (Robinson et al. 2012). *atpap26* mutants showed delayed senescence and parallelly reduced seed P concentrations. Altogether, these findings reveal significant functions of PAPs in modulating Pi status in flag leaves and seeds during grain development.

3.7 Future Perspectives

Plant PAPs are integral to Pi deficiency responses. Overexpressing many functional PAPs in Arabidopsis and major crop plants has so far proved successful in improving organic P utilization, especially phytate. Further efforts for combining actions of both secretory and intracellular PAPs may simultaneously improve Pi acquisition efficiency as well as intracellular Pi utilization. This holds the chance for improving phosphorus use efficiency (PUE) of many crop plants (e.g., rice) with very low PUE. In the case of developing nations, such efforts will also possibly relieve the economic burden lied on importing a large amount of Pi fertilizers. Moreover, targeting PAPs for improving PUE can also minimize the application of Pi fertilizers, thereby addressing several ecological concerns associated with injudicious fertilizer use.

Plant PAPs are endowed with many unique characteristic features, for instance, broad substrate specificities. Several substrates for PAP proteins have been unearthed. These substrates involve a diverse class of compounds such as phospholipids, phosphoproteins, phosphosugars, nucleotides, and phosphoinositols. Given the versatility of their substrates, plant PAPs may be associated with a range of molecular processes and signal transduction pathways. Reversible phosphorylation is one of the prime mechanisms modulating signal transduction pathways (Lan et al. 2013; Li et al. 2014). Regulation of phosphorylation status of many proteins and other compounds by PAPs may determine the course of many signaling cascades. However, the precise role of PAPs in signaling processes is still elusive and demands more investigations. Additionally, PAPs themselves are reported to undergo posttranscriptional or posttranslational modifications. The extent to which these modifications affect their structure, stability, and kinetic properties needs to be tested in details.

Over the years, plant PAPs have been largely studied from the perspective of improving low Pi tolerance. However, building evidence suggests their wider roles in multiple stresses and physiological and developmental processes (Fig. 3.2). Prominent expression of PAPs in flower parts is also intriguing and needs intense efforts to explore their possible functions in flowering. Several PAPs have been identified for their roles in remobilization of phosphate from senescing flag leaves to developing seeds (Jeong et al. 2017), while some others (e.g., AtPAP15, CaPAP7, OsPAP23) act as functional phytases. These PAPs serve as future targets for manipulating P quantity ultimately loaded into grains. This will not only reduce Pi removal from crop fields at harvest but also control seed phytate content. Notably, seed phytate constitutes around 65–80% of P in seeds (Raboy 2001). Humans and monogastric animals lack the natural ability to digest phytate. Consequently, 90% of the consumed phytate is excreted, which ultimately finds its way into rivers and lakes causing eutrophication (Raboy 2009). Therefore, minimizing seed phytate content without compromising vigor and seed germination has been suggested useful in addressing phytate promoted eutrophication (Yamaji et al. 2017). Phytate being an "anti-nutrient" also chelates many important nutrients such as iron, zinc, and

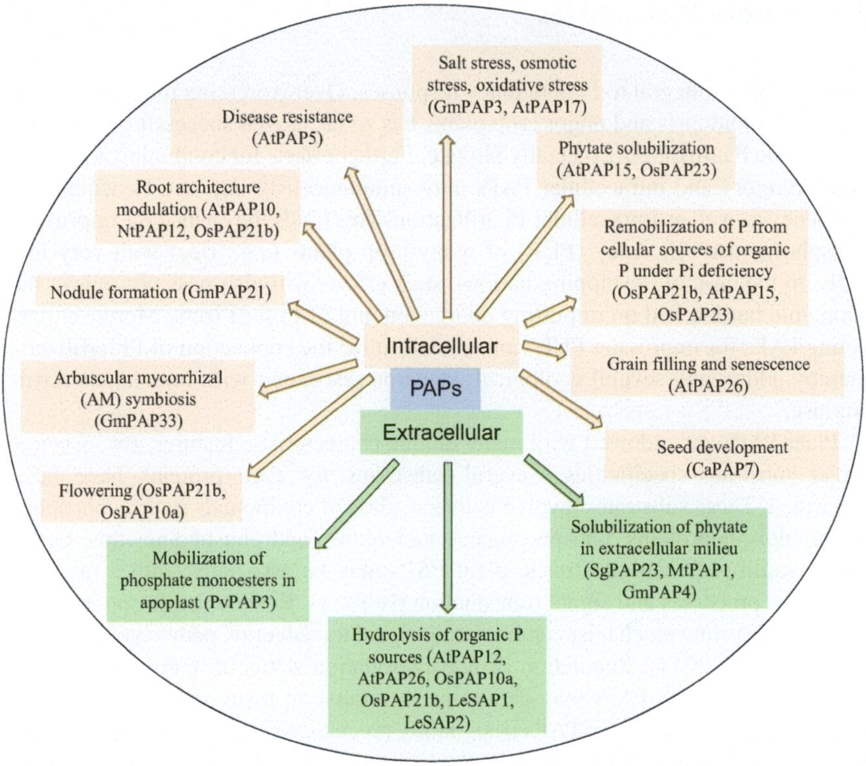

Fig. 3.2 Plant purple acid phosphatases (PAPs) are involved in diverse physiological roles. Summary of diverse functions of intracellular and/or extracellular plant PAPs

calcium, thereby reducing their bioavailability. Therefore, further research is needed to deploy highly active phytases/PAPs for minimizing seed phytate content in crops possessing high seed phytate content. Besides, "secretory" PAPs with phytase activity can also mobilize the most abundant organic P reserve, i.e., soil phytate. Thus, these PAPs can play dual roles in imparting both low P tolerance and improving seed quality.

Cumulative results from independent studies have well explored the roles of PAPs in diverse physiological processes. However, the current knowledge of their mechanisms of actions in complex metabolic networks is very limiting. Detailed investigations into their regulation, posttranscriptional modifications, structure, and kinetic properties could carve out new means for increasing their efficiency and stability. Overall, PAPs participate in multiple physiological, metabolic, and molecular processes. Therefore, a comprehensive understanding of their mechanisms and broader functions would facilitate the designing of effective breeding and biotechnological strategies for improving crop yields through multiple stress tolerance.

Acknowledgments Our research is funded by a core grant from NIPGR and DBT, India.

References

Anuradha K, Pampapathy K, Narayana N (1990) Effect of nitrogen and phosphorus on flowering, yield and quality of marigold. Indian J Hort 47:353–357

Bhadouria J, Singh AP, Mehra P, Verma L, Srivastawa R, Parida S, Giri J (2017) Identification of purple acid phosphatases in chickpea and potential roles of *CaPAP7* in seed phytate accumulation. Sci Rep 7:11012

Bozzo GG, Raghothama KG, Plaxton WC (2002) Purification and characterization of two secreted purple acid phosphatase isozymes from phosphate-starved tomato (*Lycopersicon esculentum*) cell cultures. Eur J Biochem 269:6278–6286

Bozzo GG, Singh VK, Plaxton WC (2004) Phosphate or phosphite addition promotes the proteolytic turnover of phosphate-starvation inducible tomato purple acid phosphatase isozymes. FEBS Lett 573:51–54

Bozzo GG, Dunn EL, Plaxton WC (2006) Differential synthesis of phosphate starvation inducible purple acid phosphatase isozymes in tomato (*Lycopersicon esculentum*) suspension cells and seedlings. Plant Cell Environ 29:303–313

Cashikar AG, Kumaresan R, Rao NM (1997) Biochemical characterization and subcellular localization of the red kidney bean purple acid phosphatase. Plant Physiol 114:907–915

Chiou TJ, Aung K, Lin SI, Wu CC, Chiang SF, Su CL (2006) Regulation of phosphate homeostasis by microRNA in Arabidopsis. Plant Cell 18:412–421

del Pozo JC, Allona I, Rubio V, Leyva A, de la Pena A, Aragoncillo C, Paz-Ares J (1999) A type 5 acid phosphatase gene from Arabidopsis *thaliana* is induced by phosphate starvation and by some other types of phosphate mobilizing/oxidative stress conditions. Plant J 19:579–589

Del Vecchio HA, Ying S, Park J, Knowles VL, Kanno S, Tanoi K, She YM, Plaxton WC (2014) The cell wall-targeted purple acid phosphatase AtPAP25 is critical for acclimation of Arabidopsis *thaliana* to nutritional phosphorus deprivation. Plant J 80:569–581

Devaiah BN, Karthikeyan AS, Raghothama KG (2007a) WRKY75 transcription factor is a modulator of phosphate acquisition and root development in Arabidopsis. Plant Physiol 143:1789–1801

Devaiah BN, Nagarajan VK, Raghothama KG (2007b) Phosphate homeostasis and root development in Arabidopsis are synchronized by the zinc finger transcription factor ZAT6. Plant Physiol 145:147–159

Dionisio G, Madsen CK, Holm PB, Welinder KG, Jørgensen M, Stoger E, Arcalis E, Brinch-Pedersen H (2011) Cloning and characterization of purple acid phosphatase phytases from wheat, barley, maize, and rice. Plant Physiol 156:1087–1100

Doerner P (2008) Phosphate starvation signaling: a threesome controls systemic Pi homeostasis. Curr Opin Plant Biol 11:536–540

Duff SM, Lefebvre DD, Plaxton WC (1989) Purification and characterization of a phosphoenolpyruvate phosphatase from *Brassica nigra* suspension cells. Plant Physiol 90:734–741

Duff SMG, Sarath G, Plaxton WC (1994) The role of acid phosphatases in plant phosphorus metabolism. Physiol Plant 90:791–800

Dwivedi V, Parida SK, Chattopadhyay D (2017) A repeat length variation in myo-inositol monophosphatase gene contributes to seed size trait in chickpea. Sci Rep 7:4764

Erel R, Dag A, Ben-Gal A, Schwartz A, Yermiyahu U (2008) Flowering and fruit set of olive trees in response to nitrogen, phosphorus, and potassium. J Am Soc Hortic Sci 133:639–647

Flanagan JU, Cassady AI, Schenk G, Guddat LW, Hume DA (2006) Identification and molecular modeling of a novel, plant-like, human purple acid phosphatase. Gene 377:12–20

Gao W, Lu L, Qiu W, Wang C, Shou H (2017) *OsPAP26* encodes a major purple acid phosphatase and regulates phosphate remobilization in rice. Plant Cell Physiol 58:885–892

Ghahremani M, Park J, Anderson EM, Marty-Howard NJ, Mullen RT, Plaxton WC (2018) Lectin AtGAL1 interacts with high-mannose glycoform of the purple acid phosphatase AtPAP26 secreted by phosphate-starved Arabidopsis. Plant Cell Environ. https://doi.org/10.1111/pce.13463

González-Muñoz E, Avendaño-Vázquez A-O, Chávez Montes RA, de Folter S, Andrés-Hernández L, Abreu-Goodger C, Sawers RJH (2015) The maize (*Zea mays* ssp. *mays* var. B73) genome encodes 33 members of the purple acid phosphatase family. Front Plant Sci 6:341

Gregory AL, Hurley BA, Tran HT, Valentine AJ, She YM, Knowles VL, Plaxton WC (2009) *In vivo* regulatory phosphorylation of the phosphoenolpyruvate carboxylase AtPPC1 in phosphate-starved Arabidopsis *thaliana*. Biochem J 420:57–65

Harrison AF (1987) Soil organic phosphorus—a review of world literature. CAB International, Wallingford

Hegeman CE, Grabau EA (2001) A novel phytase with sequence similarity to purple acid phosphatases is expressed in cotyledons of germinating soybean seedlings. Plant Physiol 126:1598–1608

Hurley BA, Tran HT, Marty NJ, Park J, Snedden WA, Mullen RT, Plaxton WC (2010) The dual-targeted purple acid phosphatase isozyme AtPAP26 is essential for efficient acclimation of Arabidopsis to nutritional phosphate deprivation. Plant Physiol 153:1112–1122

Jeong K, Baten A, Waters DL, Pantoja O, Julia CC, Wissuwa M, Heuer S, Kretzschmar T, Rose TJ (2017) Phosphorus remobilization from rice flag leaves during grain filling: an RNA-seq study. Plant Biotechnol J 15:15–26

Kaida R, Hayashi T, Kaneko TS (2008) Purple acid phosphatase in the walls of tobacco cells. Phytochemistry 69:2546–2551

Kaida R, Satoh Y, Bulone V, Yamada Y, Kaku T, Hayashi T, Kaneko TS (2009) Activation of β-glucan synthases by wall-bound purple acid phosphatase in tobacco cells. Plant Physiol 150:1822–1830

Kant S, Peng M, Rothstein SJ (2011) Genetic regulation by NLA and microRNA827 for maintaining nitrate-dependent phosphate homeostasis in Arabidopsis. PLoS Genet 7:e1002021

Kong Y, Li X, Ma J, Li W, Yan G, Zhang C (2014) *GmPAP4*, a novel purple acid phosphatase gene isolated from soybean (*Glycine max*), enhanced extracellular phytate utilization in Arabidopsis *thaliana*. Plant Cell Rep 33:655–667

Kong Y, Li X, Wang B, Li W, Du H, Zhang C (2018) The soybean purple acid phosphatase GmPAP14 predominantly enhances external phytate utilization in plants. Front Plant Sci 9:292

Kuang R, Chan KH, Yeung E, Lim BL (2009) Molecular and biochemical characterization of AtPAP15, a purple acid phosphatase with phytase activity, in Arabidopsis. Plant Physiol 151:199–209

Lan P, Li W, Schmidt W (2013) Genome-wide co-expression analysis predicts protein kinases as important regulators of phosphate deficiencyinduced root hair remodeling in Arabidopsis. BMC Genomics 14:210

Li D, Zhu H, Liu K, Liu X, Leggewie G, Udvardi M, Wang D (2002) Purple acid phosphatases of Arabidopsis *thaliana*. Comparative analysis and differential regulation by phosphate deprivation. J Biol Chem 277:27772–27781

Li WYF, Shao G, Lam HM (2008) Ectopic expression of *GmPAP3* alleviates oxidative damage caused by salinity and osmotic stresses. New Phytol 178:80–91

Li C, Gui S, Yang T, Walk T, Wang X, Liao H (2012a) Identification of soybean purple acid phosphatase genes and their expression responses to phosphorus availability and symbiosis. Ann Bot 109:275–285

Li RJ, Lu WJ, Guo CJ, Li XJ, Gu JT, Xiao K (2012b) Molecular characterization and functional analysis of *OsPHY1*, a purple acid phosphatase (PAP)–type phytase gene in rice (*Oryza sativa* L.). J Integr Agric 11:1217–1226

Li K, Xu C, Fan W, Zhang H, Hou J, Yang A, Zhang K (2014) Phosphoproteome and proteome analyses reveal low-phosphate mediated plasticity of root developmental and metabolic regulation in maize (*Zea mays* L.). Plant Physiol Biochem 83:232–242

Li C, Li C, Zhang H, Liao H, Wang X (2017) The purple acid phosphatase GmPAP21 enhances internal phosphorus utilization and possibly plays a role in symbiosis with rhizobia in soybean. Physiol Plant 159:215–227

Li C, Zhou J, Wang X, Liao H (2019) A purple acid phosphatase, GmPAP33, participates in arbuscule degeneration during AM symbiosis in soybean. Plant Cell Environ. https://doi.org/10.1111/pce.13530

Liang C, Tian J, Lam HM, Lim BL, Yan X, Liao H (2010) Biochemical and molecular characterization of PvPAP3, a novel purple acid phosphatase isolated from common bean enhancing extracellular ATP utilization. Plant Physiol 152:854–865

Liang C, Sun L, Yao Z, Liao H, Tian J (2012) Comparative analysis of *PvPAP* gene family and their functions in response to phosphorus deficiency in common bean. PLoS One 7:e38106

Liao H, Wong FL, Phang TH, Cheung MY, Li WYF, Shao G, Yan X, Lam HM (2003) *GmPAP3*, a novel purple acid phosphatase-like gene in soybean induced by NaCl stress but not phosphorus deficiency. Gene 318:103–111

Liu PD, Xue YB, Chen ZJ, Liu GD, Tian J (2016) Characterization of purple acid phosphatases involved in extracellular dNTP utilization in *Stylosanthes*. J Exp Bot 67:4141–4154

Liu P, Cai Z, Chen Z, Mo X, Ding X, Liang C, Liu G, Tian J (2018) A root-associated purple acid phosphatase, SgPAP23, mediates extracellular phytate-P utilization in *Stylosanthes guianensis*. Plant Cell Environ 41:2821–2834

Lu L, Qiu W, Gao W, Tyerman SD, Shou H, Wang C (2016) *OsPAP10c*, a novel secreted acid phosphatase in rice, plays an important role in the utilization of external organic phosphorus. Plant Cell Environ 39:2247–2259

Lung SC, Leung A, Kuang R, Wang Y, Leung P, Lim BL (2008) Phytase activity in tobacco (*Nicotiana tabacum*) root exudates is exhibited by a purple acid phosphatase. Phytochemistry 69:365–373

Mehra P, Pandey BK, Giri J (2016) Comparative morphophysiological analyses and molecular profiling reveal Pi-efficient strategies of a traditional rice genotype. Front Plant Sci 6:1184

Mehra P, Pandey BK, Giri J (2017) Improvement in phosphate acquisition and utilization by a secretory purple acid phosphatase (OsPAP21b) in rice. Plant Biotechnol J 15:1054–1067

Mehra P, Singh AP, Bhadouria J, Verma L, Panchal P, Giri J (2018) Phosphate homeostasis: links with seed quality and stress tolerance in chickpea. In: Wani S, Jain M (eds) Pulse improvement. Springer, Cham, pp 191–217

Menary RC, Staden JV (1976) Effect of phosphorus nutrition and cytokinins on flowering in the tomato, *Lycopersicon esculentum* Mill. Aust J Plant Physiol 3:201–205

Millar CE, Turk LM (1943) Fundamentals of soil science. Wiley, New York

Mitić N, Smith SJ, Neves A, Guddat LW, Gahan LR, Schenk G (2006) The catalytic mechanisms of binuclear metallohydrolases. Chem Rev 106:3338–3363

Mitić N, Noble CJ, Gahan LR, Hanson GR, Schenk G (2009) Metal-ion mutagenesis: conversion of a purple acid phosphatase from sweet potato to a neutral phosphatase with the formation of an unprecedented catalytically competent $Mn^{II}Mn^{II}$ active site. J Am Chem Soc 131:8173–8179

Oddie GW, Schenk G, Angel NZ, Walsh N, Guddat LW, de Jersey J, Cassady AI, Hamilton SE, Hume DA (2000) Structure, function, and regulation of tartrate-resistant acid phosphatase. Bone 27:575–584

Olczak M, Morawiecka B, Watorek W (2003) Plant purple acid phosphatases-genes, structures and biological function. Acta Biochim Pol 50:1245–1256

Petraglia A, Tomaselli M, Mondoni A, Brancaleoni L, Carbognani M (2014) Effects of nitrogen and phosphorus supply on growth and flowering phenology of the snowbed forb *Gnaphalium supinum* L. Flora Morphol Distrib Funct Ecol Plants 209:271–278

Raboy V (2001) Seeds for a better future: 'low phytate' grains help to overcome malnutrition and reduce pollution. Trends Plant Sci 6:458–462

Raboy V (2009) Approaches and challenges to engineering seed phytate and total phosphorus. Plant Sci 177:281–296

Ravichandran S, Stone S, Benkel B, Prithiviraj B (2013) Purple Acid Phosphatase5 is required for maintaining basal resistance against *Pseudomonas syringae* in Arabidopsis. BMC Plant Biol 13:107

Reddy CS, Kim KM, James D, Varakumar P, Reddy MK (2017) PgPAP18, a heat-inducible novel purple acid phosphatase 18-like gene (PgPAP18-like) from *Pennisetum glaucum*, may play a crucial role in environmental stress adaptation. Acta Physiol Plant 39:54

Robinson WD, Park J, Tran HT, Del Vecchio HA, Ying S, Zins JL, Patel K, McKnight TD, Plaxton WC (2012) The secreted purple acid phosphatase isozymes AtPAP12 and AtPAP26 play a pivotal role in extracellular phosphate-scavenging by Arabidopsis thaliana. J Exp Bot 63:6531–6542

Rubio V, Linhares F, Solano R, Martin AC, Iglesias J, Leyva A, Paz-Ares J (2001) A conserved MYB transcription factor involved in phosphate starvation signaling both in vascular plants and in unicellular algae. Genes Dev 15:2122–2133

Schenk G, Ge Y, Carrington LE, Wynne CJ, Searle IR, Carroll BJ, Hamilton S, de Jersey J (1999) Binuclear metal centers in plant purple acid phosphatases: Fe-Mn in sweet potato and Fe-Zn in soybean. Arch Biochem Biophys 370:183–189

Schenk G, Guddat LW, Ge Y, Carrington LE, Hume DA, Hamilton S, de Jersey J (2000) Identification of mammalian-like purple acid phosphatases in a wide range of plants. Gene 250:117–125

Schenk G, Gahan LR, Carrington LE, Mitic N, Valizadeh M, Hamilton SE, de Jersey J, Guddat LW (2005) Phosphate forms an unusual tripodal complex with the Fe-Mn center of sweet potato purple acid phosphatase. Proc Natl Acad Sci U S A 102:273–278

Schenk G, Miti N, Hanson GR, Comba P (2013) Purple acid phosphatase: a journey into the function and mechanism of a colorful enzyme. Coord Chem Rev 257:473–482

Sun F, Suen PK, Zhang Y, Liang C, Carrie C, Whelan J, Ward JL, Hawkins ND, Jiang L, Lim BL (2012) A dual-targeted purple acid phosphatase in Arabidopsis *thaliana* moderates carbon metabolism and its overexpression leads to faster plant growth and higher seed yield. New Phytol 194:206–219

Tian J, Liao H (2015) The role of intracellular and secreted purple acid phosphatases in plant phosphorus scavenging and recycling. In: Plaxton WC, Lambers H (eds) Phosphorus metabolism in plants, Annual plant reviews, vol 48. Wiley, Hoboken, NJ, pp 265–288

Tian J, Wang C, Zhang Q, He X, Whelan J, Shou H (2012) Overexpression of *OsPAP10a*, a root-associated acid phosphatase, increased extracellular organic phosphorus utilization in rice. J Integr Plant Biol 54:631–639

Tran HT, Hurley BA, Plaxton WC (2010a) Feeding hungry plants: the role of purple acid phosphatases in phosphate nutrition. Plant Sci 179:14–27

Tran HT, Qian W, Hurley BA, She YM, Wang D, Plaxton WC (2010b) Biochemical and molecular characterization of AtPAP12 and AtPAP26: the predominant purple acid phosphatase isozymes secreted by phosphate-starved Arabidopsis *thaliana*. Plant Cell Environ 33:1789–1803

Veljanovski V, Vanderbeld B, Knowles VL, Snedden WA, Plaxton WC (2006) Biochemical and molecular characterization of AtPAP26, a vacuolar purple acid phosphatase up-regulated in phosphate-deprived Arabidopsis suspension cells and seedlings. Plant Physiol 142:1282–1293

Venkidasamy B, Selvaraj D, Ramalingam S (2019) Genome-wide analysis of purple acid phosphatase (PAP) family proteins in *Jatropha curcas* L. Int J Biol Macromol 123:648–656

Wang XR, Wang YX, Tian J, Lim BL, Yan XL, Liao H (2009) Overexpressing *AtPAP15* enhances phosphorus efficiency in soybean. Plant Physiol 151:233–240

Wang L, Li Z, Qian W, Guo W, Gao X, Huang L, Wang H, Zhu H, Wu JW, Wang D, Liu D (2011) The Arabidopsis purple acid phosphatase AtPAP10 is predominantly associated with the root surface and plays an important role in plant tolerance to phosphate limitation. Plant Physiol 157:1283–1299

Wang L, Lu S, Zhang Y, Li Z, Du X, Liu D (2014) Comparative genetic analysis of Arabidopsis purple acid phosphatases AtPAP10, AtPAP12, and AtPAP26 provides new insights into their roles in plant adaptation to phosphate deprivation. J Integr Plant Biol 56:299–314

Wang J, Si Z, Li F, Xiong X, Lei L, Xie F, Chen D, Li Y, Li Y (2015) A purple acid phosphatase plays a role in nodule formation and nitrogen fixation in *Astragalus sinicus*. Plant Mol Biol 88:515–529

Xiao K, Harrison MJ, Wang ZY (2005) Transgenic expression of a novel *M. truncatula* phytase gene results in improved acquisition of organic phosphorus by Arabidopsis. Planta 222:27–36

Xiao K, Katagi H, Harrison M, Wang ZY (2006) Improved phosphorus acquisition and biomass production in Arabidopsis by transgenic expression of a purple acid phosphatase gene from *M. truncatula*. Plant Sci 170:191–202

Xie L, Shang Q (2018) Genome-wide analysis of purple acid phosphatase structure and expression in ten vegetable species. BMC Genomics 19:646

Yamaji N, Takemoto Y, Miyaji T, Mitani-Ueno N, Yoshida KT, Ma JF (2017) Reducing phosphorus accumulation in rice grains with an impaired transporter in the node. Nature 541:92–95

Zhang W, Gruszewski HA, Chevone BI, Nessler CL (2008) An Arabidopsis purple acid phosphatase with phytase activity increases foliar ascorbate. Plant Physiol 146:431–440

Zhang Q, Wang C, Tian J, Li K, Shou H (2011) Identification of rice purple acid phosphatases related to phosphate starvation signaling. Plant Biol 13:7–15

Zhang Y, Yu L, Yung KF, Leung DY, Sun F, Lim BL (2012) Over-expression of AtPAP2 in *Camelina sativa* leads to faster plant growth and higher seed yield. Biotechnol Biofuels 5:19

Zhang Y, Wang X, Lu S, Liu D (2014) A major root-associated acid phosphatase in Arabidopsis, AtPAP10, is regulated by both local and systemic signals under phosphate starvation. J Exp Bot 65:6577–6588

Zhu H, Qian W, Lu X, Li D, Liu X, Liu K, Wang D (2005) Expression patterns of purple acid phosphatase genes in Arabidopsis organs and functional analysis of AtPAP23 predominantly transcribed in flower. Plant Mol Biol 59:581–594

Chapter 4
Role of Serine/Threonine Phosphatase PP2A Class and Its Regulators in Salinity Stress Tolerance in Plants

Srishti Chawla, Deeksha Marothia, and Pratap Kumar Pati

4.1 Introduction

Constantly changing environmental conditions are one of the significant concerns of crop cultivators worldwide. Out of these threats, salt, drought, and temperature are the major factors affecting plant productivity drastically by changing vital plant processes, including photosynthesis, respiration, and nutrient acquisition (Gupta and Huang 2014; Fedoroff et al. 2010). A spectrum of stress signals either alone or collectively are sensed by plants especially by roots under high salt or low water conditions, which can further not just damage the plant cells but can even be detrimental to the whole plant (Vinocur and Altman 2005; Gupta and Huang 2014; Zhang and Shi 2013). Modern experimental tools such as genome/proteome sequencing, availability of whole genome sequences, genetic engineering, and microarray imaging approach have greatly aided in the development of new strategies to achieve salt tolerance in plants (Kaur and Pati 2017).

On molecular levels, the stress signal reception involves a series of biochemical phosphorylation and dephosphorylation events (Fig. 4.1). Both kinases and phosphatases exist as large families; however, plant kinases always outnumber the plant phosphatases. This numerical imbalance still maintains the phosphorylation status of the cell molecularly by regulating many vital processes of the plant cell (Wang et al. 2007). Recent developments in the understanding of molecular roles of plant phosphatases in response to salt stress and water deficiency have made a significant addition to knowledge in the field. This chapter provides insights into plant PP2A phosphatases and summarizes our current understanding of their role to combat salt stress.

S. Chawla
Department of Immunology and Microbiology, Institute of Biomedicine Sahlgrenska Academy, University of Gothenburg, Gothenburg, Sweden

D. Marothia · P. K. Pati (✉)
Department of Biotechnology, Guru Nanak Dev University, Amritsar, Punjab, India

G. K. Pandey (ed.), *Protein Phosphatases and Stress Management in Plants*,
https://doi.org/10.1007/978-3-030-48733-1_4

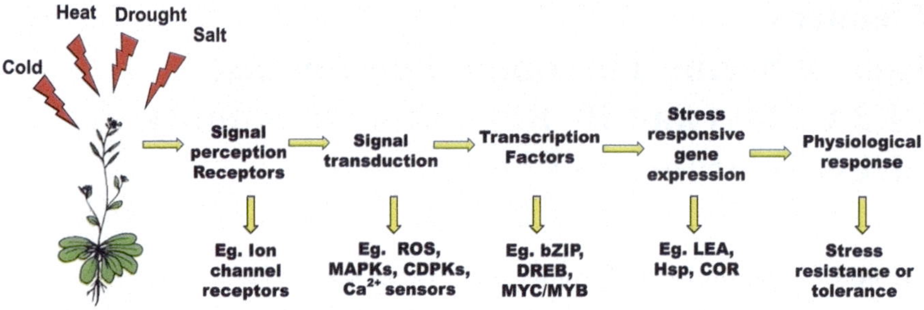

Fig. 4.1 Signal transduction pathway of plant in response to various stresses. The extracellular stress signals activate large and complex kinase phosphatase signaling cascade including the generation of secondary signal molecules

4.2 Overview of Serine-Threonine Phosphatases in Plants

4.2.1 Classification of Protein Phosphatases

The process of protein phosphorylation is very well known to regulate the critical events in the eukaryotes and prokaryotes. This phosphorylation signature appears on different amino acid substrates such as tyrosine, serine, threonine, cysteine, arginine, lysine, aspartate, glutamate, and histidine with serine, threonine, and tyrosine phosphorylation exclusively observed in plant and animal cells (Mustelin et al. 2002). In plants, the phosphatase family comprises only 150 members, whereas a sizable superfamily of 1050 members is known for protein kinases. These 150 phosphatase members were further classified into different subfamilies known as phosphoprotein phosphatases (PPPs), metal-dependent protein phosphatases (PPMs), protein tyrosine phosphatases (PTPs), and aspartate-based phosphatases based upon phospho-amino-acid-substrate specificity (Shi 2009). It is worth noting that PPPs and PPMs share common catalytic subunits (Fig. 4.2) and PPMs are monomeric enzymes that include type 2C protein phosphatases requiring Mn^{2+} or Mg^{2+} ions for their activity (Cohen 1989; Tonks 2006; Moorhead et al. 2009). Further, this class of phosphatases in the plant kingdom is metal dependent. The PPP family includes different members such as PP1, PP2A, novel phosphatases (PP4, PP5, PP7), plant-specific kelch phosphatases, and two newly discovered PPPs annotated as bacterial related SLP (*Shewanella*-like protein) and bacterial-like protein phosphatases known as the Rhizobiales/Rhodobacterales/Rhodospirillaceae-like phosphatases (RLPHs) (Andreeva and Kutuzov 2004; Uhrig and Moorhead 2011; Uhrig et al. 2013). The PPP family contains three characteristic sequence motifs within the conserved catalytic domains: GDxHG, GDxVDRG, and GNHE (Fig. 4.2). Phosphoprotein phosphatase class is highly conserved and is one of the ancient types of phosphatases found in plants. Out of the three subfamilies of Ser/Thr phosphatase classes, the PP2A class of enzymes is insensitive to mammalian inhibitors I-1 and I-2, whereas okadaic acid potentially inhibits PP2As only. However, PPP1

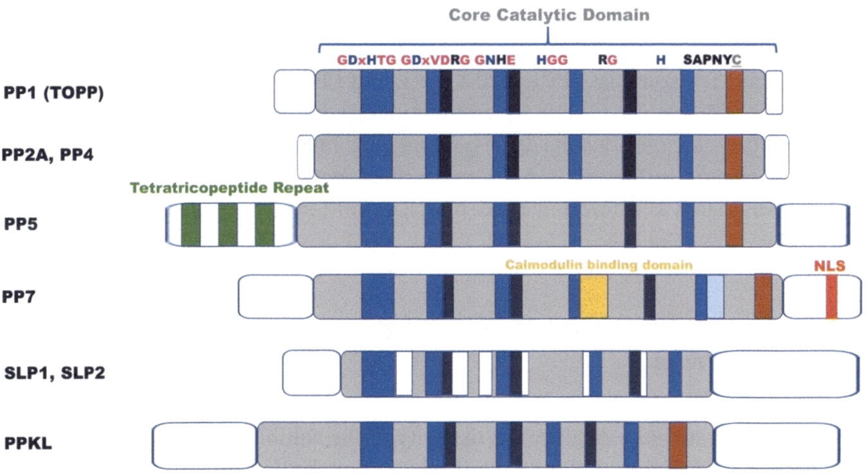

Fig. 4.2 Structural architecture of the plant phosphoprotein phosphatase (PPP) family. The highly conserved core catalytic domain of each PPP subfamily is depicted in gray with signature aspects of each motif highlighted. Dark blue and light blue represent amino acids involved in metal ion coordination and phosphate binding, respectively. The microcystin inhibition docking motif SAPNYC (brown) is also described, highlighted by a reactive cysteine (C) to which microcystin covalently attaches. PP7 maintains this motif but lacks the reactive C, whereas *Shewanella*-like protein (SLP) phosphatases completely lack this motif. Within these motifs, "x" represents any amino acid. Unique features of each subfamily are also depicted: TPR (tetratricopeptide repeat) (green), NLS (nuclear localization signal) (red), and cTP (chloroplast transit peptide)

class enzymes are sensitive to both mammalian inhibitors I and II along with okadaic acid in the nanomolar range (Wang et al. 2007).

Interestingly, in Arabidopsis *thaliana*, 26 genes encode for various PPP molecules but lack phosphatases from PPP3 class, a type of serine/threonine phosphatase subfamily known as calcineurin or PP2B. Most of PPP family-related phosphatases associate with specific regulatory subunits to direct them for a particular cellular response in a specific cellular compartment (Moorhead et al. 2009). Unlike PPPs and PPMs, the PTP subfamily is independent of any metal ion and cofactors for its activation. Phosphatases from the PTP family instead involve a cysteine from conserved signature CX5R motif to dephosphorylate its target protein (Tonks 2013). Interestingly, PTPs vary in their core domain structure as well as in their substrate preferences. Like PTPs, aspartate-based phosphatases also have a signature conserved domain (DXDXT/V) and were named after their pioneering member FCP1. Metal-dependent FCP1p removes a phosphate group explicitly from the C-terminus of RNA polymerase II and HAD (haloacid dehalogenase) family of enzymes (Archambault et al. 1997; Kamenski et al. 2004; Shi 2009).

In *A. thaliana*, various PPP molecules are encoded by 26 genes, whereas 76 different genes are known to encode multiple PP2C units (Schweighofer et al. 2004). In plants, PP2C enzymes are well known to play critical roles in stress response signal transduction, hormonal orchestration, and organ and flower development

(Luan 2003; Schweighofer et al. 2004). In the *Oryza sativa* genome, only 90 PP2C-encoding genes are known (Singh et al. 2010). The plant genome also encodes diverse members from class PP1, PP2A, PP4, and PP6 proteins. Out of these, PPPs are regulated by distinct structural and functional subunits to perform various functions on different targets via the same core subunit. Interestingly, recent studies have shown the roles of different PP2As from the plant in multiple crucial cellular signaling pathways such as saline and water deficit conditions.

4.2.2 Global Functions of PP2A: Diverse Functional Spectrum

PP2A catalytic subunits catalyze the majority of the soluble phosphatase activity at phosphoserine and phosphothreonine. It plays a vital role in the regulation of cellular metabolism. The localization of different holoenzyme forms such as PP2A-B′θ holoenzyme in peroxisomes positively affects peroxisomal β-oxidation, which was earlier responsible for consequences in hypocotyl growth. In *A. thaliana*, the B′ζ subunit of PP2A is involved in energy metabolism and is highly expressed during the senescence phase. Interestingly in an independent sugar assay, the B′ζ knockout seedlings showed hypocotyl retardation on sucrose-free medium, adding to its potential role in energy metabolism (Kataya et al. 2015). The PP2A enzyme is also required for the activation of metabolic enzymes including nitrate reductase and sucrose phosphate synthase, which plays an essential role in carbon and nitrogen metabolism. The majority of studies suggested the function of PP2A in gene expression, ion channel regulation, and developmental processes (Luan 2003).

PP2A plays a critical role during cell cycle regulation by various dephosphorylating substrates of the cell cycle pathway for survival and proper functioning of cells. It plays a significant role during cell proliferation, development and death, cell mobility, cytoskeleton dynamics, and various other signaling pathways (Ingebritsen and Cohen 1983; Janssens and Goris 2001). The PP2A enzyme is also considered as a primary cell cycle regulating enzyme because of the dynamic nature of its holoenzyme subunit, activation, and inhibition. It is involved in the regulation of various cell signaling pathways such as Wnt, mTOR, and MAP kinase. It is also involved in transcription, translation, and mitotic division, which are crucial for the proper functioning of the cell (Wlodarchak and Xing 2016).

In *A. thaliana*, the *ton2* gene encodes a c-terminal protein of PP2A in higher plants. This protein and other PP2A subunits collectively control cytoskeletal organization in plants (Luan et al. 2002). Recent reports on pathogen attack and environmental stress revealed the importance of protein phosphatases undergoing phosphorylation and dephosphorylation in various signaling pathways. Generally, reversible phosphorylation plays a vital role in the activation and inactivation of MAPKs (mitogen-activated protein kinases).

Plants undergo various cellular changes such as ion fluxes, oxygen bursts, and production of phytoalexins and salicylic acid and others in response to pathogen attack. The PP2A expression is induced in *Helianthus annuus* (sunflower) after inoculation with *Phoma macdonaldii*, a necrotrophic fungus. The activation of a defense-related protein phosphorylation cascades causes oxidative burst followed by localized cell death. Hence, the upregulation of possible negative modulators by pathogens may be a kind of protective mechanism. Therefore, the PP2As also act by desensitizing the protein phosphorylation cascades, thereby tightly regulating the defense response to prevent extensive damage to host tissues (Máthé et al. 2019).

In *A. thaliana*, protein phosphatase 2A-B′γ and B′ζ subunits help in regulation of plant tolerance to aphid infestation. Also, different PP2A subunits are involved in defense signaling in plants. In reports, B′η and B′ζ subunits of PP2A enzyme restrict the activation of BAK1 (BRI1-associated receptor kinase 1), which is a co-receptor required for the defense activation by several plasma membrane receptor kinases (Segonzac et al. 2014). Moreover, B′θ subunit PP2A localized in peroxisomes was with bacterial pathogen resistance (Kataya et al. 2015).

Over the years, PP2As have emerged as key players regulating the signal pathways to ensure the proper functioning of various development processes under normal and stressed conditions. OsPP2A-1 to PP2A-5 subunits of PP2A in *O. sativa* play a vital role in developmental stages, heat and cold stress, and drought (Bheri and Pandey 2019; Yu et al. 2003). The PP2A subunit in the tobacco plant, TaPP2Ac-1, is known to induce a drought stress response (Hu et al. 2017). Moreover, PP2As are also responsible for induction of photoprotective mechanisms and increased tolerance against abiotic stress and acclimation strategies upon environmental perturbations.

The B′η, B′θ, and B′γ subunits of PP2A in *A. thaliana* play an essential role in the regulation of flowering time. The cytoplasmic B′γ subunit of PP2A controls organellar ROS signals and SA-dependent defense responses. Also, the day length-dependent responses to intracellular oxidative stress are maintained by PP2A subunits (Bheri and Pandey 2019; Máthé et al. 2019; Li et al. 2014; Trotta et al. 2011).

4.2.3 Salinity Stress Sensor Kinases in Plants

The saline stress is sensed by a group of sensors, which detects the change in the external environment that may include mechanical stimuli, cell wall damage, or the rupture of the plasma membrane and cell wall connections (Fig. 4.2). It is still partially understood how a plant senses osmotic stress. However, during salt shock, the first line of defense is offered by the cell wall integrity pathway, which is further ensured by the cell via cell membrane including ion channels. Other signal sensors connected to cell walls are receptor-like kinases and histidine kinases, which are represented by the inner layer (Fig. 4.3) (Haswell and Verslues 2015). The *A. thaliana* genome encodes for more than 610 receptor-like kinases (RLKs), and in rice, nearly 1100 RLKs are known, which can sense hypersaline and water deficit states

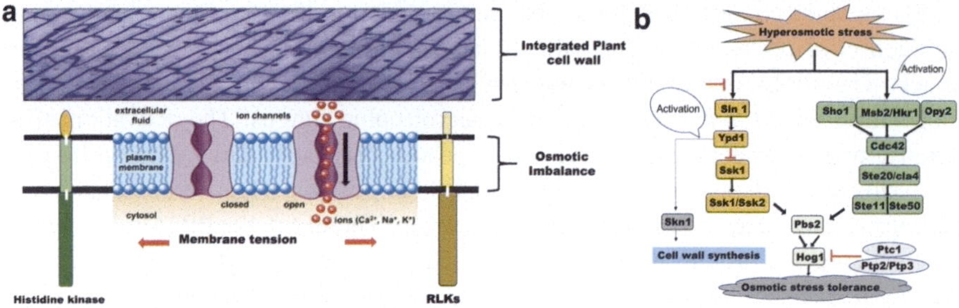

Fig. 4.3 (**a**) Potential osmosensing mechanisms in plant cells. Histidine kinases could be activated by osmotic imbalance across the plasma membrane initiating a signal transduction pathway similar to the high osmolarity glycerol response 1 pathway in *S. cerevisiae* (**b**). The cell wall is presented as a purple box. The plasma membrane is represented as a bilayer of blue lipid molecules

(Marshall et al. 2012). Other receptor like kinases includes lectin-like architecture of THESEUS and FERONIA, which monitors a cell wall (Cheung and Wu 2011). Besides lectins, there are few cell wall-associated kinases involved in maintaining the turgor of the cell wall by having a direct interaction with the pectin molecule in the cell wall (Kohorn and Kohorn 2012). Plants also have MAPKs to be linked with a range of abiotic stress and ROS generation (Morris 2001; Romeis 2001; Tena et al. 2001; Zhang and Klessig 2001; Moon et al. 2003; Nakagami et al. 2005; Mishra et al. 2006). In *A. thaliana* genome, 20 MAPKs are known to regulate osmotic homeostasis (Nakagami et al. 2005; Hamel et al. 2006). Out of these MAPKs, Sho1, a plasma membrane sensor MAPK (Serrano and Rodriguez-Navarro 2001), along with various ion transporters and SOS2 (salt overly sensitive protein kinase) comprises signal transduction pathway that becomes active upon toxic Na^+ concentrations inside and extracellular milieu and guards the plant cell (Zhu 2003). Interestingly, yeast MAPK osmosensor SLN1 (synthetic lethal of N-end rule 1) has HOG1 (High Osmolarity Glycerol) pathway downstream as a response conductor to function similarly in plant system (Saito and Posas 2012; Tran et al. 2007; Wohlbach et al. 2008). However, silencing these genes in plants does not affect the phenotypes such as the accumulation of osmoregulatory solutes and abscisic acid (ABA) (Kumar et al. 2013). Recently, the hyperosmolarity-induced increase in [Ca^{2+}], sensed by sensor OSCA1p (hyperosmolality-induced [Ca^{2+}] increase 1 protein), was proposed (Hou et al. 2014; Yuan et al. 2014). This sensor acts as a critical player in osmosensing pathways, but the mechanism by which this sensor is regulated is not yet fully understood. However, the null mutation of this sensor leads to reduced intracellular Ca^{2+} spike in the plant cell (Yuan et al. 2014).

Salt stress also enhances the biosynthesis of abscisic acid, a plant stress hormone (Koornneef et al. 1998; Taylor et al. 2000). Higher accumulation of ABA is accompanied by an improved K^+/Na^+ ratio (Maathuis and Amtamann 1999) and enhanced K^+ transport in roots (Roberts and Snowman 2000). It also attenuates some downstream pathways or upregulates osmotic stress tolerance in plants (McCourt 1999; Rock 2000; Zhu 2002). Other sensors such as the mechanosensitive channels are cell membrane proteins present on the plasma membrane which operates in response to mechanical stress (Monshausen and Haswell 2013). These are complementary to yeast homologs of MID1-CCH1 channels named as mid1-complementing activity 1 (MCA1) proteins in plants, which upregulate the Ca^{2+} influx in response to hypo-osmotic shock or mechanical stimulus (Kurusu et al. 2013). Mechanosensitive Other members of the MS channel list include plastids related MscS-like (MSL) and two-pore K^+ (TPK) channels from vacuole (Wilson et al. 2014; Maathuis 2011). Null mutation of MSLs from plastids mimics the water deficit condition for plants, which leads to higher proline and abscisic acid production inside the cell. Amino acid proline and plant hormone abscisic acid are the key molecules to identify cells under stress yet by an unknown mechanism (Wilson et al. 2014; Yoshida et al. 2014).

4.2.4 Role of PP2A Holoenzymes in Plants During Salt Stress

Various genetic rearrangements inside the cell gain molecular plasticity to combat salinity stress, which causes ionic and osmotic imbalance inside the cellular compartments and to cope against this, yet gradually in a plant system (Zhu 2001). Salinity stress causes an osmotic shock to a plant cell by extracting water from the cellular milieu and vacuolar pools, further transporting it out of the cell, which is detrimental to the cell. The biggest threat to the cell is posed by high intracellular concentrations of Na^+ and Cl^- ions, and to combat this, plant Ser/Thr phosphatase PP2A enzymes act against elevated ionic stress by deactivating the osmotic sensor kinase substrates. It is worth noting that Ser/Thr PP2A enzyme activation does not require divalent cations (either Ca^{2+} or Mg^{2+}) (Luan 2003). Structurally, PP2A enzyme class is composed of three subunits: catalytic "C," scaffolding "A," and regulatory "R." The *A. thaliana* genome encodes three As, five Cs, and 17 Bs subunits. These subunits with other various regulatory combinations could make 255 different PP2A holoenzymes (Kerk et al. 2002). Most of the physiological functions related to PP2A from plants were mostly studied by knocking out the gene in various model plant species. Protein TON2p encodes a PP2A regulatory subunit in *A. thaliana*, and loss of TON2p leads to many morphological abnormalities related to the cortical cytoskeleton in plant cells (Camilleri et al. 2002). In *O. sativa*, in response to high salinity, all five PP2A catalytic subunits, OsPP2A-1 to OsPP2A-5, get upregulated in leaves (Xu et al. 2007; Yu et al. 2003). In wheat (*Triticum aestivum*), during water deficit conditions, PP2A transcripts such as TaPP2Ac-1 catalytic subunit get accumulated in seedlings, whereas in transgenic tobacco plants, overexpression of TaPP2Ac-1 exhibits enhanced tolerance toward drought and salt stress (Xu et al. 2007). Similarly, during salt stress, potato-related PP2As such as StPP2Ac1, StPP2Ac2a, StPP2Ac2b, and StPP2Ac3 are highly transcribed in the leaves of this plant (País et al. 2009). Interestingly, mevalonate pathway biosynthetic enzyme 3-hydroxy-3-methylglutaryl-CoA reductase (HMGR), which catalyzes a key regulatory step of the mevalonate pathway for isoprenoid biosynthesis, was initially identified to interact with two B″PP2A subunits in *A. thaliana*, and it resulted in an elevated HMGR activity. This pathway represents a multilevel control of HMGR by PP2As, where B″β subunit of PP2A enzyme mediates posttranscriptional repression of HMGR during normal growth conditions. During posttranscriptional repression, B″α modulates HMGR activity by enhancing protein levels and transcript levels in response to salt challenge (Leivar and Quail 2011). In *A. thaliana*, phosphotyrosyl phosphatase activator protein (AtPTPA) is required for salinity tolerance-related functions of PP2A (Chen et al. 2014). Therefore, a compromised expression of AtPTPAp leads to altered PP2A enzyme assembly, which further hampers the responses to NaCl resulting in decreased PP2A activity eventually. The PP2A enzymes in *A. thaliana* are known to be negatively regulated by RCN1p (roots curl in naphthylphthalamic acid). The RCN1p represented "A" subunit of PP2A, having an important role in auxin transport, and was also sensitive to okadaic acid and cantharidin (Garbers et al. 1996). The holoenzyme PP2A-C5 and vacuolar membrane chloride channel proteins (CLCs) interact positively with each other to maintain vacuole ion homeostasis. This genetic interaction mediates salt tolerance in plants by upregulating the activities of CLCs to sequester Cl^- and NO^{3-} into vacuole under hypersaline conditions (Hu et al. 2017). However, this process is independent of salt overly sensitive (SOS) signaling pathway in *A. thaliana* (Zhu 2002, 2003) because double deletion of *pp2a-c5-1* and salt overly sensitive (SOS) mutants such as *sos1-1*, *sos2-2*, and *sos3-1* showed added sensitivity toward NaCl. The vacuolar CLCs in *A. thaliana* such as AtCLCa, AtCLCb, AtCLCc, and AtCLCg physically interact with PP2A-C5. Therefore, it was further questioned and later discovered that overexpressing PP2A-C5 along with *A. thaliana* vacuolar H^+-pyrophosphatase 1 (AVP1) cumulatively improved the salt and drought tolerance as compared to individual mother derivatives (Sun et al. 2018). Functionally, AVP1 is partly responsible for establishing a proton gradient across the tonoplast membrane in plants. However, AVP1 and PP2A-C5 co-overexpression in *A. thaliana* accumulates more K^+, Na^+, and Cl^- in cells under salt stress conditions. Therefore, during AVP1 and PP2A-C5 co-overexpression, a Na^+ channel, AtNHX1 in *A. thaliana* increases Na^+ sequestration into vacuoles, and its overexpression seems to prevent Cl^- accumulation in mutant plant cell where PP2A-C5 further promotes this process. Mutant plants co-overexpressing AVP1 and PP2A-C5 accumulate higher Cl^- than wild type but less than PP2A-C5 overexpressing plants. Similarly, during salt stress, PP2A overexpressing plants retained higher K^+ (Sun et al. 2018). This investigation also proved that higher K^+ retention leads to salt tolerance in *A. thaliana* ecotypes (Sun et al. 2018). These studies indicated that AVP1 and PP2A-C5 proteins act opposite to each other but at the same time implied that introducing multiple genes responsible for a peculiar phenotype can create transgenic plants with enhanced tolerance to multiple stresses.

The role of phytohormone, particularly abscisic acid, was identified to induce the expression of MsPP2A Bβ holoenzyme under drought conditions in alfalfa (*Medicago sativa*) (Tóth et al. 2000). Interestingly, catalytic PP2A-C2 subunit acts as a negative regulator of abscisic acid signaling (Pernas et al. 2007). The study showed that a mutant having inactive PP2A-C2 enzyme is hypersensitive to ABA. However, transgenic plants overexpressing PP2A-C2 were less sensitive to ABA than wild-type plants. Therefore, PP2As are the critical players of ABA signaling. Therefore, positive regulation of PP2A is dependent upon its tissue-specific localization such as in seeds or guard cells, whereas it acts as a negative regulator in roots during ABA responses (Waadt et al. 2015). Another set of regulatory PP2A subunits, Tap46p and TIP41p (TAP42-interacting protein of 41 kDa), is well known to play very crucial roles in yeast and mammals and now plants through the target of rapamycin (TOR) signaling pathway (Düvel and Broach 2004; Jacinto et al. 2001; Yan et al. 2006; Kong et al. 2004; Ahn et al. 2011, 2014; Punzo et al. 2018). It was shown that expression of Tap46p in *A. thaliana* improved growth and nitrogen-assimilating activities. Higher Tap46p expression further increased the phosphorylation of its substrate S6 kinase (S6K) involved in TOR kinase pathway signaling. After rapamycin treatment, which inhibits TOR, decreased Tap46 protein levels and higher PP2A catalytic activity confirmed that TOR and PP2A act in contrasting fashion. Therefore, association of PP2A/Tap46p is directly regulated by TOR protein kinases, and it is believed that Tap46p acts as an inhibitor of PP2A activity in plants. But the detailed mechanism needs further investigation. In another study, deletion of regulatory protein tip41Δ was found to be hypersensitive to ABA, whereas overexpressing TIP41 plants showed higher tolerance to salt, polyethylene glycol, and ABA (Punzo et al. 2018). A protein called MFP1 associated factor 1 (MAF1) is a small, soluble, serine/threonine-rich protein that is ubiquitously expressed. This protein is located at the nuclear periphery forming a component of the nuclear matrix acting as a negative regulator of the RNA Pol III (Boguta et al. 1997; Upadhya et al. 2002; Desai et al. 2005; Rollins et al. 2007; Vannini and Cramer 2012). In *A. thaliana*, AtMaf1p is phosphorylated by TOR kinases and dephosphorylated by PP2A phosphatases. Strikingly, deletion of *Maf1p* renders the plant hypersensitive toward various stresses such as oxidative stress, DNA damaging stress, and replication stress and also toward TOR inhibitors (Ahn et al. 2019). Overall, these observations strongly suggest crucial roles displayed by PP2A class of enzymes and their related regulatory associations in modulating plant growth and ABA responses.

Brassinosteroids (BRs), a class of steroidal hormones, play a significant role in salinity stress in plants (Saini et al. 2015; Sharma et al. 2017). BRs bound by receptor BRI (Brassinosteroid Insensitive 1) family of leucine-rich repeat receptor-like kinases (LRR-RLK) and PP2A enzymes dephosphorylate this kinase receptor. Brassinosteroid signaling kinase 5 (BSK5) is upregulated by BR and ABA. The abiotic stresses including salinity and drought also enhance BSK5 expression to varying extents. The BSK5 kinases are essential for salt stress and ABA-mediated drought stress tolerance. The appearance of BSK5 upregulates salt stress responses and negatively regulates salinity-induced ABA biosynthesis. Therefore, this kinase governs the process of ABA-dependent stomatal closure under salt stress (Sah et al. 2016). The PP2A class phosphatase positively regulates BRs by dephosphorylating transcriptional repressor BZR1p and negatively regulates them by dephosphorylation of the BRI1p receptor. These evidences suggest that PP2A enzymes are critical for plant survival under many abiotic stress conditions (Rahikainen et al. 2016; Hu 2016).

4.3 Concluding Remarks

The versatility of PP2As has benefited the conventional, yet essential, biochemical, molecular, and genetic methods in plant biotechnology by the integration of omics approach. However, the networking of PP2A with other phosphatases and signaling components need much exploration and validation to fully understand their functional prospects. Currently, the production of salt-tolerant varieties of plants by employing novel strategies is the top global list. In this context, understanding molecular mechanisms for the regulation of serine/threonine phosphatases, especially the PP2A class of enzymes, can play a critical role. This group of proteins like in other eukaryotic systems is involved in several complexes and crosstalk and interlinked with different pathways. Therefore, various combinatorial approaches are required to understand the interactome of plant PP2A class of holoenzymes. Phosphoproteomic approach will certainly provide more insights both quantitatively and qualitatively in physiological context of plant PP2As. Besides this, identifying more interacting partners such as target kinases and positive and negative regulators of PP2As would help us in understanding the adaption processes during or under stress for an individual pathway in which PP2As are involved. Similarly, the generation of efficient gene-editing tools such as CRISPR/Cas9 would also facilitate in overcoming many hurdles such as gene redundancy problems, which otherwise hamper the identification of stress sensors.

References

Ahn CS, Han JA, Lee HS, Lee S, Pai HS (2011) The PP2A regulatory subunit Tap46, a component of the TOR signaling pathway, modulates growth and metabolism in plants. Plant Cell 23(1):185–209

Ahn CS, Ahn HK, Pai HS (2014) Overexpression of the PP2A regulatory subunit Tap46 leads to enhanced plant growth through stimulation of the TOR signaling pathway. J Exp Bot 66(3):827–840

Ahn CS, Lee DH, Pai HS (2019) Characterization of Maf1 in Arabidopsis: function under stress conditions and regulation by the TOR signaling pathway. Planta 249(2):527–542

Andreeva AV, Kutuzov MA (2004) Widespread presence of "bacterial-like" PPP phosphatases in eukaryotes. BMC Evol Biol 4(1):47

Archambault J, Chambers RS, Kobor MS, Ho Y, Cartier M, Bolotin D et al (1997) An essential component of a C-terminal domain phosphatase that interacts with transcription factor IIF in Saccharomyces cerevisiae. Proc Natl Acad Sci 94(26):14300–14305

Bheri M, Pandey GK (2019) Protein phosphatases meet reactive oxygen species in plant signaling networks. Environ Exp Bot 161:26–40

Boguta M, Czerska K, Żołądek T (1997) Mutation in a new gene MAF1 affects tRNA suppressor efficiency in Saccharomyces cerevisiae. Gene 185(2):291–296

Camilleri C, Azimzadeh J, Pastuglia M, Bellini C, Grandjean O, Bouchez D (2002) The Arabidopsis TONNEAU2 gene encodes a putative novel protein phosphatase 2A regulatory subunit essential for the control of the cortical cytoskeleton. Plant Cell 14(4):833–845

Chen J, Hu R, Zhu Y, Shen G, Zhang H (2014) Arabidopsis PHOSPHOTYROSYL PHOSPHATASE ACTIVATOR is essential for PROTEIN PHOSPHATASE 2A holoenzyme assembly and plays important roles in hormone signaling, salt stress response, and plant development. Plant Physiol 166(3):1519–1534

Cheung AY, Wu HM (2011) THESEUS 1, FERONIA and relatives: a family of cell wall-sensing receptor kinases? Curr Opin Plant Biol 14(6):632–641

Cohen P (1989) The structure and regulation of protein phosphatases. Annu Rev Biochem 58(1):453–508

Desai N, Lee J, Upadhya R, Chu Y, Moir RD, Willis IM (2005) Two steps in Maf1-dependent repression of transcription by RNA polymerase III. J Biol Chem 280(8):6455–6462

Düvel K, Broach JR (2004) The role of phosphatases in TOR signaling in yeast. In: TOR. Springer, Berlin, pp 19–38

Fedoroff NV, Battisti DS, Beachy RN, Cooper PJ, Fischhoff DA, Hodges CN et al (2010) Radically rethinking agriculture for the 21st century. Science 327(5967):833–834

Garbers C, DeLong A, Deruere J, Bernasconi P, Söll D (1996) A mutation in protein phosphatase 2A regulatory subunit A affects auxin transport in Arabidopsis. EMBO J 15(9):2115–2124

Gupta B, Huang B (2014) Mechanism of salinity tolerance in plants: physiological, biochemical, and molecular characterization. Int J Genomics 2014:701596

Hamel LP, Nicole MC, Sritubtim S, Morency MJ, Ellis M, Ehlting J et al (2006) Ancient signals: comparative genomics of plant MAPK and MAPKK gene families. Trends Plant Sci 11(4):192–198

Haswell ES, Verslues PE (2015) The ongoing search for the molecular basis of plant osmosensing. J Gen Physiol 145(5):389–394

Hou C, Tian W, Kleist T, He K, Garcia V, Bai F et al (2014) DUF221 proteins are a family of osmosensitive calcium-permeable cation channels conserved across eukaryotes. Cell Res 24(5):632

Hu R, Zhu Y, Wei J, Chen J, Shi H, Shen G, Zhang H (2017) Overexpression of PP2A-C5 that encodes the catalytic subunit 5 of protein phosphatase 2A in Arabidopsis confers better root and shoot development under salt conditions. Plant Cell Environ 40(1):150–164

Ingebritsen TS, Cohen P (1983) Protein phosphatases: properties and role in cellular regulation. Science 221(4608):331–338

Jacinto E, Guo B, Arndt KT, Schmelzle T, Hall MN (2001) TIP41 interacts with TAP42 and negatively regulates the TOR signaling pathway. Mol Cell 8(5):1017–1026

Janssens V, Goris J (2001) Protein phosphatase 2A: a highly regulated family of serine/threonine phosphatases implicated in cell growth and signaling. Biochemical J 353(3):417–439

Kamenski T, Heilmeier S, Meinhart A, Cramer P (2004) Structure and mechanism of RNA polymerase II CTD phosphatases. Mol Cell 15(3):399–407

Kataya AR, Heidari B, Lillo C (2015) Protein phosphatase 2A regulatory subunits affecting plant innate immunity, energy metabolism, and flowering time–joint functions among B'η subfamily members. Plant Signal Behav 10(5):e1026024

Kaur N, Pati PK (2017) Integrating classical with emerging concepts for better understanding of salinity stress tolerance mechanisms in rice. Front Environ Sci 5:42

Kerk D, Bulgrien J, Smith DW, Barsam B, Veretnik S, Gribskov M (2002) The complement of protein phosphatase catalytic subunits encoded in the genome of Arabidopsis. Plant Physiol 129(2):908–925

Kohorn BD, Kohorn SL (2012) The cell wall-associated kinases, WAKs, as pectin receptors. Front Plant Sci 3:88

Kong M, Fox CJ, Mu J, Solt L, Xu A, Cinalli RM et al (2004) The PP2A-associated protein α4 is an essential inhibitor of apoptosis. Science 306(5696):695–698

Koornneef M, Alonso-Blanco C, Peeters AJ, Soppe W (1998) Genetic control of flowering time in Arabidopsis. Annu Rev Plant Biol 49(1):345–370

Kumar K, Kumar M, Kim SR, Ryu H, Cho YG (2013) Insights into genomics of salt stress response in rice. Rice 6(1):27

Kurusu T, Kuchitsu K, Nakano M, Nakayama Y, Iida H (2013) Plant mechanosensing and Ca2+ transport. Trends Plant Sci 18(4):227–233

Leivar P, Quail PH (2011) PIFs: pivotal components in a cellular signaling hub. Trends Plant Sci 16(1):19–28

Li S, Mhamdi A, Trotta A, Kangasjärvi S, Noctor G (2014) The protein phosphatase subunit PP 2A-B′ γ is required to suppress day length-dependent pathogenesis responses triggered by intracellular oxidative stress. New Phytologist 202(1):145–160
Luan S (2003) Protein phosphatases in plants. Annu Rev Plant Biol 54(1):63–92
Luan S, Kudla J, Rodriguez-Concepcion M, Yalovsky S, Gruissem W (2002) Calmodulins and calcineurin B–like proteins: calcium sensors for specific signal response coupling in plants. Plant Cell 14(suppl 1):S389–S400
Maathuis FJ, Amtmann, A. N. N. A (1999) K+ nutrition and Na+ toxicity: the basis of cellular K+/ Na+ ratios. Annals of bot 84(2):123–13
Maathuis FJ (2011) Vacuolar two-pore K+ channels act as vacuolar osmosensors. New Phytologist 191(1):84–91
Marshall A, Aalen RB, Audenaert D, Beeckman T, Broadley MR, Butenko MA et al (2012) Tackling drought stress: receptor-like kinases present new approaches. Plant Cell 24(6):2262–2278
Máthé C, Garda T, Freytag C (2019) The role of serine-threonine protein phosphatase PP2A in plant oxidative stress signaling—facts and hypotheses. Int J Mol Sci 20(12):3028
McCourt P (1999) Genetic analysis of hormone signaling. Annu Rev Plant Biol 50(1):219–243
Mishra NS, Tuteja R, Tuteja N (2006) Signaling through MAP kinase networks in plants. Arch Biochem Biophys 452(1):55–68
Monshausen GB, Haswell ES (2013) A force of nature: molecular mechanisms of mechanoperception in plants. J Exp Bot 64(15):4663–4680
Moon H, Lee B, Choi G, Shin D, Prasad DT, Lee O et al (2003) NDP kinase 2 interacts with two oxidative stress-activated MAPKs to regulate cellular redox state and enhances multiple stress tolerance in transgenic plants. Proc Natl Acad Sci 100(1):358–363
Moorhead GB, De Wever V, Templeton G, Kerk D (2009) Evolution of protein phosphatases in plants and animals. Biochem J 417(2):401–409
Morris PC (2001) MAP kinase signal transduction pathways in plants. New Phytol 151(1):67–89
Mustelin T, Abraham RT, Rudd CE, Alonso A, Merlo JJ (2002) Protein tyrosine phosphorylation in T cell signaling. Front Biosci 7:d918–d969
Nakagami H, Pitzschke A, Hirt H (2005) Emerging MAP kinase pathways in plant stress signaling. Trends Plant Sci 10(7):339–346
País SM, Téllez-Iñón MT, Capiati DA (2009) Serine/threonine protein phosphatases type 2A and their roles in stress signaling. Plant Signal Behav 4(11):1013–1015
Pernas M, García-Casado G, Rojo E, Solano R, Sánchez-Serrano JJ (2007) A protein phosphatase 2A catalytic subunit is a negative regulator of abscisic acid signaling 1. Plant J 51(5):763–778
Punzo P, Ruggiero A, Possenti M, Nurcato R, Costa A, Morelli G et al (2018) The PP 2A-interactor TIP 41 modulates ABA responses in Arabidopsis thaliana. Plant J 94(6):991–1009
Rahikainen M, Pascual J, Alegre S, Durian G, Kangasjärvi S (2016) PP2A phosphatase as a regulator of ROS signaling in plants. Antioxidants 5(1):8
Roberts SK, Snowman BN (2000) The effects of ABA on channel-mediated K+ transport across higher plant roots. J Exp Bot 51(350):1585–1594
Rock CD (2000) Tansley review no. 120: pathways to abscisic acid-regulated gene expression. New Phytol 148(3):357–396
Rollins J, Veras I, Cabarcas S, Willis I, Schramm L (2007) Human Maf1 negatively regulates RNA polymerase III transcription via the TFIIB family members Brf1 and Brf2. Int J Biol Sci 3(5):292
Romeis T (2001) Protein kinases in the plant defense response. Curr Opin Plant Biol 4(5):407–414
Sah SK, Reddy KR, Li J (2016) Abscisic acid and abiotic stress tolerance in crop plants. Front Plant Sci 7:571
Saini S, Sharma I, Pati PK (2015) Versatile roles of brassinosteroid in plants in the context of its homoeostasis, signaling and crosstalks. Front Plant Sci 6:950
Saito H, Posas F (2012) Response to hyperosmotic stress. Genetics 192(2):289–318

Schweighofer A, Hirt H, Meskiene I (2004) Plant PP2C phosphatases: emerging functions in stress signaling. Trends Plant Sci 9(5):236–243

Segonzac C, Macho AP, Sanmartín M, Ntoukakis V, Sánchez-Serrano JJ, Zipfel C (2014) Negative control of BAK 1 by protein phosphatase 2A during plant innate immunity. The EMBO journal 33(18):2069–2079

Serrano R, Rodriguez-Navarro A (2001) Ion homeostasis during salt stress in plants. Curr Opin Cell Biol 13(4):399–404

Sharma I, Kaur N, Pati PK (2017) Brassinosteroids: a promising option in deciphering remedial strategies for abiotic stress tolerance in rice. Front Plant Sci 8:2151

Shi Y (2009) Serine/threonine phosphatases: mechanism through structure. Cell 139(3):468–484

Singh A, Giri J, Kapoor S, Tyagi AK, Pandey GK (2010) Protein phosphatase complement in rice: genome-wide identification and transcriptional analysis under abiotic stress conditions and reproductive development. BMC Genomics 11(1):435

Sun L, Pehlivan N, Esmaeili N, Jiang W, Yang X, Jarrett P et al (2018) Co-overexpression of AVP1 and PP2A-C5 in Arabidopsis makes plants tolerant to multiple abiotic stresses. Plant Sci 274:271–283

Taylor IB, Burbidge A, Thompson AJ (2000) Control of abscisic acid synthesis. J Exp Bot 51(350):1563–1574

Tena G, Asai T, Chiu WL, Sheen J (2001) Plant mitogen-activated protein kinase signaling cascades. Curr Opin Plant Biol 4(5):392–400

Tonks NK (2006) Protein tyrosine phosphatases: from genes, to function, to disease. Nat Rev Mol Cell Biol 7(11):833

Tonks NK (2013) Protein tyrosine phosphatases–from housekeeping enzymes to master regulators of signal transduction. FEBS J 280(2):346–378

Tóth ÉC, Vissi E, Kovács I, Szöke A, Ariño J, Gergely P et al (2000) Protein phosphatase 2A holoenzyme and its subunits from Medicago sativa. Plant Mol Biol 43(4):527–536

Tran DQ, Ramsey H, Shevach EM (2007) Induction of FOXP3 expression in naive human CD4+ FOXP3-T cells by T-cell receptor stimulation is transforming growth factor-β–dependent but does not confer a regulatory phenotype. Blood 110(8):2983–2990

Trotta A, Wrzaczek M, Scharte J, Tikkanen M, Konert G, Rahikainen M, ... Mulo P (2011) Regulatory subunit B′ γ of protein phosphatase 2A prevents unnecessary defense reactions under low light in Arabidopsis. Plant Physiol 156(3):1464–1480

Uhrig RG, Moorhead GB (2011) Two ancient bacterial-like PPP family phosphatases from Arabidopsis are highly conserved plant proteins that possess unique properties. Plant Physiol 157(4):1778–1792

Uhrig RG, Labandera AM, Moorhead GB (2013) Arabidopsis PPP family of serine/threonine protein phosphatases: many targets but few engines. Trends Plant Sci 18(9):505–513

Upadhya R, Lee J, Willis IM (2002) Maf1 is an essential mediator of diverse signals that repress RNA polymerase III transcription. Mol Cell 10(6):1489–1494

Vannini A, Cramer P (2012) Conservation between the RNA polymerase I, II, and III transcription initiation machineries. Mol Cell 45(4):439–446

Vinocur B, Altman A (2005) Recent advances in engineering plant tolerance to abiotic stress: achievements and limitations. Curr Opin Biotechnol 16(2):123–132

Waadt R, Manalansan B, Rauniyar N, Munemasa S, Booker MA, Brandt B et al (2015) Identification of open stomata1-interacting proteins reveals interactions with sucrose non-fermenting1-related protein kinases2 and with type 2A protein phosphatases that function in abscisic acid responses. Plant Physiol 169(1):760–779

Wang Y, Liu J, Zhao H, Lü W, Zhao J, Yang L et al (2007) Human 1A6/DRIM, the homolog of yeast Utp20, functions in the 18S rRNA processing. Biochim Biophys Acta Mol Cell Res 1773(6):863–868

Wilson ME, Basu MR, Bhaskara GB, Verslues PE, Haswell ES (2014) Plastid osmotic stress activates cellular stress responses in Arabidopsis. Plant Physiol 165(1):119–128

Wohlbach DJ, Quirino BF, Sussman MR (2008) Analysis of the Arabidopsis histidine kinase ATHK1 reveals a connection between vegetative osmotic stress sensing and seed maturation. Plant Cell 20(4):1101–1117

Xu W, Zhang G, Li X, Zou S, Li P, Hu Z, Li J (2007) Occurrence and elimination of antibiotics at four sewage treatment plants in the Pearl River Delta (PRD), South China. Water Res 41(19):4526–4534

Yan G, Shen X, Jiang Y (2006) Rapamycin activates Tap42-associated phosphatases by abrogating their association with Tor complex 1. EMBO J 25(15):3546–3555

Yoshida T, Mogami J, Yamaguchi-Shinozaki K (2014) ABA-dependent and ABA-independent signaling in response to osmotic stress in plants. Curr Opin Plant Biol 21:133–139

Yu RMK, Zhou Y, Xu ZF, Chye ML, Kong RYC (2003) Two genes encoding protein phosphatase 2A catalytic subunits are differentially expressed in rice. Plant Mol Biol 51(3):295–311

Yuan F, Yang H, Xue Y, Kong D, Ye R, Li C et al (2014) OSCA1 mediates osmotic-stress-evoked Ca2+ increases vital for osmosensing in Arabidopsis. Nature 514(7522):367

Zhang S, Klessig DF (2001) MAPK cascades in plant defense signaling. Trends Plant Sci 6(11):520–527

Zhang JL, Shi H (2013) Physiological and molecular mechanisms of plant salt tolerance. Photosynth Res 115(1):1–22

Zhu JK (2001) Plant salt tolerance. Trends Plant Sci 6(2):66–71

Zhu JK (2002) Salt and drought stress signal transduction in plants. Annu Rev Plant Biol 53(1):247–273

Zhu JK (2003) Regulation of ion homeostasis under salt stress. Curr Opin Plant Biol 6(5):441–445

Chapter 5
Type 2C Protein Phosphatases in Plant Signaling Pathways under Abiotic Stress

Nguyen Nguyen Chuong, Duong Hoang Trong Nghia, Van-Anh Le Thi, Lam-Son Phan Tran, Xuan Lan Thi Hoang, and Nguyen Phuong Thao

5.1 Introduction

Protein phosphatases (PPs) form a superfamily of highly conserved enzymes from simple prokaryotes to advanced eukaryotes. These PPs are counterparts of protein kinases (PKs), and together with signal receptors, they form delicate systems for a wide range of environmental signal perception and transduction, thus playing a crucial role to the survival and development of embryophytes or land plants. Type 2C protein phosphatases (PP2Cs) are representatives of a unique class of enzymes, namely, the phosphoprotein metallophosphatase, classified by the Mg^{2+}-/Mn^{2+}-dependent characteristics (Fuchs et al. 2013). In higher plants (e.g., Arabidopsis and rice), PP2C family consists of more than 80 members, which can be divided into ten or more subgroups (A–K) with diverse functions (Singh et al. 2010; Xue et al. 2008). In this chapter, updated reviews of clades A and B PP2Cs, which have important functions under unfavorable abiotic stress conditions, particularly involved in abscisic acid (ABA)-dependent signaling pathway and mitogen-activated protein kinase (MAPK) cascade, will be our attention. Biological roles and molecular functions of PP2Cs in the signaling pathways under abiotic stresses, as well as

N. N. Chuong · D. H. T. Nghia · X. L. T. Hoang (✉) · N. P. Thao (✉)
Applied Biotechnology for Crop Development Research Unit, School of Biotechnology, International University, Ho Chi Minh City, Vietnam

Vietnam National University, Ho Chi Minh City, Vietnam
e-mail: htlxuan@hcmiu.edu.vn; npthao@hcmiu.edu.vn

V.-A. Le Thi
Graduate University of Science and Technology, Vietnam Academy of Science and Technology, Ha Noi, Vietnam

L.-S. P. Tran
Stress Adaptation Research Unit, RIKEN Center for Sustainable Resource Science, Yokohama, Japan

G. K. Pandey (ed.), *Protein Phosphatases and Stress Management in Plants*,
https://doi.org/10.1007/978-3-030-48733-1_5

PP2C-based genetic engineering approaches for crop improvement, are the major key points to be summarized below.

5.2 Regulatory Targets of PP2Cs in Plant Stress Signaling Pathways

5.2.1 Core ABA Signaling Module

The phytohormone ABA has always been recognized as the key factor in regulating plant response to disadvantageous environmental conditions. PP2Cs are known as important regulators of ABA signaling. However, it was not until the breakthrough discovery of novel ABA receptors (ABARs) in Arabidopsis genome that the crucial roles of these proteins in ABA signaling pathway were revealed (Fujii et al. 2009; Ma et al. 2009; Park et al. 2009). Since then, PP2Cs have been given more and more attention from plant scientists all over the world. Fuchs et al. (2013) categorized this plant protein family in Arabidopsis into 12 clades, in which 9 PP2Cs involved in ABA signaling module belong to clade A. These PP2Cs include ABA-insensitive (ABI) 1, ABI2 (Kuhn et al. 2006; Merlot et al. 2001; Saez et al. 2006), ABA-hypersensitive germination (AHG) 1 (Nishimura et al. 2007), AHG3/PP2CA (Kuhn et al. 2006; Yoshida et al. 2006), hypersensitive/homology to ABA (HAB) 1, HAB2 (Kuhn et al. 2006; Robert et al. 2006; Saez et al. 2004, 2006), highly ABA-induced (HAI) 1, Arabidopsis K^+ transporter 1 (AKT1)-interacting PP2C 1 (AIP1)/HAI2, and HAI3 (Fujita et al. 2009).

In majority of higher plants including Arabidopsis, there are three core protein classes that participate in the ABA signaling module in response to abiotic stress conditions, which are (1) the novel ABARs, pyrabactin resistance (PYR)/pyrabactin-like (PYL)/regulatory components of the ABA receptor (RCAR); (2) the negative regulators, clade A PP2Cs; and (3) the positive regulators, sucrose non-fermenting (SNF) 1-related protein kinases type 2 (SnRK2s) (de Zelicourt et al. 2016). Under normal condition, in the absence of ABA, PP2Cs continuously inactivate SnRK2s including SnRK2.2, SnRK2.3, and SnRK2.6 (or Open Stomata 1, OST1) by dephosphorylating their activation loop and hence preventing the kinases to phosphorylate their downstream targets (Soon et al. 2012). When plants face with environmental negative factors such as drought, salt, and cold, ABA level increases in the cytosol to initiate adaptation responses of plants (Tuteja and Sopory 2008). ABA now enters the open ligand-binding pocket of ABARs, induces the gate and latch loops to close due to the conformational change of highly conserved β-loops, and thus provides a binding surface for PP2Cs (Melcher et al. 2009). The ABA-ABAR complex interacts with the tryptophan (Trp) residue inside the PP2Cs, which leads to the inhibition of PP2C active site (Park et al. 2009). At this stage, SnRK2s are liberated from the dephosphorylation activity of PP2Cs and become activated through autophosphorylation and can then regulate a wide range of Arabidopsis downstream effectors through phosphorylation, which includes transcription factors (TFs) such as the ABA-responsive element (ABRE)-binding proteins (AREBs)/ABRE-binding factors (ABFs), ABI3 and ABI5 (Fujii and Zhu 2009; Furihata et al. 2006; Sirichandra et al. 2010), as well as plasma membrane proteins that function in controlling stomatal aperture such as slow anion channel-associated 1 (SLAC1) (Brandt et al. 2012; Geiger et al. 2009; Lee et al. 2009), quick anion channel 1/aluminum-activated malate transporter 12 (QUAC1/ALMT12) (Imes et al. 2013), the potassium channel KAT1 (Sato et al. 2009), NADPH oxidase respiratory burst oxidase homolog F, and the anion/proton exchanger CLCa (Sirichandra et al. 2009; Wege et al. 2014).

5.2.2 Chromatin Remodeling Complex

The complexities of plant response to adversities have been illustrated by various tiers of regulation. Localization analysis demonstrated the presence of Arabidopsis HAB1 in both nucleus and cytosol (Saez et al. 2008), suggesting the capacity of PP2C to interact with various ABA-signaling partners in different steps. Amino acid sequence analysis of several Arabidopsis PP2Cs revealed that at least ABI1, ABI2, HAB1, and PP2CA contain short nuclear localization signal at the end of the C-terminals (Himmelbach et al. 2002; Moes et al. 2008). Further experiments in Arabidopsis have found the nuclear-localized interaction between PP2Cs and SWI3B, which is a core subunit of the SWItch (SWI)/SNF chromatin remodeling complexes (Saez et al. 2008).

In eukaryotes in general and plants in particular, adaptation to environmental stresses requires delicate alteration in gene expression. The process of expressing a gene involves series of steps, in which the utmost requirement is the accessibility of the regulatory proteins to the gene, which requires extensive chromatin modification involving two major mechanisms—either posttranslational modification of histones or ATP-dependent reorganization of histone-DNA interactions (Han et al. 2015). Chromatin remodeling complexes are often called SWI/SNF-related ATP-dependent chromatin remodeling complexes. Their major roles in chromatin remodeling relate to the utilization of energy derived from hydrolysis of adenosine triphosphate (ATP) molecules for alteration of nucleosome occupancy or position (Archacki et al. 2016). Based on the conserved domains, chromatin remodeling ATPases have been categorized into four major subfamilies (i.e., inositol-requiring 80 (INO80)/sick with RSC/Rat1 (SWR1), chromodomain-helicase-DNA (CHD), imitation switch (ISWI), and SWI/SNF) (Han et al. 2015). Each family has a specific domain, such as chromo- or bromo-domain, or plant homeo-domain (PHD), allowing them to act in various circumstances. Among those, the yeast SWI/SNF complex was the first to be described. The complex is composed of the ATPase Swi2/Snf2 as the major catalytic subunit, and the central core consists of three additional polypeptides, Swi3, Snf5, and Swp73, which are essential for assembling and functionality of the complex (Saez et al. 2008). The ATPases of the SWI/SNF complexes were further divided into three types, which are encoded by all land plants and share significant similarities with the metazoan counterparts: BRAHMA (BRM), SPLAYED (SYD), and MINUSCULE (MINU) (Han et al. 2015).

HAB1-SWI1B is the first PP2C interaction with the SWI/SNF chromatin remodeling complex to be discovered in Arabidopsis, emphasizing the consistent role of PP2C as a negative regulator of ABA signaling under various tiers of gene expressing regulation (Saez et al. 2008). Strong interaction between HAB1 catalytic domain and SWI3B was revealed by a yeast two-hybrid assay (Saez et al. 2008). Swi3p, Rsc8p, Moira (also known as SWIRM), and ZZ zinc finger domains of SWI3B protein were required for the interaction as deletion of either of them results in abolition of HAB1-SWI3B interaction (Saez et al. 2008). Moreover, nuclear localization signals are present at the C-terminal of the HAB1, ABI2, and PP2CA, illustrating the capacity of nuclear protein interactions of PP2C and SWI3. Meanwhile, the *swi3b* mutants displayed reduced ABA sensitivity and ABA-dependent gene expression, as well as the HAB1-SWI3B interaction (Saez et al. 2008). These findings strongly implied the role of SWI3B as a positive regulator of ABA signaling and the ability of PP2Cs, HAB1 in particular, to regulate the putative SWI/SNF complex to repress activities of some ABA-dependent promoters.

Nevertheless, a recent study on ATPase BRM in Arabidopsis has revealed a direct linkage of the chromatin remodeling to ABA signaling pathway (Peirats-Llobet et al. 2016). Bimolecular fluorescence complementation (BiFC) assay images confirmed positive BRM physical interactions with either OST1/SnRK2.2/2.3/2.6 or HAB1/PP2CA clade A PP2Cs in nucleus at both N- and C-terminals (Peirats-Llobet et al. 2016). Further mass spectrometry experiments demonstrated the capacity of phosphorylation/dephosphorylation of the C-terminal region of BRM by OST1/PP2CA *in vitro*. The major phosphorylation sites were identified to be around the AT hook and bromo-domain of BRM (Peirats-Llobet et al. 2016). Such regions have been reported to play crucial roles in BRM functionality, enabling interactions with linker and nucleosomal DNA as well as the histone octamer (Farrona et al. 2007). However, no phosphorylation sites have been detected in other regions of BRM required for ATPase activities, including the active site of ATP hydrolysis and the Snf2-ATP coupling (SnAC) domain, both of which utilize energy from ATP hydrolysis into nucleosome rearrangement. Hence, such "hotspots" of phosphorylation in the C-terminal, starting from the AT-hook domain and the bromo-domain, are considered as the core regulatory sites via phosphorylation/dephosphorylation activities. The Arabidopsis *brm* loss-of-function mutants exhibited ABA-hypersensitive phenotypes, further indicating that SnRK2 phosphorylation releases the repression of BRM in ABA signaling. In contrast, consistent role of PP2C phosphatases as negative regulators of ABA responses, such as PP2CA and HAB1, is emphasized as the dephosphorylation of BRM returning its activities in the absence of ABA and enabling plant growth in normal condition (Peirats-Llobet et al. 2016).

It was shown that *A. thaliana brm* knock-out mutant shared the same ABA-hypersensitive germination phenotype as the *swi3c* mutant, one of the plant's four SWI3 homologues, suggesting the notion that both BRM and SWI3C belong to the same complex (Sarnowska et al. 2016). While both BRM and SWI3B exhibit direct interactions to at least one PP2C in Arabidopsis (i.e., HAB1), the *swi3b* defective mutant was found to have opposite phenotype to *brm* mutant, whereby the former displayed lower ABA sensitivity during germination and reduced expression of the ABA-responsive genes, such as *responsive to ABA 18* (*RAB18*) and *response to desiccation 29B* (*RD29B*) (Saez et al. 2008). In other words, BRM and SWI3B are negative and positive regulators of the ABA signaling pathway, respectively. In the Asensi-Fabado et al. 2017 review, Asensi-Fabado suggested two possibilities, which are either SWI3B competing with BRM for HAB1 binding thereby dephosphorylating BRM or SWI3B being associated with a different complex with distinct function to the BRM/SWI3C complex (Sarnowski et al. 2005; Sarnowska et al. 2016). The exact mode by which either BRM or SWI3B regulates the target loci remains to be elucidated, but nucleosome repositioning involvement might be the best explanation.

5.2.3 *MAPK Cascades*

MAPK signaling pathway is one of the most well-studied signaling mechanisms that is evolutionary conserved throughout eukaryotic organisms such as plants, insects, yeast, and mammals (Hamel et al. 2012). In plant kingdom, MAPK cascades consist of proteins from a large family with the ability to sense and transduce stress signals for appropriate responses during plant adaptation (Danquah et al. 2014). A MAPK cascade comprises at least three protein kinases at three levels, which are MAPK kinase kinases (MAPKKK/MKKK/MEKK, MAP3K), MAPK kinases (MAPKK/MKK/MEK, MAP2K), and MAPK (MPK). These proteins stimulate each other in a sequential manner through phosphorylation (Colcombet and Hirt 2008). The large number of MAPK pathway components in plants allows them to form thousands of different MAPK cascades. During the first step, activated MAPKKK phosphorylates two threonine (Thr/T)/or serine (Ser/S) residues located within the activation loop of MAPKK. MAPK is then activated in the next step via dual phosphorylation along T-X-Tyrosine (Tyr/Y) motif by the activated MAPKK (Hamel et al. 2012). This consecutive activation results in the phosphorylation of specific targets and the regulation of TF activities as well as the expression of different sets of genes that function in response to various environmental stresses (Popescu et al. 2009; Taj et al. 2010). In 2008, Colcombet and Hirt suggest that there may be a fourth level of kinases, MAP4Ks (MAP3K kinases), as mediators to link upstream signaling steps to core MAPK modules. Data from different studies have shown the involvement of MAPKs in signal transduction of plant adaptions to divergent stimuli, both biotic and abiotic (de Zelicourt et al. 2016).

Clade B PP2Cs have been categorized as regulators of MAPK activities and consist of six genes in Arabidopsis, namely, phosphatases type 2C (AP2Cs) which are orthologous to *Medicago sativa* protein phosphatases 2C (MP2Cs) (Fuchs et al. 2013). Although most of research focuses on the role of clade B PP2Cs in biotic stress responses, this group of proteins also displays potential association with abiotic stress. Four members of this cluster (AP2C1–4) have been identified to contain the MAPK interaction motif known as kinase interaction motif (KIM), which regulates interaction activities of MAPKs with MAPK phosphatases (MKPs, phosphatase enzymes responsible for downregulation of MAPK signaling), MAP2K, or TFs in animals and plants (Fuchs et al. 2013). All MAPKs, except for the most distant group D, carry and are activated through a T-glutamic acid (Glu/E)-Y phosphorylation motif (Ichimura et al. 2002). AP2C/MP2C deactivates MAPK activities through the dephosphorylation of the pT (phosphorylated threonine residue) in the pTEpY activation loop of MAPK (Schweighofer et al. 2007; Umbrasaite et al. 2010). All four AP2Cs contain KIM domain and interact with Arabidopsis MPK3, MPK4, and MPK6, the three MAPKs which are involved in various stress signaling pathways (Moustafa et al. 2008; Hoang et al. 2012), suggesting a potential role of clade B PP2Cs in plant adaptation.

In addition to clade B PP2Cs, members of clade A are also believed to be involved in the regulation of MAPK cascades in plants due to close relation between MAPKs and ABA. Recently, Mitula et al. (2015) have successfully identified a member of Arabidopsis MAPK cascade, MKKK18, which is regulated by ABI1, a clade A PP2C. MKKK18 functions in an MAPK module comprised of MKKK18-MKK3-MPK1/2/7/14 (Danquah et al. 2015) and acts as a positive regulator of stomata density and ABA-induced stomatal closure (Mitula et al. 2015). In Arabidopsis, ABI1 was also found to interact with MAPK6 (Leung et al. 2006), suggesting that this clade A PP2C, and possibly some other members in the cluster, might also be involved in different tiers to regulate MAPK cascades in plant stress adaptation (Fig. 5.1).

5.2.4 *Other Targets*

Aside from their inhibition with the calcium-independent kinase SnRK2 family, PP2Cs are also found to interact with other kinase families such as calcium-dependent protein kinases (CDPKs/CPKs) or SnRK3s/calcineurin B-like protein (CBL)-interacting protein kinases (CIPKs). However, the effects of these interactions are quite different. Zhao et al. (2011) have reported the phosphorylation activity of Arabidopsis CPK12 with ABI2, a clade A PP2C, which results in stimulating catalytic activity of this phosphatase, suggesting that CPK12 could be negatively involve in ABA signaling pathways. Interestingly, two clade A PP2Cs, ABI1 and ABI2, have been reported to inhibit Arabidopsis CIPK26 in a similar manner in which they inactivate SnRK2s (Lyzenga et al. 2013, 2017). These findings may indicate an antagonistic correlation between these two kinase families, yet more studies should be conducted to clarify this assumption.

In Arabidopsis, ABI1, along with PP2CA, was also found to interact with and dephosphorylate SnRK1.1, a member of the subgroup 1 of SnRK-type protein kinases that is involved in sugar responses under stress controlled through ABA signaling (Rodrigues et al. 2013). In addition to this finding, Chen et al. (2016) also

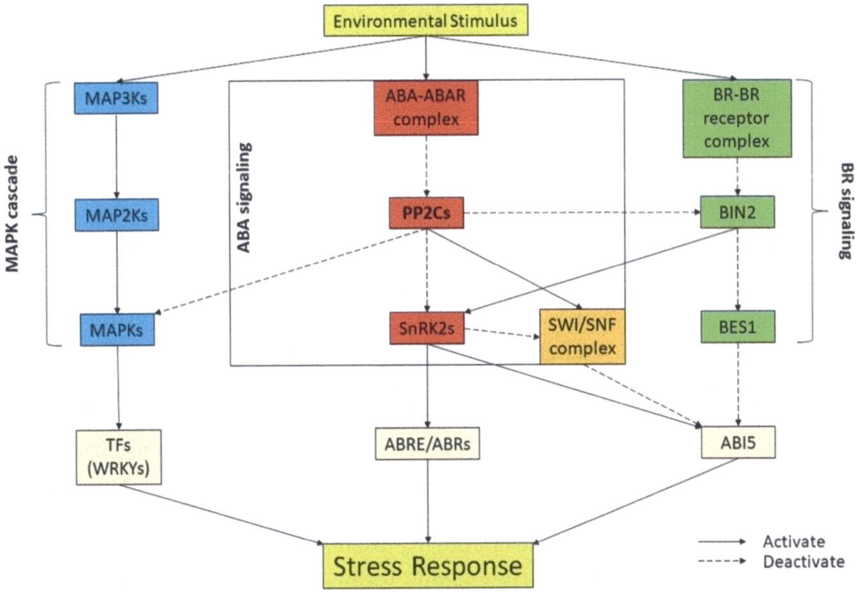

Fig. 5.1 Suggested model of regulatory functions of type 2C protein phosphatases (PP2Cs) in different pathways relating to abscisic acid (ABA)- and brassinosteroid (BR) signaling pathways and mitogen-activated protein kinase (MAPK) cascade in plant response to abiotic stress conditions such as drought, salinity, and cold. Under normal condition, PP2Cs regulate components of different signaling pathways through dephosphorylation. Upon exposing to stress, plants receive and transmit signal and initiate ABA-ABA receptor (ABAR) binding which will inhibit the active site of PP2Cs and in turn alter activities of downstream components including various kinase enzymes, resulting in appropriate responses (Ojolo et al. 2018; Opdenakker et al. 2012; Planas-Riverola et al. 2019; Singh et al. 2016)

reported the direct interaction between Arabidopsis ABI2 and the receptor-like kinase FERONIA (FER), suggesting the coordinated function of ABA in different pathways during plant stress responses.

Interestingly, Wang et al. (2018) reported the regulatory activities of Arabidopsis ABI1 and ABI2 in brassinosteroid (BR) signaling pathway through the BR-negative regulator brassinosteroid insensitive 2 (BIN2) kinase, which can serve as a bridge for constructing the crosstalk between two phytohormones, ABA and BR, under abiotic stress. Moreover, Arabidopsis PP2Cs, including ABI1 and ABI2, also interact with salicylic acid (SA) in an antagonistic manner with respect to ABA (Manohar et al. 2017). These findings indicate the cooperative correlation between/among phytohormones in balancing plant growth and development during stress conditions.

In general, PP2Cs have appeared as essential factors of ABA signaling pathway in regulating plant reactions to abiotic stresses. Through interaction and modulation of different target proteins, this protein family has provided significant roles in adaptive response of several plant species. Latest findings have further revealed some novel and unique functions of PP2Cs in ABA signaling pathway and uncovered a whole new area for future research.

5.3 Current Studies on PP2Cs in Plant under Abiotic Stress

5.3.1 Clade A

Due to the important role of PP2Cs in ABA signaling pathway, several investigations under different abiotic stress conditions have been reported to aid further understanding of their molecular function. A common characteristic that is shared between different clade A PP2Cs is that their expression levels are induced by ABA and stressful conditions (Fujita et al. 2009, 2011). Expression analyses in different plant species, such as Arabidopsis (Xue et al. 2008), rice (Singh et al. 2010), tomato (Sun et al. 2011), maize (Wei and Pan 2014), Chinese cabbage (Kong et al. 2018), and banana (Hu et al. 2017), have consolidated this fact. Many studies also discovered the fact that *PP2C* genes from different species have been found to exhibit overlapping expression profile under various abiotic stress treatments, including drought, salt, cold, and heat stresses (Cao et al. 2016; Li et al. 2018; Singh et al. 2010; Xue et al. 2008; Yang et al. 2018). These findings may indicate that plants respond to different abiotic stresses through a resembling molecular mechanism, in which PP2Cs can be important cross-talking factors between/among different signaling pathways.

Among the nine clade A PP2C members that involve in ABA signaling pathway identified in Arabidopsis, ABI1 and ABI2 are two best-studied proteins, which control the full range of ABA responses under abiotic stresses and during development (Fuchs et al. 2013; Singh and Pandey 2012). These two PP2Cs were found to physically interact with various cytosolic and nuclear-localized proteins. Such interactions are commonly seen in various PP2Cs such as interaction with homeodomain TF (e.g., HB6, CIPK24) or the more selective, specific interaction with preprotein of fibrillin and CIPK8 (Fuchs et al. 2013). In addition to being the regulators of activities of SnRK2s, ABI1 and its homolog, ABI2, have been identified to be associated with other regulatory activities in plant response to environmental changes, such as the CBL1/CBL9-CIPK23 pathway in stomata aperture regulation (Mao et al. 2016) and nitrate sensing (Léran et al. 2015), and the proteasome degradation through the ubiquitin/26 s proteasome system (UPS) (Ludwików 2015).

Besides ABI1 and ABI2, the three "HAI" PP2Cs, HAI1, AKT1-interacting PP2C1/HAI2, and HAI3, also show interesting behaviors under abiotic stress conditions. Under polyethylene glycol (PEG) treatment condition, *HAI* PP2C mutants showed enhancing proline and osmoregulatory solute accumulation, whereas these features were not apparently seen in Arabidopsis carrying mutation in other clade A PP2Cs (Bhaskara et al. 2012). While Arabidopsis PP2C *HAI* single mutants did not produce ABA-responsive phenotype (Yoshida et al. 2006), double and triple mutation in genes encoding HAI PP2C showed different ABA sensitivity levels at different stages of plant development. During germination stage, Arabidopsis *HAI* double and triple mutants were found to be ABA insensitive, which is in contrast with the hypersensitive phenotypes of other clade A *PP2C* mutants (Yoshida et al. 2006). However, when entering post-germination stage, mutants of various *HAI* genes in this cluster showed similar hypersensitivity characteristics to ABA (Bhaskara et al. 2012). Furthermore, the *pp2ca-1hai1-1* (or *ahg3-1hai1-1*) double mutant plants displayed enhanced ABA-mediated growth inhibition, increased ABA-responsive gene induction, and diminished water loss compared with the *pp2ca*-1 single mutants (Antoni et al. 2012), suggesting that *hai1-1* mutation enhanced the ABA sensitivity of *pp2ca-1/ahg3-1* mutant. These results indicate that HAI PP2Cs may have a greater role in ABA-independent pathway rather than in ABA-dependent pathway in response to drought stress.

Several *in planta* studies have been conducted to further understand the functional roles of PP2Cs under abiotic stresses. This protein family has shown to possess potential candidates for producing abiotic stress-tolerant transgenic plants. Singh et al. (2015) reported that transgenic Arabidopsis overexpressing rice clade A PP2C *OsPP108* confers high tolerance under salt, mannitol, and drought stresses. Similarly, transgenic Arabidopsis ectopically expressing *Brachypodium distachyon BdPP2CA6* displayed enhanced stomatal closure and salinity tolerance (Zhang et al. 2017a). Singh et al. (2016) also summarized studies of other PP2Cs on their expression level and *in planta* functional roles in response to stress conditions (Table 5.1).

5.3.2 Clade B

Among clade B PP2Cs, AP2C1 and AP2C3 are the two best-characterized members and were found to regulate stomatal developmental pathway in Arabidopsis (Schweighofer et al. 2007; Umbrasaite et al. 2010). In the absence of AP2C1 and AP2C3, stomatal closure was impaired, indicating the involvement of these two proteins in regulating water loss rate of plants, especially under adverse conditions. In 2007, Schweighofer et al. discovered that AP2C1 controlled wound-induced MAPK activities and stress-induced ethylene responses. The early expression of *AP2C1* at the site of wounding indicates its involvement in order to antagonize effect of the aforementioned stress conditions. Moreover, AP2C1 also negatively controls production of jasmonate, a phytohormone that is believed to play a role in plant responses to abiotic stresses such as drought, salt, and heat. Results from investigation revealed that *ap2c1* knock-out plants showed enhanced jasmonate production upon wounding and better tolerance to herbivory effects (Schweighofer et al. 2007). On the other hand, AP2C3, which is closely related to AP2C1, shows unique expression that differs from those of other members of the family. Distinct expression pattern in stomata and stomatal lineage cells along with the ability to interact with/downregulate signaling activity of MAPKs consolidates the participation of AP2C3 in regulating MPK3 and MPK6 during stomatal development (Umbrasaite et al. 2010). Recently, AP2C1 was also found to negatively regulate CIPK9 under K^+ deficiency condition, which is also a kind of abiotic stress experienced by the plants (Singh et al. 2018).

Table 5.1 Summary of recent studies on identification of PP2Cs that mediate plant response to abiotic stress conditions

Plant species	Gene	Type of study	Findings	References
Arabidopsis *thaliana*	Clade D—*AtPP2Cs*	RT-qPCR	Expression of *AtPP2Cs* was significantly influenced by alkali and salt stresses, suggesting possible involvement or direct interaction	Chen et al. (2018)
Soybean (*Glycine soja*)	Clade D—*GsPP2Cs*	RT-qPCR	Similar expression pattern in clade D—*PP2Cs* of *G. soja* and in clade D—*PP2Cs* of *A. thaliana*, suggesting conserved functions of clade D—PP2Cs between plant species	
Rice (*Oryza sativa*)	Clade F—*OsPP18*	Molecular, genetic, and physiological analyses	Overall expression of *OsPP18* led to osmotic and oxidative stress tolerance. *ospp18* mutants and suppressed *OsPP18*-RNAi exhibit drought-hypersensitive phenotypes, with lower reactive oxygen species (ROS)-scavenging gene expression, suggesting potential role of OsPP18 in drought tolerance mediation	You et al. (2014)
	Clade A—*OsPP108*	Molecular and genetic analyses using heterologous system	Arabidopsis ectopically expressing *OsPP108* showed enhanced abscisic acid (ABA) insensitivity and high tolerance to salt, mannitol, and drought stresses at various stages of development	Singh et al. (2015)
Brachypodium distachyon	Various clades—*BdPP2Cs*	Transcriptome/ RT-qPCR	50–80% of *BdPP2C* genes displayed upregulation in response to abiotic stresses (cold, heat, PEG and NaCl treatments), suggesting possible involvement of BdPP2Cs in *B. distachyon* resistance to abiotic stresses	Cao et al. (2016)

(continued)

Table 5.1 (continued)

Plant species	Gene	Type of study	Findings	References
Barrel clover (*Medicago truncatula*)	Various clades—*MtPP2Cs*	Microarray/ RT-qPCR	Most of *MtPP2C* genes showed differential expression patterns in response to abiotic stresses (i.e., cold, drought, and ABA stress), aiding the identification of stress-related *MtPP2C* genes	Yang et al. (2018)
Wheat (*Triticum aestivum*)	Clade F—*TaPP2C1*	Expression, molecular, biochemical, and physiological analyses using heterologous system	Ectopic expression of *TaPP2C1* in tobacco resulted in reduced ABA sensitivity and increased salt tolerance of the transgenic seedlings	Hu et al. (2015)
Tomato (*Solanum lycopersicum*)	Clade A— *SlPP2C1*	Expression, molecular, and biochemical analyses using homologous system	*SlPP2C1*-RNAi tomato plants displayed hypersensitivity to ABA and increased drought stress tolerance	Zhang et al. (2017b)

5.3.3 *Other Clades*

Apart from clades A and B, there is limited information about function of other clades in plants under abiotic stress conditions. However, some lines of evidence also indicate the involvement of these proteins in plant adaptation. For example, Chen et al. (2018) found that expression level of members of clade D PP2Cs was significantly altered upon alkali and salt stress treatments in soybean and Arabidopsis, suggesting the direct or indirect association of this class of PP2Cs in stress signaling pathways.

5.4 Conclusion

Plant genomes code for larger number of PP2Cs than other groups of organisms including yeast, mouse, or human, which indicates the important role of this protein family in various cellular processes in plants. Recent analyses of plant PP2Cs have revealed the novel regulatory modes and functions of this class of protein phosphatases in different signaling pathways. Due to the large number of members in this gene family, many functions and activities of PP2Cs in plants remain unknown. However, results obtained from performed studies and analyses on PP2Cs have proven them to be potential targets for further investigation to thoroughly understand their role in mediating plant adaptation to environmental stimuli and serve as base to develop appropriated methods for overcoming stress conditions.

Acknowledgments This research is funded by Vietnam National University Ho Chi Minh City (VNU-HCM) under grant numbers B2017-28-02 and C2018-28-04.

References

Antoni R, Gonzalez-Guzman M, Rodriguez L, Rodrigues A, Pizzio GA, Rodriguez PL (2012) Selective inhibition of clade A phosphatases type 2C by PYR/PYL/RCAR abscisic acid receptors. Plant Physiol 158(2):970–980

Archacki R, Yatusevich R, Buszewicz D, Krzyczmonik K, Patryn J, Iwanicka-Nowicka R, Biecek P, Wilczynski B, Koblowska M, Jerzmanowski A, Swiezewski S (2016) Arabidopsis SWI/SNF chromatin remodeling complex binds both promoters and terminators to regulate gene expression. Nucleic Acids Res 45(6):3116–3129

Asensi-Fabado MA, Amtmann A, Perrella G (2017) Plant responses to abiotic stress: the chromatin context of transcriptional regulation. Biochim Biophys Acta Gene Regul Mech 1860(1):106–122

Bhaskara GB, Nguyen TT, Verslues PE (2012) Unique drought resistance functions of the highly ABA-induced clade A protein phosphatase 2Cs. Plant Physiol 160(1):379–395

Brandt B, Brodsky DE, Xue S, Negi J, Iba K, Kangasjärvi J, Ghassemian M, Stephan AB, Hu H, Schroeder JI (2012) Reconstitution of abscisic acid activation of SLAC1 anion channel by CPK6 and OST1 kinases and branched ABI1 PP2C phosphatase action. Proc Natl Acad Sci 109(26):10593–10598

Cao J, Jiang M, Li P, Chu Z (2016) Genome-wide identification and evolutionary analyses of the PP2C gene family with their expression profiling in response to multiple stresses in Brachypodium distachyon. BMC Genomics 17(1):175

Chen J, Yu F, Liu Y, Du C, Li X, Zhu S, Wang X, Lan W, Rodriguez PL, Liu X, Li D (2016) FERONIA interacts with ABI2-type phosphatases to facilitate signaling cross-talk between abscisic acid and RALF peptide in Arabidopsis. Proc Natl Acad Sci 113(37):E5519–E5527

Chen C, Yu Y, Ding X, Liu B, Duanmu H, Zhu D, Sun X, Cao L, Zaib-un-Nisa, Li Q, Zhu Y (2018) Genome-wide analysis and expression profiling of PP2C clade D under saline and alkali stresses in wild soybean and Arabidopsis. Protoplasma 255(2):643–654

Colcombet J, Hirt H (2008) Arabidopsis MAPKs: a complex signaling network involved in multiple biological processes. Biochem J 413(2):217–226

Danquah A, de Zelicourt A, Colcombet J, Hirt H (2014) The role of ABA and MAPK signaling pathways in plant abiotic stress responses. Biotechnol Adv 32(1):40–52

Danquah A, de Zélicourt A, Boudsocq M, Neubauer J, Frei dit Frey N, Leonhardt N, Pateyron S, Gwinner S, Tamby JP, Ortiz-Masia D, Marcote MJ (2015) Identification and characterization of an ABA-activated MAP kinase cascade in Arabidopsis *thaliana*. Plant J 82(2):232–244

de Zelicourt A, Colcombet J, Hirt H (2016) The role of MAPK modules and ABA during abiotic stress signaling. Trends Plant Sci 21(8):677–685

Farrona S, Hurtado L, Reyes JC (2007) A nucleosome interaction module is required for normal function of Arabidopsis *thaliana* BRAHMA. J Mol Biol 373(2):240–250

Fuchs S, Grill E, Meskiene I, Schweighofer A (2013) Type 2C protein phosphatases in plants. FEBS J 280(2):681–693

Fujii H, Zhu JK (2009) Arabidopsis mutant deficient in 3 abscisic acid-activated protein kinases reveals critical roles in growth, reproduction, and stress. Proc Natl Acad Sci 106(20):8380–8385

Fujii H, Chinnusamy V, Rodrigues A, Rubio S, Antoni R, Park SY, Cutler SR, Sheen J, Rodriguez PL, Zhu JK (2009) *In vitro* reconstitution of an abscisic acid signaling pathway. Nature 462(7273):660

Fujita Y, Nakashima K, Yoshida T, Katagiri T, Kidokoro S, Kanamori N, Umezawa T, Fujita M, Maruyama K, Ishiyama K, Kobayashi M (2009) Three SnRK2 protein kinases are the main positive regulators of abscisic acid signaling in response to water stress in Arabidopsis. Plant Cell Physiol 50(12):2123–2132

Fujita Y, Fujita M, Shinozaki K, Yamaguchi-Shinozaki K (2011) ABA-mediated transcriptional regulation in response to osmotic stress in plants. J Plant Res 124(4):509–525

Furihata T, Maruyama K, Fujita Y, Umezawa T, Yoshida R, Shinozaki K, Yamaguchi-Shinozaki K (2006) Abscisic acid-dependent multisite phosphorylation regulates the activity of a transcription activator AREB1. Proc Natl Acad Sci 103(6):1988–1993

Geiger D, Scherzer S, Mumm P, Stange A, Marten I, Bauer H, Ache P, Matschi S, Liese A, Al-Rasheid KAS, Herdrich R, Romeis T (2009) Activity of guard cell anion channel SLAC1 is controlled by drought-stress signaling kinase-phosphatase pair. Proc Natl Acad Sci 106(50):21425–21430

Hamel LP, Nicole MC, Duplessis S, Ellis BE (2012) Mitogen-activated protein kinase signaling in plant-interacting fungi: distinct messages from conserved messengers. Plant Cell 24(4):1327–1351

Han SK, Wu MF, Cui S, Wagner D (2015) Roles and activities of chromatin remodeling ATP ases in plants. Plant J 83(1):62–77

Himmelbach A, Hoffmann T, Leube M, Höhener B, Grill E (2002) Homeodomain protein ATHB6 is a target of the protein phosphatase ABI1 and regulates hormone responses in Arabidopsis. EMBO J 21(12):3029–3038

Hoang MHT, Nguyen XC, Lee K, Kwon YS, Pham HTT, Park HC, Yun DJ, Lim CO, Chung WS (2012) Phosphorylation by AtMPK6 is required for the biological function of AtMYB41 in Arabidopsis. Biochem Biophys Res Commun 422(1):181–186

Hu W, Yan Y, Hou X, He Y, Wei Y, Yang G, He G, Peng M (2015) TaPP2C1, a group F2 protein phosphatase 2C gene, confers resistance to salt stress in transgenic tobacco. PLoS One 10(6):e0129589

Hu W, Yan Y, Shi H, Liu J, Miao H, Tie W, Ding Z, Ding X, Wu C, Liu Y, Wang J (2017) The core regulatory network of the abscisic acid pathway in banana: genome-wide identification and expression analyses during development, ripening, and abiotic stress. BMC Plant Biol 17(1):145

Ichimura K, Shinozaki K, Tena G, Sheen J, Henry Y, Champion A, Kreis M, Zhang S, Hirt H, Wilson C, Heberle-Bors E (2002) Mitogen-activated protein kinase cascades in plants: a new nomenclature. Trends Plant Sci 7(7):301–308

Imes D, Mumm P, Böhm J, Al-Rasheid KA, Marten I, Geiger D, Hedrich R (2013) Open stomata 1 (OST 1) kinase controls R–type anion channel QUAC 1 in Arabidopsis guard cells. Plant J 74(3):372–382

Kong L, Deng H, Hu S, Wang F, Miao L, Chen C, Zhao K, Yu X (2018) Isolation, expression, and evolution analysis of the type 2C protein phosphatase gene *BcABI1* involved in abiotic and biotic stress in *Brassica campestris* ssp. chinensis. Plant Growth Regul 85(2):317–327

Kuhn JM, Boisson-Dernier A, Dizon MB, Maktabi MH, Schroeder JI (2006) The protein phosphatase AtPP2CA negatively regulates abscisic acid signal transduction in Arabidopsis, and effects of abh1 on AtPP2CA mRNA. Plant Physiol 140(1):127–139

Lee SC, Lan W, Buchanan BB, Luan S (2009) A protein kinase-phosphatase pair interacts with an ion channel to regulate ABA signaling in plant guard cells. Proc Natl Acad Sci 106(50):21419–21424

Léran S, Edel KH, Pervent M, Hashimoto K, Corratgé-Faillie C, Offenborn JN, Tillard P, Gojon A, Kudla J, Lacombe B (2015) Nitrate sensing and uptake in Arabidopsis are enhanced by ABI2, a phosphatase inactivated by the stress hormone abscisic acid. Sci Signal 8(375):ra43

Leung J, Orfanidi S, Chefdor F, Mészaros T, Bolte S, Mizoguchi T, Shinozaki K, Giraudat J, Bögre L (2006) Antagonistic interaction between MAP kinase and protein phosphatase 2C in stress recovery. Plant Sci 171(5):596–606

Li J, Jia H, Zhang J, Sun J, Zhang Y, Lu M, Xin X, Hu J (2018) Genome-wide characterization of protein phosphatase 2C genes in *Populus euphratica* and their expression profiling under multiple abiotic stresses. Tree Genet Genomes 14(6):80

Ludwików A (2015) Targeting proteins for proteasomal degradation—a new function of Arabidopsis ABI1 protein phosphatase 2C. Front Plant Sci 6:310

Lyzenga WJ, Liu H, Schofield A, Muise-Hennessey A, Stone SL (2013) Arabidopsis CIPK26 interacts with KEG, components of the ABA signaling network and is degraded by the ubiquitin–proteasome system. J Exp Bot 64(10):2779–2791

Lyzenga WJ, Sullivan V, Liu H, Stone SL (2017) The kinase activity of calcineurin B-like interacting protein kinase 26 (CIPK26) influences its own stability and that of the ABA-regulated ubiquitin ligase, Keep on Going (KEG). Front Plant Sci 8:502

Ma Y, Szostkiewicz I, Korte A, Moes D, Yang Y, Christmann A, Grill E (2009) Regulators of PP2C phosphatase activity function as abscisic acid sensors. Science 324(5930):1064–1068

Manohar M, Wang D, Manosalva PM, Choi HW, Kombrink E, Klessig DF (2017) Members of the abscisic acid co-receptor PP 2C protein family mediate salicylic acid–abscisic acid crosstalk. Plant Direct 1(5):e00020

Mao J, Manik S, Shi S, Chao J, Jin Y, Wang Q, Liu H (2016) Mechanisms and physiological roles of the CBL-CIPK networking system in Arabidopsis *thaliana*. Genes 7(9):62

Melcher K, Ng LM, Zhou XE, Soon FF, Xu Y, Suino-Powell KM, Park SY, Weiner JJ, Fujii H, Chinnusamy V, Kovach A (2009) A gate–latch–lock mechanism for hormone signaling by abscisic acid receptors. Nature 462(7273):602

Merlot S, Gosti F, Guerrier D, Vavasseur A, Giraudat J (2001) The ABI1 and ABI2 protein phosphatases 2C act in a negative feedback regulatory loop of the abscisic acid signaling pathway. Plant J 25(3):295–303

Mitula F, Tajdel M, Cieśla A, Kasprowicz-Maluśki A, Kulik A, Babula-Skowrońska D, Michalak M, Dobrowolska G, Sadowski J, Ludwików A (2015) Arabidopsis ABA-activated kinase MAPKKK18 is regulated by protein phosphatase 2C ABI1 and the ubiquitin–proteasome pathway. Plant Cell Physiol 56(12):2351–2367

Moes D, Himmelbach A, Korte A, Haberer G, Grill E (2008) Nuclear localization of the mutant protein phosphatase abi1 is required for insensitivity towards ABA responses in Arabidopsis. Plant J 54(5):806–819

Moustafa K, Vos LD, Leprince AS, Savouré A, Lauriere C (2008) Analysis of the Arabidopsis mitogen-activated protein kinase families: organ specificity and transcriptional regulation upon water stresses. Scholarly Res Exchange 2008:143656

Nishimura N, Yoshida T, Kitahata N, Asami T, Shinozaki K, Hirayama T (2007) ABA-hypersensitive Germination1 encodes a protein phosphatase 2C, an essential component of abscisic acid signaling in Arabidopsis seed. Plant J 50(6):935–949

Ojolo SP, Cao S, Priyadarshani SVGN, Li W, Yan M, Aslam M, Zhao H, Qin Y (2018) Regulation of plant growth and development: a review from a chromatin remodeling perspective. Front Plant Sci 9:1232

Opdenakker K, Remans T, Vangronsveld J, Cuypers A (2012) Mitogen-activated protein (MAP) kinases in plant metal stress: regulation and responses in comparison to other biotic and abiotic stresses. Int J Mol Sci 13(6):7828–7853

Park SY, Fung P, Nishimura N, Jensen DR, Fujii H, Zhao Y, Lumba S, Santiago J, Rodrigues A, Chow TF, Alfred SE (2009) Abscisic acid inhibits type 2C protein phosphatases via the PYR/PYL family of START proteins. Science 324(5930):1068–1071

Peirats-Llobet M, Han SK, Gonzalez-Guzman M, Jeong CW, Rodriguez L, Belda-Palazon B, Wagner D, Rodriguez PL (2016) A direct link between abscisic acid sensing and the chromatin-remodeling ATPase BRAHMA via core ABA signaling pathway components. Mol Plant 9(1):136–147

Planas-Riverola A, Gupta A, Betegón-Putze I, Bosch N, Ibañes M, Caño-Delgado AI (2019) Brassinosteroid signaling in plant development and adaptation to stress. Development 146(5):dev151894

Popescu SC, Popescu GV, Bachan S, Zhang Z, Gerstein M, Snyder M, Dinesh-Kumar SP (2009) MAPK target networks in Arabidopsis *thaliana* revealed using functional protein microarrays. Genes Dev 23(1):80–92

Robert N, Merlot S, N'Guyen V, Boisson-Dernier A, Schroeder JI (2006) A hypermorphic mutation in the protein phosphatase 2C HAB1 strongly affects ABA signaling in Arabidopsis. FEBS Lett 580(19):4691–4696

Rodrigues A, Adamo M, Crozet P, Margalha L, Confraria A, Martinho C, Elias A, Rabissi A, Lumbrebras V, Gonzaléz-Guzmán M, Antoni R (2013) ABI1 and PP2CA phosphatases are negative regulators of Snf1-related protein kinase1 signaling in Arabidopsis. Plant Cell 25(10):3871–3884

Saez A, Apostolova N, Gonzalez-Guzman M, Gonzalez-Garcia MP, Nicolas C, Lorenzo O, Rodriguez PL (2004) Gain-of-function and loss-of-function phenotypes of the protein phosphatase 2C HAB1 reveal its role as a negative regulator of abscisic acid signaling. Plant J 37(3):354–369

Saez A, Robert N, Maktabi MH, Schroeder JI, Serrano R, Rodriguez PL (2006) Enhancement of abscisic acid sensitivity and reduction of water consumption in Arabidopsis by combined inactivation of the protein phosphatases type 2C ABI1 and HAB1. Plant Physiol 141(4):1389–1399

Saez A, Rodrigues A, Santiago J, Rubio S, Rodriguez PL (2008) HAB1–SWI3B interaction reveals a link between abscisic acid signaling and putative SWI/SNF chromatin-remodeling complexes in Arabidopsis. Plant Cell 20(11):2972–2988

Sarnowska E, Gratkowska DM, Sacharowski SP, Cwiek P, Tohge T, Fernie AR, Siedlecki JA, Koncz C, Sarnowski TJ (2016) The role of SWI/SNF chromatin remodeling complexes in hormone crosstalk. Trends Plant Sci 21(7):594–608

Sarnowski TJ, Ríos G, Jásik J, Świeżewski S, Kaczanowski S, Li Y, Kwiatkowska A, Pawlikowska K, Koźbiał M, Koźbiał P, Koncz C (2005) SWI3 subunits of putative SWI/SNF chromatin-remodeling complexes play distinct roles during Arabidopsis development. Plant Cell 17(9):2454–2472

Sato A, Sato Y, Fukao Y, Fujiwara M, Umezawa T, Shinozaki K, Hibi T, Taniguchi M, Miyake H, Goto DB, Uozumi N (2009) Threonine at position 306 of the KAT1 potassium channel is essential for channel activity and is a target site for ABA-activated SnRK2/OST1/SnRK2.6 protein kinase. Biochem J 424(3):439–448

Schweighofer A, Kazanaviciute V, Scheikl E, Teige M, Doczi R, Hirt H, Schwanninger M, Kant M, Schuurink R, Mauch F, Buchala A (2007) The PP2C-type phosphatase AP2C1, which negatively regulates MPK4 and MPK6, modulates innate immunity, jasmonic acid, and ethylene levels in Arabidopsis. Plant Cell 19(7):2213–2224

Singh A, Pandey GK (2012) Protein phosphatases: a genomic outlook to understand their function in plants. J Plant Biochem Biotechnol 21(1):100–107

Singh A, Giri J, Kapoor S, Tyagi AK, Pandey GK (2010) Protein phosphatase complement in rice: genome-wide identification and transcriptional analysis under abiotic stress conditions and reproductive development. BMC Genomics 11(1):435

Singh A, Jha SK, Bagri J, Pandey GK (2015) ABA inducible rice protein phosphatase 2C confers ABA insensitivity and abiotic stress tolerance in Arabidopsis. PLoS One 10(4):e0125168

Singh A, Pandey A, Srivastava AK, Tran LSP, Pandey GK (2016) Plant protein phosphatases 2C: from genomic diversity to functional multiplicity and importance in stress management. Crit Rev Biotechnol 36(6):1023–1035

Singh A, Yadav AK, Kaur K, Sanyal SK, Jha SK, Fernandes JL, Sharma P, Tokas I, Pandey A, Luan S, Pandey GK (2018) A protein phosphatase 2C, AP2C1, interacts with and negatively regulates the function of CIPK9 under potassium-deficient conditions in Arabidopsis. J Exp Bot 69(16):4003–4015

Sirichandra C, Gu D, Hu HC, Davanture M, Lee S, Djaoui M, Valot B, Zyvy M, Leung J, Merlot S, Kwak JM (2009) Phosphorylation of the Arabidopsis AtrbohF NADPH oxidase by OST1 protein kinase. FEBS Lett 583(18):2982–2986

Sirichandra C, Davanture M, Turk BE, Zivy M, Valot B, Leung J, Merlot S (2010) The Arabidopsis ABA-activated kinase OST1 phosphorylates the bZIP transcription factor ABF3 and creates a 14-3-3 binding site involved in its turnover. PLoS One 5(11):e13935

Soon FF, Ng LM, Zhou XE, West GM, Kovach A, Tan ME, Suino-Powell KM, He Y, Xu Y, Chalmers MJ, Brunzelle JS (2012) Molecular mimicry regulates ABA signaling by SnRK2 kinases and PP2C phosphatases. Science 335(6064):85–88

Sun L, Wang YP, Chen P, Ren J, Ji K, Li Q, Li P, Dai SJ, Leng P (2011) Transcriptional regulation of SlPYL, SlPP2C, and SlSnRK2 gene families encoding ABA signal core components during tomato fruit development and drought stress. J Exp Bot 62(15):5659–5669

Taj G, Agarwal P, Grant M, Kumar A (2010) MAPK machinery in plants: recognition and response to different stresses through multiple signal transduction pathways. Plant Signal Behav 5(11):1370–1378

Tuteja N, Sopory SK (2008) Chemical signaling under abiotic stress environment in plants. Plant Signal Behav 3(8):525–536

Umbrasaite J, Schweighofer A, Kazanaviciute V, Magyar Z, Ayatollahi Z, Unterwurzacher V, Choopayak C, Boniecka J, Murray JAH, Bogre L, Meskiene I (2010) MAPK phosphatase AP2C3 induces ectopic proliferation of epidermal cells leading to stomata development in Arabidopsis. PLoS One 5(12):e15357

Wang H, Tang J, Liu J, Hu J, Liu J, Chen Y, Cai Z, Wang X (2018) Abscisic acid signaling inhibits brassinosteroid signaling through dampening the dephosphorylation of BIN2 by ABI1 and ABI2. Mol Plant 11(2):315–325

Wege S, De Angeli A, Droillard MJ, Kroniewicz L, Merlot S, Cornu D, Gambale F, Martinoia E, Barbier-Brygoo H, Thomine S, Filleur S, Leonhardt N (2014) Phosphorylation of the vacuolar anion exchanger AtCLCa is required for the stomatal response to abscisic acid. Sci Signal 7(333):ra65

Wei K, Pan S (2014) Maize protein phosphatase gene family: identification and molecular characterization. BMC Genomics 15(1):773

Xue T, Wang D, Zhang S, Ehlting J, Ni F, Jakab S, Zheng C, Zhong Y (2008) Genome-wide and expression analysis of protein phosphatase 2C in rice and Arabidopsis. BMC Genomics 9(1):550

Yang Q, Liu K, Niu X, Wang Q, Wan Y, Yang F, Li G, Wang Y, Wang R (2018) Genome-wide identification of *PP2C* genes and their expression profiling in response to drought and cold stresses in *Medicago truncatula*. Sci Rep 8(1):12841

Yoshida T, Nishimura N, Kitahata N, Kuromori T, Ito T, Asami T, Shinozaki K, Hirayama T (2006) ABA-hypersensitive germination3 encodes a protein phosphatase 2C (AtPP2CA) that strongly regulates abscisic acid signaling during germination among Arabidopsis protein phosphatase 2Cs. Plant Physiol 140(1):115–126

You J, Zong W, Hu H, Li X, Xiao J, Xiong L (2014) A STRESS-RESPONSIVE NAC1-regulated protein phosphatase gene rice protein phosphatase18 modulates drought and oxidative stress tolerance through abscisic acid-independent reactive oxygen species scavenging in rice. Plant Physiol 166(4):2100–2114

Zhang F, Wei Q, Shi J, Jin X, He Y, Zhang Y, Luo Q, Wang Y, Chang J, Yang G, He G (2017a) *Brachypodium distachyon* BdPP2CA6 interacts with BdPYLs and BdSnRK2 and positively regulates salt tolerance in transgenic Arabidopsis. Front Plant Sci 8:264

Zhang Y, Li Q, Jiang L, Kai W, Liang B, Wang J, Du Y, Zhai X, Wang J, Zhang Y, Sun Y (2017b) Suppressing type 2C protein phosphatases alters fruit ripening and the stress response in tomato. Plant Cell Physiol 59(1):142–154

Zhao R, Sun HL, Mei C, Wang XJ, Yan L, Liu R, Zhang XF, Wang XF, Zhang DP (2011) The Arabidopsis $Ca2^{+}$-dependent protein kinase CPK12 negatively regulates abscisic acid signaling in seed germination and post-germination growth. New Phytol 192(1):61–73

Chapter 6
Plant Protein Phosphatase 2C: Critical Negative Regulator of ABA Signaling

Lokesh K. Saini, Nidhi Singh, and Girdhar K. Pandey

6.1 Introduction

Plants are commonly encountered with various environmental stresses in their life span, and they thrive due to well-developed and efficient sensory system to the constantly changing environment. Environmental stress includes both biotic such as bacteria, fungi, viruses, and insects and abiotic stresses like cold salinity, and drought. Abiotic stress is becoming a major worldwide threat to food security due to climate change, which hampers the growth and development in plants as well as in animals (Huang et al. 2013). Abiotic stress influences the crop productivity and yield by affecting the plant health. Plants have developed complex signaling network to sense environmental cues and respond by changing their physiological and biochemical processes (Droillard et al. 2002; Franz et al. 2011).

Plants synthesize numerous diffusible hormonal signals like abscisic acid, brassinosteroids, and methyl jasmonate which work together to maintain growth, development, and cellular physiology and also to respond against environmental stresses (Lumba et al. 2010). Abscisic acid (ABA) is the carotenoid derivative, which functions as the key abiotic stress signal in plants. ABA is a ubiquitous phytohormone, which regulates growth and development of plants such as maturation of embryo, inhibition of seed germination via maintaining dormancy, and inhibition of post-germination growth and transition from vegetative growth to reproductive growth of plants (Cutler et al. 2010). ABA has a significant role to play in responding to environmental stimuli and in triggering modifications in a number of plant physiological and developmental processes like stomatal development and function, resulting in adaptation to stress circumstances (Melotto et al. 2006). In response to environmental stimuli, ABA biosynthesis takes place in the cell which leads to the activation of the signaling pathway. There are three major components of ABA signaling in

L. K. Saini · N. Singh · G. K. Pandey (✉)
Department of Plant Molecular Biology, University of Delhi South Campus, New Delhi, India
e-mail: gkpandey@south.du.ac.in

G. K. Pandey (ed.), *Protein Phosphatases and Stress Management in Plants*,
https://doi.org/10.1007/978-3-030-48733-1_6

plants: ABA receptors (PYR/PYL/RCAR), type 2C protein phosphatases (PP2Cs), and SNF1-related protein kinases 2 (SnRK2s). These core components of ABA signaling function in a double negative regulatory mechanism in which ABA receptors inhibit protein phosphatase and protein phosphatases inhibit protein kinases (Ikegami et al. 2009). Here, we will attempt to provide a basic understanding toward the core ABA signaling, mainly focusing onto the role of protein phosphatases in ABA signaling.

6.2 ABA Signaling: Major Stress Signaling Pathway of Plants

Abscisic acid (ABA) is one of the most important plant hormones for the regulation of various elements of plant life. ABA is a significant phytohormone that is essential in plant reaction to abiotic stress variables like cold, drought, salinity, and heat. Abscisic acid was identified to control the water status and acclimatization to a variety of stresses in crop plants (Koornneef et al. 1984). In Arabidopsis, its core signaling pathway is comprised of ABA receptor family, protein phosphatase 2Cs (PP2Cs), and SnRK2 (SnRK2 family members are plant-specific serine/threonine kinases) protein kinases (Fig. 6.1). In normal condition, the physiologically active concentration is in sub-micromolar range, while its working concentration increases to low micromolar range during stress (Ikegami et al. 2009). At the early stage of ABA signaling, ABA is sensed by PYR/PYL/RCAR (Pyrabactin Resistance/ Pyrabactin Resistance Like/Regulatory Component of ABA Receptors) receptor family which is comprised of 14 members in Arabidopsis (Table 6.1) (Park et al. 2009; Ma et al. 2009).

Protein phosphatase 2Cs (PP2Cs) are the upstream component of this pathway which negatively regulates ABA signaling via dephosphorylating the downstream component like SnRK2s that hence keep ABA-responsive genes and transcription factors silent in the absence of ABA (Fig. 6.3). The interaction between ABA receptors and PP2Cs contributes to negative feedback regulation of PP2Cs at the moment of ABA perception. In the presence of sufficient concentration of ABA, SnRK2s act as a critical positive regulator in the ABA signaling, generally activated by autophosphorylation mechanism (Fujii and Zhu 2009; Fujita et al. 2009).

ABA signaling in plant cell leads to change in the gene expression of approximately 5–10% genes of genome mainly involved in stress tolerance (Nakashima et al. 2009b). Promoters of these genes possess many *cis*-regulatory elements named as ABA-responsive elements (ABREs). ABRE-binding protein (AREB), also known as ABRE-binding factor (ABF), was identified by using yeast 1-hybrid (Y1H) screening (Uno et al. 2000; Choi et al. 2000). The ABFs belong to group A subfamily of basic leucine zipper (bZIP) domain transcription factors. This family is comprised of nine homologues in Arabidopsis that share conserved C-terminal bZIP domain and N-terminal regions (Jakoby et al. 2002). Several reports suggest that

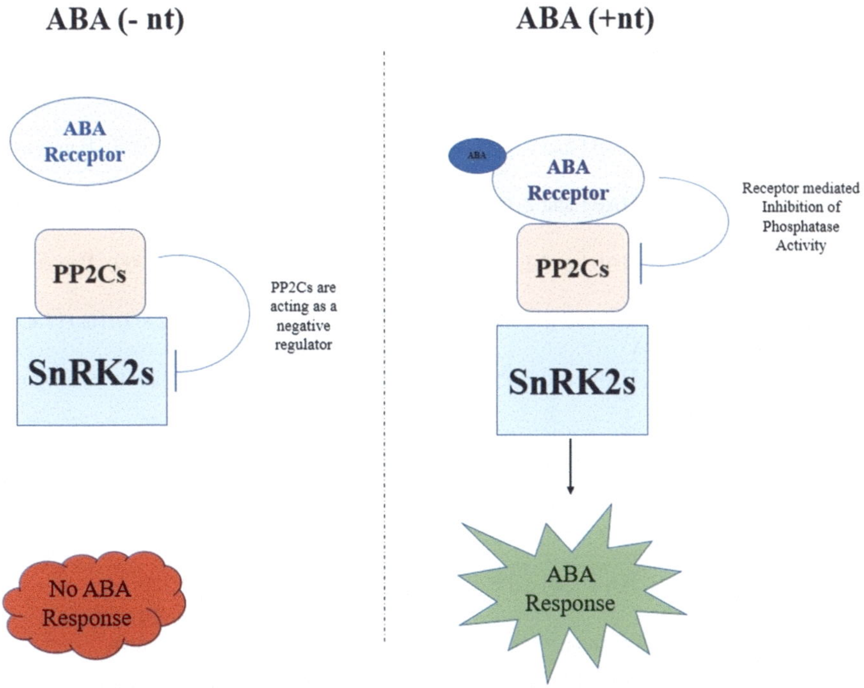

Fig. 6.1 Core component of ABA signaling pathway in plants. In the absence of ABA, PP2Cs prevent the accumulation of active SnRK2s by dephosphorylating them, but in the presence of ABA, receptor-mediated inhibition of PP2Cs results in the activation or accumulation of SnRK2s and ultimately leads to ABA responses

ABFs require ABA-dependent phosphorylation at the conserved domain for their full activation (Furihata et al. 2006; Fujii et al. 2007). So, phosphorylation and activation of ABFs by SnRK2s are essential for the ABA-induced gene expression.

6.2.1 ABA Receptors: Site for ABA Perception

In 1990s, studies on implication of ABA analogs have suggested that ABA-binding proteins are generally present either on plasma membrane or inside the cell (Allan et al. 1994; Schwartz et al. 1994). Isolation of ABA-binding protein in plants was considered to build a key step toward the identification of ABA receptors. The protein having ability to bind with ABA was identified from broad bean (*Vicia faba*) epidermal protein extracts via affinity chromatography (Zhang et al. 2002). In Arabidopsis, it was named as ABAR (ABA receptor) having ABA binding ability. Protein sequence studies revealed that ABAR was a component of Mg-chelatase, which is a multi-subunit complex present on plastid. T-DNA insertion mutants of ABAR were responsive to ABA (Shen et al. 2006).

Table 6.1 Identified interaction between ABA receptors and PP2Cs

ABA receptors	Interactive PP2Cs	References
PYL1/RCAR12	ABI1 and 2, HAB1 and 2, PP2CA	Fujii et al. (2009), Park et al. (2009), Ma et al. (2009), Hao et al. (2011)
PYL2/RCAR14	ABI1, HAB1 and 2, PP2CA	Fujii et al. (2009), Park et al. (2009), Hao et al. (2011)
PYL3/RCAR13	ABI1, HAB1 and 2, PP2CA	Fujii et al. (2009), Park et al. (2009), Hao et al. (2011)
PYL4/RCAR10	ABI1, HAB1 and 2, PP2CA	Fujii et al. (2009), Park et al. (2009), Hao et al. (2011), Pizzio et al. (2013)
PYL5/RCAR8	ABI1 and 2, HAB1 and 2, PP2CA	Fujii et al. (2009), Ma et al. (2009), Hao et al. (2011), Santiago et al. (2009)
PYL6/RCAR9	ABI1, HAB1 and 2, PP2CA	Fujii et al. (2009), Hao et al. (2011), Santiago et al. (2009)
PYL7/RCAR2	ABI1	Fujii et al. (2009)
PYL8/RCAR3	ABI1 and 2, HAB1 and 2, PP2CA	Fujii et al. (2009), Ma et al. (2009), Hao et al. (2011), Santiago et al. (2009)
PYL9/RCAR1	ABI1 and 2, HAB1 and 2, PP2CA	Fujii et al. (2009), Park et al. (2009), Ma et al. (2009), Hao et al. (2011)
PYL10/RCAR4	ABI1, HAB1 and 2, PP2CA	Fujii et al. (2009), Hao et al. (2011)
PYL11/RCAR5	ABI1	Fujii et al. (2009), Hao et al. (2011)
PYL12/RCAR6	PP2CA/AHG3	Fujii et al. (2009), Park et al. (2009)
PYL13/RCAR7	ABI1 and 2, PP2CA/AHG3	Zhao et al. (2013), Fuchs et al. (2014)
PYR1/RCAR11	ABI1 and 2, HAB1 and 2, PP2CA/AHG3	Fujii et al. (2009), Park et al. (2009), Hao et al. (2011)

Many groups have worked to isolate and identify the PYR/PYL/RCAR proteins via different approaches (Ma et al. 2009; Park et al. 2009; Santiago et al. 2009; Nishimura et al. 2010). The characterization of a synthetic ABA agonist named Pyrabactin has made the connection between PYR1 and ABA signaling. *In vitro* studies revealed that both Pyrabactin and ABA help PYR1 to interact and inhibit the clade A protein phosphatase 2C (PP2C) such as *ABA Insensitive 1* (*ABI1*), *ABI2*, and *Hypersensitive to ABA 1* (*HAB1*) (Park et al. 2009). Regulatory component of ABA response 1 (RCAR1), identical to PYL9, was discovered by using Y2H method. This investigation was leading to the identification of remaining 13 members of ABA receptor family (Ma et al. 2009). By using HAB1 as bait in Y2H screening, Rodriguez group identified PYL5, PYL6, and PYL8 (Santiago et al. 2009). In another *in planta* study, nine PYR/PYL/RACR proteins were identified as interactor proteins of ABI1 by using affinity chromatography followed by mass spectrometry (Nishimura et al. 2010).

PYR/PYL/RCAR proteins are total 14 in number which are the members of soluble ligand binding superfamily named as START (steroidogenic acute regulatory protein (StAR)-related lipid transfer)-domain superfamily (Iyer et al. 2001). The crystal structure of PYR/PYL/RCAR ABA receptors has revealed that all PYLs share similar helix grip structure which is composed of a seven-stranded β-sheet flanked by two α-helices. The characterization of START-domain proteins is based on the presence of "helix grip" which is required for the formation of central hydrophobic ligand-binding pocket (Iyer et al. 2001). Crystal structures of ternary complexes have revealed that six members of the PYR/PYL/RCAR family of proteins (PYR1/RCAR11, PYL1/RCAR12, PYL2/RCAR14, PYL3/RCAR13, PYL8/RCAR3, PYL9/RCAR1) have been shown to bind to protein phosphatase 2C in the presence of ABA (Ma et al. 2009; Melcher et al. 2009; Park et al. 2009; Zhang et al. 2012b; Antoni et al. 2013).

Genetic evidences have confirmed the ABA receptors as a central regulator of ABA signaling. Since *pyr1* mutant did not show sensitivity to ABA, maybe because of functional redundancy of other family members of ABA receptor family. However, the reduced ABA sensitivity was observed in the triple and quadruple mutant of *pyr1pyl1pyl4* and *pyr1pyl1pyl2pyl4*, respectively (Park et al. 2009). Overexpression lines of RCAR1, PYL5, and PYL8 showed enhanced ABA sensitivity and stress tolerance to drought (Ma et al. 2009; Santiago et al. 2009). Thus, ABA receptors can be targeted for the better tolerance for abiotic stresses such as drought in crop plants.

6.2.2 Protein Kinases: Positive Regulator of ABA Signaling

Protein phosphorylation is the major event in almost all signal transduction pathways. Many protein kinases have been isolated and characterized as an important component of ABA signaling pathway (Hirayama and Shinozaki 2007, 2010). Major protein kinases involved in ABA signaling belong to SNF1-related kinase 2 (SnRK2) kinase family. Some other kinase families like SNF1-related kinase 3 (SnRK3; CBL-interacting protein kinases (CIPKs)), calcium-dependent protein kinase (CDPK), and mitogen-activated protein kinase (MAPK) family were also found to be involved in ABA signaling (Colcombet and Hirt 2008; Jammes et al. 2009; Cutler et al. 2010; Kudla et al. 2010). In Arabidopsis *thaliana*, there are ten members in SnRK2 family which are characterized into three subclasses (I, II, and III). SnRK2 family is conserved in all land plants, and their involvement has been shown in ABA signaling in barley, maize, pea, and rice (Li et al. 2000; Shen et al. 2001; Kobayashi et al. 2004; Huai et al. 2008). SnRK2.2/3/6/7/8 were found to be activated by ABA. Interestingly, members of subclass III of SnRK2 family, SnRK2.2/3/6, have been exhibiting strongest activation by ABA. These groups of kinases are activated within 30 mins of ABA treatment which suggests their involvement in early signal transduction for initiation of ABA signaling (Mustilli et al. 2002; Yoshida et al. 2002, 2006a; Boudsocq et al. 2004). Phenotypic analysis of mutants and interaction capability of CIPKs including CIPK1/3/8/14/15/20/23/24 with PP2Cs revealed their involvement in ABA signaling (Ohta et al. 2003; Kudla et al. 2010). Calcium-dependent protein kinases (CDPKs) such as CPK3/4/6/11/32 were also found to be involved in ABA signaling. *CPK4* and *CPK11* have sequence similarity, and both phosphorylate ABF1 and ABF4, which are known as

ABA-responsive transcription factors (Choi et al. 2005; Mori et al. 2006; Zhu et al. 2007).

Several Ca^{2+}-independent (SnRK2s) and Ca^{2+}-dependent kinases (SnRK3s/CIPKs and CDPKs/CPKs) have been characterized and known to regulate ABA signaling. The first identified kinase was an SnRK2 named PKABA1 (Wheat Abscisic Acid-Responsive Protein Kinase), which was originally isolated from wheat and was highly upregulated by ABA (Gomez-Cadenas et al. 1999). PKABA1 phosphorylates TaABF1 (a transcription factor of AREB family which binds on ABRE sequence) and mediates ABA suppression in the gibberellic acid-induced gene expression (Gomez-Cadenas et al. 1999). SnRK2.6/Open Stomata 1 (OST1) was the first functionally characterized kinase in guard cells, which led to closing of stomata in response to ABA (Mustilli et al. 2002).

Accumulation of ABA in plant cells leads to the activation of protein kinases, which regulate the phosphorylation status of downstream component of the ABA signaling. Although the molecular mechanism underlying autoactivation of SnRK2s is not very well known, a very recent report suggests that SnRK2 is phosphorylated at a specific OST1 site by MAPKK kinase (M3Ks) for ABA-induced activation (Takahashi et al. 2020).

Some reports using genetic approach have deciphered the importance of subclass III SnRKs in ABA signaling in Arabidopsis. The triple mutant of SnRK2.2/2.3/2.6 has shown the ABA-insensitive phenotype on higher concentration of ABA in the context of seed germination, seed dormancy, post-germination growth, and stomatal movement. These triple mutants did not show any significant expression of ABA-responsive genes (Fujii and Zhu 2009; Fujita et al. 2009; Nakashima et al. 2009a; Umezawa et al. 2009). Based on above studies, it can be concluded that subclass III of SnRK2 kinases functions as a central hub in ABA signaling.

6.2.3 Protein Phosphatase 2C: Negative Regulator of ABA Signaling

Reversible phosphorylation process is a well-understood event which is comprised of protein phosphorylation and dephosphorylation catalyzed by the protein kinases and protein phosphatases, respectively. Previous research was more focused on protein kinases, but nowadays, extensive research is ongoing on protein phosphatases to decipher the mechanism of reversible phosphorylation. Reversible phosphorylation regulates many biological processes in eukaryotes by modulating stability, activity, conformation, and localization of the target substrate in growth, development, and signal transduction pathways (Hunter 1995). Protein kinases and phosphatases are the key players of cellular phosphorylation processes in almost all the organisms under normal and stressed conditions. Protein kinases have been thoroughly studied for their structure and evolution in many organisms (Manning et al. 2002; Kerk et al. 2008).

6.2.4 Classification and Evolution of Protein Phosphatases in Plants

Protein phosphatases are evolutionary conserved from bacteria to complex organism. They have been classified into different subclasses on the basis of structural similarity (Cohen 1989). In higher plants like Arabidopsis and rice, there are 76 and 90 members of protein phosphatases 2C, respectively, which are further subdivided into ten or more subgroups (Xue et al. 2008; Singh et al. 2010). Protein phosphatases are basically divided into two major categories on the basis of their substrate specificity: serine/threonine (Ser/Thr) phosphatases and tyrosine (Tyr)-specific phosphatases. Ser/Thr phosphatases dephosphorylate the phosphoserine/phosphothreonine residue, whereas Tyr-specific phosphatases dephosphorylate the phosphotyrosine residue of substrate protein. Ser/Thr phosphatases are further classified into two families on the basis of molecular, biochemical and genomic analysis: (1) phosphoprotein phosphatases (PPs) consisting of PP1, PP2A, PP2B, and other distantly related phosphatases like PP4, PP5, PP6, and PP7 and (2) metallo-dependent phosphatase (PPM) family. PPM family requires metal ion for their activation or catalysis process, comprising PP2C and other Mg^{2+}-dependent phosphatases (Singh et al. 2010). However, a Ca^{2+}-dependent phosphatase, PP2B, is found in animals and also known as calcineurin A (CNA) but could not be found in plants so far (Uhrig et al. 2013). On the other hand, plant protein Tyr phosphatases (PTPs) are also classified into Tyr-specific phosphatases (PTPs) and dual-specificity phosphatases (DSPs). DSPs have ability to dephosphorylate both phosphotyrosine and phosphoserine/phosphothreonine (de la Fuente van Bentem and Hirt 2009; Shankar et al. 2015) (Fig. 6.2).

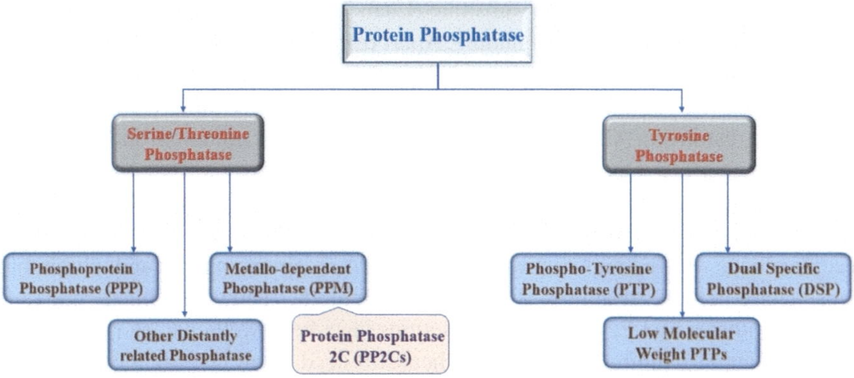

Fig. 6.2 Classification of plant protein phosphatase based on its substrate specificity. Protein phosphatases are categorized into Ser/Thr and Tyr phosphatase. Ser/Thr phosphatases are further subcategorized into three families: PPP, PPM, and other distantly related phosphatases. PPM family is also referred to as PP2Cs. Tyr-specific phosphatases are also subcategorized into PTP, DSP, and LMWPs

Plant genomes have a large number of protein phosphatase-encoding genes. Arabidopsis genome encodes 126 protein phosphatases (Kerk et al. 2008), while other plant species like rice (*Oryza sativa*), tomato (*Solanum lycopersicum*), and hot pepper (*Capsicum annuum*) have 132, 113, and 102 protein phosphatases, respectively (Singh et al. 2010; Kim et al. 2014). The major group of protein phosphatase is PP2Cs having 80, 88, 90, and 91 genes in the Arabidopsis, hot pepper, rice, and tomato genome, respectively (Kim et al. 2014; Singh et al. 2010). In Arabidopsis, PP2Cs are further subdivided into 11 sub-clades, clade A to clade K (Singh et al. 2016). The presence of clade A PP2Cs in unicellular algae (e.g., *Chlamydomonas reinhardtii*) and clade B PP2Cs in *Selaginella moellendorffii* suggests that these PP2Cs are conserved throughout the plant kingdom (Fuchs et al. 2013). Interestingly, six out of nine clade A PP2Cs have been found to be involved in ABA signaling and have been characterized as negative regulators of ABA signaling.

6.2.5 PP2Cs as a Fine Modulator of ABA Signaling

Protein phosphatases are the largest efficient group of proteins in all eukaryotes suggesting their involvement in almost all of the signaling processes in eukaryotic organism. Through genetic studies, several Arabidopsis mutants have been identified which showed insensitivity to ABA. These ABA-insensitive (*abi*) mutants have mutations in numerous genes, which are essential for ABA signaling. Interestingly, instead of kinases, two protein phosphatases were also found to be a critical component in transmission of ABA signal. These findings illustrated a paradigm that protein phosphatase also plays an important role in regulation of protein kinases by dephosphorylating them or their substrate (Luan 1998). Clade A PP2Cs have been very well documented in ABA signaling. Apart from these PP2Cs, a dual-specificity protein phosphatase (DSP), Propyzamide Hypersensitive 1 (PHS1), has been identified and characterized as a critical regulator of ABA signaling, since the mutation in PHS1 gene resulted in ABA hypersensitivity during seed germination and reduction in stomatal aperture (Quettier et al. 2006).

Earlier, we have discussed that there is receptor-mediated inhibition of PP2Cs that takes place in ABA signaling in the presence of ABA. Now, most of the ABA receptors have shown to interact with PP2Cs. ABA receptors interact and inhibit the phosphatase activity of almost all the PP2Cs. Table 6.1 shows the interactions of ABA receptors with PP2Cs. Some interactions have been well characterized in the context of activity inhibition, whereas others still need detail investigation.

6.2.6 Clade A PP2Cs and ABA Signaling

There are nine PP2C members in clade A; most of them have been characterized as modulators of ABA signaling. The first identified ABA-responsive PP2C was ABA Insensitive 1 (ABI1). In Arabidopsis, ABI1 and its homologue regulate

ABA-dependent responses like transpiration, growth, and seed germination (Leung et al. 1994, 1997; Meyer et al. 1994). ABI1 and ABI2 play an important role during early events in ABA signaling (Koornneef et al. 1984). Consequently, six out of nine members of clade A PP2Cs have been emerged as negative regulators of ABA signaling and regulate physiological responses such as germination, stomatal conductance, and root growth (Merlot et al. 2001; Saez et al. 2004; Umezawa et al. 2009). In addition to ABI1 and ABI2, clade A consists of PP2CA/AHG3 (ABA hypersensitive germination 3) (Yoshida et al. 2006b), HAB1 (Homology to ABI1 1) (Rodriguez et al. 1998b), HAB2 (Homology to ABI1 2) (Umezawa et al. 2009), HAI1 (Highly ABA-induced PP2C 1), HAI2 (Highly ABA-induced PP2C 2), HAI3 (Highly ABA-induced PP2C 3), and AHG1 (ABA hypersensitive germination 1) phosphatases (Nishimura et al. 2007).

Under stressed condition or in response to high ABA, PP2Cs stimulate the ABA biosynthesis in the plant cell (Saez et al. 2006; Rubio et al. 2009). It has been postulated that this upregulation of PP2Cs desensitizes the plant to high level of ABA in a negative feedback loop (Szostkiewicz et al. 2010).

The various PP2C members physically interact with cytosolic or nuclear localized proteins suggesting their multiple role and multiple targets in ABA response (Chérel et al. 2002; Saez et al. 2008; Brandt et al. 2012). Among the interactors of clade A PP2Cs, a group of proteins, which is structurally related to pollen allergen Bet V 1, was identified as soluble ABA-binding proteins and named as ABA co-receptors regulatory component of ABA receptors (RCAR) and ABA receptor proteins named Pyrabactin resistance 1 (PYR1/PYR1-like). RCARs/PYR1-like proteins inhibit the function of protein phosphatase activity in the presence of ABA (Ma et al. 2009; Nishimura et al. 2010). ABA receptors require co-receptors ABI1 or other clade A PP2Cs to sense the sub-micromolar concentration. This trimeric complex formation between ABA ligand and holoenzyme leads to inhibition of PP2C activity in ABA signaling. SnRK2 family (subfamily 2 of SNF1-related kinases) member OST1 (Open Stomata 1) functions as a positive regulator downstream to PP2Cs in ABA signaling (Mustilli et al. 2002; Rubio et al. 2009; Kulik et al. 2011). ABI1 and HAB1 dephosphorylate pSer176 of OST1, which is present in activation loop and necessary for autoactivation of this kinase (Ng et al. 2011; Yunta et al. 2011). Some studies also show that RCARs serve as pseudo-substrate for PP2Cs, which can be locked in the presence of ABA (Klingler et al. 2010). It has been already shown by biochemical and protein interaction analysis that RCAR proteins interact with eight clade A PP2Cs to inhibit the activity of PP2Cs (Nishimura et al. 2010; Antoni et al. 2012; Zhang et al. 2012a).

6.2.7 ABA Insensitive 1 and 2 (ABI1 and ABI2)

The *abi1-1* and *abi2-1* mutants have been isolated from ethyl methane sulfonate-mutagenized Arabidopsis seeds through a genetic screening (Koornneef et al. 1984). These mutations lead to phenotypic alterations such as ABA-tolerant seed germination and seedling growth, abnormality in stomatal regulation, and altered drought stress responses (Koornneef et al. 1984; Finkelstein and Somerville 1990). Both of these loci encode homologous proteins and are transcriptionally induced by ABA. Substitution from Gly to Asp in the catalytic domain of ABI1 and ABI2 results in significant reduction in phosphatase activity and Mg^{2+} binding affinity in both. Due to this missense mutation, Gly is changed to Asp in both *abi1-1* (G180D) and *abi2-1* (G168D), which leads to a dominant insensitive phenotype (Leube et al. 1998; Rodriguez et al. 1998a). Further studies have revealed that this mutated protein ABI1-1 is unable to bind with ABA receptors (PYL) in the presence of ABA (Ma et al. 2009; Park et al. 2009) but can still bind to SnRK2s (Umezawa et al. 2009; Vlad et al. 2009). Crystal structure of ternary complex (PYLs-ABA-PP2Cs) also showed that the conserved Gly in active site and conserved Ser in the "gate" loop of PP2Cs form a hydrogen bond, but in mutated protein ABI1-1, hydrogen bond is disrupted due to G to D substitution (Yin et al. 2009). So, it can be concluded that receptor-mediated inhibition of PP2Cs is not happening in the *abi1-1* and *abi2-1* mutants; hence, these mutants show ABA insensitivity.

ABI1-1 was found to inhibit ABA signal transduction in isolated protoplasts using ABA-responsive promoters HAV1-GUS and RBCS-GUS (Sheen 1998). The similar type mutation in the alfalfa MP2C results in diminished phosphatase activity of recombinant protein *in vitro* but could not affect its dephosphorylation activity for MAPK substrate *in vivo* (Meskiene et al. 2003). Isolation and analysis of mutants of *abi1-1* and *abi2-1* provided evidence that these both PP2Cs negatively regulate the ABA signaling in plants. Mutations in the catalytic domain of the ABI1 or ABI2 cause loss of function with very low phosphatase activity and also make plants supersensitive to ABA. Near 50% of the ABA-induced PP2C activity is contributed by ABI1 and ABI2 that indicates that other PP2Cs may also involve in ABA signaling (Merlot et al. 2001). Overexpression and microinjection approach has shown that mutant protein abi1-1 but not wild type ABI1 can block ABA, cyclic ADP-ribose, and Ca^{2+}-induced activation of promoters KIN2-GUS and RD29A-GUS (Wu et al. 2003). As per the model suggested by authors, both PP2Cs are functioning at different levels; *abi1-1* acts upstream, whereas *abi2-1* acts downstream of ROS production induced by ABA in guard cells (Murata et al. 2001). Interaction of ABI1 with ABA-inducible transcription factor ATHB6 and promoter reporter expression of ATHB6 was inhibited in *abi1-1* mutant plants, suggesting that ABI1 acts upstream of ATHB6 (Himmelbach et al. 2002).

6.2.8 ABA Hypersensitive Germination 3 (AHG3/AtPP2CA)

Another clade A type PP2C named *PP2CA/AHG3* was found to block ABA signaling when it is expressed transiently in protoplasts (Sheen 1998). Transcriptional upregulation of *PP2CA* is induced in cold, drought, salt, and ABA. Reduction in mRNA level of *PP2CA* in cold and drought in the *aba1-1* mutant suggested that expression of *PP2CA* is ABA dependent. Using antisense approach, downregulation of *AtPP2CA* resulted in better plant development and led to freezing tolerance, suggesting its negative role in ABA responses during cold (Tähtiharju and Palva 2001). Interestingly, interaction between AtPP2C5 and AKT2/AKT3 in yeast was specific for catalytic domain of phosphatase compared with other potassium (K^+) shaker channels. This interaction enables *AtPP2C5* to regulate inward rectifying K^+ channel AKT2 in *Xenopus* oocytes. The *AtPP2CA* and *AKT2* genes are regulated by ABA and show similar tissue-specific expression with highest expression level in phloem vasculature. This study suggests the regulation of AKT2 channel by AtPP2CA enables K^+ uptake and also maintains membrane potential during stress condition (Vranová et al. 2001; Chérel et al. 2002).

6.2.9 Hypersensitive to ABA (HAB1)

One of the very close relatives of *ABI1* and *ABI2* is *HAB1* (previously known as AtPP2CA), upregulated by ABA and expressed in almost all part of Arabidopsis plant including root, stem, leaf, flower, and silique (Rodriguez 1998; Rodriguez et al. 1998b). ABA insensitivity in seeds and vegetative tissues was seen in constitutive overexpression lines of this gene, which suggests that *HAB1* is also a negative regulator of ABA signaling. Overexpression of *HAB1* results in impaired stomatal activity, ABA insensitive root growth, and reduced ABA-inducible gene expression (Saez et al. 2004).

Previous data suggests that clade A PP2Cs function as co-receptors of ABA, and functional analysis of other PP2Cs revealed their involvement in ABA-dependent stress responses (Jia et al. 2009). Moreover, clade A protein phosphatase (PP2C) is the negative regulator of ABA signaling, but a few reports also suggest that some PP2Cs can also regulate positively to ABA signaling. Splice variant HAB1.2 positively regulates ABA signaling by interacting with OST1 but not inhibiting the kinase activity (Wang et al. 2015). In another study, overexpression of beech (*Fagus sylvatica*) PP2C 2 (FsPP2C2) in Arabidopsis showed the enhanced expression of ABA-responsive genes, suggesting its role as a positive regulator of ABA signaling (Reyes et al. 2006).

6.3 Role of PP2Cs in Various Signaling Pathways in Plants

6.3.1 PP2Cs in Developmental Signaling

Integration of different signaling pathway is required for proper growth and development of plants. Now, it has been established that protein kinases and protein phosphatases play a vital role in plant developmental pathways. MAPK (mitogen-activated protein kinase) signaling pathway is a major developmental pathway, which is also activated in biotic and abiotic stress condition. Clade B protein phosphatase 2C, named AP2C3/PP2C5, interacts with MPK3, MPK4, and MPK6 and dephosphorylates MPK6. These data suggest that AP2C3 inhibits the MAPK signaling pathway to maintain the balance between stomatal differentiation and pavement cell differentiation (Umbrasaite et al. 2010).

Protein phosphatases are found to play an important role in developmental signaling during different stages of plant development (Singh and Pandey 2012). Most significant PP2Cs include *KAPP* (kinase-associated protein phosphatase) and *POL* (POLTERGEIST); both of these are involved in *CLAVATA 1* (CLV1, a receptor-like kinase) signaling, regulating the Arabidopsis flower development (Luan 2003). KAPP interacts and dephosphorylates the CLV1 *in vitro*, and POL modulates the activity of a transcription factor WUSCHEL (WUS) (Yu et al. 2003). However, genetic analysis of double mutant of *clv/wus* and triple mutant of *pol/clv/wus* showed that POL functions in WUS-dependent and independent pathway (Yu et al. 2003). Besides, double mutant of *POL* and *POL-Like* (PLL) has shown seedling lethality suggesting their implication in shoot and root meristem as well as embryo development (Song and Clark 2005; Wang et al. 2007). Later on, it was also found that POL and PLL1 regulate the stem cell fate by affecting the WUS expression

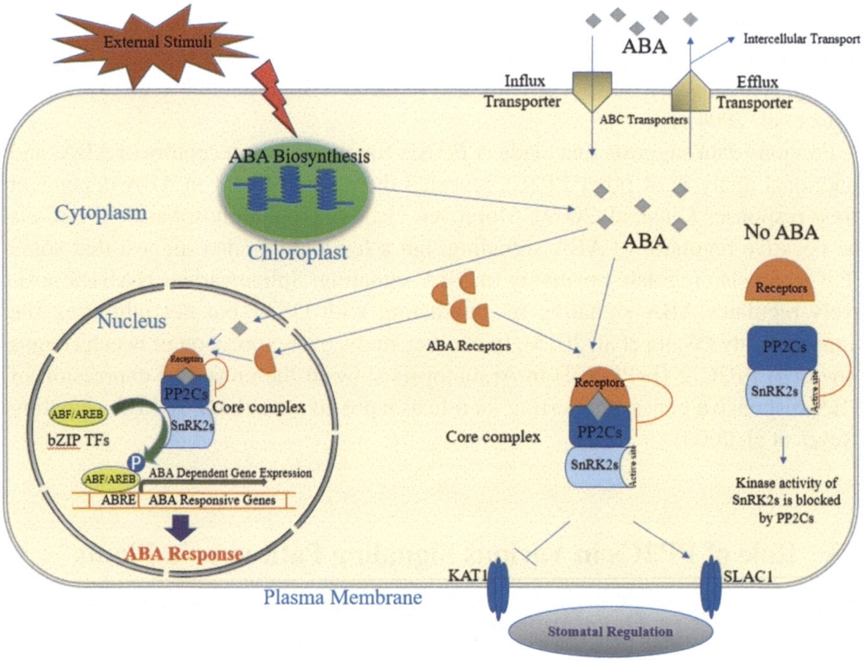

Fig. 6.3 Model of the core ABA signaling pathway in plants. Under normal conditions, PP2Cs remain bound to SnRK2s and inhibit their kinase activity and hence negatively regulate ABA-mediated responses. When ABA is perceived by PYR/PYL/RCAR ABA receptor family, these receptors bind with PP2Cs and block the phosphatase activity. This receptor-mediated inhibition of PP2Cs leads to activation of SnRK2s. These SnRK2s further phosphorylate the downstream components of the signaling pathway to activate or deactivate them like KAT1 and SLAC1 for stomatal regulation

downstream to CLV1. POL and PLL1 modulate the development of early embryo and root meristem by mediating the CLE40 (CLV-related 40)/WOX5 (WUSCHEL-related homeobox-5) pathway (Gagne and Clark 2010; Song et al. 2008). Therefore, POL and PLL1 involved in asymmetric stem cell division and in maintenance of stem cell polarity by controlling the early embryo and root meristem development (Fuchs et al. 2013). Moreover, genetic analysis of PLL4 and PLL5 has suggested their role in leaf development (Song and Clark 2005). On the basis of above data, it can be concluded that PP2Cs regulate the plant growth and development via mediating the receptor kinase signaling pathway.

6.3.2 PP2Cs in Abiotic Stress Signaling

Activation of Ca^{2+} signaling network, which basically consists of calcineurin B-like proteins (CBLs) and CBL-interacting protein kinases (CIPKs), leads to triggering an adaptive mechanism to maintain K^+ homeostasis during K^+ deficiency. Clade A PP2C AIP1 (AKT1-interacting PP2C 1)/HAI2 was found to interact with CIPK23 and also AKT1 (Arabidopsis K^+ transporter 1). CIPK23 promotes the activity of AKT1, and AIP1 downregulates the activity of AKT1, suggesting that *AIP1* is a negative regulator of this pathway and negatively regulating the K^+ uptake during K^+ deficiency condition (Lee et al. 2007). Another clade B PP2C named AP2C1 was also found to interact and dephosphorylate the CIPK9 under K^+ deficiency. *CIPK9* positively regulates whereas *AP2C1* negatively regulates Arabidopsis root growth and seedling development under K^+ deficiency and hence acts as phosphorylation-dephosphorylation-based switch under low K^+ condition (Singh et al. 2018).

6.3.3 PP2Cs in Biotic Stress Signaling

AP2C1, clade B PP2C, is also known to regulate wound and biotic stress response by dephosphorylating/inactivating MAPK4 and MAPK6, which are known as positive regulators of wound and pathogen-triggered signaling (Schweighofer et al. 2007). *AP2C1* expression is upregulated in response to tissue injury and fungal pathogen *Botrytis cinerea*. Overexpression of AP2C1 phosphatase has been shown to reduce the kinase activity of MAPK in response to wounding. Overexpression lines of AP2C1 have shown low level of ethylene and hampered innate immunity against *B. cinerea*. Moreover, *ap2c1* mutants exhibited higher accumulation of jasmonic acid induced by wounding and also resistance to phytophagous herbivore (*Tetranychus urticae*) (Galletti et al. 2011; Fuchs et al. 2013). These findings suggest that *AP2C1* negatively regulates MAPK signaling. Thus, AP2C1 regulates plants innate immunity during pathogen attack via controlling the defense hormone level (Schweighofer et al. 2007; Galletti et al. 2011; Fuchs et al. 2013).

6.4 Conclusions and Future Perspectives

Plants encode a larger number of PP2Cs than yeast, mouse, and human, suggesting PP2Cs as a major group of protein phosphatase, regulating diverse physiological processes in plants. The phytohormone ABA activates a complex signaling that regulates numerous cellular and physiological processes in plants. Recent in-depth studies on ABA receptor uncovered the novel functions of PP2Cs as hormone co-receptors in plants. Now, it has been understood that PP2Cs negatively regulate the kinase function in most of the signaling pathways. But some reports also suggest that PP2Cs can regulate positively to ABA signaling for better adaptive responses against stresses. According to latest paradigm, ABA receptor, PP2Cs, and SnRK2s come together in different combination and interact via specific domain. This coordination regulates the ABA signaling in response to stress and developmental stimuli and enables plants to acclimatize in stress conditions.

Based on functional genomic approaches, novel uncharacterized PP2C candidates can be identified followed by their characterization by various genetic, cellular, and molecular approaches. As PP2Cs are involved in many signal transduction pathways, they can serve as connecting link between different pathways. Also, these can be targeted as potential candidate to decipher the crosstalk between different stress and hormone signaling pathways. Another important aspect for plant biologist is to decipher the role of PP2Cs in regulation of downstream targets of ABA signaling as well as in stress and developmental signaling pathways. Information collected from model plants like Arabidopsis could also be used for generating tools to improve the crop productivity under stress conditions. Functional characterization of PP2Cs should also be explored in other crop plants such as cereal and non-cereal crops to overcome the worldwide food security for constantly growing population.

Acknowledgments We are thankful to the Department of Biotechnology (DBT), Science and Engineering Research Board (SERB), and Council for Scientific and Industrial Research (CSIR), Delhi University, India, for research funding in GKP's lab.

References

Allan AC, Fricker MD, Ward JL, Beale MH, Trewavas AJ (1994) Two transduction pathways mediate rapid effects of abscisic acid in commelina guard cells. Plant Cell 6:1319–1328

Antoni R, Gonzalez-Guzman M, Rodriguez L, Rodrigues A, Pizzio GA, Rodriguez PL (2012) Selective inhibition of clade A phosphatases type 2C by PYR/PYL/RCAR abscisic acid receptors. Plant Physiol 158:970

Antoni R, Gonzalez-Guzman M, Rodriguez L, Peirats-Llobet M, Pizzio GA, Fernandez MA, De Winne N, De Jaeger G, Dietrich D, Bennett MJ, Rodriguez PL (2013) PYRABACTIN RESISTANCE1-LIKE8 plays an important role for the regulation of abscisic acid signaling in root. Plant Physiol 161:931

Boudsocq M, Barbier-Brygoo H, Laurière C (2004) Identification of nine sucrose nonfermenting 1-related protein kinases 2 activated by hyperosmotic and saline stresses in Arabidopsis thaliana. J Biol Chem 279:41758

Brandt B, Brodsky DE, Xue S, Negi J, Iba K, Kangasjarvi J, Ghassemian M, Stephan AB, Hu H, Schroeder JI (2012) Reconstitution of abscisic acid activation of SLAC1 anion channel by CPK6 and OST1 kinases and branched ABI1 PP2C phosphatase action. Proc Natl Acad Sci 109:10593

Chérel I, Michard E, Platet N, Mouline K, Alcon C, Sentenac H, Thibaud J-B (2002) Physical and functional interaction of the Arabidopsis K(+) channel AKT2 and phosphatase AtPP2CA. Plant Cell 14:1133–1146

Choi HI, Hong JH, Ha JO, Kang JY, Kim SY (2000) ABFs, a family of ABA-responsive element binding factors. J Biol Chem 275:1723–1730

Choi H, Park H-J, Park JH, Kim S, Im M-Y, Seo H-H, Kim Y-W, Hwang I, Kim SY (2005) Arabidopsis calcium-dependent protein kinase AtCPK32 interacts with ABF4, a transcriptional regulator of abscisic acid-responsive gene expression, and modulates its activity. Plant Physiol 139:1750

Cohen P (1989) The structure and regulation of protein phosphatases. Annu Rev Biochem 58:453

Colcombet J, Hirt H (2008) Arabidopsis MAPKs: a complex signaling network involved in multiple biological processes. Biochem J 413:217

Cutler SR, Rodriguez PL, Finkelstein RR, Abrams SR (2010) Abscisic acid: emergence of a core signaling network. Annu Rev Plant Biol 61:651–679

de la Fuente van Bentem S, Hirt H (2009) Protein tyrosine phosphorylation in plants: more abundant than expected? Trends Plant Sci 14:71–76

Droillard MJ, Boudsocq M, Barbier-Brygoo H, Laurière C (2002) Different protein kinase families are activated by osmotic stresses in Arabidopsis thaliana cell suspensions: involvement of the MAP kinases AtMPK3 and AtMPK6. FEBS Lett 527:43

Finkelstein RR, Somerville CR (1990) Three classes of abscisic acid (ABA)-insensitive mutations of arabidopsis define genes that control overlapping subsets of ABA responses. Plant Physiol 94:1172–1179

Franz S, Ehlert B, Liese A, Kurth J, Cazalé AC, Romeis T (2011) Calcium-dependent protein kinase CPK21 functions in abiotic stress response in Arabidopsis thaliana. Mol Plant 4:83

Fuchs S, Grill E, Meskiene I, Schweighofer A (2013) Type 2C protein phosphatases in plants. FEBS J 280:681–693

Fuchs S, Tischer SV, Wunschel C, Christmann A, Grill E (2014) Abscisic acid sensor RCAR7/PYL13, specific regulator of protein phosphatase coreceptors. Proc Natl Acad Sci 111:5741

Fujii H, Zhu J-K (2009) Arabidopsis mutant deficient in 3 abscisic acid-activated protein kinases reveals critical roles in growth, reproduction, and stress. Proc Natl Acad Sci 106:8380

Fujii H, Verslues PE, Zhu J-K (2007) Identification of two protein kinases required for abscisic acid regulation of seed germination, root growth, and gene expression in Arabidopsis. Plant Cell 19:485–494

Fujii H, Chinnusamy V, Rodrigues A, Rubio S, Antoni R, Park SY, Cutler SR, Sheen J, Rodriguez PL, Zhu JK (2009) *In vitro* reconstitution of an abscisic acid signaling pathway. Nature 462:660

Fujita Y et al (2009) Three SnRK2 protein kinases are the main positive regulators of abscisic acid signaling in response to water stress in arabidopsis. Plant Cell Physiol 50:2123

Furihata T, Maruyama K, Fujita Y, Umezawa T, Yoshida R, Shinozaki K, Yamaguchi-Shinozaki K (2006) Abscisic acid-dependent multisite phosphorylation regulates the activity of a transcription activator AREB1. Proc Natl Acad Sci U S A 103:1988–1993

Gagne JM, Clark SE (2010) The Arabidopsis stem cell factor POLTERGEIST is membrane localized and phospholipid stimulated. Plant Cell 22:729–743

Galletti R, Ferrari S, De Lorenzo G (2011) Arabidopsis MPK3 and MPK6 play different roles in basal and oligogalacturonide- or flagellin-induced resistance against Botrytis cinerea. Plant Physiol 157:804

Gomez-Cadenas A, Verhey SD, Holappa LD, Shen Q, Ho T-HD, Walker-Simmons MK (1999) An abscisic acid-induced protein kinase, PKABA1, mediates abscisic acid-suppressed gene expression in barley aleurone layers. Proc Natl Acad Sci 96:1767–1772

Hao Q, Yin P, Li W, Wang L, Yan C, Lin Z, Wu JZ, Wang J, Yan SF, Yan N (2011) The molecular basis of ABA-independent inhibition of PP2Cs by a subclass of PYL proteins. Mol Cell 42:662

Himmelbach A, Hoffmann T, Leube M, Höhener B, Grill E (2002) Homeodomain protein ATHB6 is a target of the protein phosphatase ABI1 and regulates hormone responses in Arabidopsis. EMBO J 21:3029

Hirayama T, Shinozaki K (2007) Perception and transduction of abscisic acid signals: keys to the function of the versatile plant hormone ABA. Trends Plant Sci 12:343

Hirayama T, Shinozaki K (2010) Research on plant abiotic stress responses in the post-genome era: past, present and future. Plant J 61:1041

Huai J, Wang M, He J, Zheng J, Dong Z, Lv H, Zhao J, Wang G (2008) Cloning and characterization of the SnRK2 gene family from Zea mays. Plant Cell Rep 27:1861–1868

Huang J, Levine A, Wang Z (2013) Plant abiotic stress. Sci World J 2013:1–2

Hunter T (1995) Protein kinases and phosphatases: the yin and yang of protein phosphorylation and signaling. Cell 80:225–236

Ikegami K, Okamoto M, Seo M, Koshiba T (2009) Activation of abscisic acid biosynthesis in the leaves of Arabidopsis thaliana in response to water deficit. J Plant Res 122:235

Iyer LM, Koonin EV, Aravind L (2001) Adaptations of the helix-grip fold for ligand binding and catalysis in the START domain superfamily. Proteins 43:134–144

Jakoby M, Weisshaar B, Dröge-Laser W, Vicente-Carbajosa J, Tiedemann J, Kroj T, Parcy F (2002) bZIP transcription factors in Arabidopsis. Trends Plant Sci 7:106–111

Jammes F et al (2009) MAP kinases MPK9 and MPK12 are preferentially expressed in guard cells and positively regulate ROS-mediated ABA signaling. Proc Natl Acad Sci 106:20520

Jia F, Gampala SSL, Mittal A, Luo Q, Rock CD (2009) Cre-lox univector acceptor vectors for functional screening in protoplasts: analysis of Arabidopsis donor cDNAs encoding ABSCISIC ACID INSENSITIVE1-like protein phosphatases. Plant Mol Biol 70:693

Kerk D, Templeton G, Moorhead GBG (2008) Evolutionary radiation pattern of novel protein phosphatases revealed by analysis of protein data from the completely sequenced genomes of humans, green algae, and higher plants. Plant Physiol 146:351–367

Kim S et al (2014) Genome sequence of the hot pepper provides insights into the evolution of pungency in Capsicum species. Nat Genet 46:270

Klingler JP, Batelli G, Zhu JK (2010) ABA receptors: the START of a new paradigm in phytohormone signaling. J Exp Bot 61:3199

Kobayashi Y, Yamamoto S, Minami H, Kagaya Y, Hattori T (2004) Differential activation of the rice sucrose nonfermenting1–related protein kinase2 family by hyperosmotic stress and abscisic acid. Plant Cell 16:1163

Koornneef M, Reuling G, Karssen CM (1984) The isolation and characterization of abscisic acid-insensitive mutants of Arabidopsis thaliana. Physiol Plant 61:377

Kudla J, Batistic O, Hashimoto K (2010) Calcium signals: the Lead currency of plant information processing. Plant Cell 22:541

Kulik A, Wawer I, Krzywińska E, Bucholc M, Dobrowolska G (2011) SnRK2 protein kinases—key regulators of plant response to abiotic stresses. OMICS 15:859–872

Lee SC, Lan W-Z, Kim B-G, Li L, Cheong YH, Pandey GK, Lu G, Buchanan BB, Luan S (2007) A protein phosphorylation/dephosphorylation network regulates a plant potassium channel. Proc Natl Acad Sci U S A 104:15959

Leube M, Grill E, Amrhein N (1998) ABI1 of Arabidopsis is a protein serine/threonine phosphatase highly regulated by the proton and magnesium ion concentration. FEBS Lett 424:100–104

Leung J, Bouvier-Durand M, Morris PC, Guerrier D, Chefdor F, Giraudat J (1994) Arabidopsis ABA response gene ABI1: features of a calcium-modulated protein phosphatase. Science 264:1448–1452

Leung J, Merlot S, Giraudat J (1997) The Arabidopsis ABSCISIC ACID-INSENSITIVE2 (ABI2) and ABI1 genes encode homologous protein phosphatases 2C involved in abscisic acid signal transduction. Plant Cell 9:759–771

Li J, Wang XQ, Watson MB, Assmann SM (2000) Regulation of abscisic acid-induced stomatal closure and anion channels by guard cell AAPK kinase. Science 287:300–303

Luan S (1998) Protein phosphatases and signaling cascades in higher plants. Trends Plant Sci 3:271

Luan S (2003) Protein phosphatases in plants. Annu Rev Plant Biol 54:63

Lumba S, Cutler S, McCourt P (2010) Plant nuclear hormone receptors: a role for small molecules in protein-protein interactions. Annu Rev Cell Dev Biol 26:445

Ma Y, Szostkiewicz I, Korte A, Moes D, Yang Y, Christmann A, Grill E (2009) Regulators of PP2C phosphatase activity function as abscisic acid sensors. Science 324:1064–1068

Manning G, Plowman GD, Hunter T, Sudarsanam S (2002) Evolution of protein kinase signaling from yeast to man. Trends Biochem Sci 27:514

Melcher K et al (2009) A gate-latch-lock mechanism for hormone signaling by abscisic acid receptors. Nature 462:602

Melotto M, Underwood W, Koczan J, Nomura K, He SY (2006) Plant stomata function in innate immunity against bacterial invasion. Cell 126:969–980

Merlot S, Guerrier D, Vavasseur A, Gosti F, Giraudat J (2001) The ABI1 and ABI2 protein phosphatases 2C act in a negative feedback regulatory loop of the abscisic acid signaling pathway. Plant J 25:295–303

Meskiene I, Baudouin E, Schweighofer A, Liwosz A, Jonak C, Rodriguez PL, Jelinek H, Hirt H (2003) Stress-induced protein phosphatase 2C is a negative regulator of a mitogen-activated protein kinase. J Biol Chem 278:18945–18952

Meyer K, Leube MP, Grill E (1994) A protein phosphatase 2C involved in ABA signal transduction in Arabidopsis thaliana. Science 264:1452–1455

Mori IC, Murata Y, Yang Y, Munemasa S, Wang YF, Andreoli S, Tiriac H, Alonso JM, Harper JF, Ecker JR, Kwak JM, Schroeder JI (2006) CDPKs CPK6 and CPK3 function in ABA regulation of guard cell S-type anion- and Ca2+-permeable channels and stomatal closure. PLoS Biol 4:e327

Murata Y, Pei Z-M, Mori IC, Schroeder J (2001) Abscisic acid activation of plasma membrane Ca^{2+} channels in guard cells requires cytosolic NAD(P)H and is differentially disrupted upstream and downstream of reactive oxygen species production in *abi1-1* and *abi2-1* protein phosphatase 2C mutants. Plant Cell 13:2513–2523

Mustilli A-C, Merlot S, Vavasseur A, Fenzi F, Giraudat J (2002) Arabidopsis OST1 protein kinase mediates the regulation of stomatal aperture by abscisic acid and acts upstream of reactive oxygen species production. Plant Cell 14:3089–3099

Nakashima K, Fujita Y, Kanamori N, Katagiri T, Umezawa T, Kidokoro S, Maruyama K, Yoshida T, Ishiyama K, Kobayashi M, Shinozaki K, Yamaguchi-Shinozaki K (2009a) Three arabidopsis SnRK2 protein kinases, SRK2D/SnRK2.2, SRK2E/SnRK2.6/OST1 and SRK2I/SnRK2.3, involved in ABA signaling are essential for the control of seed development and dormancy. Plant Cell Physiol 50:1345–1363

Nakashima K, Ito Y, Yamaguchi-Shinozaki K (2009b) Transcriptional regulatory networks in response to abiotic stresses in Arabidopsis and grasses. Plant Physiol 149:88–95

Ng L-M et al (2011) Structural basis for basal activity and autoactivation of abscisic acid (ABA) signaling SnRK2 kinases. Proc Natl Acad Sci U S A 108:21259–21264

Nishimura N, Yoshida T, Kitahata N, Asami T, Shinozaki K, Hirayama T (2007) ABA-hypersensitive germination1 encodes a protein phosphatase 2C, an essential component of abscisic acid signaling in Arabidopsis seed. Plant J 50:935

Nishimura N, Sarkeshik A, Nito K, Park SY, Wang A, Carvalho PC, Lee S, Caddell DF, Cutler SR, Chory J, Yates JR, Schroeder JI (2010) PYR/PYL/RCAR family members are major *in-vivo* ABI1 protein phosphatase 2C-interacting proteins in Arabidopsis. Plant J 61:290–299

Ohta M, Guo Y, Halfter U, Zhu J-K (2003) A novel domain in the protein kinase SOS2 mediates interaction with the protein phosphatase 2C ABI2. Proc Natl Acad Sci 100:11771

Park SY et al (2009) Abscisic acid inhibits type 2C protein phosphatases via the PYR/PYL family of START proteins. Science 324:1068–1071

Pizzio GA, Rodriguez L, Antoni R, Gonzalez-Guzman M, Yunta C, Merilo E, Kollist H, Albert A, Rodriguez PL (2013) The PYL4 A194T mutant uncovers a key role of PYR1-LIKE4/PROTEIN PHOSPHATASE 2CA interaction for abscisic acid signaling and plant drought resistance. Plant Physiol 163:441

Quettier AL, Bertrand C, Habricot Y, Miginiac E, Agnes C, Jeannette E, Maldiney R (2006) The phs1-3 mutation in a putative dual-specificity protein tyrosine phosphatase gene provokes hypersensitive responses to abscisic acid in Arabidopsis thaliana. Plant J 47:711

Reyes D, Rodríguez D, González-García MP, Lorenzo O, Nicolás G, García-Martínez JL, Nicolás C (2006) Overexpression of a protein phosphatase 2C from beech seeds in Arabidopsis shows phenotypes related to abscisic acid responses and gibberellin biosynthesis. Plant Physiol 141:1414–1424

Rodriguez PL (1998) Protein phosphatase 2C (PP2C) function in higher plants. Plant Mol Biol 38:919–927

Rodriguez PL, Benning G, Grill E (1998a) ABI2, a second protein phosphatase 2C involved in abscisic acid signal transduction in Arabidopsis. FEBS Lett 421:185–190

Rodriguez PL, Leube MP, Grill E (1998b) Molecular cloning in Arabidopsis thaliana of a new protein phosphatase 2C (PP2C) with homology to ABI1 and ABI2. Plant Mol Biol 38:879–883

Rubio S, Rodrigues A, Saez A, Dizon MB, Galle A, Kim T-H, Santiago J, Flexas J, Schroeder JI, Rodriguez PL (2009) Triple loss of function of protein phosphatases type 2C leads to partial constitutive response to endogenous abscisic acid. Plant Physiol 150:1345–1355

Saez A, Apostolova N, Gonzalez-Guzman M, Gonzalez-Garcia MP, Nicolas C, Lorenzo O, Rodriguez PL (2004) Gain-of-function and loss-of-function phenotypes of the protein phosphatase 2C HAB1 reveal its role as a negative regulator of abscisic acid signaling. Plant J 37:354–369

Saez A, Robert N, Maktabi MH, Schroeder JI, Serrano R, Rodriguez PL (2006) Enhancement of abscisic acid sensitivity and reduction of water consumption in Arabidopsis by combined inactivation of the protein phosphatases type 2C ABI1 and HAB1. Plant Physiol 141:1389

Saez A, Rodrigues A, Santiago J, Rubio S, Rodriguez PL (2008) HAB1-SWI3B interaction reveals a link between abscisic acid signaling and putative SWI/SNF chromatin-remodeling complexes in Arabidopsis. Plant Cell 20:2972–2988

Santiago J, Rodrigues A, Saez A, Rubio S, Antoni R, Dupeux F, Park SY, Márquez JA, Cutler SR, Rodriguez PL (2009) Modulation of drought resistance by the abscisic acid receptor PYL5 through inhibition of clade A PP2Cs. Plant J 60:575

Schwartz A, Wu WH, Tucker EB, Assmann SM (1994) Inhibition of inward K+ channels and stomatal response by abscisic acid: an intracellular locus of phytohormone action. Proc Natl Acad Sci 91:4019–4023

Schweighofer A et al (2007) The PP2C-type phosphatase AP2C1, which negatively regulates MPK4 and MPK6, modulates innate immunity, jasmonic acid, and ethylene levels in Arabidopsis. Plant Cell Online 19:2213

Shankar A, Agrawal N, Sharma M, Pandey A, Pandey K, G. (2015) Role of protein tyrosine phosphatases in plants. Curr Genomics 16:224–236

Sheen J (1998) Mutational analysis of protein phosphatase 2C involved in abscisic acid signal transduction in higher plants. Proc Natl Acad Sci U S A 95:975

Shen Q, Gomez-Cadenas A, Zhang P, Walker-Simmons MK, Sheen J, Ho TH (2001) Dissection of abscisic acid signal transduction pathways in barley aleurone layers. Plant Mol Biol 47:437–448

Shen Y-Y, Wang X-F, Wu F-Q, Du S-Y, Cao Z, Shang Y, Wang X-L, Peng CC, Yu X-C, Zhu S-Y, Fan R-C, Xu Y-H, Zhang D-P (2006) The Mg-chelatase H subunit is an abscisic acid receptor. Nature 443(7113):823–826

Singh A, Pandey GK (2012) Protein phosphatases: a genomic outlook to understand their function in plants. J Plant Biochem Biotechnol 21:100

Singh A, Giri J, Kapoor S, Tyagi AK, Pandey GK (2010) Protein phosphatase complement in rice: genome-wide identification and transcriptional analysis under abiotic stress conditions and reproductive development. BMC Genomics 11:435

Singh A, Pandey A, Srivastava AK, Tran LS, Pandey GK (2016) Plant protein phosphatases 2C: from genomic diversity to functional multiplicity and importance in stress management. Crit Rev Biotechnol 36:1023–1035

Singh A, Yadav AK, Kaur K, Sanyal SK, Jha SK, Fernandes JL, Sharma P, Tokas I, Pandey A, Luan S, Pandey GK (2018) A protein phosphatase 2C, AP2C1, interacts with and negatively regulates the function of CIPK9 under potassium-deficient conditions in Arabidopsis. J Exp Bot 69:4003–4015

Song SK, Clark SE (2005) POL and related phosphatases are dosage-sensitive regulators of meristem and organ development in Arabidopsis. Dev Biol 285:272

Song SK, Hofhuis H, Lee MM, Clark SE (2008) Key divisions in the early Arabidopsis embryo require POL and PLL1 phosphatases to establish the root stem cell organizer and vascular axis. Dev Cell 15:98–109

Szostkiewicz I, Richter K, Kepka M, Demmel S, Ma Y, Korte A, Assaad FF, Christmann A, Grill E (2010) Closely related receptor complexes differ in their ABA selectivity and sensitivity. Plant J 61:25

Tähtiharju S, Palva T (2001) Antisense inhibition of protein phosphatase 2C accelerates cold acclimation in Arabidopsis thaliana. Plant J 26:461–470

Takahashi Y, Zhang J, Hsu PK, Ceciliato PHO, Zhang L, Dubeaux G, Munemasa S, Ge C, Zhao Y, Hauser F, Schroeder JI (2020) MAP3Kinase-dependent SnRK2-kinase activation is required for abscisic acid signal transduction and rapid osmotic stress response. Nat Commun 11:1–12

Uhrig RG, Labandera AM, Moorhead GB (2013) Arabidopsis PPP family of serine/threonine protein phosphatases: many targets but few engines. Trends Plant Sci 18:505

Umbrasaite J, Schweighofer A, Kazanaviciute V, Magyar Z, Ayatollahi Z, Unterwurzacher V, Choopayak C, Boniecka J, Murray JAH, Bogre L, Meskiene I (2010) MAPK phosphatase AP2C3 induces ectopic proliferation of epidermal cells leading to stomata development in Arabidopsis. PLoS One 5:e15357

Umezawa T, Sugiyama N, Mizoguchi M, Hayashi S, Myouga F, Yamaguchi-Shinozaki K, Ishihama Y, Hirayama T, Shinozaki K (2009) Type 2C protein phosphatases directly regulate abscisic acid-activated protein kinases in Arabidopsis. Proc Natl Acad Sci 106:17588

Uno Y, Furihata T, Abe H, Yoshida R, Shinozaki K, Yamaguchi-Shinozaki K (2000) Arabidopsis basic leucine zipper transcription factors involved in an abscisic acid-dependent signal transduction pathway under drought and high-salinity conditions. Proc Natl Acad Sci U S A 97:11632–11637

Vlad F, Rubio S, Rodrigues A, Sirichandra C, Belin C, Robert N, Leung J, Rodriguez PL, Laurière C, Merlot S (2009) Protein phosphatases 2C regulate the activation of the Snf1-related kinase OST1 by abscisic acid in Arabidopsis. Plant Cell 21:3170–3184

Vranová E, Tähtiharju S, Sriprang R, Willekens H, Heino P, Palva ET, Inzé D, Van Camp W (2001) The AKT3 potassium channel protein interacts with the AtPP2CA protein phosphatase 2C. J Exp Bot 52:181–182

Wang H, Chevalier D, Larue C, Cho SK, Walker JC (2007) The protein phosphatases and protein kinases of Arabidopsis thaliana. Arabidopsis Book 5:e0106

Wang Z, Ji H, Yuan B, Wang S, Su C, Yao B, Zhao H, Li X (2015) ABA signaling is fine-tuned by antagonistic HAB1 variants. Nat Commun 6:8138

Wu Y, Sanchez JP, Lopez-Molina L, Himmelbach A, Grill E, Chua N-H (2003) The abi1-1 mutation blocks ABA signaling downstream of cADPR action. Plant J 34:307–315

Xue T, Wang D, Zhang S, Ehlting J, Ni F, Jakab S, Zheng C, Zhong Y (2008) Genome-wide and expression analysis of protein phosphatase 2C in rice and Arabidopsis. BMC Genomics 9:550

Yin P, Fan H, Hao Q, Yuan X, Wu D, Pang Y, Yan C, Li W, Wang J, Yan N (2009) Structural insights into the mechanism of abscisic acid signaling by PYL proteins. Nat Struct Mol Biol 16:1230–1236

Yoshida R, Hobo T, Ichimura K, Mizoguchi T, Takahashi F, Aronso J, Ecker JR, Shinozaki K (2002) ABA-activated SnRK2 protein kinase is required for dehydration stress signaling in Arabidopsis. Plant Cell Physiol 43:1473

Yoshida R, Umezawa T, Mizoguchi T, Takahashi S, Takahashi F, Shinozaki K (2006a) The regulatory domain of SRK2E/OST1/SnRK2.6 interacts with ABI1 and integrates abscisic acid (ABA) and osmotic stress signals controlling stomatal closure in Arabidopsis. J Biol Chem 281:5310–5318

Yoshida T, Nishimura N, Kitahata N, Kuromori T, Ito T, Asami T, Shinozaki K, Hirayama T (2006b) ABA-hypersensitive germination3 encodes a protein phosphatase 2C (AtPP2CA) that strongly regulates abscisic acid signaling during germination among Arabidopsis protein phosphatase 2Cs. Plant Physiol 140:115–126

Yu LP, Miller AK, Clark SE (2003) POLTERGEIST encodes a protein phosphatase 2C that regulates CLAVATA pathways controlling stem cell identity at Arabidopsis shoot and flower meristems. Curr Biol 13:179–188

Yunta C, Martínez-Ripoll M, Zhu J-K, Albert A (2011) The structure of Arabidopsis thaliana OST1 provides insights into the kinase regulation mechanism in response to osmotic stress. J Mol Biol 414:135–144

Zhang D-P, Wu Z-Y, Li X-Y, Zhao Z-X (2002) Purification and identification of a 42-kilodalton abscisic acid-specific-binding protein from epidermis of broad bean leaves. Plant Physiol 128:714–725

Zhang K, Xia X, Zhang Y, Gan SS (2012a) An ABA-regulated and Golgi-localized protein phosphatase controls water loss during leaf senescence in Arabidopsis. Plant J 69:667

Zhang X et al (2012b) Complex structures of the abscisic acid receptor PYL3/RCAR13 reveal a unique regulatory mechanism. Structure 20:780

Zhao Y, Chan Z, Xing L, Liu X, Hou YJ, Chinnusamy V, Wang P, Duan C, Zhu JK (2013) The unique mode of action of a divergent member of the ABA-receptor protein family in ABA and stress signaling. Cell Res 23:1380

Zhu S-Y et al (2007) Two calcium-dependent protein kinases, CPK4 and CPK11, regulate abscisic acid signal transduction in Arabidopsis. Plant Cell 19:3019

Chapter 7
Protein Phosphatases at the Interface of Sugar and Hormone Signaling Pathways to Balance Growth and Stress Responses in Plants

Harshita B. Saksena, Dhriti Singh, Manvi Sharma, Muhammed Jamsheer K., Sunita Jindal, Mohan Sharma, Archna Tiwari, Prakhar, Sanjay Singh Rawat, and Ashverya Laxmi

7.1 Introduction

Plants are often exposed to fluctuating environmental conditions, and being sessile, they require intricate mechanisms for their survival. Plants rely on various stress sensing and signaling networks to respond to abiotic and biotic stresses. Posttranslational modifications through phosphorylation of proteins to regulate cellular functions are a phenomenon which is ubiquitous to all the organisms (Smith and Walker 1996). Protein kinases and phosphatases catalyze phosphorylation and dephosphorylation of cellular proteins, respectively (Smith and Walker 1996). Phosphorylation leads to conformational changes in proteins, thereby regulating their activity and interaction with other proteins to form a complex (Pawson 1995; Luan 2003). Intracellular signaling in response to external stimulus involves phosphorylation of many proteins in order to migrate to the destination site in the cell and regulate the target proteins. Reversible phosphorylation of cellular proteins controls a spectrum of biological functions like growth and development processes, metabolism, cell cycle control, and stress responses (Luan 2003). Protein phospha-

---

Harshita B. Saksena, Dhriti Singh, and Manvi Sharma contributed equally to this work.

Muhammed Jamsheer K., Sunita Jindal, Mohan Sharma, Archna Tiwari, Prakhar, and Sanjay Singh Rawat contributed equally to this work.

H. B. Saksena · D. Singh · M. Sharma · S. Jindal · M. Sharma · A. Tiwari · Prakhar · S. S. Rawat · A. Laxmi (✉)
National Institute of Plant Genome Research, New Delhi, India
e-mail: ashverya_laxmi@nipgr.ac.in

M. Jamsheer K.
Amity Food and Agriculture Foundation, Amity University Uttar Pradesh, Noida, India

G. K. Pandey (ed.), *Protein Phosphatases and Stress Management in Plants*,
https://doi.org/10.1007/978-3-030-48733-1_7

tases target hydroxyl group of specific amino acid residues, majorly of serine and threonine and seldom of tyrosine (Shenolikar 1994; Smith and Walker 1996). Several studies also report the evidence of phosphohistidine phosphorylation in eukaryotic organisms (Ota and Varshavsky 1993; Huber et al. 1994; Crovello et al. 1995; Smith and Walker 1996). Localization of protein phosphatase occurs majorly in subcellular organelles including nuclei, chloroplast, mitochondria, cytosol, and some membrane fractions (MacKintosh et al. 1991; Huber et al. 1994; Smith and Walker 1996). Protein phosphatases have been classified into type I (PP1) and type II (PP2) in mammals and plants on the basis of their specificity to certain substrates and sensitivity to the inhibitors, inhibitor-1 (I-1) and inhibitor-2 (I-2) (Ingebritsen and Cohen 1983; Cohen 1989; Smith and Walker 1996). PP1 targets the β-subunit of mammalian phosphorylase kinase as it is a substrate for dephosphorylation and is sensitive to inhibitor-1 (I-1) and inhibitor-2 (I-2). In contrast, PP2 dephosphorylates α subunit of phosphorylase kinase, and its activity is not inhibited by inhibitor-1 (I-1) and inhibitor-2 (I-2) (Cohen 1989; Smith and Walker 1996). On the basis of structure, substrate specificity, and requirement of divalent cation as a cofactor, PP2 type phosphatases have been further divided into three subgroups: protein phosphatase 2A (PP2A), protein phosphatase 2B (PP2B), and protein phosphatase 2C (PP2C). PP2A is heterotrimeric consisting of a catalytic C-subunit and two distinct regulatory A- and B-subunits and does not require any cofactor for its activity. In contrast, PP2B (also known as calcineurin) functions as a heterodimer comprising a catalytic A-subunit and a regulatory B-subunit and requires Ca^{2+} for its activity. A third subclass consists of PP2C, which exists as a monomer and requires Mg^{2+} for its activity (Shenolikar and Nairn 1991; Smith and Walker 1996). Sugars are not only a source of energy but also act as signaling molecules. Photosynthesis in plants generates sucrose, which further splits into fructose and glucose. Glucose, a monosaccharide sugar, has been reported to function as a signaling molecule through various pathways including hexokinase 1 (HXK1)-dependent pathway, regulator of G-protein signaling 1 (RGS1)-dependent pathway, and target of rapamycin (TOR)-sucrose non-fermenting 1 (SNF)-related kinase 1 (SnRK1) pathway (Li and Sheen 2016). Phosphorylation and dephosphorylation of these glucose sensors are an essential way to regulate their activity. A recent report on yeast HXK2 by Barbosa et al. (2016) revealed that it is dephosphorylated by PP2A-like phosphatase to modulate certain growth-related aspects such as cell cycle regulation and life span. Similarly in yeast, PP2C has also been reported to regulate TOR signaling (González et al. 2009). PP2C also represses SnRK1-mediated energy starvation and stress responses through regulating SnRK1 activity (Rodrigues et al. 2013). PP2A class of phosphatases is known to repress ABA responses in plants via negatively regulating SnRK2 (Waadt et al. 2015). Several reports in mammals, yeast, and plants have discovered the role of TOR in phosphorylating type 2A-phosphatase-associated protein 42 kDa (TAP42), which is a regulatory-associated protein of PP2A (Murata et al. 1997; Jacinto et al. 2001; Di Como and Jiang 2006; Prickett and Brautigan 2006; Ahn et al. 2011, 2015). PP2A and PP2C have been reported to function in stress responses by regulating ABA signaling. Pernas et al. (2007) studied a mutant of a catalytic subunit of PP2A known as PP2AC-2 and identified distinct drought and salt stress responses suggesting a crosstalk of ABA signaling and PP2C. Further, two members of PP2C family including PP2CA and ABI1 are known to dephosphorylate SnRK1 to negatively regulate its activity; ABA promotes SnRK1 signaling by suppressing PP2Cs, thereby collectively suggesting an interconnection between PP2 type phosphatases, ABA signaling, and stress responses (Rodrigues et al. 2013). It is noteworthy that although the function of individual sugar sensing pathways is explored in great detail, how these pathways interact at the molecular level with protein phosphatases to optimize growth needs further investigation.

7.2 Structure, Subunit Composition, and General Functions of Protein Phosphatases

7.2.1 PP2A

Protein phosphatase 2A are ubiquitous serine/threonine phosphatases and are highly conserved among eukaryotes. These phosphatases regulate a plethora of cellular processes including cell cycle, gene regulation, and translation by dephosphorylation of many kinases and other cellular molecules (Wlodarchak and Xing 2016). Protein phosphatases 2A are predominant among members in plant protein phosphatases. The primary structure of plant protein phosphatases is composed of heterotrimeric complexes comprising a catalytic C subunit, structural A subunit consisting of 15 HEAT repeats that acts as scaffold, and regulatory B-type subunits. The B-type subunits are required for targeting and substrate specificity, and therefore, appropriate B-type subunit is crucial in maintaining holoenzyme assembly for PP2A target specificity and their regulation. The C-subunit exhibits high sequence and structural similarity to the catalytic subunits of other related phosphatases such as PP4, PP6, and PP1 (Moorhead et al. 2008). The A-subunit possesses a series of conserved alpha helical repeats and serves as protein scaffold for B- and C-subunits to form the holoenzyme (Mumby 2007). Phylogenetic analysis among the members of PPP family shows that PP2A, PP4, and PP6 form a separate cluster suggesting a common ancestral relationship (Moorhead et al. 2008). There are five genes which encode for C-subunit in rice, tomato, and Arabidopsis *thaliana*. However, one to several genes encode for A-subunit in plants (Delong 2006). Unlike A and C, B-subunit is encoded by phylogenetically and structurally unrelated gene families. High conservation of B-subunit gene families exists in plants and animals. These are B55 gene family that encodes for beta-propeller proteins, the B56 that encodes for Huntingtin, EF3, A subunit of PP2A, TOR (HEAT) repeat proteins, and the B72 that encodes EF-hand-containing proteins (Farkas et al. 2007). There are two genes encoding B55, nine encoding B56, and six encoding B72 family. Moreover, eukaryotic catalytic subunit of PP2A, PP4, and PP6 contains C-terminal YFL motif for potential leucine methylation. In Arabidopsis, PP2A binds the target of rapamycin substrate TAP46 and through RNA-induced gene silencing regulates cell growth and survival, autophagy, and translation (Ahn et al. 2011). Recent reports showed that PP2A associated with histone deacetylase HDA14, which was shown to be involved in deacetylation of alpha-tubulin in regulating microtubule function (Tran et al. 2012). Yeast two-hybrid and *in vitro* pull down assays identified the direct physical interaction of Arabidopsis PP2A-A1 subunit [ROOTS CURL IN NAPHTHYLPHTHALAMIC ACID 1 (RCN1)] with phototropin2 (PHOT2). The binding of PP2A-A1 and PHOT2 downregulates phototropism and stomatal opening through direct dephosphorylation of PHOT2 in response to blue light signaling (Tseng and Briggs 2010). PP2A-C together with protein kinase PINOID regulates the phosphorylation state of auxin efflux PIN proteins in fine-tuning auxin transport in roots (Ballesteros et al. 2013). In the past few years, several groups have identified the significant role of PP2A in modulating brassinosteroid (BR) signaling. PP2A regulates the dephosphorylation and protein turnover of BR receptor BRI1 (Wu et al. 2011). Besides, PP2A also dephosphorylates the BR signaling transcription factors BZR1 and BES1 (BZR2) in regulating BR signaling cascade (Tang et al. 2011). For a detailed review on the roles of PP2A phosphatases, see Bheri and Pandey (2019).

7.2.2 PP2C

PP2C (protein phosphatase 2C) is a Ser/Thr phosphatase, which belongs to the Mn^{2+}/Mg^{2+}-dependent PPM (metal-dependent protein phosphatase) family (Schweighofer and Meskiene 2008). In contrast to the PPP (phosphoprotein phosphatase) family phosphatase, PP2C is insensitive to inhibition by known phosphatase inhibitors such as okadaic acid and microcystin (Rogers et al. 2006). PP2C represents a large family of highly conserved protein phosphatases, with 16 definite PP2C genes in the human genome that are known to form 22 different isoforms (Lammers and Lavi 2007). Arabidopsis and rice possess 80 and 78 PP2C genes, respectively (Xue et al. 2008; Singh et al. 2010, 2015). This remarkable expansion of PP2C in plants indicates the possibility of neo-functionalization and functional specialization. A prominent function of plant PP2Cs is the modulation of ABA-mediated stress signaling (Rodriguez 1998; Schweighofer et al. 2004; Moorhead et al. 2007; Singh et al. 2010, 2015). PP2Cs work as monomers. The conserved catalytic core domain of human PP2C carries a central β sandwich, and each β sheet is flanked by a pair of α helices. This arrangement forms a cleft between two β sheets, in which two metal ions are located at the base of the cleft. Three additional α helices, unique to PP2C, connect with the core domain on one side and contribute to substrate specificity or regulation (Das et al. 1996). No regulatory subunit is known for PP2C enzymes. Compared to human PP2Cs, the plant PP2Cs display a characteristic structural pattern. They consist of a basic PP2C core with N terminal domains of variable length (Meskiene et al. 1998). Variations in N terminal extensions specify the function of PP2Cs (Fuchs et al. 2013; Stone et al. 1994). For example, distinct N terminals are present in ABI1 and ABI2 PP2Cs, which are involved in ABA signaling (Leung et al. 1994, 1997; Meyer et al. 1994). KAPP (kinase-associated protein phosphatase) is another type of PP2C, which regulates RLK (receptor-like kinase) signaling pathway in plants (Braun et al. 1997; Wang et al. 2007). MP2C (*Medicago* PP2C) and its other orthologues are known to be involved in MAPK (mitogen-activated protein kinase) signaling pathway. MP2C was identified through genetic screening in yeast. MP2C acts as a negative regulator of the stress-activated MAPK (SMAPK) pathway, which is transiently activated in response to cold, drought, touch, and wounding in plants (Jonak et al. 1996; Bogre et al. 1997; Meskiene et al. 1998; Schweighofer et al. 2007; Brock et al. 2010; Umbrasaite et al. 2010). Taken together, these results suggest that specific members of the PP2C family are specialized to act on specific kinases.

7.3 Sugar as a Signaling Molecule

Plants being autotrophs synthesize sugars, which act as structural, energy, and signaling molecule to sustain life on earth. Sugar status in plant acts as an internal cue to regulate developmental transition, nutrient homeostasis, and stress responses. In plants, sucrose, the byproduct of photosynthesis, is further split into hexoses (glucose and fructose). Glucose receptors and signaling pathways are well characterized in the plant (Urano et al. 2012; Sheen 2014). Studies have shown plant-specific extracellular (RGS1) and intracellular (HXK1) receptors for glucose. A highly conserved energy sensing module includes two serine/threonine kinases, SnRK and TOR, to regulate growth accordingly. RGS1 is a plasma membrane-bound hybrid receptor protein having G protein-coupled receptor (GPCR) and C-terminal RGS box (Sakr et al. 2018). Glucose mediates the phosphorylation of RGS1 by With No lysine Kinases (WNKs). WNK8-dependent phosphorylation promotes endocytosis of RGS1, thus activating G protein α subunit 1 (GPA) (Urano et al. 2012). GPA1 regulates different aspects of plant development such as cell division and elongation, organ development, and hormonal responses (Urano et al. 2013). Hexokinase (HXK1) catalyzes phosphorylation of glucose into glucose-6-phosphate in glycolysis. Glucose sensor activity of HXK1 was found to be independent of its catalytic activity (Moore et al. 2003). HXK1 interacts with the 19S regulatory particle of proteasome subunit (RPT5B) and vacuolar H^+-ATPase B1 (VHA-B1) in nucleus. This multimeric complex directly binds with promoters of glucose-inducible genes to regulate their transcriptional activity (Cho et al. 2006). In eukaryotes, cellular respiration, which depends on sugar availability, produces energy in the form of ATP. TOR is a conserved pathway of energy, which acts as an integrator of exogenous (environmental) and endogenous (nutrient and sugar availability) signals to regulate growth (Sakr et al. 2018). In plants, target of rapamycin (TOR) complex is composed of three subunits, TOR, RAPTOR, and LST8, which positively govern energy-dependent promotion of metabolism, cell growth, and protein synthesis (Xiong and Sheen 2015). Experimental evidence has supported the upregulation of TORC1 activity in the presence of nutrients and growth factors, while inactivation of TORC1 was found in energy deprivation, stress, and starvation in both animals and plants (Saxton and Sabatini 2017; Shi et al. 2018). In plants, glucose as photosynthesis product drives TOR signaling via glycolysis and controls root meristem activity, growth hormone signaling, and stem cell maintenance (Xiong et al. 2013). Glucose-TOR signaling also dictates transcription regulation of genes involved in primary and secondary metabolism, cell cycle, transcription, signaling, and protein folding (Xiong et al. 2013). Glucose modulates TOR activity, but the underlying mechanism of TOR upregulation is still not well known; on the other hand, the activity of SnRK1 gets inhibited in the presence of glucose (Baena-González et al. 2007). SnRK1α1 is a serine/threonine kinase that belongs to CDPK-SnRK superfamily (Hrabak et al. 2003). In Arabidopsis, SnRK1 is composed of a heterotrimeric complex with one catalytic α-subunit (kinase) and regulatory β and βγ-subunit (Broeckx et al. 2016). The activity of SnRK1 was found to be lowered in the presence of glucose; similarly, a negative correlation was found between the expression profile of genes upregulated by SnRK1 and the genes upregulated by glucose (Baena-González et al. 2007). SnRK1α1 interacts and phosphorylates RAPTOR1b, which is a regulatory subunit of TOR kinase (Nukarinen et al. 2016). This suggests that SnRK1 and TOR both antagonistically regulate nutrient and energy availability to control metabolism, transcriptome, cell growth, and development; however, further studies are required for better assessment of this yin-yang module in the plant (Fig. 7.1).

The following section focuses on how the various sugar signaling pathways described in detail above are linked to the two protein phosphatase subgroups, PP2A and PP2C. Few reports suggest a possible connection of hexokinase-dependent pathway with protein phosphatases. However, a considerable amount of studies report a crosstalk between protein phosphatases and components of energy signaling including TOR and SnRK1 in mediating a trade-off between stress and growth.

7.4 Crosstalk Between Protein Phosphatases and Sugar Signaling

The glucose sensor RGS1 and energy sensors TOR and SnRK1 are regulated by specific phosphorylation events in plants (Urano et al. 2012; Jamsheer et al. 2019). Thus, dephosphorylation mediated by protein phosphatases would be an important regulatory mechanism for modulating the functions of these proteins. It is yet to be seen whether glucose sensor HXK1 in plants is also regulated by phosphorylation-dephosphorylation-dependent mechanisms. In yeast, HXK2 is dephosphorylated by PP2A-like phosphatase to regulate cell cycle progression, mitochondrial function, and life span (Barbosa et al. 2016). In response to glucose, RGS1 is phosphorylated at the C-terminal by WITH NO LYSINE KINASEs (WNKs), which leads to endocytosis of RGS1 and thus activation of G-protein-mediated glucose signaling (Urano et al. 2012; Fu et al. 2014). Similarly, pathogen attack triggers RGS1 phosphoryla-

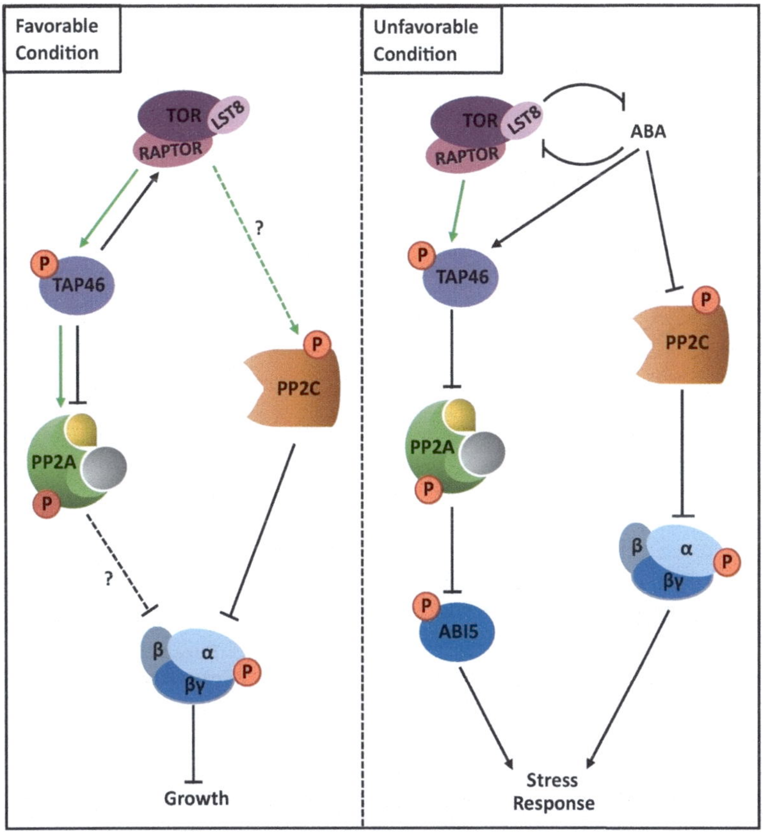

Fig. 7.1 Protein phosphatases and sugar signaling interaction. Under favorable growth conditions, activated TOR phosphorylates TAP46 and activates it which in turn promotes TOR signaling. TAP46 can regulate PP2A in both positive and negative manner. TOR might be utilizing this module to suppress SnRK1. Similarly, under nutrient-rich condition, PP2C suppresses SnRK1 activity and thereby stress-responsive changes in plants and promotes growth and development. TOR might also be involved in promoting PP2C to promote growth, whereas under stress conditions, TOR activates TAP46, which in turn negatively regulates PP2A. In the absence of activated TAP46, PP2A negatively regulates ABI5 and expression of its downstream target genes, which carry out stress responses. Exogenous ABA also upregulates TAP46. PP2C is a major negative regulator of ABA signaling. PP2C also negatively regulates SnRK1, thus playing a key role in cross talk of ABA signaling and energy status

tion by immune receptors which leads to its dissociation from GPA1 resulting in the activation of G-protein-mediated immune signaling (Liang et al. 2018). Taken together, these results suggest that phosphorylation status is a crucial regulator of RGS1 activity. However, phosphatases acting on RGS1 to regulate sugar signaling are yet to be identified (Fig. 7.2).

The TOR and SnRK1 are conserved serine-threonine kinases whose activity is regulated by phosphorylation at specific residues (Jamsheer et al. 2019). Intriguingly,

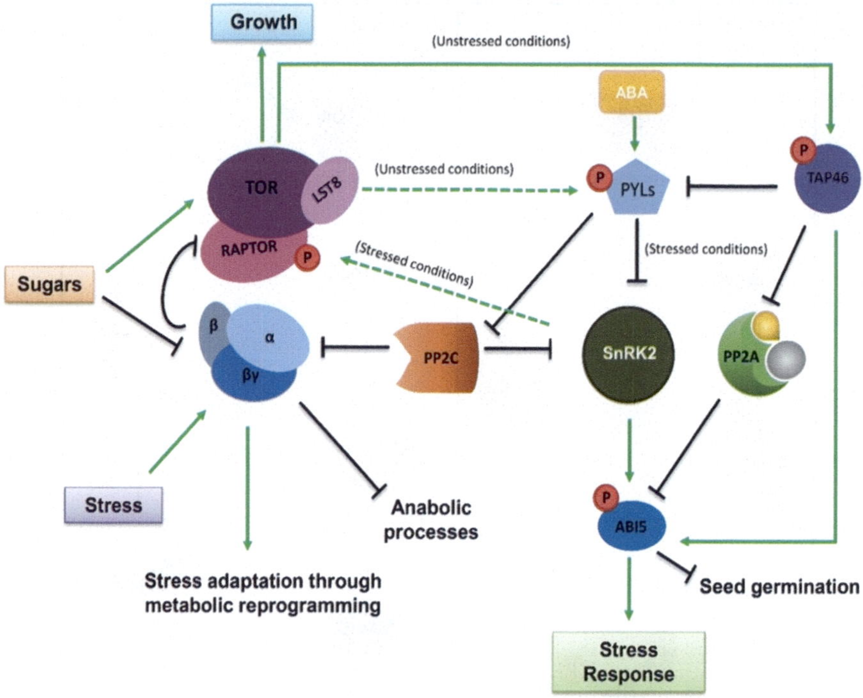

Fig. 7.2 Schematic representation of the crosstalk among the components of TOR, SnRKs, and ABA signaling pathways in Arabidopsis. TOR downregulates ABA receptor PYLs under normal conditions. However, SnRK2 inactivates TOR as soon as the plant perceives abiotic stress conditions, leading to activation of ABA signal transduction. TOR also regulates the expression of TAP46 by phosphorylation which interacts with the ABA signal component ABI5

the T-loop phosphorylation of SnRK1α1 (the kinase subunit of SnRK1) was found to be insensitive to the phosphatases of its mammalian homologue (Emanuelle et al. 2015). This could be due to the structural changes that occurred in plant lineage (Broeckx et al. 2016; Emanuelle et al. 2016). SnRK1 is negatively regulated by PP2C class of phosphatases to suppress energy starvation and abiotic stress-responsive transcriptome changes during favorable growth conditions (Rodrigues et al. 2013). Intriguingly, SnRK2 and SnRK3 kinases in plants, which are closely related to SnRK1, are also negatively regulated by PP2C class of phosphatases indicating that PP2C module works as a common negative regulator of SnRKs in favorable growth conditions (Jamsheer et al. 2019). Thus, it is possible that PP2C is also linked with TOR in promoting growth under favorable conditions. In line with this hypothesis, PP2C signaling was found to be required for TOR function in yeast indicating the communication between these pathways (González et al. 2009). However, direct communication between TOR and PP2C signaling is yet to be identified in plants.

PP2A class of phosphatases negatively regulates the activity of mammalian homologue of SnRK1 through dephosphorylation (Pernas et al. 2007; Joseph et al. 2015). PP2A negatively regulates SnRK2 family member to suppress ABA responses in plants (Waadt et al. 2015); however, it is still unknown whether PP2A similarly regulates SnRK1. TOR phosphorylates the type 2A phosphatase-associated protein 42 kDa (TAP42), a regulatory-associated protein of PP2A, to modulate the activity of PP2A in mammals, yeast, and plants (Murata et al. 1997; Jacinto et al. 2001; Di Como and Jiang 2006; Prickett and Brautigan 2006; Ahn et al. 2011, 2015). The plant homologue of Tap42/α4 known as Tap46 was found to be a positive regulator of TOR functions (Ahn et al. 2011, 2015). Thus, in favorable growth conditions, active TOR might utilize the PP2A module to suppress SnRK1 functions in plants. Studies in this direction will reveal more intricacies on growth control in response to environmental fluctuations.

Studies so far identified that PP2A and PP2C class of protein phosphatases is intimately connected to energy sensors to fine-tune nutrient-dependent growth control in plants. SnRK family kinases are dephosphorylated by these phosphatases in normal growth conditions to suppress stress signaling and promote growth. A member of myotubularin phosphatidylinositol (PI) 3-phosphatase suppresses TOR function in mammals (Hao et al. 2016). Similarly, plants may have specific phosphatases to suppress TOR function in specific conditions. Taken together, studies so far identified many interesting molecular leads between the components of sugar signaling and protein phosphatases. More studies are needed to establish many of these connections.

7.5 Evolutionary Dynamics of Nutrient and Energy Sensing Among Eukaryotes

The TOR-AMPK module is a well-conserved fundamental signaling pathway that is implicated in regulating growth according to the nutrient status of the organism and the changes in its internal and external environmental milieu. TOR, which belongs to the PIKK family (phosphatidylinositol kinase-related kinase) of serine/threonine kinases, was first identified in budding yeast (Heitman et al. 1991) and is found to be present in vast majority of organisms including invertebrates (Soulard et al. 2009), plants (Crespo et al. 2005; Deprost et al. 2007), and animals (Hall 2008) except for some intracellular parasites (Serfontein et al. 2010). Similarly, AMPK kinases and their orthologues in other eukaryotes are highly conserved serine/threonine kinases and regulate several aspects of metabolism in response to energy deprivation (Crozet et al. 2014). Through a mutualistic antagonism with the starvation-induced AMPK pathway, TOR orchestrates a myriad of cell and tissue developmental programs in response to amino acids, hormones, carbon, nitrogen, light regimes, and other nutrient-derived signals for efficient growth, metabolism, and survival of the organism (Dobrenel et al. 2016; Li and Sheen 2016).

TOR kinases are high molecular weight complexes (~250 kDa) which are involved in varied cellular and developmental processes including cell division, translation, embryogenesis, autophagy, ribosome biogenesis, senescence, etc., among several others. Recent phylogenetic profiling provides novel insights into the complex evolution of energy sensing, homeostasis, and maintenance by TOR and AMPK signaling (Van Dam et al. 2011; Roustan et al. 2016). Except yeast and certain other fungi which contain two *TOR* genes, almost all other eukaryotes possess single *TOR* gene (Crespo et al. 2005; Lee et al. 2005). TOR kinases in the plant lineage, except that of green alga *Chlamydomonas reinhardtii* and *Zea mays* (Crespo et al. 2005; Sotelo et al. 2010), are insensitive toward rapamycin treatment at lower physiological conditions (Xiong and Sheen 2012), while mammalian and yeast TOR are rapamycin sensitive even at low concentrations (nanomolar) (Heitman et al. 1991). However, in plants, the susceptibility toward rapamycin was restored following heterologous expression of the *FKBP12* protein in Arabidopsis (Sormani et al. 2007) and is enhanced following FKBP12 overexpression (Xiong and Sheen 2012). Yeast and mammals contain two TOR multiprotein subunits, TORCI and TORC2, which are involved in diverse yet distinct developmental regimes and have different subunit compositions, in contrast to just one TORC1 in plants (Tatebe and Shiozaki 2017). TORCI in mammals comprises TOR, RAPTOR, and LST8, and TORC2 is composed of TOR, LST8, and RICTOR (Hara et al. 2002; Kim et al. 2003). The presence of TORC2 in plants is rather questionable as no plant RICTOR orthologue has been found till date (Tatebe and Shiozaki 2017). Among plants, TOR has also been shown to have conserved functions and overall domain structure (John et al. 2011). The plant kinase domain has around 75% similarity to the kinase domain present in animals, implying that TOR is functionally conserved among eukaryotes (Xiong and Sheen 2015). In fact, certain modules of the TOR pathway such as LST8 (lethal with SEC13 protein 8) and TOR which form the core are conserved throughout evolution and were most likely present in the last eukaryotic common ancestor (LECA). Additional inputs/segments were probably lost and/or accommodated around this highly complex pathway during both plant and animal evolution, indicating that the energy sensing transduction pathway is both highly conserved yet amenable to incorporate changes across phylogenetic groups which might cater to the specific needs of the organism (Van Dam et al. 2011).

Several TOR substrates, including the ribosomal protein S6K and the PP2A phosphatase-associated protein TAP46, are also conserved in plants (Dobrenel et al. 2016). TOR activates the downstream effector S6K by phosphorylation to regulate translation and protein synthesis (Xiong et al. 2013). TOR has also been implicated in regulating physiological outcomes in times of stress. Under unstressed conditions, TOR phosphorylates ABA receptor PYLs (PYR1/PYL/RCAR family of proteins) and inactivates SnRK2, thus leading to negative regulation of stress-induced ABA signaling (Wang et al. 2018). However, during the onset of unfavorable conditions, SnRK2 inactivates TOR via phosphorylation of RAPTOR (Wang et al. 2018). Apart from its function in regulating PP2C's activity through ABA stress signaling pathway, TOR has also been shown to directly phosphorylate the downstream protein effector TAP46 in plants, a conserved regulatory subunit of protein phosphatase

2A (PP2A) that controls plant growth and development including seed germination and maturation via regulating several stress-related genes (Ahn et al. 2011, 2015). As discussed above, TAP46 interacts with the transcription factor ABI5 (ABA-Insensitive 5) *in vivo* and stabilizes it, thereby positively regulating core ABA signaling (Hu et al. 2014). Another PP2A protein interactor TIP41 (TAP42-interacting protein of 41 kDa) has been implicated in the modulation of ABA-mediated responses (Punzo et al. 2018). Thus, a reciprocal interaction of key components of TOR-ABA pathway results in equilibrium between induction of stress responses and plant growth.

Starvation-induced AMP-activated kinase (AMPK) equivalents have also been thoroughly characterized in yeast (SNF1, sucrose non-fermenting 1) and plants (SnRKs, SNF1-related kinase) (Broeckx et al. 2016; Hardie 2018) which inhibit several anabolic pathways while promoting induction of several catabolic processes and are known to regulate a common set of regulatory proteins during stressful conditions (Baena-González et al. 2007; Roustan et al. 2016). In all organisms evaluated, the AMPK module is rather conserved and comprises a heterotrimeric protein complex composed of one catalytic subunit, α, and two regulatory subunits, β (plants possess atypical βγ and β subunits) and γ, and is encoded by multiple genes (Polge and Thomas 2007; Ramon et al. 2008; Emanuelle et al. 2015; Hardie 2018) and is regulated via multiple posttranslational modifications and upstream kinases including PP2As and PP2Cs (Crozet et al. 2014). Across all eukaryotes, the SNF1/AMPKα/SnRK1α members require phosphorylation at a conserved residue involving threonine for their activation (Polge and Thomas 2007; Hardie 2018). SNF1/AMPK modules have been shown to regulate several aspects of cell cycle progression and division (Carling et al. 2012) and are specifically regulated by AMP/ADP, while SnRK1 is insensitive toward this regulatory mechanism (Emanuelle et al. 2015). Similarly, SnRKs are also critical for fine-tuning growth and physiological responses in times of unfavorable conditions. SnRKs have been implicated in the adjustment of plant responses to several biotic and abiotic environmental stimuli including nutrient limitation, starvation, dehydration, cold, salt, and osmotic stresses (Coello et al. 2011; Wurzinger et al. 2018). SnRK1 regulates several aspects of cell metabolism and is probably involved in cell cycle progression (Guérinier et al. 2013; Crozet et al. 2014). SnRK1 in plants is repressed by sugars such as glucose-6-phosphate (G6P) and sucrose and indirectly by trehalose-6-phosphate (T6P) and high energy status (Zhang et al. 2009; Rodrigues et al. 2013). The SnRK1 subfamily comprises three catalytic subfamilies, AKIN10/AKIN11/AKIN12 (Baena-González et al. 2007). Genome duplication and neo-functionalization in plants have led to an increase in the number of SnRK domains and their highly complex structures (Emanuelle et al. 2015). Plants possess two additional homologues, namely, SnRK2 and SnRK3/CIPK, both of which have additional subclasses, signaling intermediates, and key components that might be the result of gene duplication events to fulfill broader spectrum of functions in plants (Broeck et al. 2016). SnRK2, as discussed above, are activated under stress conditions and are categorized into three subcategories depending upon their affinity toward ABA and regulation via clade A PP2Cs (Kulik et al. 2011). Emerging roles of SnRK3 family of kinases in plant adaptation to stress have revealed that these calcineurin B-like (CBL) calcium-binding proteins mediate responses to drought, cold, ABA, sugar, salinity, and pH changes (Coello et al. 2011; Kulik et al. 2011). Not only do the SNF1/AMPK/SnRK1 modules share significant homology in terms of structure and functions, but their upstream activating kinases also exhibit sequence similarity to one another indicating the conservation of activation mechanisms in the eukaryotic lineage (Glab et al. 2017). In fact, certain kinases and phosphatases including PP2As regulating AMPK activity have also been traced back to majority of eukaryotic species (Roustan et al. 2016). An exhaustive phylogenetic analysis (Roustan et al. 2016) on the evolutionary aspects of energy sensing network has revealed that the AMPK/TOR pathway-related kinases were already functional in the prokaryotic lineage. A link between mTOR pathway and retrograde signaling involving mitochondria has also been speculated (Komeili et al. 2000; Morita et al. 2015). For a detailed overview on evolutionary kinetics of SnRK and TOR pathway, see Roustan et al. (2016) and Jamsheer et al. (2019).

Thus, TOR/AMPK module works at the interface of major plant physiological responses which are governed by several developmental and environmental inputs and are prevalent in all photosynthetic eukaryotes. Although elucidation of TOR/AMP kinases as the central integrators of diverse metabolic pathways has just begun in plants, several areas concerned with translational and metabolic reprogramming via related kinases need to be explored. A complete and rigorous overview would require high-end tools like phospho-proteomics, plant TOR chemical inhibitors, and experimental designs to unravel yet newer components and effectors of this ancient central signal transduction hub.

7.6 Protein Phosphatases and Interaction with Sugar and ABA in Managing Stress

Plants activate signal transduction pathways under stress conditions. These pathways begin with signal perception leading to activation of different protein kinases and phosphatases that finally change the activity of target proteins or transcription factors. Thus, protein phosphatases act as a regulatory hub for different stress responses. On the other hand, sugar signaling integrates various internal and external cues so it plays a vital role in modulating various stress responses. There are different studies suggesting the crosstalk between sugar signaling and protein phosphatases in regulation of stress responses.

In Arabidopsis, the bifunctional Lys-ketoglutarate reductase (LKR)/saccharopine dehydrogenase (SDH) is an important enzyme is lysine catabolism. The expression of this gene is altered during various biotic and abiotic stresses in different plants, suggesting its involvement in stress-related metabolism. LDR/SDH protein expression was upregulated by ABA signaling in which two homologous PP2Cs, ABI1 and ABI2, play a vital role. In addition to this, sugar starvation also stimulated the expression of LDR/SDH protein via HXK-dependent pathway (Stepansky and Galili 2003).

In an attempt to understand the function of PP2A catalytic subunits (PP2A-C), a large number of T-DNA mutant lines were generated and screened. A specific catalytic subunit of PP2A (PP2A-C2) was identified as a negative regulator of ABA signaling. Loss-of-function *pp2ac-2* showed hypersensitive response to inhibition of primary and lateral roots as well as inhibition of seed germination in the presence of ABA, whereas plants overexpressing *PP2A-C* were less sensitive to ABA-mediated inhibition of these responses. In addition to this, *PP2A-C* mutant plants showed altered salt and drought stress response. *PP2A-C* mutants were hypersensitive to glucose-mediated inhibition of cotyledon greening and leaf development. And, in this case, *PP2A-C* overexpression plants showed response similar to wild type. All these results collectively suggested that this specific catalytic subunit of PP2A (PP2A-C2) is involved in ABA, stress, and glucose signaling (Pernas et al. 2007).

Rice MAPK kinase kinase gene (*ENHANCED DISEASE RESISTANCE 1*) *OsERD1* expression was significantly upregulated by sugar and protein phosphatase inhibitors, CANTHARIDIN (CN), and okadaic acid (Kim et al. 2003), thus suggesting that it might be antagonistically regulated by sugar and protein phosphatases. The transcript level of *OsERD1* was upregulated by drought, high salt, and heavy metal as well as by different hormones, ABA, SA, JA, and ethylene, suggesting its involvement in stress signaling pathways and development (Kim et al. 2003).

Arabidopsis UDP-glucose pyrophosphorylase encoding gene *UGP* is an important enzyme of sucrose synthesis which produces UDP-glucose. *UGP* expression was found to be upregulated in the presence of sucrose and light and inhibited in the presence of okadaic acid. The transcript levels of *UGP* were significantly increased in response to cold stress via ABA-independent pathway. All these observations suggested that UGP is a sugar-responsive and okadaic acid-sensitive enzyme involved in plant responses to environmental signals.

A study conducted in developing pea embryos where ABA levels were downregulated by immunomodulation showed a decreased hexose sugar levels as well as expression of PP2A was decreased. ABA deficiency also caused downregulation of many stress-related genes. This study indicated a connection between sugar, PP2A, stress signaling, and ABA (Radchuk et al. 2010).

As discussed earlier, TAP46 is a positive regulator of TOR signaling. TAP46 is also involved in ABA signaling. In Arabidopsis, TAP46 overexpressing transgenic plants showed hypersensitivity to ABA responses, whereas the mutant *tap46* lines showed hyposensitive response, thus suggesting that it is a positive regulator of ABA signaling. Conversely, these transgenic line studies showed that TAP46 is a negative regulator of PP2A. Furthermore, TAP46 has been shown to interact with ABA Insensitive 5 (ABI5) and stabilize it. This binding may interfere with PP2A activity on ABI5, thereby maintaining ABI5 in its active form. TAP46 also modulated some of stress-related gene expression (Hu et al. 2014). In addition to TAP46, TIP41 (TAP42-interacting protein of 41 kDa) is also involved in ABA, TOR, and stress pathways. ABA and stress perturbed expression of TIP41. TIP41 seems to negatively regulate ABA signaling, while it is positively regulated by TOR signaling (Punzo et al. 2018).

ABI1 and PP2CA have been shown to facilitate coordinated action of ABA signaling, energy signaling, and stress responses. In a study, it was found that these two PP2Cs, ABI1 and PP2CA, interact with the catalytic subunit of SnRK1 and dephosphorylate and inactivate it (Rodrigues et al. 2013). In double and quadruple knockout mutants of *pp2c*, SnRK1 inactivation was blocked. In addition to this, these knockout mutants showed sugar hypersensitivity similar to the phenotype of *SnRK1* overexpression (Rodrigues et al. 2013). ABA could promote SnRK1 signaling by repressing PP2Cs as suggested by reporter gene assays. This was further confirmed by *SnRK1* target gene expression in the presence of ABA. When transcriptional profiles associated with *SnRK1.1* activation were compared to the transcriptional profile of ABA treated seedlings, which were already available in literature, there was a large overlap between these two profiles despite the tissue difference. All these results implied that PP2Cs play a vital role in crosstalk between ABA signaling and sugar/energy signaling strengthening the stress response (Rodrigues et al. 2013; Pizzio et al. 2013). On the other hand, ABA signaling and TOR signaling antagonistically regulate each other depending upon the condition (Wang et al. 2018).

7.7 Conclusions and Future Perspective

Reversible protein phosphorylation is recognized as the major switch in regulation of various cellular and developmental processes in eukaryotes. There has been a remarkable progress in understanding the functions of protein phosphatases in recent years. Reversible phosphorylation-dephosphorylation mechanisms have been shown to regulate sugar signaling events at multiple levels (Urano et al. 2012; Barbosa et al. 2016; Jamsheer et al. 2019). Although a large number of PKs are known to phosphorylate energy sensors like SnRK1 and RGS1, their partner protein phosphatases are scarcely studied (Urano et al. 2012; Emanuelle et al. 2015; Barbosa et al. 2016; Jamsheer et al. 2019). Thus, due to lack in knowledge of the phosphatases involved in sugar signaling hub, the clear picture on the regulation of sugar signaling cascade and how it influences plant's response toward stress is missing and therefore needs further exploration. Moreover, it is yet to be studied whether HXK1 is regulated by the reversible phosphorylation-dephosphorylation events.

Emerging evidence has suggested a connection between protein phosphatases and sugar signaling and their role in managing stress responses. However, most of the links between the signaling pathways are poorly understood. Since protein phosphatases, mainly PP2A and PP2C, mediate both sugar signaling and ABA signaling in stress responses and developmental signaling, they might also act as a nodal point of connection between sugar and ABA signaling cascades and their role in mitigating stress responses. Also, extensive research is necessary in understanding how these proteins interact with hormones and sugars in combating biotic stress responses.

It is just the beginning to elucidate the complex networks involving protein phosphatases, sugars, and hormones in cellular functions. The cornerstone of understanding the functions of protein phosphatases will be identifying the protein kinases that protein phosphatases counteract. Research in this direction in the future will be required to understand how these protein phosphatases are integrated into regulatory networks and how the dynamics of these networks balance growth and stress responses and their eventual utilization for stress management in crop plants.

Acknowledgments The research in AL laboratory is supported by project grants from the Department of Biotechnology, Government of India, and a Core Grant from the National Institute of Plant Genome Research. M.S., H.B.S., D.S., S.J., M.J.K., M.S., A.T., and P.A. duly acknowledge research fellowships from the Department of Biotechnology, Department of Science and Technology, Council of Scientific and Industrial Research, University Grants Commission, and National Institute of Plant Genome Research, Government of India. M.J.K. acknowledges research grant and fellowship from INSPIRE faculty program, Department of Science and Technology, Government of India (Grant no. IFA18-LSPA110). The authors acknowledge DBT-eLibrary Consortium (DeLCON) for providing access to e-resources.

References

Ahn CS, Han JA, Lee HS, Lee S, Pai HS (2011) The PP2A regulatory subunit Tap46, a component of the TOR signaling pathway, modulates growth and metabolism in plants. Plant Cell 23(1):185–209. https://doi.org/10.1105/tpc.110.074005

Ahn CS, Ahn HK, Pai HS (2015) Overexpression of the PP2A regulatory subunit Tap46 leads to enhanced plant growth through stimulation of the TOR signaling pathway. J Exp Bot 66(3):827–840. https://doi.org/10.1093/jxb/eru438

Baena-González E, Rolland F, Thevelein JM, Sheen J (2007) A central integrator of transcription networks in plant stress and energy signaling. Nature 448(7156):938–942. https://doi.org/10.1038/nature06069

Ballesteros I, Dom T, Sauer M, Paredes P, Duprat A, Rojo E, Sanmart M (2013) Specialized functions of the PP2A subfamily II catalytic subunits PP2A-C3 and PP2A-C4 in the distribution of auxin fluxes and development in Arabidopsis. Plant J 73(5):862–872. https://doi.org/10.1111/tpj.12078

Barbosa AD, Pereira C, Osório H, Moradas-Ferreira P, Costa V (2016) The ceramide-activated protein phosphatase Sit4p controls lifespan, mitochondrial function and cell cycle progression by regulating hexokinase 2 phosphorylation. Cell Cycle 15(12):1620–1630. https://doi.org/10.1080/15384101.2016.1183846

Bheri M, Pandey GK (2019) PP2A Phosphatases take a giant leap in the post-genomics era. Curr Genomics 20(3):154–171. https://doi.org/10.2174/1389202920666190517110605

Bogre L, Ligterink W, Meskiene I, Barker PJ, Heberle-Bors E, Huskisson NS, Hirt H (1997) Wounding induces the rapid and transient activation of a specific MAP kinase pathway. Plant Cell 9(1):75–83. https://doi.org/10.1105/tpc.9.1.75

Braun DM, Stone JM, Walker JC (1997) Interaction of the maize and Arabidopsis kinase interaction domains with a subset of receptor-like protein kinases: implications for transmembrane signaling in plants. Plant J 12(1):83–95

Brock AK, Willmann R, Kolb D, Grefen L, Lajunen HM, Bethke G, Lee J, Nürnberger T, Gust AA (2010) The Arabidopsis mitogen-activated protein kinase phosphatase PP2C5 affects seed germination, stomatal aperture, and abscisic acid-inducible gene expression. Plant Physiol 153(3):1098–1111. https://doi.org/10.1104/pp.110.156109

Broeckx T, Hulsmans S, Rolland F (2016) The plant energy sensor: evolutionary conservation and divergence of SnRK1 structure, regulation, and function. J Exp Bot 67(22):6215–6252. https://doi.org/10.1093/jxb/erw416

Carling D, Thornton C, Woods A, Sanders MJ (2012) AMP-activated protein kinase: new regulation, new roles?. Biochem J 445(1):11–27. https://doi.org/10.1042/BJ20120546

Cho YH, Yoo SD, Sheen J (2006) Regulatory functions of nuclear hexokinase1 complex in glucose signaling. Cell 127(3):579–589. https://doi.org/10.1016/j.cell.2006.09.028

Coello P, Hey SJ, Halford NG (2011) The sucrose non-fermenting-1-related (SnRK) family of protein kinases: potential for manipulation to improve stress tolerance and increase yield. J Exp Bot 62(3):883–893. https://doi.org/10.1093/jxb/erq331

Cohen P (1989) The structure and regulation of protein phosphatases. Annu Rev Biochem 58(1):453–508. https://doi.org/10.1146/annurev.bi.58.070189.002321

Crespo JL, Díaz-Troya S, Florencio FJ (2005) Inhibition of target of rapamycin signaling by rapamycin in the unicellular green alga Chlamydomonas reinhardtii. Plant Physiol 139(4):1736–1749. https://doi.org/10.1104/pp.105.070847

Crovello CS, Furie BC, Furie B (1995) Histidine phosphorylation of P-selectin upon stimulation of human platelets: a novel pathway for activation-dependent signal transduction. Cell 82(2):279–286

Crozet P, Margalha L, Confraria A, Rodrigues A, Martinho C, Adamo M, Elias CA, Baena-González E (2014) Mechanisms of regulation of SNF1/AMPK/SnRK1 protein kinases. Front Plant Sci 5:190. https://doi.org/10.3389/fpls.2014.00190

Das AK, Helps NR, Cohen PT, Barford D (1996) Crystal structure of the protein serine/threonine phosphatase 2C at 2.0 A° resolution. EMBO J 15(24):6798–6809

Delong A (2006) Switching the flip: protein phosphatase roles in signaling pathways. Curr Opin Plant Biol 9:470–477. https://doi.org/10.1016/j.pbi.2006.07.015

Deprost D, Yao L, Sormani R, Moreau M, Leterreux G, Nicolaï M, Bedu M, Robaglia C, Meyer C (2007) The Arabidopsis TOR kinase links plant growth, yield, stress resistance and mRNA translation. EMBO Rep 8(9):864–870. https://doi.org/10.1038/sj.embor.7401043

Di Como CJ, Jiang Y (2006) The association of Tap42-phosphatase complexes with TORC1—another level of regulation in TOR signaling. Cell Cycle 5(23):2729–2732. https://doi.org/10.4161/cc.5.23.3516

Dobrenel T, Caldana C, Hanson J, Robaglia C, Vincentz M, Veit B, Meyer C (2016) TOR signaling and nutrient sensing. Annu Rev Plant Biol 29;67:261–285. https://doi.org/10.1146/annurev-arplant-043014-114648

Emanuelle S, Hossain MI, Moller IE, Pedersen HL, van de Meene AM, Doblin MS, Koay A, Oakhill JS, Scott JW, Willats WG, Kemp BE, Bacic A, Gooley PR, Stapleton DI (2015) SnRK1 from Arabidopsis *thaliana* is an atypical AMPK. Plant J 82(2):183–192. https://doi.org/10.1111/tpj.12813

Emanuelle S, Doblin MS, Stapleton DI, Bacic A, Gooley PR (2016) Molecular insights into the enigmatic metabolic regulator, SnRK1. Trends Plant Sci 21(4):341–353. https://doi.org/10.1016/j.tplants.2015.11.001

Farkas I, Dombra V, Koncz C (2007) Arabidopsis PPP family of serine/threonine phosphatases. Trends Plant Sci 12(4):169–176. https://doi.org/10.1016/j.tplants.2007.03.003

Fu Y, Lim S, Urano D, Tunc-Ozdemir M, Phan NG, Elston TC, Jones AM (2014) Reciprocal encoding of signal intensity and duration in a glucose-sensing circuit. Cell 156(5):1084–1095. https://doi.org/10.1016/j.cell.2014.01.013

Fuchs S, Grill E, Meskiene I, Schweighofer A (2013) Type 2C protein phosphatases in plants. FEBS J 280(2):681–693. https://doi.org/10.1111/j.1742-4658.2012.08670.x

Glab N, Oury C, Guérinier T, Domenichini S, Crozet P, Thomas M, Vidal J, Hodges M (2017) The impact of Arabidopsis thaliana SNF1-related-kinase 1 (SnRK1)-activating kinase 1 (SnAK1) and SnAK2 on SnRK1 phosphorylation status: characterization of a SnAK double mutant. Plant J 89(5):1031–1041. https://doi.org/10.1089/omi.2011.0091

González A, Ruiz A, Casamayor A, Ariño J (2009) Normal function of the yeast TOR pathway requires the type 2C protein phosphatase Ptc1. Mol Cell Biol 29(10):2876–2888. https://doi.org/10.1128/MCB.01740-08

Guérinier T, Millan L, Crozet P, Oury C, Rey F, Valot B, Mathieu C, Vidal J, Hodges M, Thomas M, Glab N (2013) Phosphorylation of p27(KIP1) homologs KRP6 and 7 by SNF1-related protein kinase-1 links plant energy homeostasis and cell proliferation. Plant J 75(3):515–525. https://doi.org/10.1111/tpj.12218

Hall MN (2008) mTOR-what does it do?. Transplant Proc 40(10 Suppl):S5–S8. https://doi.org/10.1016/j.transproceed.2008.10.009

Hara K, Maruki Y, Long X, Yoshino K, Oshiro N, Hidayat S, Tokunaga C, Avruch J, Yonezawa K (2002) Raptor, a binding partner of target of rapamycin (TOR), mediates TOR action. Cell 110(2):177–189. https://doi.org/10.1016/s0092-8674(02)00833-4

Hardie DG (2018) Keeping the home fires burning: AMP-activated protein kinase. J R Soc Interface 15(138):20170774. https://doi.org/10.1098/rsif.2017.0774

Hao F, Itoh T, Morita E, Shirahama-Noda K, Yoshimori T, Noda T (2016) The PtdIns3-phosphatase MTMR3 interacts with mTORC1 and suppresses its activity. FEBS Lett 590(1):161–173. https://doi.org/10.1002/1873-3468.12048

Heitman J, Movva NR, Hall MN (1991) Targets for cell cycle arrest by the immunosuppressant rapamycin in yeast. Science 253(5022):905–909. https://doi.org/10.1126/science.1715094

Hrabak EM, Chan CW, Gribskov M, Harper JF, Choi JH, Halford N, Kudla J, Luan S, Nimmo HG, Sussman MR, Thomas M, Walker-Simmons K, Zhu JK, Harmon AC (2003) The Arabidopsis CDPK-SnRK superfamily of protein kinases. Plant Physiol 132(2):666–680. https://doi.org/10.1104/pp.102.011999

Hu R, Zhu Y, Shen G, Zhang H (2014) TAP46 plays a positive role in the ABSCISIC ACID INSENSITIVE5-regulated gene expression in Arabidopsis. Plant Physiol 164(2):721–734. https://doi.org/10.1104/pp.113.233684

Huber SC, Huber JL, McMichael RW (1994) Control of plant enzyme activity by reversible protein phosphorylation. Int Rev Cytol 149:47–98. https://doi.org/10.1016/S0074-7696(08)62086-0

Ingebritsen TS, Cohen P (1983) Protein phosphatases: properties and role in cellular regulation. Science 221(4608):331–338

Jacinto E, Guo B, Arndt KT, Schmelzle T, Hall MN (2001) TIP41 interacts with TAP42 and negatively regulates the TOR signaling pathway. Mol Cell 8(5):1017–1026. https://doi.org/10.1016/S1097-2765(01)00386-0

Jamsheer KM, Jindal S, Laxmi, A. (2019) Evolution of TOR-SnRK dynamics in green plants and their integration with phytohormone signaling networks. J Exp Bot (in press)

John F, Roffler S, Wicker T, Ringli C (2011) Plant TOR signaling components. Plant Signal Behav 6(11):1700–1705. https://doi.org/10.4161/psb.6.11.17662

Jonak C, Kiegerl S, Ligterink W, Barker PJ, Huskisson NS, Hirt H (1996) Stress signaling in plants: a mitogen-activated protein kinase pathway is activated by cold and drought. Proc Natl Acad Sci 93(20):11274–11279. https://doi.org/10.1073/PNAS.93.20.11274

Joseph BK, Liu HY, Francisco J, Pandya D, Donigan M, Gallo-Ebert C, Giordano C, Bata A, Nickels JT Jr (2015) Inhibition of AMP kinase by the protein phosphatase 2A heterotrimer, PP2APpp2r2d. J Biol Chem 290(17):10588–10598. https://doi.org/10.1074/jbc.M114.626259

Kim JA, Agrawal GK, Rakwal R, Han KS, Kim KN, Yun CH, Heu S, Park SY, Lee YH, Jwa NS (2003) Molecular cloning and mRNA expression analysis of a novel rice (Oryza sativa L.) MAPK kinase kinase, OsEDR1, an ortholog of Arabidopsis AtEDR1, reveal its role in defense/stress signaling pathways and development. Biochem Biophys Res Commun 300(4):868–876. https://doi.org/10.1016/S0006-291X(02)02944-3

Komeili A, Wedaman KP, O'Shea EK, Powers T (2000) Mechanism of metabolic control. Target of rapamycin signaling links nitrogen quality to the activity of the Rtg1 and Rtg3 transcription factors. J Cell Biol 13;151(4):863–878. https://doi.org/10.1083/jcb.151.4.863

Kulik A, Wawer I, Krzywińska E, Bucholc M, Dobrowolska G (2011) SnRK2 protein kinases–key regulators of plant response to abiotic stresses. OMICS (12):859–872. https://doi.org/10.1089/omi.2011.0091

Lammers T, Lavi S (2007) Role of type 2C protein phosphatases in growth regulation and in cellular stress signaling. Crit Rev Biochem Mol Biol 42(6):437–461. https://doi.org/10.1080/10409230701693342

Lee S, Comer FI, Sasaki A, McLeod IX, Duong Y, Okumura K, Yates JR 3rd, Parent CA, Firtel RA (2005) TOR complex 2 integrates cell movement during chemotaxis and signal relay in Dictyostelium. Mol Biol Cell 16(10):4572–4583. https://doi.org/10.1091/mbc.e05-04-0342

Leung J, Bouvier-Durand M, Morris PC, Guerrier D, Chefdor F, Giraudat J (1994) Arabidopsis ABA response gene ABI1: features of a calcium-modulated protein phosphatase. Science 264(5164):1448–1452

Leung J, Merlot S, Giraudat J (1997) The Arabidopsis ABSCISIC ACID-INSENSITIVE2 (ABI2) and ABI1 genes encode homologous protein phosphatases 2C involved in abscisic acid signal transduction. Plant Cell 9(5):759–771

Li L, Sheen J (2016) Dynamic and diverse sugar signaling. Curr Opin Plant Biol 33:116–125. https://doi.org/10.1016/j.pbi.2016.06.018

Liang X, Ma M, Zhou Z, Wang J, Yang X, Rao S, Bi G, Li L, Zhang X, Chai J, Chen S, Zhou JM (2018) Ligand-triggered de-repression of Arabidopsis heterotrimeric G proteins coupled to immune receptor kinases. Cell Res 28(5):529–543. https://doi.org/10.1038/s41422-018-0027-5

Luan S (2003) Protein phosphatases in plants. Annu Rev Plant Biol 54(1):63–92. https://doi.org/10.1146/annurev.arplant.54.031902.134743

MacKintosh C, Coggins J, Cohen P (1991) Plant protein phosphatases: subcellular distribution, detection of protein phosphatase 2C and identification of protein phosphatase 2A as the major quinate dehydrogenase phosphatase. Biochem J 273(3):733–738

Meskiene I, Bögre L, Glaser W, Balog J, Brandstötter M, Zwerger K, Ammerer G, Hirt H (1998) MP2C, a plant protein phosphatase 2C, functions as a negative regulator of mitogen-activated protein kinase pathways in yeast and plants. Proc Natl Acad Sci U S A 95(4):1938–1943

Meyer K, Leube MP, Grill E (1994) A protein phosphatase 2C involved in ABA signal transduction in Arabidopsis *thaliana*. Science 264(5164):1452–1455

Moore B, Zhou L, Rolland F, Hall Q, Cheng WH, Liu YX, Hwang I, Jones T, Sheen J (2003) Role of the Arabidopsis glucose sensor HXK1 in nutrient, light, and hormonal signaling. Science 300(5617):332–336. https://doi.org/10.1126/science.1080585

Moorhead GBG, Trinkle-Mulcahy L, Ulke-Lemée A (2007) Emerging roles of nuclear protein phosphatases. Nat Rev Mol Cell Biol 8(3):234–244. https://doi.org/10.1038/nrm2126

Moorhead GBG, Wever VDE, Templeton G, Kerk D (2008) Evolution of protein phosphatases in plants and animals. Biochem J 417(2):401–409. https://doi.org/10.1042/BJ20081986

Morita M, Gravel SP, Hulea L, Larsson O, Pollak M, St-Pierre J, Topisirovic I (2015) mTOR coordinates protein synthesis, mitochondrial activity and proliferation. Cell Cycle 14(4):473–480. https://doi.org/10.4161/15384101.2014.991572

Mumby M (2007) The 3D structure of protein phosphatase 2A: new insights into a ubiquitous regulator of cell signaling. ACS Chem Biol 2(2):99–103

Murata K, Wu J, Brautigan DL (1997) B cell receptor-associated protein alpha4 displays rapamycin-sensitive binding directly to the catalytic subunit of protein phosphatase 2A. Proc Natl Acad Sci U S A 94(20):10624–10629. https://doi.org/10.1073/PNAS.94.20.10624

Nukarinen E, Nägele T, Pedrotti L, Wurzinger B, Mair A, Landgraf R, Börnke F, Hanson J, Teige M, Baena-Gonzalez E, Dröge-Laser W, Weckwerth W (2016) Quantitative phosphoproteomics reveals the role of the AMPK plant ortholog SnRK1 as a metabolic master regulator under energy deprivation. Sci Rep 6(1):31697. https://doi.org/10.1038/srep31697

Ota IM, Varshavsky A (1993) A yeast protein similar to bacterial two-component regulators. Science 262(5133):566–569

Pawson T (1995) Protein modules and signaling networks. Nature 373(6515):573–580. https://doi.org/10.1038/373573a0

Pernas M, García-Casado G, Rojo E, Solano R, Sánchez-Serrano JJ (2007) A protein phosphatase 2A catalytic subunit is a negative regulator of abscisic acid signaling. Plant J 51(5):763–778. https://doi.org/10.1111/j.1365-313X.2007.03179.x

Pizzio GA, Rodriguez L, Antoni R, Gonzalez-Guzman M, Yunta C, Merilo E, Kollist H, Albert A, Rodriguez PL (2013) The PYL4 A194T mutant uncovers a key role of PYR1-LIKE4/PROTEIN PHOSPHATASE 2CA interaction for abscisic acid signaling and plant drought resistance. Plant Physiol 163(1):441–455. https://doi.org/10.1104/pp.113.224162

Prickett TD, Brautigan DL (2006) The α4 regulatory subunit exerts opposing allosteric effects on protein phosphatases PP6 and PP2A. J Biol Chem 281(41):30503–30511. https://doi.org/10.1074/jbc.M601054200

Polge C, Thomas M (2007) SNF1/AMPK/SnRK1 kinases, global regulators at the heart of energy control?. Trends Plant Sci 12(1):20–28. https://doi.org/10.1016/j.tplants.2006.11.005

Punzo P, Ruggiero A, Possenti M, Nurcato R, Costa A, Morelli G, Grillo S, Batelli G (2018) The PP2A-interactor TIP41 modulates ABA responses in Arabidopsis *thaliana*. Plant J, pp. 94(6):991–1009. https://doi.org/10.1111/tpj.139134

Radchuk R, Conrad U, Saalbach I, Giersberg M, Emery RJ, Küster H, Nunes-Nesi A, Fernie AR, Weschke W, Weber H (2010) Abscisic acid deficiency of developing pea embryos achieved by immunomodulation attenuates developmental phase transition and storage metabolism. Plant J 64(5):715–730. https://doi.org/10.1111/j.1365-313X.2010.04376.x

Ramon M, Rolland F, Sheen J (2008) Sugar sensing and signaling, The Arabidopsis book. Am Soc Plant Biol 6:e0117. https://doi.org/10.1199/tab.0117

Rodrigues A, Adamo M, Crozet P, Margalha L, Confraria A, Martinho C, Elias A, Rabissi A, Lumbreras V, González-Guzmán M, Antoni R, Rodriguez PL, Baena-González E (2013) ABI1 and PP2CA phosphatases are negative regulators of Snf1-Related Protein Kinase1 signaling in Arabidopsis. Plant Cell 25(10):3871–3884. https://doi.org/10.1105/tpc.113.114066

Rodriguez PL (1998) Protein phosphatase 2C (PP2C) function in higher plants. Plant Mol Biol 38(6):919–927. https://doi.org/10.1023/A:1006054607850

Rogers JP, Beuscher AE 4th, Flajolet M, McAvoy T, Nairn AC, Olson AJ, Greengard P (2006) Discovery of protein phosphatase 2C inhibitors by virtual screening. J Med Chem 49(5):1658–1667. https://doi.org/10.1021/jm051033y

Roustan V, Jain A, Teige M, Ebersberger I, Weckwerth W (2016) An evolutionary perspective of AMPK-TOR signaling in the three domains of life. J Exp Bot 67(13):3897–3907. https://doi.org/10.1093/jxb/erw211

Sakr S, Wang M, Dédaldéchamp F, Perez-Garcia MD, Ogé L, Hamama L, Atanassova R (2018) The sugar-signaling hub: overview of regulators and interaction with the hormonal and metabolic network. Int J Mol Sci 19(9):2506. https://doi.org/10.3390/ijms19092506

Saxton RA, Sabatini DM (2017) mTOR signaling in growth, metabolism, and disease. Cell 168(6):960–976. https://doi.org/10.1016/J.CELL.2017.02.004

Serfontein J, Nisbet RE, Howe CJ, de Vries PJ (2010) Evolution of the TSC1/TSC2-TOR signaling pathway. Sci Signal 2010;3(128):ra49. https://doi.org/10.1126/scisignal.2000803

Schweighofer A, Meskiene I (2008) Protein phosphatases in plant growth signaling pathways, in plant growth signaling. Springer, Berlin, pp 277–297. https://doi.org/10.1007/7089_2007_155

Schweighofer A, Hirt H, Meskiene I (2004) Plant PP2C phosphatases: emerging functions in stress signaling. Trends Plant Sci 9(5):236–243. https://doi.org/10.1016/j.tplants.2004.03.007

Schweighofer A, Kazanaviciute V, Scheikl E, Teige M, Doczi R, Hirt H, Schwanninger M, Kant M, Schuurink R, Mauch F, Buchala A, Cardinale F, Meskiene I (2007) The PP2C-type phosphatase AP2C1, which negatively regulates MPK4 and MPK6, modulates innate immunity, jasmonic acid, and ethylene levels in Arabidopsis. Plant Cell 19(7):2213–2224. https://doi.org/10.1105/tpc.106.049585

Sheen J (2014) Master regulators in plant glucose signaling networks. J Plant Biol 57(2):67–79. https://doi.org/10.1007/s12374-014-0902-7

Shenolikar S (1994) Protein serine/threonine phosphatases—new avenues for cell regulation. Annu Rev Cell Biol 10(1):55–86. https://doi.org/10.1146/annurev.cb.10.110194.000415

Shenolikar S, Nairn AC (1991) Protein phosphatases: recent progress. Adv Second Messenger Phosphoprotein Res 23:1–121
Shi L, Wu Y, Sheen J (2018) TOR signaling in plants: conservation and innovation. Development 145(13):dev160887. https://doi.org/10.1242/dev.160887
Singh A, Giri J, Kapoor S, Tyagi AK, Pandey GK (2010) Protein phosphatase complement in rice: genome-wide identification and transcriptional analysis under abiotic stress conditions and reproductive development. BMC Genomics 11:435. https://doi.org/10.1186/1471-2164-11-435
Singh A, Jha SK, Bagri J, Pandey GK (2015) ABA inducible rice protein phosphatase 2C confers ABA insensitivity and abiotic stress tolerance in Arabidopsis. PLoS One 10(4):e0125168. https://doi.org/10.1371/journal.pone.0125168
Smith RD, Walker JC (1996) Plant protein phosphatases. Annu Rev Plant Physiol Plant Mol Biol 47(1):101–125. https://doi.org/10.1146/annurev.arplant.47.1.101
Sotelo R, Garrocho-Villegas V, Aguilar R, Calderón ME, de Jiménez ES (2010) Coordination of cell growth and cell division in maize (Zea mays L.) relevance of the conserved TOR signal transduction pathway. *In Vitro* Cell Dev Biol-Plant 46:578–586. https://doi.org/10.1007/s11627-010-9293-8
Soulard A, Cohen A, Hall MN (2009) TOR signaling in invertebrates. Curr Opin Cell Biol 21(6):825–836. https://doi.org/10.1016/j.ceb.2009.08.007
Stepansky A, Galili G (2003) Synthesis of the Arabidopsis bifunctional lysine-ketoglutarate reductase/saccharopine dehydrogenase enzyme of lysine catabolism is concertedly regulated by metabolic and stress-associated signals. Plant Physiol 133:1407–1415. https://doi.org/10.1104/pp.103.026294
Stone JM, Collinge MA, Smith RD, Horn MA, Walker JC (1994) Interaction of a protein phosphatase with an Arabidopsis serine-threonine receptor kinase. Science 266(5186):793–795. https://doi.org/10.1126/SCIENCE.7973632
Tang W, Yuan M, Wang R, Yang Y, Wang C, Oses-prieto JA, Kim T, Zhou H, Deng Z, Gampala SS, Gendron JM, Jonassen EM, Lillo C, Delong A, Burlingame AL, Sun Y, Wang Z (2011) PP2A activates brassinosteroid-responsive gene expression and plant growth by dephosphorylating BZR1. Nat Cell Biol 13(2):124–131. https://doi.org/10.1038/ncb2151
Tatebe H, Shiozaki K (2017) Evolutionary conservation of the components in the TOR Signaling Pathways. Biomolecules 7(4):77. https://doi.org/10.3390/biom7040077
Tran HT, Nimick M, Uhrig RG, Templeton G, Morrice N, Gourlay R, Delong A, Moorhead GBG (2012) Arabidopsis *thaliana* histone deacetylase 14 (HDA14) is an a-tubulin deacetylase that associates with PP2A and enriches in the microtubule fraction with the putative histone acetyltransferase ELP3. Plant J 71:263–272. https://doi.org/10.1111/j.1365-313X.2012.04984.x
Tseng T, Briggs WR (2010) The Arabidopsis *rcn1-1* mutation impairs dephosphorylation of Phot2, resulting in enhanced blue light responses. Plant Cell 22:392–402. https://doi.org/10.1105/tpc.109.066423
Umbrasaite J, Schweighofer A, Kazanaviciute V, Magyar Z, Ayatollahi Z, Unterwurzacher V, Choopayak C, Boniecka J, Murray JA, Bogre L, Meskiene I (2010) MAPK phosphatase AP2C3 induces ectopic proliferation of epidermal cells leading to stomata development in Arabidopsis. PLoS ONE 5(12):e15357. https://doi.org/10.1371/journal.pone.0015357
Urano D, Phan N, Jones JC, Yang J, Huang J, Grigston J, Taylor JP, Jones AM (2012) Endocytosis of the seven-transmembrane RGS1 protein activates G-protein-coupled signaling in Arabidopsis. Nat Cell Biol 14(10):1079–1088. https://doi.org/10.1038/ncb2568
Urano D, Chen JG, Botella JR, Jones AM (2013) Heterotrimeric G protein signaling in the plant kingdom. Open Biol 3(3):120186–120186. https://doi.org/10.1098/rsob.120186
Van Dam TJ, Zwartkruis FJ, Bos JL, Snel B (2011) Evolution of the TOR pathway. J Mol Evol 73(3–4):209–220. https://doi.org/10.1007/s00239-011-9469-9
Waadt R, Manalansan B, Rauniyar N, Munemasa S, Booker MA, Brandt B, Waadt C, Nusinow DA, Kay SA, Kunz HH, Schumacher K, DeLong A, Yates JR 3rd, Schroeder JI (2015) Identification of open stomata1-interacting proteins reveals interactions with Sucrose Non-fermenting1-

Related Protein Kinases2 and with type 2A protein phosphatases that function in abscisic acid responses. Plant Physiol 169(1):760–779. https://doi.org/10.1104/pp.15.00575

Wang H, Ngwenyama N, Liu Y, Walker JC, Zhang S (2007) Stomatal development and patterning are regulated by environmentally responsive mitogen-activated protein kinases in Arabidopsis. Plant Cell 19(1):63–73. https://doi.org/10.1105/tpc.106.048298

Wang P, Zhao Y, Li Z, Hsu CC, Liu X, Fu L, Hou YJ, Du Y, Xie S, Zhang C, Gao J, Cao M, Huang X, Zhu Y, Tang K, Wang X, Tao WA, Xiong Y, Zhu JK (2018) Reciprocal regulation of the TOR kinase and ABA receptor balances plant growth and stress response. Mol Cell 69(1):100–112.e6. https://doi.org/10.1016/j.molcel.2017.12.002

Wlodarchak N, Xing Y (2016) PP2A as a master regulator of the cell cycle. Crit Rev Biochem Mol Biol 51(3):162–184. https://doi.org/10.3109/10409238.2016.1143913

Wu G, Wang X, Li X, Kamiya Y, Otegui MS, Chory J (2011) Methylation of a phosphatase specifies dephosphorylation and degradation of activated brassinosteroid receptors. Sci Signal 4(172):1–12

Wurzinger B, Nukarinen E, Nägele T, Weckwerth W, Teige M (2018) The SnRK1 kinase as central mediator of energy signaling between different organelles. Plant Physiol 176(2):1085–1094. https://doi.org/10.1104/pp.17.01404

Xiong Y, Sheen J (2012) Rapamycin and glucose-target of rapamycin (TOR) protein signaling in plants. J Biol Chem 287(4):2836–2842. https://doi.org/10.1074/jbc.M111.300749

Xiong Y, Sheen J (2015) Novel links in the plant TOR kinase signaling network. Curr Opin Plant Biol 28:83–91. https://doi.org/10.1016/J.PBI.2015.09.006

Xiong Y, McCormack M, Li L, Hall Q, Xiang C, Sheen J (2013) Glucose–TOR signaling reprograms the transcriptome and activates meristems. Nature 496(7444):181–186. https://doi.org/10.1038/nature12030

Xue T, Wang D, Zhang S, Ehlting J, Ni F, Jakab S, Zheng C, Zhong Y (2008) Genome-wide and expression analysis of protein phosphatase 2C in rice and Arabidopsis. BMC Genomics 9(1):550. https://doi.org/10.1186/1471-2164-9-550

Zhang Y, Primavesi LF, Jhurreea D, Andralojc PJ, Mitchell RA, Powers SJ, Schluepmann H, Delatte T, Wingler A, Paul MJ (2009) Inhibition of SNF1-related protein kinase1 activity and regulation of metabolic pathways by trehalose-6-phosphate. Plant Physiol 149(4):1860–1871. https://doi.org/10.1104/pp.108.133934

Chapter 8
Protein Phosphatases in Guard Cells: Key Role in Stomatal Closure and Opening

Shashibhushan Gahir, Vaidya Sunitha, Pulimamidi Bharath, and Agepati S. Raghavendra

8.1 Introduction

8.1.1 Importance of Stomata

Plants are sessile organisms and cannot move away from the harsh environments. Abiotic stresses that are encountered by plants are changes in temperature (high or low), drought (water deficit), waterlogging, salinity, and metal toxicity, while biotic stresses are due to pathogens and herbivores. Plants strive to adapt to different stress conditions by employing diverse mechanisms from morphological to molecular levels. Among these adaptations against stress conditions, the stomata, minute pores present on the epidermis of leaves, are crucial. Being the exit and entry points for water and carbon dioxide (CO_2), respectively, stomata not only play an important role in the regulation of photosynthesis and transpiration but also restrict the pathogen entry by closing themselves. The role of stomata during stress adaptations has been reviewed extensively (Murata et al. 2015; Agurla and Raghavendra 2016; Melotto et al. 2017; Agurla et al. 2018a; Zoulias et al. 2018; Buckley 2019).

Stomatal pores are formed by a pair of specialized guard cells. Closing and opening of stomata are determined by flaccidity and turgidity of guard cells. These events are triggered by abiotic/biotic stress signals, and guard cells have intricate signal transduction network and metabolisms to regulate stomatal movements (Medeiros et al. 2019). Under stress conditions, abscisic acid (ABA), a stress hormone produced in plants, is a key player that regulates stomatal function and modulates gene expression (Nakashima and Yamaguchi-Shinozaki 2013).

S. Gahir · V. Sunitha · P. Bharath · A. S. Raghavendra (✉)
Department of Plant Sciences, School of Life Sciences, University of Hyderabad, Hyderabad, India
e-mail: asrsl@uohyd.ernet.in

G. K. Pandey (ed.), *Protein Phosphatases and Stress Management in Plants*,
https://doi.org/10.1007/978-3-030-48733-1_8

8.1.2 Signals That Induce Closure/Opening

Among abiotic factors, elevated CO_2 levels, darkness, moisture deficit, and high ABA or ethylene concentration promote closing of stomata. In contrast high light, low CO_2, waterlogging conditions, and cytokinin promote opening (Araujo et al. 2011). Similarly, during pathogen attack (biotic stress), stomata sense the microbe-associated molecular patterns (MAMPs) or elicitors and try to close themselves (Melotto et al. 2017). Even polyamines can induce stomatal closure (Agurla et al. 2018b). These abiotic and biotic stress conditions provoke various alterations in the signaling components, i.e., nitric oxide (NO), ROS, pH, and calcium (Ca^{2+}), which lead to ion efflux, turgor loss, and finally stomatal closure (Raghavendra and Murata 2017). Hence, these abiotic and biotic stress factors contribute to advancement of stomata and signaling complex of guard cells.

8.1.3 Events During Stomatal Closure by ABA

Stomatal closure is a complex process comprising of multiple steps involving various signaling components. When guard cells take up ions, their turgor increases making stomata open. In contrast, ion efflux decreases the guard cell turgor making stomata close. The rise in pH of guard cell is an early event and is followed by an increase in ROS, NO, and cytosolic Ca^{2+}. Both ROS and NO increase cytosolic Ca^{2+}, which stimulates calcium-dependent protein kinases (CDPKs) (Brandt et al. 2015). The active CDPKs facilitate ion efflux through guard cell-associated slow anion channels (SLAC) as well as outward K^+ ion channels (GORK). During the stomatal closure, the signaling events converge at ROS, NO, and cytosolic Ca^{2+}, which later diverge their actions (Agurla and Raghavendra 2016; Agurla et al. 2018a). All the events ultimately modulate the ion levels in guard cells. The signaling components involved during stomatal closure by ABA have been described in detail (Lee and Luan 2012; Gayatri et al. 2013; Kollist et al. 2014; Murata et al. 2015; Agurla and Raghavendra 2016; Singh et al. 2017; Agurla et al. 2018a).

Upon microbial pathogen challenge also the stomata close. The signals under such microbial attack are pathogen-associated molecular patterns (PAMPs) or MAMPs that are recognized by PRRs (pattern recognition receptors). These PAMPs or MAMPs stimulate mitogen-activated protein kinase (MAPK) activity and upregulate SLAC (Rasmussen et al. 2012). Upon perception of PAMP, ROS is produced by NADPH oxidase in guard cells followed by other components such as NO, Ca^{2+}, K^+, and anion channels in that order (Arnaud and Hwang 2015). Salicylic acid (SA) signaling pathway involving non-expresser of pathogenesis related 1 (NPR1), receptor of SA, can lead to stomatal closure, as defense against pathogen (Seyfferth and Tsuda 2014; Manohar et al. 2017). These events of stomatal closure are further explained below.

8.2 Signal Transduction in Guard Cells

8.2.1 Signal Perception and Transmission

Changes in the stomatal aperture are initiated by different abiotic or biotic stress signals. For example, ABA or methyl jasmonate (MJ) can act as abiotic stress signals, while microbial elicitors and SA are biotic stress signals. The presence of ABA is sensed by ABA receptors (also called as PYR/PYL/RCAR family proteins), type 2C protein phosphatases (PP2Cs), and SnRK2-type protein kinases (Raghavendra et al. 2010; Lim et al. 2015). These events are further explained in Sect. 8.3.2.

The PYR/PYL/RCAR family is often perceived as PYLs. Apart from ABA, there are several compounds that can cause stomatal closure such as MJ, SA, chitosan, cryptogein, and flagellin22 (flg22). However, it is not clear how these microbial elicitors of cryptogein, MJ, or chitosan are perceived (Agurla and Raghavendra 2016). There are suggestions that the receptor for flg22 is flagellin-sensitive 2 (FLS2) (Chinchilla et al. 2006) and SA is perceived by NPR1 in the guard cell (Wu et al. 2012).

8.2.2 ABA-Receptor-PP2C Complex Formation

In default conditions, the active PP2C downregulates SnRK2s by dephosphorylation and thus arrests the ABA response. When present, ABA combines to PYL/PYR/RCAR receptor and then to PP2C, making PP2C ineffective. The inhibition of PP2C results in phosphorylated SnRK2s, which in turn modulate further downstream signaling in guard cells (Zhang et al. 2015). The complex of PYL-ABA-PP2C is stabilized by the binding of ABA to PYR and then with protein phosphatase 2C (PP2C) making PP2C ineffective and nonoperational (Gonzalez-Guzman et al. 2012). As a result of blocking PP2C action, the SnRK2 remains phosphorylated and active (Kline et al. 2010; Fujii and Zhu 2009). Thus, PYR/PYL/RCAR protein family and SnRKs are involved in positive regulation of ABA signaling, whereas PP2C acts as a negative regulator of ABA.

An example of SnRK2s is OST1 kinase, a serine/threonine protein kinase (Zhang et al. 2014; Ye and Murata 2016). The phosphorylated SnRK2 or OST1 phosphorylates and activates NADPH oxidases (respiratory burst oxidase homologs RBOH D/F) to generate reactive oxygen species (ROS), an important second messenger involved in the stomatal closure (Munemasa et al. 2015). The rise in ROS is associated with the rise in cytoplasmic pH, NO, and Ca^{2+} (Gonugunta et al. 2008; Agurla and Raghavendra 2016). ABA-induced cascade of signaling components finally targets ion channels at the plasma membrane and tonoplast. The presence of ABA activates K^+ outward channel while inhibiting inward K^+ channels and S-type anion

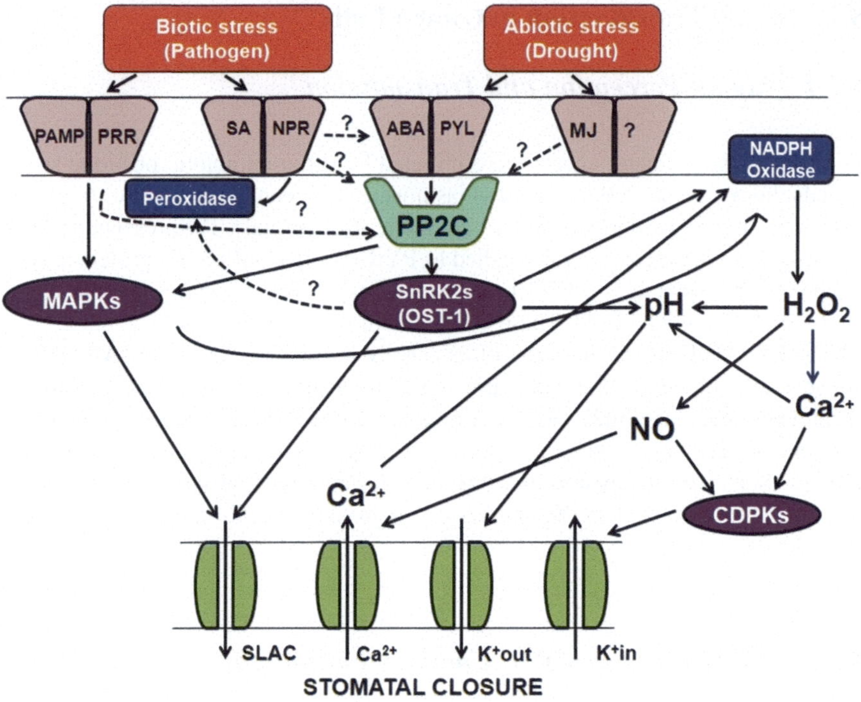

Fig. 8.1 Schematic illustration of signal transduction during stomatal closure under biotic and abiotic stress conditions. Abiotic stress, e.g., drought, leads to the rise in levels of ABA and MJ. Then PYL receptors sense the ABA and bind to PP2C making PP2C nonfunctional and unable to exert its phosphatase activity. As a result, SnRK2s (e.g., OST-1) become active, phosphorylate, and stimulate SLAC1 releasing the anions from guard cells, leading to stomatal closure. The activated SnRK2s can also upregulate NADPH oxidase and increase ROS production. Other signaling components, such as ROS, NO, and Ca^{2+} in stomatal guard cells, all lead to loss of K^+ anions from guard cells causing turgor loss and stomatal closure. Under biotic stress (e.g., microbial pathogens) conditions, PRRs recognize PAMPs released by pathogens and stimulate MAPKs. The activated MAPKs facilitate ROS production by NADPH oxidase. The hypothetical steps are indicated by broken arrows. The solid lines represent actions that are supported by experimental evidences. (Adapted from Agurla and Raghavendra (2016) and Agurla et al. (2018a))

channels, leading to ion efflux, loss of turgor in guard cells, and stomatal closure (Lee and Luan 2012; Edel and Kudla 2016). Figure 8.1 represents graphically the spectrum of events that occur during stomatal closure by abiotic or biotic stress factors. PP2C is a key component in these signal transduction systems.

8.2.3 *ABA Analogues Used as Interacting Partners of PP2C/ ABI1*

Efforts are always on to identify ABA analogues, in view of their potential use to conserve water loss and improve drought tolerance of crops. Many ABA analogues have been tested for their interactions with PYL receptors. Pyrabactin is an analogue of ABA and can induce stomatal closure by interacting with ABA-insensitive 1 (ABI1) (Puli and Raghavendra 2012). Three compounds, xanthoxin (ABA precursor) and two of its catabolites (ABA alcohol and ABA aldehyde), were found to mediate the reduction in stomatal aperture which were also interacted with ABI1 (Kepka et al. 2011).

Many other analogues of ABA have been found to cause closure of the stomatal guard cells, but their role and interactions with PP2Cs are not completely studied. Examples are biotinylated ABA and RCA-7a (Yamazaki et al. 2003), 8′ acetylene ABA methyl-ester (Weaver and Iersel 2014), 2′,3′-dihydro-ABA (Yamamoto and Oritani 1995), and compounds 1–4 described by Orton and Mansfield (1974). There is an evidence of interaction of PP2C with the ABA analogue 3′-hexylsulfanyl-ABA, which specifically inhibits PYL-PP2C interaction by steric hindrance (Takeuchi et al. 2014). During chemical screening studies, two compounds SCL1 and SCL2 were found to cause stomatal closure. These two compounds suppressed stomatal opening by blue light due to complete suppression of phosphorylation of H^+-ATPases by SCL1 and 50% reduction in phosphorylation status by SCL2 (Toh et al. 2018).

8.3 Protein Phosphatases (PPs)

8.3.1 *Different Forms of PPs*

Protein phosphatases are called as "housekeeping" enzymes as they perform central functions by coordinating with protein kinases in plant signaling pathways during abiotic and biotic stress conditions (Schweighofer and Meskiene 2015). Based on primary sequence, specificity to substrates, and catalytic reaction mechanisms, protein phosphatases are classified into four groups. These are phosphoprotein phosphatase (PPP), metal ion-dependent (Mg^{2+} or Mn^{2+}) protein phosphatase (PPM)/protein phosphatase 2C (PP2C), phosphotyrosine phosphatase (PTP), and dual specificity phosphatase (Ser/Thr and Tyr-specific)/aspartate (Asp)-dependent (Uhrig et al. 2013; Farkas et al. 2007).

The major plant protein phosphatases belong to the PPP family, and they play a central role in cellular signaling (Uhrig et al. 2013). The PPPs include Ser/Thr protein phosphatase types: PP1, PP2A, and PP2B (Shi 2009). PP1 plays an important role in cell cycle regulation, embryo development, cell differentiation, and salt tolerance (Farkas et al. 2007). Only PP2As are reported in plants. These PP2As participate in various stress cellular signaling and metabolisms and provide tolerance to plants during biotic and abiotic stress conditions. The role of PP2A in various abiotic and biotic stresses has been reviewed (País et al. 2009; Durian et al. 2016; Sun et al. 2018). Among other PPs, the role of protein phosphatase 4 (PP4) is not clear. However, protein phosphatase 5 (PP5) is important for disease resistance and thermotolerance, while protein phosphatase 6 (PP6) plays a role in auxin efflux and phosphorylation of pin-formed (PIN) proteins. The literature on PPs other than PP2C and PP2A is quite limited. Readers interested in diversity of PPs in plants may consult recent reviews (Rodriguez 1998; Luan 2003; Fuchs et al. 2013; Uhrig et al. 2013; Farkas et al. 2007; Singh et al. 2016; Bheri and Pandey 2019).

The protein phosphatases 2C, which are present in Arabidopsis, are ABI1, ABA-insensitive 2 (ABI2), AtP2C-HA, AtPP2CA/AtPP2C5/AP2C3, KAPP proteins, and alfalfa MP2C protein. ABI1 and ABI2 display a similar architecture; i.e., C-terminal domain is 86% identical and contains the PP2C core (Wu et al. 2003). They play a key role in ABA signal transduction and control transpiration, vegetative growth, seed germination, stomatal closure, and gene expressions (Meyer et al. 1994; Allen et al. 1999). They are expressed in different parts of plants including stomatal guard cells and function as negative regulators of ABA signaling (Saez et al. 2004). A summary of different forms of PPs and their involvement in stomatal opening/closure is given in Table 8.1.

8.3.2 PP2C: Essential for Stomatal Closure

A second major group of PPs are protein serine/threonine phosphatases called PP2C enzymes (Singh et al. 2016). They are Mg^{2+}- or Mn^{2+}-dependent protein phosphatase (PPM) and insensitive to okadaic acid. A major function of PP2C is promotion of stomatal opening. In default condition, when plants are not exposed to any stress, PP2C is active and keeps the downstream targets, e.g., SnRK2 in dephosphorylated and inactive form. PP2Cs have multiple actions in the signaling pathway of ABA-induced stomatal closure. Stomatal closure induced by ABA signaling is regulated by interaction of a protein kinase SnRK2-type kinase and PP2C with ion channel SLAC1 (Lee and Luan 2012). SnRK2.6 (OST1), a serine/threonine protein kinase, is a key positive regulator of NADPH oxidases in guard cells of Arabidopsis and modulates stomatal closure (Acharya et al. 2013). Although PP2C phosphatases and SnRK2 protein kinases constitute the hub of ABA pathway, MAPKs also are involved in ABA signaling (Lee et al. 2016). In brief, sensing of ABA signaling steps is initiated by the binding of ABA and PYL receptors to PP2C, thus blocking its phosphatase activity. As a result, the downstream components of SnRK2-type kinases are activated and lead to a series of steps. These are all described further.

The PP2Cs that are involved in stomatal closure during ABA signaling are ABI1, ABI2, AtP2C-HA, and homology to ABI1 (HAB1) in Arabidopsis, and they all negatively regulate ABA signaling (Saez et al. 2006; Ma et al. 2009). The only member of PP2C that acts as positive regulator of ABA is AKT1 interacting protein

Table 8.1 The spectrum of protein phosphatases (PPs) reported in guard cells and their role in stomatal movements

Role in stomatal function		References
Phosphoprotein phosphatase (PPP)		
PP1	Promotes stomatal opening by activating H^+-ATPase in response to blue light receptors, e.g., phototropin 1/2 (Phot1/2)	Takemiya et al. (2006)
	A subunit of PP1, PRSL1 (PP1 regulatory subunit 2-like protein1), modulates blue light signaling in stomatal guard cells	Takemiya et al. (2012)
PP2A	Stomatal closure by activating slow anion channels and Ca^{2+} increase	Kwak et al. (2002)
	Closure by upregulation of ROS production	Saito et al. (2008)
	Stomatal closure by restricting the activity of Phot2	Tseng and Briggs (2010)
	Closure by interaction with SnRK2-type protein kinases	Waadt et al. (2015)
	Stomatal closure by ROS production and reduced polymerization of microtubules	Lijun et al. (2018)
PP4, PP5, and PP6	No report on the role in stomatal movement	
PP7	Promotes stomatal opening by interacting with HRB1 (hypersensitive to red and blue 1)	Sun et al. (2012)
Protein phosphatase 2C (PP2C)/metal-ion dependent PP (PPM)		
ABI1	Promotes stomatal closure by activation of NADPH oxidase and ROS production	Murata et al. (2001)
	Interacts with cGMP and modulates H_2O_2/no production through Ca^{2+}	Dubovskaya et al. (2011)
	Stomatal closure by polymerization of actin filaments in guard cell	Eun et al. (2001)
ABI2	Promotes stomatal closure mediated by ROS and Ca^{2+}	Murata et al. (2001)
AtP2C-HA	Mutation leads to ABA-hypersensitive regulation of stomatal closing	Leonhardt et al. (2004)
AtPP2CA	Promotes stomatal closure by activating MAPKs	Brock et al. (2010)
	Downregulates MAPKs during stomatal guard cell differentiation	Umbrasaite et al. (2010)
KAPP	Interacts with flagellin 22 and activates protein kinase	Gómez-Gómez et al. (2001)
Phosphotyrosine phosphatase (PTP)		
	Mediates closure by H_2O_2	Shi et al. (2004)
	Promotes closure by increasing intracellular Ca^{2+}	MacRobbie (2002)
	Promotes stomatal closure by triggering K^+ efflux from guard cells	Luan (2002)
	Promotes closure by starch degradation	Qin et al. (2015)
Dual specificity phosphatase (DSP)		
	No report on the role in stomatal movement	

phosphatase 1 (*AIP1*) (Lim et al. 2012). ABI1 and ABI2 are closely related protein phosphatase 2C as they carry the same amino acid substitution on ABI1 and ABI2 PP2C domains (Leung et al. 1997). They are identified by mutations *abi1-1* and *abi2-1* in Arabidopsis, and they contribute 50% of ABA-induced PP2C activity and they have overlapping functions (Merlot et al. 2001). The *abi1* and *abi2* mutants show reduced sensitivity of root growth to ABA and impaired regulation of the ABA-dependent stomatal closure (Rodriguez 1998). Stomatal closure is induced in *abi1-1* and *abi2-1* mutants. In *abi1-1*, ABA acts at upstream of ROS production mediated by NAD(P)H oxidase, and in *abi2-1*, ABA acts downstream of ROS production and mediates stomatal closure by activating Ca^{2+} channel (Murata et al. 2001). The *abi1-1* interacts with cADPR and cGMP (Wu et al. 2003) and promotes release of Ca^{2+} (Leckie et al. 1998; Dubovskaya et al. 2011).

8.3.3 PP2A: Key Component of Stomatal Closure

PP2A is involved in stomata movements in plants, particularly during stress conditions. Kwak et al. (2002) reported that RCN1 (roots curl in naphthylphthalamic acid 1), a regulatory unit of PP2A, is expressed in guard cells and positively regulates ABA signal transduction. The RCN1 functions upstream of Ca^{2+} rise. The *rcn1* mutant had reduced cytosolic Ca^{2+}, inactivation of anion channels resulting in reduced stomatal closure (Kwak et al. 2002). RCN1 may modulate ROS production, an upstream event of ROS in guard cells (Saito et al. 2008). PP2A regulatory subunits interact with activated SnRK2type protein kinases and form complexes. The *pp2a* double mutants had reduced stomatal closure due to lack of interaction with SnRK2-type protein kinases (Waadt et al. 2015). When the PP2A activity is reduced, the activity of Phot2 increases and leads to opening of stomata (Tseng and Briggs 2010). Similarly, when the PP2A activity is increased due to ALA (5-aminolevulinic acid), ROS production reduces and polymerization of microtubules increases in guard cells leading to stomatal closure (Lijun et al. 2018).

8.3.4 PP1: Promotes Stomatal Opening

Protein phosphatases are involved not only in promoting stomatal closure but also stomatal opening (Table 8.1). The importance and essentiality of PP2C for ABA-induced stomatal closure is well established. Similarly PP1, another phosphoprotein phosphatase-type protein phosphatase, is essential for blue light-induced stomatal opening (Kinoshita and Shimazaki 1997). Inhibitors of protein phosphatases, like calyculin A and okadaic acid, restricted the blue light-induced opening, suggesting that PP1 is essential for stomatal function. Phots and Crys, especially the former, are considered as the major receptors of blue light. Activation of Phot, by a serine/threonine kinase by autophosphorylation, is triggered by blue light. As a result,

H^+-ATPase is activated and membrane hyperpolarized to allow K^+ influx into guard cells and stomatal opening (Zhang et al. 2014). Phot1 and Phot2 are the major phototropins, involved in such blue light responses. After blue light perception, Phot1 phosphorylates a guard cell-specific BLUe light Signaling 1 (BLUS1) kinase that can activate H^+-ATPase on the membrane. This phosphorylated H^+-ATPase is stabilized by the presence of 14-3-3 protein (Fig. 8.2). Phototropin-activated PP1 ensures the downward regulation of PP and inactivation of H^+-ATPase (Takemiya et al. 2015; Zhang et al. 2014).

The activation of Phot1 and Phot2 by blue light converges to PP1C, a serine/threonine phosphatase (Takemiya et al. 2012). The promotion of stomatal opening

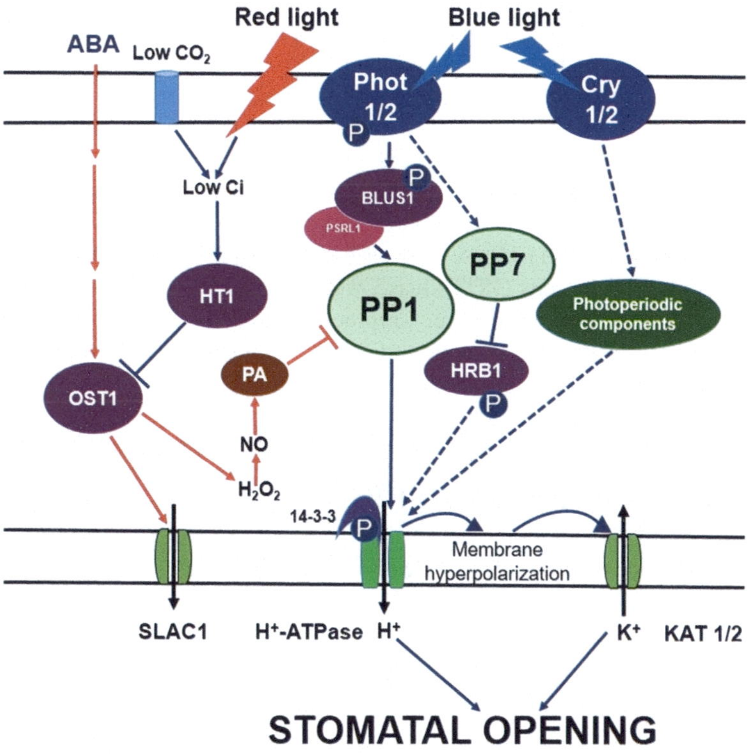

Fig. 8.2 Scheme of events during blue light promoted stomatal opening, emphasizing the role PP1 and PP7 in guard cells. Blue light is perceived by Phots or Crys, and autophosphorylation of Phot activates PSRL1 through BLUS1 kinase and then activates PP1. Phototropin can upregulate PP7, which in turn dephosphorylates HRB1, and together can stimulate H^+-ATPase activity. Crys also acts as light receptors and modulates photoperiodic components to activate H^+-ATPase. When active H^+-ATPase pumps protons out, the membrane hyperpolarization results in influx of K^+ ions and stomatal opening. The events with obscure evidences are represented by broken arrows. Signal transduction events during stomatal opening by blue light are represented in dark blue lines. Red colored lines are symbolized to show restriction of opening, favoring closure. Further detailed description of these events can be found in reviews of Shimazaki et al. (2007), Inoue et al. (2010), and Inoue and Kinoshita (2017)

by blue or red light can be inhibited by ABA in two ways (Takemiya and Shimazaki 2010). One of them is the ABA-induced production of phosphatidic acid (PA), which can impair stomatal opening. The second possibility is the activation by ABA of OST1 kinase that activates SLAC and stimulates efflux of anions, thus promoting stomatal closure. It has been found that the activation of H^+-ATPase is ensured by phosphorylated 14-3-3 proteins. After blue light perception, phototropins phosphorylate BLUS1, which in turn stabilizes the binding of 14-3-3 to H^+-ATPase and activates H^+ efflux, leading to membrane hyperpolarization, influx of K^+ via K^+ channel of Arabidopsis *thaliana* 1/2 (KAT1/2), and thereby stomatal closure (Assmann and Jegla 2016; Kang et al. 2018). In addition, two master regulators, hypersensitive to red and blue light 1 (HRB1) and PP7, also mediate the blue light-dependent stomatal opening. PP7 is an interacting partner of HRB1 during signaling, as PP7 dephosphorylates HRB1 and both promote stomatal opening (Sun et al. 2012). These events are illustrated in Fig. 8.2.

Low CO_2 and red light along with blue light induce stomatal opening in plants (Fig. 8.2). Phots sense the blue light, autophosphorylate itself, and mediate phosphorylation of BLUS1, which activates H^+-ATPase and subsequently causes membrane hyperpolarization leading to the opening of stomata. On the other hand low CO_2 or low intracellular concentration of CO_2 under red light activates the HT1 (High leaf Temperature 1) kinase and ensures stomatal opening (Hiyama et al. 2017). Further, HT1 inhibits OST1 kinase, which activates S-type anion channel in guard cells, thus restricting closure. Cryptochromes facilitate blue light-induced stomatal opening, with the help of photoperiodic components like Flowering Locus T (FLT) and COnstans (CO) (Zhang et al. 2014).

8.4 Interacting Partners of PP2C

8.4.1 ABA Receptors: PYR/PYL/RCAR Proteins

In Arabidopsis, receptors for ABA are called PYR/PYL/RCAR (Raghavendra et al. 2010; Rodriguez et al. 2014). The PYR/PYL/RCAR family proteins are identified as ABA-binding proteins located both in the cytosol and the nucleus. The PYR/PYL family represents the initial step of ABA signaling pathway and perceives intracellular ABA. These interactions are described earlier in Sect. 8.3.2.

The number of PYR/PYL genes is variable among the different species, e.g., 14 in Arabidopsis, 11 in *Zea mays*, 12 in *Oryza sativa*, 14 in *Hevea brasiliensis*, and 46 in *Brassica napus* (Guo et al. 2017). The PYR/PYL family in Arabidopsis (14 genes) has nine PYR/PYL proteins reported as ABA-binding signal transduction proteins. These nine are AtPYL1, 2, 3, 4, 5, 6, 7, 8, and 9. Quadruple mutant plants pyr1/pyl1/pyl2/pyl4 exhibited a strong ABA-insensitive phenotype, suggesting that PYR1/PYL1/PYL2/PYL4 function in ABA signal transduction pathway (Nishimura et al. 2010). Such PYR1/PYL1/PYL2/PYL4 quadruple mutant-mediated ABA signaling is found not only in guard cells but also in seed germination and root growth. In conclusion, ABA and PYR/PYL receptors are essential for sensing ABA and subsequent regulation of stomatal aperture.

8.4.2 Protein Kinases Involved in ABA-Mediated Signaling in Guard Cells

ABA induces stomatal closure in response to drought stress by regulating the ion fluxes through Ca^{2+}-dependent protein kinases. Transient phosphorylation and dephosphorylation of key signaling components act as "on-and-off" switches (Liang et al. 2015), to regulate downstream targets involved in ABA-mediated stomatal closure. During ABA signaling, four types of protein kinases operate. These are SnRKs, MAPKs, CDPKs, and RLKs (receptor-like kinases). All the four kinases phosphorylate the serine/threonine residues in the downstream components to regulate the ABA-mediated signaling.

The SnRKs (serine-threonine kinases) belong to the CDPK-SnRK superfamily, containing 38 genes in Arabidopsis. Unlike the variability in PYR/PYL genes, the number of SnRK proteins is similar in all the plant species, examined so far. These SnRK protein kinases (plant-specific serine/threonine kinases) are divided into three subfamilies: SnRK1, SnRK2, and SnRK3 with 3, 10, and 25 members of proteins, respectively (Hrabak et al. 2003). Only SnRK2 and SnRK3 subfamilies have plant-specific kinase actions. The SnRK2 family consists of ten proteins: SnRK2.1 to SnRK2.10 in Arabidopsis. Among these SnRK2.2, SnRK2.3 and SnRK2.6 are involved in ABA-mediated guard cell signaling. One of the SnRK2 types, SnRK2.6 mutant (Open STomata 1, *ost1*) is a classic example of impaired ABA-mediated stomatal closure, confirming the role of SnRK 2.6/*ost1* during ABA-mediated stomatal closure (Mustilli et al. 2002; Yoshida et al. 2006). Downstream targets of SnRK2s play a major role during stomatal closure mediated by ABA.

CDPKs are the major members of CDPK-SnRK superfamily and CDPKs are abbreviated with three letter word CPK (Hrabak et al. 2003). In Arabidopsis 34 genes are expected to code CDPKs (Li et al. 2008; Ma et al. 2013). Among these AtCPK10, AtCPK4, and AtCPK11 members in Arabidopsis function are positive regulators of ABA actions in an intracellular Ca^{2+}-dependent manner. Double mutant of *cpk4cpk11* showed strong ABA-insensitive responses. ABA-responsive transcription factors (ABFs) are the other downstream components of CDPKs in Arabidopsis to mediate ABA signaling.

Guard cell Hydrogen peroxide-Resistant1 (GHR1) is a kinase belonging to leucine-rich repeat receptor-like kinases (LRR-RLK). GHR acts as a signaling component involved in downstream regulation of stomatal closure in response to H_2O_2 and ABA. GHR acts downstream of ABI1 and ABI2 to phosphorylate and activate the SLAC1 anion channel. In addition to GHR, another kinase, MPK3 (a MAP Kinase), acts as downstream signaling component of ROS during ABA inhibition of stomatal opening (Gudesblat et al. 2007). Two MAP kinases, AtMPK9 and AtMPK12, are highly expressed MAP kinases in guard cell and function downstream of ROS in response to ABA or elicitors. A detailed list of kinases reported in guard cells and their possible function is presented in Table 8.2.

Table 8.2 Protein kinases in guard cells involved in signaling events leading to stomatal closure or opening

Abbreviation (name)	Opening/ closure	Remarks	References
OST1 (Open STomata 1)	Closure	OST1 upregulated as a consequence of PYL/ABA/PP2C interaction leading to activation of NADPH oxidase and production of ROS	Mustilli et al. (2002), Acharya et al. (2013)
BAK1 (Brassinosteroid-insensitive 1-Associated receptor Kinase 1)	Closure	Forms complex with OST1 and required also for OST1 expression	Shang et al. (2016)
BIK1 (*Botrytis*-Induced Kinase 1)	Closure	Phosphorylates NADPH oxidase RbohD to generate H_2O_2 in response to flg22	Li et al. (2014)
MPK3/MPK6 (mitogen-activated protein kinase)	Closure	Pathogen responsive; mediates ABA-independent stomatal closure	Su et al. (2017)
MAPK9/MAPK12 (myosin-activated protein kinase)	Closure	Acts downstream of ROS and Ca^{2+} and upstream of S-type anion channels	Khokon et al. (2015), Jammes et al. (2009)
AAPK (ABA-Activated Protein Kinase)	Closure	Guards cell-specific response regulator. Activates slow anion channels (SLAC) in response to ABA and implicated in stomatal closure	Li et al. (2000)
GHR1 (Guard cell Hydrogen peroxide-Resistant1)	Closure	Activates SLAC1	Hua et al. (2012)
CDPK6/CDPK3 (calcium-dependent protein kinase)	Closure	Activation of S-type anion channels by modulating cytosolic Ca^{2+}	Mori et al. (2006)
CBL1 and CBL9 (Calcineurin B-Like proteins)	Closing	Phosphorylates RBOHF and enhances ROS production	Drerup et al. (2013)
HT1 (High leaf Temperature kinase1)	Opening	Highly expressed in guard cells and inhibits OST1 and S-type anion channels	Hiyama et al. (2017), Hashimoto et al. (2006)
BLUS1 (BLUe light Signaling1)	Opening	Phosphorylated BLUS1 activates H^+-ATPase	Zhang et al. (2014)
CIPK23 (Calcineurin B-like (CBL)-Interacting Protein Kinase 23)	Opening	Complex of CIPK23 and CBL1/9 activates K^+ uptake	Nieves-Cordones et al. (2011)

8.4.3 Molecular Interactions of PP2C with PYLs and Kinases

Whenever an abiotic stress or even biotic stress develops, levels of ABA increase and modulate developmental and physiological processes like stomatal movement and seed germination. The PP2Cs act as the central component in the ABA signaling, during such stomatal closure. In the absence of stress signal, such as ABA, the responsive genes are downregulated due to the inactive protein kinases (Yoon et al. 2018). Under stress, the elevated ABA binds to the PYLs and then PP2C. As a consequence of nonavailability of PP2C, SnRK2s are allowed to be in phosphorylated state and target ion channels, like of SLAC1 and K^+ channels. These kinases also activate NADPH oxidases, which otherwise are dephosphorylated and inactive (Zhang et al. 2015). The two crucial PP2Cs are encoded by ABA-insensitive 1 and 2 genes (Rodriguez et al. 1998), and the mutants *abi1-1* and *abi2-1* show insensitivity to ABA during germination and guard cell function (Umezawa et al. 2010). These interactions are represented graphically in Fig. 8.3.

The structure and molecular interactions of ABA receptors (PYL proteins) and protein phosphatases (PP2C) have been studied, using the crystal structures (Hubbard et al. 2010; Melcher et al. 2010b; Umezawa et al. 2010; Dupeux et al. 2011; Miyakawa and Tanokura 2011; Zhang et al. 2015). The 14 members in PYL family, i.e., PYR1 and PYL1 to PYL13, are all from START/Bet v 1 large superfamily. The PYLs are structurally similar in architecture of helix-grip structure, consisting of seven antiparallel ß-strands supported by two α-helices on either side. The PYL1, PYL2, PYL3, PYL5, PYL9, PYL10, and PYL13 proteins have been crystallized, and their structures showing bound ABA are documented (Melcher et al. 2010b; Zhang et al. 2015).

When ABA molecule binds to the apo-receptor PYLs, changes in the conformation occur in two highly conserved ß-loops to make them act as gate and latch. Subsequently the closure of gate induced by ABA creates a docking surface for PP2C and turns the receptor into closed conformation. This process is termed gate-latch-lock mechanism (Melcher et al. 2009, 2010b). PP2Cs along with PYLs and ABA form the ternary complexes (Zhang et al. 2015). PP2C activity is inhibited by blockage of substrate entrance to catalytic site due to presence of complex at bottom of active site and the protruding domains (Miyazono et al. 2009). Replacement of amino acid glycine with aspartate in PP2Cs' active site impairs the interaction of receptor and PP2Cs in ABA-insensitive plants (Sheen 1998). Pyrabactin, an analogue of ABA (Puli and Raghavendra 2012), has been shown to bind PYR1 and PYL1 (which can inhibit PP2C) but failed to couple with PYL2 and PYL3 (Melcher et al. 2010a).

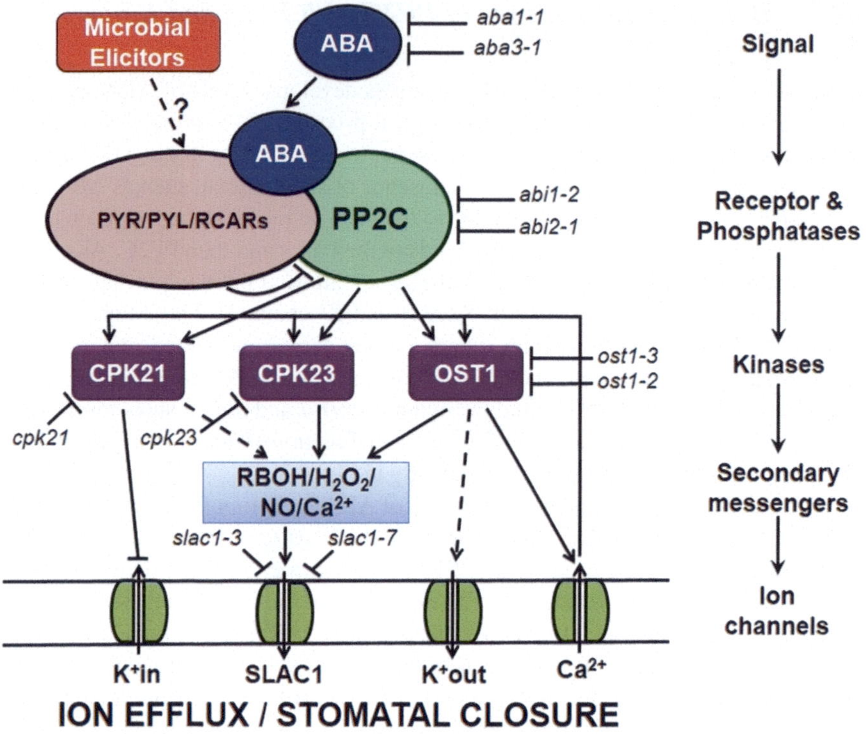

Fig. 8.3 Events during stomatal closure by ABA, emphasizing the role of ABA receptors and PP2C. When present, ABA is sensed intracellularly by PYR/PYL/RCAR receptor and forms PYR/ABA/PP2C ternary complex, which blocks the normal function of PP2C. As a result, PP2C cannot dephosphorylate the kinases acting downstream. As a consequence, the phosphorylated forms of kinases like CPK21, CPK23, and OST1 activate NADPH oxidase (RBOH) and lead to an increase in ROS, NO, and cytosolic Ca^{2+}. These events promote ion efflux from guard cells through SLAC and K^{+}_{out} channels. At the same time CPK21 ensures inactivation of K^{+}_{in} channel, restricting K^{+} influx into guard cells. The ion efflux from guard cells results in stomatal closure. Further description can be found in the reviews of Kollist et al. (2014), Agurla and Raghavendra (2016), Agurla et al. (2018a), and Buckley (2019)

8.4.4 Arabidopsis *Mutants: Versatile Tools to Study the Role of PP2C and PP1 in Stomatal Function*

Steady and stupendous progress in plant biology has been made possible with use of mutants. Arabidopsis *thaliana* mutants have been excellent model systems and versatile tools to study the structure and function of plant cells. Further, stomatal guard cells are ideal to study the mechanism of signal transduction in plants.

These investigations on stomatal function in Arabidopsis mutants got a boost with the discovery and optimization of infrared thermal imaging for identifying plants, with altered stomatal function (Merlot et al. 2002). Since then the search and use of Arabidopsis mutants to study stomatal guard cells has intensified (Agurla et al. 2017). A list of Arabidopsis mutants, which have been useful in validating the role of PPs in stomatal functions, is given in Table 8.3.

Table 8.3 Arabidopsis mutants deficient in signaling components involved in stomatal closure, particularly in relation to ABA perception and PP2C function

Mutant name (deficiency)	Effects/remarks	References
pyr1/pyl1/pyl2/pyl4 quadruple mutant (receptors of ABA)	Formation of the RCAR-ABA-PP2C complex cannot take place in mutants. ABA-induced closure is restricted	Nishimura et al. (2010)
abi1-2 and *abi2-1* (ABA-insensitive 1 and 2 type 2C protein phosphatases)	ABI1 and ABI2 dephosphorylate and inactivate kinases involved in ABA signaling and stomatal closure	Merlot et al. (2001)
abi1-2/abi2-2 double mutant	Showed enhanced ABA-induced stomatal closure and reduced water loss	Rubio et al. (2009)
hab1-1 (hypersensitive to ABA 1, a PP2C)	Mutants exhibit enhanced ABA-mediated stomatal closure	Zhang et al. (2013)
pp2ca-1 (protein phosphatase type 2C-A)	Impaired the stomatal closure by ABA	Lefoulon et al. (2016)
hab1-1/pp2ca-1 and *abi1-2/pp2ca-1* double mutant	Enhanced stomatal closure by ABA	Rubio et al. (2009)
pp2ca-1/hai1-1 double mutant (PP2CA/HAI1, highly ABA-induced-1)	Extreme sensitivity to ABA	Antoni et al. (2012)
ost1-1/snrk2.6 (Open STomata 1 kinases)	Impaired phosphorylation by SnRK2 and disrupted ABA induction of stomatal closure as well as ABA inhibition of light-induced stomatal opening	Mustilli et al. (2002)
hab1-1abi1-2abi2-2 and *hab1-1abi1-2pp2ca-1* triple mutant	Extreme response to exogenous ABA	Rubio et al. (2009)
srk2d/srk2e/srk2i (srk2d/e/i) triple mutant	Reduced tolerance of plants to drought stress and highly enhanced insensitivity to ABA	Nakashima et al. (2009)
areb1/areb2/abf3 triple mutant (ABA-responsive elements)	These elements are partially associated with stomatal closure	Yoshida et al. (2010)
ait1/nrt1.2 (ABA-importing transporter, AIT/nitrate transporter, NRT1.2)	AIT1/NRT1.2 mediates ABA uptake into guard cells, necessary for stomatal closure. Mutants unable to take up ABA resulting in wider stomatal aperture and excess water loss	Kanno et al. (2012)
hrb1 and *pp7* mutants (hypersensitive to red and blue 1 HRB1 and protein phosphatase 7, PP7)	Expressed in the guard cells and required for stomatal opening in response to a light-to-dark or dark-to-light transition	Sun et al. (2012)

8.5 Concluding Remarks

The phenomenon of phosphorylation and dephosphorylation of proteins/enzymes is an essential mode of modulating key steps in plant growth, development, and stress adaptation. Plants employ a diverse spectrum of protein kinases and protein phosphatases. There is considerable diversity of protein kinases in plants and their distribution. In contrast, the protein phosphatases are poorly understood. It is imperative that detailed work on PPs is to be further intensified. The interest on PP2C got a boost with the path-breaking discovery of ternary complex formation by PP2C, PYL, and ABA and its function as an on-off switch of protein kinases. The downstream regulation of SnRK2s (like ABA-stimulated protein kinases) is studied at multiple levels of organization (proteins, genes) and in different organs, starting from leaf to intracellular and intra-organellar level. On the other hand, the available information on the diversity and substrate specificity of PPs is scattered and incomplete.

It is necessary to check if PPs other than PP2C act as receptors of other hormones or microbial elicitors and MAMPs. There could be also other PPs, occurring specifically in guard cells. Mutants of Arabidopsis would be a suitable platform to assess the specificity and roles of PPs in different plant organs, particularly stomatal guard cells. A critical examination of homology of the protein phosphatases from different taxons of plant kingdom may reveal clues about their evolution and function.

Acknowledgments Our work on stomatal guard cell signaling is supported by grants to A.S.R. of a JC Bose National Fellowship (No. SR/S2/JCB-06/2006) from the Department of Science and Technology and another from the Council of Scientific and Industrial Research (CSIR) (No. 38 (1404)/15/EMR-II), both in New Delhi. S.G. is supported by a Junior Research Fellowship from University Grants Commission. V.S. is supported by SERB-NPDF fellowship and BP is supported by BBL fellowship (UoH). We also thank DBT-CREBB, DST-FIST, and UGC-SAP for support of infrastructure in department/school.

References

Acharya BR, Jeon BW, Zhang W, Assmann SM (2013) Open stomata 1 (OST1) is limiting in abscisic acid responses of Arabidopsis guard cells. New Phytol 200:1049–1063

Agurla S, Raghavendra AS (2016) Convergence and divergence of signaling events in guard cells during stomatal closure by plant hormones or microbial elicitors. Front Plant Sci 7:1332

Agurla S, Gayatri G, Raghavendra AS (2017) Signal transduction components in guard cells during stomatal closure by plant hormones and microbial elicitors. In: Pandey G (ed) Mechanism of plant hormone signaling under stress, vol 2, 1st edn. Wiley, Hoboken, NJ, pp 355–387

Agurla S, Gahir S, Munemasa S, Murata Y, Raghavendra AS (2018a) Mechanism of stomatal closure in plants exposed to drought and cold stress. Adv Exp Med Biol 1081:15–232

Agurla S, Gayatri G, Raghavendra AS (2018b) Polyamines increase nitric oxide and reactive oxygen species in guard cells of Arabidopsis *thaliana* during stomatal closure. Protoplasma 255:153–162

Allen GJ, Kuchitsu K, Chu SP, Murata Y, Schroeder JI (1999) Arabidopsis *abi1-1* and *abi2-1* phosphatase mutations reduce abscisic acid-induced cytoplasmic calcium rises in guard cells. Plant Cell 11:1785–1798

Antoni R, Gonzalez-Guzman M, Rodriguez L, Rodrigues A, Pizzio GA, Rodriguez PL (2012) Selective inhibition of clade A phosphatases type 2C by PYR/PYL/RCAR abscisic acid receptors. Plant Physiol 158:970–980

Araujo WL, Fernie AR, Nunes-Nesi A (2011) Control of stomatal aperture: a renaissance of the old guard. Plant Signal Behav 6:1305–1311

Arnaud D, Hwang I (2015) A sophisticated network of signaling pathways regulates stomatal defenses to bacterial pathogens. Mol Plant 8:566–581

Assmann SM, Jegla T (2016) Guard cell sensory systems: recent insights on stomatal responses to light, abscisic acid, and CO_2. Curr Opin Plant Biol 33:157–167

Bheri M, Pandey GK (2019) PP2A take a giant leap in the post genomic era. Curr Genomics 20:154–171

Brandt B, Munemasa S, Wang C, Nguyen D, Yong T, Yang PG, Poretsky E, Belknap TF, Waadt R, Alem F, Schroeder JI (2015) Calcium specificity signaling mechanisms in abscisic acid signal transduction in Arabidopsis guard cells. eLife 4:1–25

Brock AK, Willmann R, Kolb D, Grefen L, Lajunen HM, Bethke G, Lee J, Nürnberger T, Gust AA (2010) The Arabidopsis mitogen-activated protein kinase phosphatase PP2C5 affects seed germination, stomatal aperture, and abscisic acid-inducible gene expression. Plant Physiol 153:1098–1111

Buckley TN (2019) How do stomata respond to water status? New Phytol 224:21–36

Chinchilla D, Bauer Z, Regenass M, Boller T, Felix G (2006) The Arabidopsis receptor kinase FLS2 binds flg22 and determines the specificity of flagellin perception. Plant Cell 18:465–476

Drerup MM, Schlücking K, Hashimoto K, Manishankar P, Steinhorst L, Kuchitsu K, Kudla J (2013) The Calcineurin B-like calcium sensors CBL1 and CBL9 together with their interacting protein kinase CIPK26 regulate the Arabidopsis NADPH oxidase RBOHF. Mol Plant 6:559–569

Dubovskaya LV, Bakakina YS, Kolesneva EV, Sodel DL, McAinsh MR, Hetherington AM, Volotovski ID (2011) cGMP-dependent ABA-induced stomatal closure in the ABA-insensitive Arabidopsis mutant *abi1-1*. New Phytol 191:57–69

Dupeux F, Santiago J, Betz K, Twycross J, Park SY, Rodriguez L, Gonzalez-Guzman M, Jensen MR, Krasnogor N, Blackledge M, Holdsworth M (2011) A thermodynamic switch modulates abscisic acid receptor sensitivity. EMBO J 30:4171–4184

Durian G, Rahikainen M, Alegre S, Brosché M, Kangasjärvi S (2016) Protein phosphatase 2A in the regulatory network underlying biotic stress resistance in plants. Front Plant Sci 7:812

Edel KH, Kudla J (2016) Integration of calcium and ABA signaling. Curr Opin Plant Biol 33:83–91

Eun SO, Bae SH, Lee Y (2001) Cortical actin filaments in guard cells respond differently to abscisic acid in wild-type and *abi1-1* mutant Arabidopsis. Planta 212:466–469

Farkas I, Dombrádi V, Miskei M, Szabados L, Koncz C (2007) Arabidopsis PPP family of serine/threonine phosphatases. Trends Plant Sci 12:169–176

Fuchs S, Grill E, Meskiene I, Schweighofer A (2013) Type 2C protein phosphatases in plants. FEBS J 280:681–693

Fujii H, Zhu JK (2009) Arabidopsis mutant deficient in 3 abscisic acid-activated protein kinases reveals critical roles in growth, reproduction, and stress. Proc Natl Acad Sci U S A 106:8380–8385

Gayatri G, Agurla S, Raghavendra AS (2013) Nitric oxide in guard cells as an important secondary messenger during stomatal closure. Front Plant Sci 4:1–11

Gómez-Gómez L, Bauer Z, Boller T (2001) Both the extracellular leucine-rich repeat domain and the kinase activity of FLS2 are required for flagellin binding and signaling in Arabidopsis. Plant Cell 13:1155–1163

Gonugunta VK, Srivastava N, Puli MR, Raghavendra AS (2008) Nitric oxide production occurs after cytosolic alkalinization during stomatal closure induced by abscisic acid. Plant Cell Environ 31:1717–1724

Gonzalez-Guzman M, Pizzio GA, Antoni R, Vera-Sirera F, Merilo E, Bassel GW, Rodriguez PL (2012) Arabidopsis PYR/PYL/RCAR receptors play a major role in quantitative regulation of stomatal aperture and transcriptional response to abscisic acid. Plant Cell 24:2483–2496

Gudesblat GE, Iusem ND, Morris PC (2007) Guard cell-specific inhibition of Arabidopsis MPK3 expression causes abnormal stomatal responses to abscisic acid and hydrogen peroxide. New Phytol 173:713–721

Guo D, Zhou Y, Li HL, Zhu JH, Wang Y, Chen XT, Peng SQ (2017) Identification and characterization of the abscisic acid (ABA) receptor gene family and its expression in response to hormones in the rubber tree. Sci Rep 7:45157

Hashimoto M, Negi J, Young J, Israelsson M, Schroeder JI, Iba K (2006) Arabidopsis HT1 kinase controls stomatal movements in response to CO_2. Nat Cell Biol 8:391–397

Hiyama A, Takemiya A, Munemasa S, Okuma E, Sugiyama N, Tada Y, Murata Y, Shimazaki KI (2017) Blue light and CO_2 signals converge to regulate light-induced stomatal opening. Nat Commun 8:1284

Hrabak EM, Chan CW, Gribskov M, Harper JF, Choi JH, Halford N, Thomas M (2003) The Arabidopsis CDPK-SnRK superfamily of protein kinases. Plant Physiol 132:666–680

Hua D, Wang C, He J, Liao H, Duan Y, Zhu Z, Guo Y, Chen Z, Gong Z (2012) A plasma membrane receptor kinase, GHR1, mediates abscisic acid-and hydrogen peroxide-regulated stomatal movement in Arabidopsis. Plant Cell 24:2546–2561

Hubbard KE, Nishimura N, Hitomi K, Getzoff ED, Schroeder JI (2010) Early abscisic acid signal transduction mechanisms: newly discovered components and newly emerging questions. Genes Dev 24:1695–1708

Inoue SI, Kinoshita T (2017) Blue light regulation of stomatal opening and the plasma membrane H^+-ATPase. Plant Physiol 174:531–538

Inoue SI, Takemiya A, Shimazaki KI (2010) Phototropin signaling and stomatal opening as a model case. Curr Opin Plant Biol 13:587–593

Jammes F, Song C, Shin D, Munemasa S, Takeda K, Gu D, Cho D, Lee S, Giordo R, Sritubtim S, Leonhardt N (2009) MAP kinases MPK9 and MPK12 are preferentially expressed in guard cells and positively regulate ROS-mediated ABA signaling. Proc Natl Acad Sci U S A 106:20520–20525

Kang X, Xu G, Lee B, Chen C, Zhang H, Kuang R, Ni M (2018) HRB2 and BBX21 interaction modulates Arabidopsis ABI5 locus and stomatal aperture. Plant Cell Environ 41:1912–1925

Kanno Y, Hanada A, Chiba Y, Ichikawa T, Nakazawa M, Matsui M, Koshiba T, Kamiya Y, Seo M (2012) Identification of an abscisic acid transporter by functional screening using the receptor complex as a sensor. Proc Natl Acad Sci U S A 109:9653–9658

Kepka M, Benson CL, Gonugunta VK, Nelson KM, Christmann A, Grill E, Abrams SR (2011) Action of natural abscisic acid precursors and catabolites on abscisic acid receptor complexes. Plant Physiol 157:2108–2119

Khokon MA, Salam MA, Jammes F, Ye W, Hossain MA, Uraji M, Nakamura Y, Mori IC, Kwak JM, Murata Y (2015) Two guard cell mitogen-activated protein kinases, MPK 9 and MPK 12, function in methyl jasmonate-induced stomatal closure in Arabidopsis *thaliana*. Plant Biol 17:946–952

Kinoshita T, Shimazaki KI (1997) Involvement of calyculin A-and okadaic acid-sensitive protein phosphatase in the blue light response of stomatal guard cells. Plant Cell Physiol 38:1281–1285

Kline KG, Sussman MR, Jones AM (2010) Abscisic acid receptors. Plant Physiol 154:479–482

Kollist H, Nuhkat M, Roelfsema MR (2014) Closing gaps: linking elements that control stomatal movement. New Phytol 203:44–62

Kwak JM, Moon JH, Murata Y, Kuchitsu K, Leonhardt N, DeLong A, Schroeder JI (2002) Disruption of a guard cell-expressed protein phosphatase 2A regulatory subunit, RCN1, confers abscisic acid insensitivity in Arabidopsis. Plant Cell 14:2849–2861

Leckie CP, McAinsh MR, Allen GJ, Sanders D, Hetherington AM (1998) Abscisic acid-induced stomatal closure mediated by cyclic ADP-ribose. Proc Natl Acad Sci U S A 95:15837–15842
Lee SC, Luan S (2012) ABA signal transduction at the crossroad of biotic and abiotic stress responses. Plant Cell Environ 35:53–60
Lee Y, Kim YJ, Kim MH, Kwak JM (2016) MAPK cascades in guard cell signal transduction. Front Plant Sci 7:80
Lefoulon C, Boeglin M, Moreau B, Véry AA, Szponarski W, Dauzat M, Michard E, Gaillard I, Chérel I (2016) The Arabidopsis AtPP2CA protein phosphatase inhibits the GORK K+ efflux channel and exerts a dominant suppressive effect on phosphomimetic-activating mutations. J Biol Chem 291:6521–6533
Leonhardt N, Kwak JM, Robert N, Waner D, Leonhardt G, Schroeder JI (2004) Microarray expression analyses of Arabidopsis guard cells and isolation of a recessive abscisic acid hypersensitive protein phosphatase 2C mutant. Plant Cell 16:596–615
Leung J, Merlot S, Giraudat J (1997) The Arabidopsis ABSCISIC ACID–INSENSITIVE2 (ABI2) and ABI1genes encode redundant protein phosphatases 2C involved in abscisic acid signal transduction. Plant Cell 9:759–771
Li J, Wang XQ, Watson MB, Assmann SM (2000) Regulation of abscisic acid-induced stomatal closure and anion channels by guard cell AAPK kinase. Science 287:300–303
Li AL, Zhu YF, Tan XM, Wang X, Wei B, Guo HZ, Zhang ZL, Chen XB, Zhao GY, Kong XY, Jia JZ, Mao L (2008) Evolutionary and functional study of the CDPK gene family in wheat (*Triticum aestivum* L.). Plant Mol Biol 66:429–443
Li L, Li M, Yu L, Zhou Z, Liang X, Liu Z, Cai G, Gao L, Zhang X, Wang Y, Chen S (2014) The FLS2-associated kinase BIK1 directly phosphorylates the NADPH oxidase RbohD to control plant immunity. Cell Host Microbe 15:329–338
Liang S, Lu K, Wu Z, Jiang SC, Yu YT, Bi C, Zhang DP (2015) A link between magnesium-chelatase H subunit and sucrose nonfermenting 1 (SNF1)-related protein kinase SnRK2. 6/OST1 in Arabidopsis guard cell signaling in response to abscisic acid. J Exp Bot 66:6355–6369
Lijun X, Yuyan AN, Liangju W (2018) The role of microtubule skeleton and PP1/PP2A protein phosphatase in ALA-ABA regulating stomatal movement in apple leaves. Acta Hort Sin 45:2073–2088
Lim CW, Kim JH, Baek W, Kim BS, Lee SC (2012) Functional roles of the protein phosphatase 2C, AtAIP1, in abscisic acid signaling and sugar tolerance in Arabidopsis. Plant Sci 187:83–88
Lim C, Baek W, Jung J, Kim JH, Lee S (2015) Function of ABA in stomatal defense against biotic and drought stresses. Int J Mol Sci 16:15251–15270
Luan S (2002) Tyrosine phosphorylation in plant cell signaling. Proc Natl Acad Sci U S A 9:11567–11569
Luan S (2003) Protein phosphatases in plants. Annu Rev Plant Biol 54:63–92
Ma Y, Szostkiewicz I, Korte A, Moes D, Yang Y, Christman A, Grill E (2009) Regulators of PP2C phosphatase activity function as abscisic acid sensors. Science 324:1064–1068
Ma P, Liu J, Yang X, Ma R (2013) Genome-wide identification of the maize calcium-dependent protein kinase gene family. Appl Biochem Biotechnol 169:2111–2125
MacRobbie EA (2002) Evidence for a role for protein tyrosine phosphatase in the control of ion release from the guard cell vacuole in stomatal closure. Proc Natl Acad Sci U S A 99:11963–11968
Manohar M, Wang D, Manosalva PM, Choi HW, Kombrink E, Klessig DF (2017) Members of the abscisic acid co receptor PP2C protein family mediate salicylic acid–abscisic acid crosstalk. Plant Direct 1:14
Medeiros DB, Luz LM, Oliveira HO, Araújo WL, Daloso DM, Fernie AR (2019) Metabolomics for understanding stomatal movements. Theor Exp Plant Physiol 31:91–102
Melcher K, Ng LM, Zhou XE, Soon FF, Xu Y, Suino-Powell KM, Park SY, Weiner JJ, Fujii H, Chinnusamy V, Kovach A (2009) A gate–latch–lock mechanism for hormone signaling by abscisic acid receptors. Nature 462:602

Melcher K, Xu Y, Ng LM, Zhou XE, Soon FF, Chinnusamy V, Suino-Powell KM, Kovach A, Tham FS, Cutler SR, Li J (2010a) Identification and mechanism of ABA receptor antagonism. Nat Struct Mol Biol 17:1102

Melcher K, Zhou XE, Xu HE (2010b) Thirsty plants and beyond: structural mechanisms of abscisic acid perception and signaling. Curr Opin Struct Biol 20:722–729

Melotto M, Zhang L, Oblessuc PR, He SY (2017) Stomatal defense a decade later. Plant Physiol 174:561–571

Merlot S, Gosti F, Guerrier D, Vavasseur A, Giraudat J (2001) The ABI1 and ABI2 protein phosphatases 2C act in a negative feedback regulatory loop of the abscisic acid signaling pathway. Plant J 25:295–303

Merlot S, Mustilli AC, Genty B, North H, Lefebvre V, Sotta B, Vavasseur A, Giraudat J (2002) Use of infrared thermal imaging to isolate Arabidopsis mutants defective in stomatal regulation. Plant J 30:601–609

Meyer K, Leube MP, Grill E (1994) A protein phosphatase 2C involved in ABA signal transduction in Arabidopsis *thaliana*. Science 264:1452–1455

Miyakawa T, Tanokura M (2011) Regulatory mechanism of abscisic acid signaling. Biophysics 7:123–128

Miyazono KI, Miyakawa T, Sawano Y, Kubota K, Kang HJ, Asano A, Miyauchi Y, Takahashi M, Zhi Y, Fujita Y, Yoshida T (2009) Structural basis of abscisic acid signaling. Nature 462:609

Mori IC, Murata Y, Yang Y, Munemasa S, Wang YF, Andreoli S, Tiriac H, Alonso JM, Harper JF, Ecker JR, Kwak JM (2006) CDPKs CPK6 and CPK3 function in ABA regulation of guard cell S-type anion-and Ca^{2+}-permeable channels and stomatal closure. PLoS Biol 4:e327

Munemasa S, Hauser F, Park J, Waadt R, Brandt B, Schroeder JI (2015) Mechanisms of abscisic acid-mediated control of stomatal aperture. Curr Opin Plant Biol 28:154–162

Murata Y, Pei ZM, Mori IC, Schroeder J (2001) Abscisic acid activation of plasma membrane Ca2+ channels in guard cells requires cytosolic NAD(P)H and is differentially disrupted upstream and downstream of reactive oxygen species production in *abi1-1* and *abi2-1* protein phosphatase 2C mutants. Plant Cell 13:2513–2523

Murata Y, Mori IC, Munemasa S (2015) Diverse stomatal signaling and the signal integration mechanism. Annu Rev Plant Biol 66:369–392

Mustilli AC, Merlot S, Vavasseur A, Fenzi F, Giraudat J (2002) Arabidopsis OST1 protein kinase mediates the regulation of stomatal aperture by abscisic acid and acts upstream of reactive oxygen species production. Plant Cell 14:3089–3099

Nakashima K, Yamaguchi-Shinozaki K (2013) ABA signaling in stress-response and seed development. Plant Cell Rep 32:959–970

Nakashima K, Fujita Y, Kanamori N, Katagiri T, Umezawa T, Kidokoro S, Maruyama K, Yoshida T, Ishiyama K, Kobayashi M, Shinozaki K (2009) Three Arabidopsis SnRK2 protein kinases, SRK2D/SnRK2.2, SRK2E/SnRK2.6/OST1 and SRK2I/SnRK2.3, involved in ABA signaling are essential for the control of seed development and dormancy. Plant Cell Physiol 50:1345–1363

Nieves-Cordones M, Caballero F, Martínez V, Rubio F (2011) Disruption of the Arabidopsis *thaliana* inward-rectifier K^+ channel AKT1 improves plant responses to water stress. Plant Cell Physiol 53:423–432

Nishimura N, Sarkeshik A, Nito K, Park SY, Wang A, Carvalho PC, Lee S, Caddell DF, Cutler SR, Chory J, Yates JR (2010) PYR/PYL/RCAR family members are major *in-vivo* ABI1 protein phosphatase 2C-interacting proteins in Arabidopsis. Plant J 61:290–299

Orton PJ, Mansfield TA (1974) The activity of abscisic acid analogues as inhibitors of stomatal opening. Planta 121:263–272

País SM, Téllez-Iñón MT, Capiati DA (2009) Serine/threonine protein phosphatases type 2A and their roles in stress signaling. Plant Signal Behav 4:1013–1015

Puli MR, Raghavendra AS (2012) Pyrabactin, an ABA agonist, induced stomatal closure and changes in signaling components of guard cells in abaxial epidermis of *Pisum sativum*. J Exp Bot 63:1349–1356

Qin XM, Bian MD, Yang ZM, Shi WL (2015) Tyrosine phosphorylation mediates starch metabolism in guard cell of *Vicia faba*. Biologia 70:574–580
Raghavendra AS, Murata Y (2017) Editorial: signal transduction in stomatal guard cells. Front Plant Sci 7:114
Raghavendra AS, Gonugunta VK, Christmann A, Grill E (2010) ABA perception and signaling. Trends Plant Sci 15:395–401
Rasmussen MW, Roux M, Petersen M, Mundy J (2012) MAP kinase cascades in Arabidopsis innate immunity. Front Plant Sci 3:169
Rodriguez PL (1998) Protein phosphatase 2C (PP2C) function in higher plants. Plant Mol Biol 38:919–927
Rodriguez PL, Leube MP, Grill E (1998) Molecular cloning in Arabidopsis *thaliana* of a new protein phosphatase 2C (PP2C) with homology to ABI1 and ABI2. Plant Mol Biol 38:879–883
Rodriguez L, Gonzalez-Guzman M, Diaz M, Rodrigues A, Izquierdo-Garcia AC, Peirats-Llobet M, Mulet JM (2014) C2-domain abscisic acid-related proteins mediate the interaction of PYR/PYL/RCAR abscisic acid receptors with the plasma membrane and regulate abscisic acid sensitivity in Arabidopsis. Plant Cell 26:4802
Rubio S, Rodrigues A, Saez A, Dizon MB, Galle A, Kim TH, Santiago J, Flexas J, Schroeder JI, Rodriguez PL (2009) Triple loss of function of protein phosphatases type 2C leads to partial constitutive response to endogenous abscisic acid. Plant Physiol 150:1345–1355
Saez A, Apostolova N, Gonzalez-Guzman M, Gonzalez-Garcia MP, Nicolas C, Lorenzo O, Rodriguez PL (2004) Gain-of-function and loss-of-function phenotypes of the protein phosphatase 2C HAB1 reveal its role as a negative regulator of abscisic acid signaling. Plant J 37:354–369
Saez A, Robert N, Maktabi MH, Schroeder JI, Serrano R, Rodriguez PL (2006) Enhancement of abscisic acid sensitivity and reduction of water consumption in Arabidopsis by combined inactivation of the protein phosphatases type 2C ABI1 and HAB1. Plant Physiol 141:1389–1399
Saito N, Munemasa S, Nakamura Y, Shimoishi Y, Mori IC, Murata Y (2008) Roles of RCN1, regulatory A subunit of protein phosphatase 2A, in methyl jasmonate signaling and signal crosstalk between methyl jasmonate and abscisic acid. Plant Cell Physiol 49:1396–1401
Schweighofer A, Meskiene I (2015) Phosphatases in plants. Methods Mol Biol 1306:25–46
Seyfferth C, Tsuda K (2014) Salicylic acid signal transduction: the initiation of biosynthesis, perception and transcriptional reprogramming. Front Plant Sci 5:697
Shang Y, Dai C, Lee MM, Kwak JM, Nam KH (2016) BRI1-associated receptor kinase 1 regulates guard cell ABA signaling mediated by open stomata 1 in Arabidopsis. Mol Plant 9:447–460
Sheen J (1998) Mutational analysis of protein phosphatase 2C involved in abscisic acid signal transduction in higher plants. Proc Natl Acad Sci U S A 95:975–980
Shi Y (2009) Serine/threonine phosphatases: mechanism through structure. Cell 139:468–484
Shi W, Jia W, Liu X, Zhang S (2004) Protein tyrosine phosphatases involved in signaling of the ABA-induced H_2O_2 generation in guard cells of *Vicia faba* L. Chin Sci Bull 49:1841–1846
Shimazaki KI, Doi M, Assmann SM, Kinoshita T (2007) Light regulation of stomatal movement. Annu Rev Plant Biol 58:219–247
Singh A, Pandey A, Srivastava AK, Tran LS, Pandey GK (2016) Plant protein phosphatases 2C: from genomic diversity to functional multiplicity and importance in stress management. Crit Rev Biotechnol 36:1023–1035
Singh R, Parihar P, Singh S, Mishra RK, Singh VP, Prasad SM (2017) Reactive oxygen species signaling and stomatal movement: current updates and future perspectives. Redox Biol 11:213–218
Su J, Zhang M, Zhang L, Sun T, Liu Y, Lukowitz W, Xu J, Zhang S (2017) Regulation of stomatal immunity by interdependent functions of a pathogen-responsive MPK3/MPK6 cascade and abscisic acid. Plant Cell 29:526–542
Sun X, Kang X, Ni M (2012) Hypersensitive to red and blue 1 and its modification by protein phosphatase 7 are implicated in the control of Arabidopsis stomatal aperture. PLoS Genet 8:e1002674

Sun L, Pehlivan N, Esmaeili N, Jiang W, Yang X, Jarrett P, Mishra N, Zhu X, Cai Y, Herath M, Shen G, Zhang H (2018) Co-overexpression of AVP1 and PP2A-C5 in Arabidopsis makes plants tolerant to multiple abiotic stresses. Plant Sci 274:271–283
Takemiya A, Shimazaki KI (2010) Phosphatidic acid inhibits blue light-induced stomatal opening via inhibition of protein phosphatase 1. Plant Physiol 153:1555–1562
Takemiya A, Kinoshita T, Asanuma M, Shimazaki K (2006) Protein phosphatase 1 positively regulates stomatal opening in response to blue light in *Vicia faba*. Proc Natl Acad Sci U S A 103:13549–13554
Takemiya A, Yamauchi S, Yano T, Ariyoshi C, Shimazaki KI (2012) Identification of a regulatory subunit of protein phosphatase 1 which mediates blue light signaling for stomatal opening. Plant Cell Physiol 54:24–35
Takemiya A, Doi A, Yoshida S, Okajima K, Tokutomi S, Shimazaki KI (2015) Reconstitution of an initial step of phototropin signaling in stomatal guard cells. Plant Cell Physiol 57:152–159
Takeuchi J, Okamoto M, Akiyama T, Muto T, Yajima S, Sue M, Seo M, Kanno Y, Kamo T, Endo A, Nambara E (2014) Designed abscisic acid analogs as antagonists of PYL-PP2C receptor interactions. Nat Chem Biol 10:477
Toh S, Inoue S, Toda Y, Yuki T, Suzuki K, Hamamoto S, Fukatsu K, Aoki S, Uchida M, Asai E, Uozumi N (2018) Identification and characterization of compounds that affect stomatal movements. Plant Cell Physiol 59:1568–1580
Tseng TS, Briggs WR (2010) The Arabidopsis *rcn1-1* mutation impairs dephosphorylation of Phot2, resulting in enhanced blue light responses. Plant Cell 22:392–402
Uhrig RG, Labandera AM, Moorhead GB (2013) Arabidopsis PPP family of serine/threonine protein phosphatases: many targets but few engines. Trends Plant Sci 18:505–513
Umbrasaite J, Schweighofer A, Kazanaviciute V, Magyar Z, Ayatollahi Z, Unterwurzacher V, Choopayak C, Boniecka J, Murray JA, Bogre L, Meskiene I (2010) MAPK phosphatase AP2C3 induces ectopic proliferation of epidermal cells leading to stomata development in Arabidopsis. PLoS One 5:e15357
Umezawa T, Nakashima K, Miyakawa T, Kuromori T, Tanokura M, Shinozaki K, Yamaguchi-Shinozaki K (2010) Molecular basis of the core regulatory network in ABA responses: sensing, signaling and transport. Plant Cell Physiol 51:1821–1839
Waadt R, Manalansan B, Rauniyar N, Munemasa S, Booker MA, Brandt B, Waadt C, Nusinow DA, Kay SA, Kunz HH, Schumacher K, DeLong A, Yates JR, Schroeder JI (2015) Identification of open stomata1-interacting proteins reveals interactions with sucrose non-fermenting1-related protein kinases2 and with type 2A protein phosphatases that function in abscisic acid responses. Plant Physiol 169:760–779
Weaver GM, Iersel MW (2014) Antitranspirational efficacy and longevity of abscisic acid and synthetic abscisic acid analog in pansies (*Viola X wittrockiana*). Hort Sci 49:779–784
Wu Y, Sanchez JP, Lopez-Molina L, Himmelbach A, Grill E, Chua NH (2003) The abi1-1 mutation blocks ABA signaling downstream of cADPR action. Plant J 34:307–315
Wu Y, Zhang D, Chu JY, Boyle P, Wang Y, Brindle ID, De Luca V, Després C (2012) The Arabidopsis NPR1 protein is a receptor for the plant defense hormone salicylic acid. Cell Rep 1:639–647
Yamamoto H, Oritani T (1995) Chiral synthesis and stomatal closure activities of γ-pseudo-and dihydro-abscisic acids. Phytochemistry 40:1033–1039
Yamazaki D, Yoshida S, Asami T, Kuchitsu K (2003) Visualization of abscisic acid-perception sites on the plasma membrane of stomatal guard cells. Plant J 35:129–139
Ye W, Murata Y (2016) Microbe associated molecular pattern signaling in guard cells. Front Plant Sci 7:583
Yoon JT, Ahn HK, Pai HS (2018) The subfamily II catalytic subunits of protein phosphatase 2A (PP2A) are involved in cortical microtubule organization. Planta 248:1551–1567
Yoshida R, Umezawa T, Mizoguchi T, Takahashi S, Takahashi F, Shinozaki K (2006) The regulatory domain of SRK2E/OST1/SnRK2. 6 interacts with ABI1 and integrates abscisic acid (ABA) and osmotic stress signals controlling stomatal closure in Arabidopsis. J Biol Chem 281:5310–5318

Yoshida T, Fujita Y, Sayama H, Kidokoro S, Maruyama K, Mizoi J, Shinozaki K, Yamaguchi-Shinozaki K (2010) AREB1, AREB2, and ABF3 are master transcription factors that cooperatively regulate ABRE-dependent ABA signaling involved in drought stress tolerance and require ABA for full activation. Plant J 61:672–685

Zhang J, Li X, He Z, Zhao X, Wang Q, Zhou B, Yu D, Huang X, Tang D, Guo X, Liu X (2013) Molecular character of a phosphatase 2C (PP2C) gene relation to stress tolerance in Arabidopsis *thaliana*. Mol Biol Rep 40:2633–2644

Zhang T, Chen S, Harmon AC (2014) Protein phosphorylation in stomatal movement. Plant Signal Behav 9:e972845

Zhang XL, Jiang L, Xin Q, Liu Y, Tan JX, Chen ZZ (2015) Structural basis and functions of abscisic acid receptors PYLs. Front Plant Sci 19:88

Zoulias N, Harrison EL, Casson SA, Gray JE (2018) Molecular control of stomatal development. Biochem J 475:441–454

Chapter 9
Deciphering the Roles of Protein Phosphatases in the Regulation of Salt-Induced Signaling Responses in Plants

Aditya Banerjee and Aryadeep Roychoudhury

9.1 Introduction

Salinity is one of the most prevalent kinds of abiotic stress which drastically deteriorates crop productivity and negatively affects the overall growth physiology. It has been reported that salt stress reduces the mean yield of susceptible plant species by about 50% (Wang et al. 2003; Banerjee and Roychoudhury 2018a, b). This is a serious problem because large stretches of land used for agricultural pursuits are largely inflicted with high salt concentration. As a result, the conventional crop production programs experience a tremendous pressure in order to satiate the hunger of the ever-growing population (Banerjee et al. 2018, 2019). Transgenic approaches have hence been undertaken on experimental basis to improve crop yield under suboptimal conditions including salt stress. In order to achieve significant success in this field, proper understanding of the plant metabolome is extremely crucial. It is quite well known that phosphorylation at Ser, Thr, or Tyr residues is largely responsible for mediating a major part of the plant signalosome (Roychoudhury and Banerjee 2017). Such phosphorylations are catalyzed by specialized enzymes known as the kinases. However, the removal of the phosphoryl group is also essential for terminating the signaling pathway/cascade or even as a separate posttranslational modification (Schweighofer and Meskiene 2015). The reversal of protein phosphorylation is mediated by a group of enzymes known as the phosphatases. These enzymes were largely considered to be housekeeping enzymes; however, recently, their specific and crucial roles have been undermined during important cellular and physiological processes in plants (Schweighofer and Meskiene 2015). In this chapter, we have briefly highlighted the general roles of plant phosphatases and their involvement in salt-induced responses.

A. Banerjee · A. Roychoudhury (✉)
Post Graduate Department of Biotechnology, St. Xavier's College (Autonomous), Kolkata, West Bengal, India

G. K. Pandey (ed.), *Protein Phosphatases and Stress Management in Plants*,
https://doi.org/10.1007/978-3-030-48733-1_9

9.2 Phosphatases: A Brief Outlook

Compared to the plant system, protein phosphatases have been better characterized in animal models with respect to their structure and function. In plants, phosphatases cleaving the phosphoryl group at Ser and Thr residues have been broadly classified under two classes: type 1 (PP-1) and type 2 (PP-2A, 2B, and 2C) depending on the specificity to substrates, inhibitors, and ionic requirements (Brautigan 2013).

9.2.1 PP-1

This group of phosphatase has been partially purified from *Brassica napus*, *Triticum aestivum*, *Zea mays*, Arabidopsis *thaliana*, etc. (Sopory and Munshi 1998). A 35 kDa phosphatase isolated from wheat specifically acted upon phosphotyrosines instead of phosphoserines or phosphothreonines (Li et al. 2007). It was also reported that PP-1 from *Pisum sativum* was localized in the cytoplasm, whereas that in wheat was associated with the microsomes (Gonzalez Besteiro and Ulm 2013). Another phosphotyrosine-specific phosphatase was isolated from the nuclei of *Pisum sativum* plants which exhibited its activity irrespective of Ca^{2+}, Mg^{2+}, or Mn^{2+}. The activity was in turn stimulated in the presence of divalent cation chelators like EDTA and EGTA (Guo and Roux 1995). This clearly illustrated the presence of phosphotyrosyl phosphatases (PTPases) in plant systems similar to the mammalian systems.

9.2.2 PP-2

This class of enzymes is further subdivided into smaller groups based on the cationic requirement of the proteins for executing optimal enzymatic activity. PP-2A does not require any cation for its functioning. However, Ca^{2+}/calmodulin and Mg^{2+} are necessary for PP-2B and 2C, respectively (Fuchs et al. 2013).

PP-2A has been reported to exist as a holoenzyme. The C-subunit (36 kDa) is the catalytic structure complexed with the regulatory A-subunit (65 kDa). The specific properties of the phosphatase are determined by the variable B-subunit with associates with the core dimer (Fuchs et al. 2013). PP-2A enzymes have been isolated across several plant species from various subcellular fractions, nuclei, insoluble fractions, and even plasma membranes (Uhrig et al. 2013). Experimental evidences regarding the participation of PP-2A proteins in regulating the activity of other enzymes like quinate dehydrogenase, sucrose-phosphate synthase, phosphoenolpyruvate carboxylase, and nitrate reductase have been established (Sopory and Munshi 1998).

The PP-2B class of phosphatases contain a 19 kDa subunit which exhibits strong affinity for Ca^{2+}. Hence, the activity of these enzymes is stimulated in the presence of Ca^{2+}/calmodulin (DeLong 2006). This class of phosphatases has very restricted substrate specificity and is often themselves phosphorylated by cAMP-dependent protein kinases (PKAs). The enzymes are involved in triggering Ca^{2+}-mediated secondary signaling. Allen and Sanders (1995) reported the roles of PP-2B protein, calcineurin, in regulating the Ca^{2+}-permeable slow vacuolar ion channel in the stomatal guard cells of broad bean plants.

The PP-2C enzymes dephosphorylate at phosphoserine and phosphothreonine residues. The protein also exhibits strong affinity toward phosphocasein. A PP-2C enzyme isolated from *Daucus carota* was able to dephosphorylate the α-subunit present in phosphorylase kinases (Sopory and Munshi 1998). Kinase-associated protein phosphatase (KAPP) also belongs to the PP-2C subfamily and binds to the phosphorylated receptor Ser-Thr kinase (RLK5) (Park et al. 2012). Alongside a typical PP-2C catalytic domain, this phosphatase contains a kinase interaction (KI) domain and also an N-terminal signal anchor. A number of positively charged residues follow the N-terminal anchor. These residues act as a type I signal anchor in KAPP (Manabe et al. 2008). PP-2Cs also fall under the PP-M subfamily of phosphatases. The PP-2C class of phosphatases is widely considered as the negative regulator of abscisic acid (ABA, the major stress hormone) signaling pathway (Merlot et al. 2001; Zhang et al. 2013). Singh et al. (2015) reported that overexpression of group A PP-2C, *OsPP108*, in rice resulted in enhanced salt tolerance phenotype in the transgenics. This was generally because the group A PP-2Cs usually regulate the activity of SnRK2 group of kinases, viz., SnRK2.2, SnRK2.3, and SnRK2.6, and positively regulate ABA-dependent signaling which essentially dictates salt tolerance (Danquah et al. 2014).

9.2.3 PP-4 to PP-7

Close structural and phylogenetic resemblance has been observed among PP-2A, PP-4/PPX, and PP-6 (Uhrig et al. 2013). Cloning of *PPX-1* and *PPX-2* genes in Arabidopsis revealed their differential expression in almost all organisms (Farkas et al. 2007). Unlike in animals, PP-5 is less characterized in plants. The *PP-5* gene undergoes variable pre-mRNA splicing in tomato and Arabidopsis plants, thus indicating at a genus-specific variation in the alternative splicing pattern (Van Bentem et al. 2003). Park et al. (2011) reported heat susceptibility in *pp-5* T-DNA insertion mutants of Arabidopsis. Farkas et al. (2007) inferred the intricate interactions of PP-6 with phytochrome complexes to regulate flowering. The PP-6 of pea seedlings was observed to bind with phosphorylated phytochrome A (Phy A) and far red derivatives of Phy A and Phy B (Farkas et al. 2007). The regulation of flowering was mediated via the functional interactions between phytochrome kinase and phosphatase (Kim et al. 2002). The PP-7 subfamily is unique among the other subfamilies in the sense that it lacks the EF-hand motifs (which binds to Ca^{2+}) and also the

N- and C-terminal adjuncts (Uhrig et al. 2013). However, PP-7 activity is still regulated in the presence of Ca^{2+} since it binds to calmodulin moieties in a Ca^{2+}-dependent fashion and also mediates flowering by regulating the functions of phytochromes and cryptochromes (Uhrig et al. 2013).

9.2.4 Protein Tyrosine Phosphatases (PTPs)

The PTPs are specific group of proteins which dephosphorylate usually at the tyrosine residues. Among the PTPs, the tyrosine-specific PTPs specifically dephosphorylate tyrosine residues, whereas the dual-specificity PTPs can act upon phosphotyrosine, phosphoserine, and also phosphothreonine residues (Chae et al. 2009). These two groups of proteins have retained close similarity in their respective crystal structures. Based on localization of the proteins, the PTPs consist of two groups, viz. (1) the receptor-like PTPs which are extracellular and contain a ligand-binding domain and (2) the cytoplasmic PTPs (PTP1B) possessing a catalytic domain and extensions at the 5′ and 3′ termini (Chae et al. 2009). PTPs have been widely known to regulate the elaborate signal transduction processes in protozoans, animals and plants as well (Singh et al. 2010). PTPs have been identified in Arabidopsis, *Pinus*, tomato, and rice plants (Shankar et al. 2015).

9.2.5 Phosphatases Which Regulate Inositol Signaling

Inositol phosphatases are an emerging group of enzymes which have been intricately tagged with abiotic stress responses (Vollmer et al. 2011). Several phosphatases regulate the inositol phosphate (IP) and phosphoinositide (PI) pathways. The polyphosphate-5-phosphatases (5PTases) consist of the largest family of inositol phosphatases comprising of 15 members in Arabidopsis, 21 in rice, and 39 in soybean (Zhang et al. 2019). Based on their substrate specificity, these have been subdivided into four groups which together are believed to terminate the inositol (1,4,5) P_3 pathway and regulate ABA signaling and Ca^{2+} release (Burnette 2003). The suppressor of actin (SAC) phosphatases is polyphosphoinositide phosphatases containing the catalytic SAC domain with limited knowledge regarding their substrate specificity. In Arabidopsis, almost all SACs are expressed ubiquitously except SAC6 which is localized only in the floral parts (Jia et al. 2019). The SAL1 phosphatase/FIERY1 (FRY1) and its homologues, inositol monophosphatase (IMP), and phosphatase and tensin homologue deleted on chromosome 10 (PTEN)-related phosphatases also behave as bifunctional enzymes regulating the inositol-mediated signaling (Jia et al. 2019). The general roles of the inositol pathway and the substrate specificity of the phosphatases operative in this pathway have been highlighted in Fig. 9.1a, b (extracted from Jia et al. 2019).

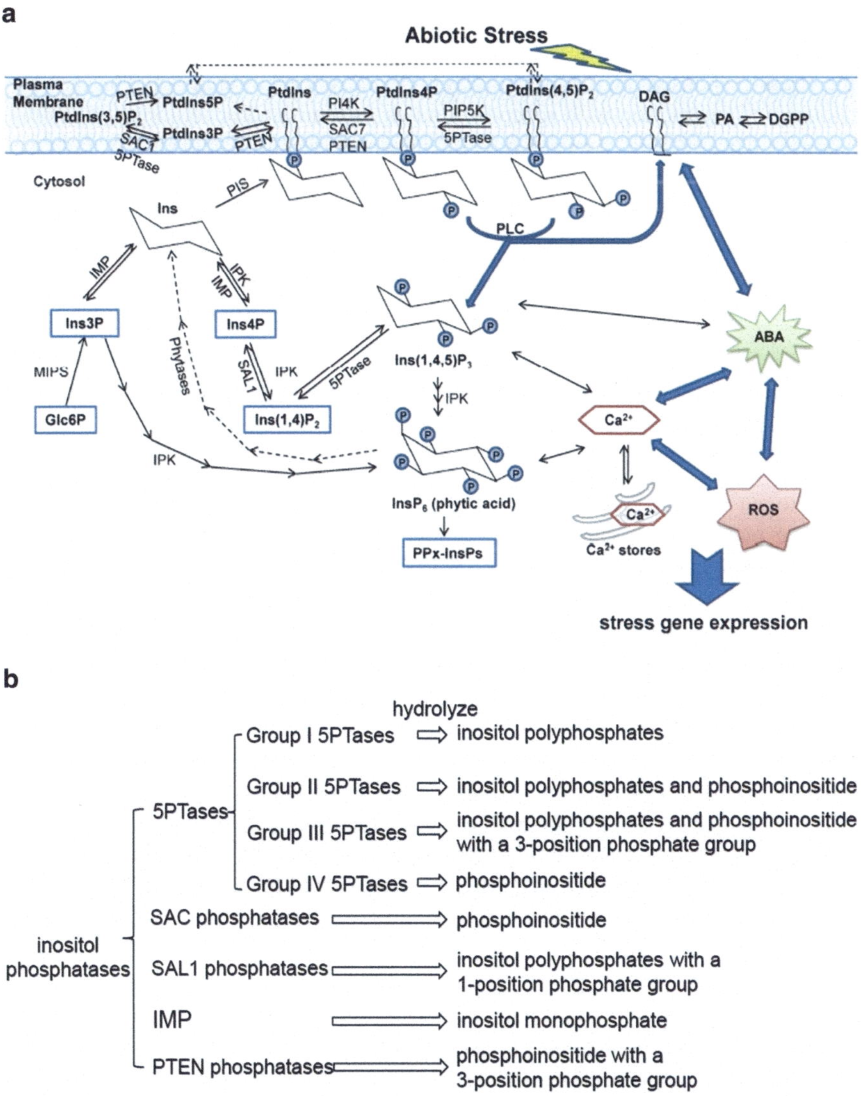

Fig. 9.1 The roles of inositol phosphatases during stress signaling in plants (**a**). *ABA* abscisic acid, *DGPP* diacylglycerol pyrophosphate, *Glc6P* glucose-6-phosphate, *IMP* inositol monophosphatase, *IPK* inositol polyphosphate multi-kinase, *MIPS* myo-inositol-3-phosphate synthase, *P* phosphate, *PIP5K* PtIns4P 5-kinase, *PI4K* phosphatidylinositol 4-kinase, *PIS* phosphatidylinositol synthase, *PKC* protein kinase C, *PLC* phospholipase C, *PPx-InsPs* pyrophosphates, *PTEN* phosphatase and tensin homologue deleted on chromosome 10, *PtdIns* phosphatidylinositol, *ROS* reactive oxygen species, *SAC* suppressor of actin, *5PTases* inositol polyphosphate 5-phosphatases; the related substrates of the inositol phosphatases (**b**). (The figures have been extracted from Jia et al. 2019)

The family of protein phosphatases in plants is variable in their structural assembly and functional characteristics. This enables them to largely regulate multiple stress-responsive signaling pathways. Dephosphorylation at Ser, Thr, and Tyr residues and interaction with receptor kinases mediate and/or terminate signaling responses. These actions forwarded by the phosphatases ultimately signify the overall survival and health of the plants considering the physiological perspectives. The roles of these crucial enzymes are less documented during salinity stress, and in the next section, we shall focus on the recent and more significant roles of protein phosphatases in regulating signaling responses induced by high soil salinity.

9.3 The Roles of Phosphatases in Regulating Salt Stress in Plants

The mitogen-activated protein kinase (MAPK) cascades importantly dictate several signaling pathways in plants (Roychoudhury and Banerjee 2017). The mitogen-activated protein kinase phosphatases (MKPs) negatively regulate such signaling cascades, thus maintaining the cellular metabolite homeostasis. The effects of transgenic plants overexpressing or showing the modulation of various representative phosphatase-encoding genes have been represented in Table 9.1. Zaidi et al. (2016) identified MKP1 in durum wheat. Overexpression of the gene encoding this phosphatase resulted in improved salt tolerance phenotype in transformed yeast cells. Arabidopsis seedlings overexpressing *MKP1* from durum wheat also exhibited increased germination rate compared to the wild-type plants. The transgenics were tolerant to salt stress due to increased activity of antioxidant enzymes like superoxide dismutase (SOD), catalase (CAT), and peroxidases (POX) (Zaidi et al. 2016). Ghorbel et al. (2019) showed that MKP1 interacted with calmodulin, 14-3-3, and MAPK3/6 proteins in durum wheat. It was observed that the differential association of MKP1 with its substrates could be responsible for regulating the cellular responses to multiple abiotic stresses (Ghorbel et al. 2019). Overexpression of *PP2A-C5* increased salt and drought tolerance in the transgenic Arabidopsis plants by stimulating the overall salt-responsive pathways (Sun et al. 2018). Bradai et al. (2018) reported induced expression of *PP1a* in *Triticum durum* exposed to salt stress. This suggests a potential role of this protein in regulating salt stress responses in wheat. Loss of function of *calcineurin B-like protein 10* (*CBL10*) in *Solanum lycopersicum* distorted the structure of the shoot apex and reproductive organs during salt stress (Egea et al. 2018). It was found that CBL10 effectively maintained an optimum and low Na^+/Ca^{2+} ratio in developing organs during salt stress. The tomato mutants of *cbl10* exhibited impaired Ca^{2+} transport via abnormal expression of *cation exchanger 1* and *two-pore channel 1* (*TPC1*). The tomato CBL10 was also found to be a close homologue of the Arabidopsis CBL10. This protein effectively regulated Na^+ and Ca^{2+} fluxes within the vacuole via cooperation with vacuolar cation

Table 9.1 List of genes encoding phosphatases which have been modulated to generate salt-tolerant phenotype in plants

Phosphatase gene	Mechanism of salt tolerance	Target organism	Reference
MKP1	Overexpression of the gene enhanced germination under stress	Arabidopsis	Zaidi et al. (2016)
OsPP108	Overexpression of the gene led to ABA insensitivity and enhanced tolerance under salt, mannitol, and drought stresses	Arabidopsis	Singh et al. (2015)
PP2A-C5	Overexpression stimulated the overall salt-responsive pathways	Arabidopsis	Sun et al. (2018)
PP1a	The gene was induced in response to salt stress in the wild-type plants	*Triticum durum*	Bradai et al. (2018)
CBL10	Mutation of the gene led to impaired shoot apex and reproductive development due to impaired Na^+/Ca^{2+} ratio	*Solanum lycopersicum*	Egea et al. (2018)
PP-2C and *ABI1*	Negative interaction between PP-2C and ABI1 regulated SnRK2.4 function in response to salinity	Arabidopsis	Krzywinska et al. (2016)
ABI2	Interacts with SOS2 to activate the SOS pathway	Arabidopsis	Ohta et al. (2003)
PP-2C1	Overexpression induced salt tolerance	*Nicotiana tabacum*	Hu et al. (2015)
AtPP-2C group 1 (AtPP-2CG1)	Overexpression of the gene led to salt tolerance by activating the ABA-dependent signaling	Arabidopsis	Liu et al. (2012)
ZmPP-2C	Constitutive expression of the gene resulted in lowered salt tolerance since the transgenics were hyposensitive to ABA	Arabidopsis	Liu et al. (2009)
At5PTase7 and *At5PTase9*	Overexpression promoted salt tolerance	Arabidopsis	Kaye et al. (2011); Golani et al. (2013)
GmSAL1	Ectopic expression promoted salt tolerance	*Nicotiana tabacum* BY-2 cells	Ku et al. (2013)

channel TPC1 and the vacuolar proton pumps, viz., high-affinity K^+ transporter 1;2 (AVP1) and V-ATPase (VHA-A1) in salt-stressed tomato plants (Egea et al. 2018).

Wu et al. (2017) performed de novo assembly and analysis of the *Fagopyrum tataricum* transcriptome to identify phosphatases involved in salt adaptation. In another study, Chen et al. (2018) analyzed the expression profile of PP2C clade D proteins in wild soybean (*Glycine soja*) and Arabidopsis seedlings exposed to salt and alkali stresses. Thirteen PP2C orthologs were detected in the wild soybean genome, whereas several PP2C-encoding genes were found to be responsive to salt stress in both the plant species. The research established a scaffold for future functional studies on the genes belonging to PP2C clade D (Chen et al. 2018). In a recent report, a novel family of ABA-induced transcription repressors (AITRs) involved in feedback response of ABA signaling was found to negatively regulate the expression of selective *PP2C* genes. However, the *aitr* mutants of Arabidopsis exhibited ABA hyposensitivity along with increased susceptibility to salt and drought (Tian et al. 2017). Creighton et al. (2017) highlighted the importance of posttranslational modification on PP2A proteins during salt stress. It was shown that the enzymes LCMT1 and PME1 catalyze the methylation and demethylation of PP2A catalytic subunit (PP2A-c), respectively. Interestingly salt stress was found to induce the demethylation of PP2A-c. The physiological significance of such modifications is still enigmatic (Creighton et al. 2017).

Han et al. (2017) suggested the involvement of ABCB transporters, PP-2A, and auxin metabolism during halotropic growth in model plants. Overexpression of the catalytic subunit 5 of PP-2A (*PP2A-c5*) improved salt tolerance in the transgenic Arabidopsis plants (Hu et al. 2017). It was reported that the PP-2Ac5 did function similar to the Salt Overly Sensitive (SOS) proteins. The association of PP-2Ac5 with multiple membrane-bound chloride channels was observed. It was inferred that these membrane channels might act as substrates of PP-2Ac5. As a result, the *PP-2Ac5*-overexpressing plants could increase the channelization of Na^+ and Cl^- ions to the vacuoles, thereby promoting salt tolerance (Hu et al. 2017). Salt stress-induced methylation was detected in Ser/Thr protein phosphatase encoding genes in foxtail millet via methylation-sensitive amplified polymorphism (MSAP) (Pandey et al. 2017). This indicates at salinity-induced epigenetic alterations in phosphatase-encoding genes. This phenomenon is less characterized.

The myristoylated Ca^{2+}-binding protein, SOS3, acts as the cellular Ca^{2+} sensor during salt stress and activates SOS2 which functions as a Ser/Thr protein kinase (belonging to the SnRK3 family) (Hrabak et al. 2003). In Arabidopsis, the calcineurin B-like 10 (CBL10) has been reported as another regulator of SOS2 in the shoot tissues (Quan et al. 2007). Du et al. (2011) inferred that the CBL/SOS3-like calcium binding protein (SCaBP)-CBL interacting protein kinase (CIPK) regulates the phosphorylation of SOS3-like proteins. As a result, the downstream Na^+/H^+ antiporter, namely, SOS1, is activated upon successful interactions between SOS3-SOS2 and SCaBP8-SOS2, following which SOS2 is translocated to the cell membrane to activate SOS1 (Quintero et al. 2011). Halophytes like *Thellungiella salsuginea* also regulate Na^+ entry by maintaining high activity of SOS1. Knockdown studies showed that 50% reduction in SOS1 activity led to loss of the halophytic phenotype in the *T. salsuginea* plants due to excess Na^+ accumulation within the tissues (Oh et al. 2009). An overview of the SOS pathway in plants has been represented in Fig. 9.2 (extracted from Gupta and Huang 2014).

The clade A PP-2Cs inhibit the action of SNF1-related protein kinases 2 (SnRK2s), which in turn are activated in presence of the universal stress phytohormone, ABA. Krzywinska et al. (2016) reported that the PP-2C named ABA-insensitive 1 (ABI1) negatively regulated the activity of SnRK2.4 during salt stress response in Arabidopsis. The negative interaction of these proteins actually determined the fate of root development under saline conditions (Krzywinska et al.

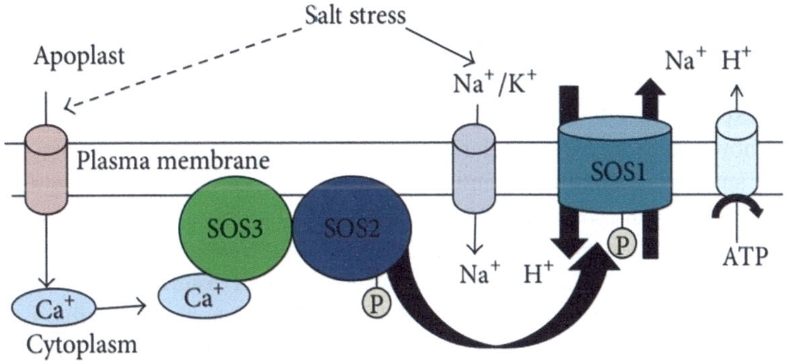

Fig. 9.2 An overview of the SOS pathway in plants in response to salt stress. (Extracted from Gupta and Huang 2014)

2016). Rapid alkalinization factor (RALF) is a peptide signal which determines some responses during abiotic stresses. The protein phosphatase, ABI2, negatively regulated the receptor-like kinase, FERONIA (FER), by removing the phosphoryl groups added in presence of ABA and RALF. Thus, ABI2 negatively regulates ABA and RALF-induced signaling during abiotic stresses (Chen et al. 2016). Ohta et al. (2003) reported the interaction between SOS2 and ABI2 proteins in Arabidopsis. A 37-amino-acid-long domain was identified via deletion analysis that was named as the protein phosphatase interaction (PPI) motif essential for the SOS2-ABI2 interaction (Ohta et al. 2003). The PPI motif has been observed to be conserved in SOS2 family kinases, DNA repair kinases, and cell cycle checkpoint kinases across organisms including humans. It was also observed that protein kinase S (PKS belonging to the SOS2 family) strongly interacted with ABI2, whereas others associated with ABI1 (Ohta et al. 2003).

Xu et al. (2016) performed a comparative expression analysis of calcineurin B-like family gene *CBL10A* between a salt-tolerant and a salt-sensitive cultivar of *Brassica oleracea*. It was observed that the expression of this gene was significantly higher in the tolerant cultivar compared to the susceptible variety. The gene also exhibited longer period of induction and shorter time of response in the tolerant cultivar. Such differences were attributed to substantial variation in the coding region of the gene. The gene was found to be 741 bp in the sensitive cultivar and 829 bp in the tolerant *B. oleracea* variety (Xu et al. 2016). Dong et al. (2015) reported that the overexpression of the *CBL10* gene from *Nicotiana sylvestris* promoted salt tolerance in transgenic Arabidopsis plants via efficient maintenance of Na^+ homeostasis. In another study, it was found that overexpression of *PP-2C1* from *Triticum aestivum* improved the salt-sensitive character in transgenic tobacco plants (Hu et al. 2015). The transgenics exhibited ABA hyposensitivity, reduced accumulation of reactive oxygen species (ROS), and activated antioxidant machinery to ameliorate the salt-induced injuries (Hu et al. 2015).

The salt susceptibility of Arabidopsis was increased in the mutants of *At5PTase7* and *At5PTase9*, whereas the overexpression of the genes promoted tolerance, thus showing the positive regulatory action of these phosphatases during salinity (Kaye et al. 2011; Golani et al. 2013). The expression of osmotic stress-responsive genes like *responsive to dehydration 29A* (*rd29A*) and *rd22* was not stimulated in the mutants exposed to salt stress (Jia et al. 2019). Jia et al. (2019) also reviewed the widely known roles of AtSAL1 during multiple abiotic stresses including salinity. High salt stress was ameliorated in the tobacco BY-2 cells overexpressing the soybean SAL1 (Ku et al. 2013). However, in case of Arabidopsis, the overexpression of *AtSAL1* could not increase salt tolerance, whereas the transgenic lines behaving as loss-of-function mutants of *AtSAL1* showed enhanced tolerance to desiccation stress, indicating that in Arabidopsis, SAL1 behaves as a negative regulator of abiotic stress (Ku et al. 2013). Incorporation of the META motif from the SAL1 homologue (HAL2) in black yeast *Aureobasidium pullulans* increased salinity tolerance in transgenic Arabidopsis plants (Gasparic et al. 2013).

9.4 Conclusion

Phosphatases are crucial molecular components in plant signaling pathway culminating multiple physiological responses. Some of these proteins especially those involved in the ABA signaling pathway negatively regulate the cascade by deactivating SnRKs in absence of ABA. Such regulation in fact is necessary to properly coordinate the switching on and switching off of the signaling processes. Phosphatases thus dictate a major portion of global systemic responses. Salt stress is one of the most prevalent forms of abiotic stress, responsible for large-scale crop losses worldwide. It affects the plant by creating osmotic shocks and by triggering uncontrolled production of ROS. ABA is largely related to abiotic stress tolerance in multiple crop species. Hence, the involvement of protein phosphatases in salt stress regulation is very evident. The chapter highlights the most recent reports on the involvement of phosphatases and even Ca^{2+}-responsive calcineurins in dictating salt stress tolerance in crop species. Posttranslational alterations and even epigenetic modifications in phosphatases and encoding genes would indicate at newer avenues of research which are still not well understood.

9.5 Future Perspectives

The research on the involvement of phosphatases in generating salt tolerance is limiting in the current literature. Genome-wide studies based on next-generation sequencing platforms can be designed to identify novel phosphatases operative exclusively during salt stress. Cloning and overexpression of such genes might generate multiple stress-tolerant crops. Epigenomic studies and bisulfite sequencing can be adopted to identify the global epigenetic modifications occurring in phosphatase-encoding genes. Studies exclusively based on mass spectrometry can be performed to identify and report the posttranslational changes occurring in phosphatases as a result of salt stress. Studying any change in activity due to such modifications will also open newer avenues of research.

Acknowledgments Financial assistance from Science and Engineering Research Board, Government of India, through the grant EMR/2016/004799 and Department of Higher Education, Science and Technology and Biotechnology, Government of West Bengal, through the grant 264(Sanc.)/ST/P/S&T/1G-80/2017 to Dr. Aryadeep Roychoudhury is gratefully acknowledged. Mr. Aditya Banerjee is thankful to University Grants Commission, Government of India, for providing Senior Research Fellowship in course of this work.

References

Allen GJ, Sanders D (1995) Calcineurin, a type 2B protein phosphatase, modulates the Ca^{2+}-permeable slow vacuolar ion channel of stomatal guard cells. Plant Cell 7:1473–1483

Banerjee A, Roychoudhury A (2018a) Seed priming technology in the amelioration of salinity stress in plants. In: Rakshit A, Singh HB (eds) Advances in seed priming. Springer Nature, Singapore, pp 81–93

Banerjee A, Roychoudhury A (2018b) Effect of salinity stress on growth and physiology of medicinal plants. In: Ghorbanpour M et al (eds) Medicinal plants and environmental challenges. Springer International Publishing AG, Cham, pp 177–188

Banerjee A, Tripathi DK, Roychoudhury A (2018) Hydrogen sulphide trapeze: environmental stress amelioration and phytohormone crosstalk. Plant Physiol Biochem 132:46–53

Banerjee A, Tripathi DK, Roychoudhury A (2019) The karrikin 'callisthenics': can compounds derived from smoke help in stress tolerance? Physiol Plant 165:290–302

Bradai M, Mahjoubi H, Chini A, Chaboute ME, Hanin M, Ebel C (2018) Genome wide identification of wheat and *Brachypodium* type one protein phosphatases and functional characterization of durum wheat TdPP1a. PLoS One 13:e0191272

Brautigan DL (2013) Protein Ser/Thr phosphatases—the ugly ducklings of cell signaling. FEBS J 280:324–345

Burnette RN (2003) An Arabidopsis inositol 5-phosphatase gain-of-function alters abscisic acid signaling. Plant Physiol 132:1011–1019

Chae L, Pandey GK, Luan S, Cheong YH, Kim KN (2009) Protein kinases and phosphatases for stress signal transduction in plants. In: Abiotic stress adaptation in plants. Springer, Dordrecht, pp 123–163

Chen J, Yu F, Liu Y, Du C, Li X, Zhu S et al (2016) FERONIA interacts with ABI2-type phosphatases to facilitate signaling cross-talk between abscisic acid and RALF peptide in Arabidopsis. Proc Natl Acad Sci U S A 113:E5519–E5527

Chen C, Yu Y, Ding X, Liu B, Duanmu H et al (2018) Genome-wide analysis and expression profiling of PP2C clade D under saline and alkali stresses in wild soybean and Arabidopsis. Protoplasma 255:643–654

Creighton MT, Kolton A, Kataya ARA, Maple-Grodem J et al (2017) Methylation of protein phosphatase 2A-influence of regulators and environmental stress factors. Plant Cell Environ 40:2347–2358

Danquah A, de Zelicourt A, Colcombet J, Hirt H (2014) The role of ABA and MAPK signaling pathways in plant abiotic stress responses. Biotechnol Adv 32:40–52

DeLong A (2006) Switching the flip: protein phosphatase roles in signaling pathways. Curr Opin Plant Biol 9:470–477

Dong L, Wang Q, Manik SM, Song Y, Shi S et al (2015) *Nicotiana sylvestris* calcineurin B-like protein NsylCBL10 enhances salt tolerance in transgenic Arabidopsis. Plant Cell Rep 34:2053–2063

Du W, Lin H, Chen S, Wu Y, Zhang J, Fuglsang AT, Palmgren MG, Wu W, Guo Y (2011) Phosphorylation of SOS3-like calcium-binding proteins by their interacting SOS2-like protein kinases is a common regulatory mechanism in Arabidopsis. Plant Physiol 156:2235–2243

Egea I, Pineda B, Ortiz-Atienza A, Plasencia FA, Drevensek S et al (2018) The SlCBL10 calcineurin B-like protein endures plant growth under salt stress by regulating Na^+ and Ca^{2+} homeostasis. Plant Physiol 176:1676–1693

Farkas I, Dombradi V, Miskei M, Szabados L, Koncz C (2007) Arabidopsis PPP family of serine/threonine phosphatases. Trends Plant Sci 12:169–176

Fuchs S, Grill E, Meskiene I, Schweighofer A (2013) Type 2C protein phosphatases in plants. FEBS J 280:681–693

Gasparic MB, Lenassi M, Gostincar C, Rotter A, Plemenitaš A, Gunde-Cimerman N, Gruden K, Žel J (2013) Insertion of a specific fungal 30-phosphoadenosine-50-phosphatase motif into a plant homologue improves halotolerance and drought tolerance of plants. PLoS One 8:e81872

Ghorbel M, Zaidi I, Ebel C, Hanin M (2019) Differential regulation of the durum wheat MAPK phosphatase 1 by calmodulin, bivalent cations and possibly mitogen activated protein kinase 3. Plant Physiol Biochem 135:242–252

Golani Y, Kaye Y, Gilhar O, Ercetin M, Gillaspy G, Levine A (2013) Inositol polyphosphate phosphatidylinositol 5-phosphatase9 (At5PTase9) controls plant salt tolerance by regulating endocytosis. Mol Plant 6:1781–1794

Gonzalez Besteiro MA, Ulm R (2013) Phosphorylation and stabilization of Arabidopsis MAP kinase phosphatase 1 in response to UV-B stress. J Biol Chem 288:480–486

Guo YL, Roux SJ (1995) Partial purification and characterisation of an enzyme from pea nuclei with protein tyrosine phosphatase activity. Plant Physiol 107:167–175

Gupta B, Huang B (2014) Mechanism of salinity tolerance in plants: physiological, biochemical, and molecular characterization. Int J Genomics 2014:701596

Han EH, Petrella DP, Blakeslee JJ (2017) 'Bending' models of halotropism: incorporating protein phosphatase 2A, ABCB transporters, and auxin metabolism. J Exp Bot 68:3071–3089

Hrabak EM, Chan CW, Gribskov M, Harper JF, Choi JH, Halford N, Kudla J, Luan S, Nimmo HG, Sussman MR et al (2003) The Arabidopsis CDPK-SnRK superfamily of protein kinases. Plant Physiol 132:666–680

Hu W, Yan Y, Hou X, He Y, Wei Y, Yang G, et al. (2015) TaPP2C1, a group F2 protein phosphatase 2C gene, confers resistance to salt stress in transgenic tobacco. PLoS ONE 10:e0129589.

Hu R, Zhu Y, Shen G, Zhang H (2017) Overexpression of the *PP2A-C5* gene confers increased salt tolerance in Arabidopsis *thaliana*. Plant Signal Behav 12:e1276687

Jia Q, Kong D, Li Q, Sun S, Song J, Zhu Y, Liang K et al (2019) The function of inositol phosphatases in plant tolerance to abiotic stress. Int J Mol Sci 20:3999

Kaye Y, Golani Y, Singer Y, Leshem Y, Cohen G, Ercetin M, Gillaspy G, Levine A (2011) Inositol polyphosphate 5-phosphatase7 regulates the production of reactive oxygen species and salt tolerance in Arabidopsis. Plant Physiol 157:229–241

Kim DH, Kang JG, Yang SS, Chung KS, Song PS, Park CM (2002) A phytochrome-associated protein phosphatase 2A modulates light signals in flowering time control in Arabidopsis. Plant Cell 14:3043–3056

Krzywinska E, Kulik A, Bucholc M, Fernandez MA et al (2016) Protein phosphatase type 2C PP2CA together with ABI1 inhibits SnRK2.4 activity and regulates plant responses to salinity. Plant Signal Behav 11:e1253647

Ku Y-S, Koo NS-C, Li FW-Y, Li M-W, Wang H, Tsai S-N, Sun F, Lim BL, Ko W-H, Lam H-M (2013) GmSAL1 hydrolyzes inositol-1,4,5-trisphosphate and regulates stomatal closure in detached leaves and ion compartmentalization in plant cells. PLoS One 8:e78181

Li Y, Yang S, Yang H, Hua J (2007) The TIR-NB-LRR gene SNC1 is regulated at the transcript level by multiple factors. Mol Plant Microbe Interact 20:1449–1456

Liu L, Hu X, Song J, Zong X, Li D, Li D (2009) Over-expression of a *Zea mays L. protein phosphatase 2C* gene (*ZmPP2C*) in Arabidopsis *thaliana* decreases tolerance to salt and drought. J Plant Physiol 166:531–542

Liu X, Zhu Y, Zhai H, Cai H, Ji W, Luo X, Li J, Bai X (2012) AtPP2CG1, a protein phosphatase 2C, positively regulates salt tolerance of Arabidopsis in abscisic acid-dependent manner. Biochem Biophys Res Commun 422:710–715

Manabe Y, Bressan RA, Wang T, Li F, Koiwa H et al (2008) The Arabidopsis kinase-associated protein phosphatase regulates adaptation to Na^+ stress. Plant Physiol 146:612–622

Merlot S, Gosti F, Guerrier D, Vavasseur A, Giraudat J (2001) The ABI1 and ABI2 protein phosphatases 2C act in a negative feedback regulatory loop of the abscisic acid signaling pathway. Plant J 25:295–303

Oh DH, Leidi E, Zhang Q, Hwang SM, Li Y, Quintero FJ, Jiang X, D'Urzo MP, Lee SY, Zhao Y et al (2009) Loss of halophytism by interference with SOS1 expression. Plant Physiol 151:210–222

Ohta M, Guo Y, Halfter U, Zhu J-K (2003) A novel domain in the protein kinase SOS2 mediates interaction with the protein phosphatase 2C ABI2. Proc Natl Acad Sci U S A 100:11771–11776

Pandey G, Yadav CB, Sahu PP, Muthamilarasan M, Prasad M (2017) Salinity induced differential methylation patterns in contrasting cultivars of foxtail millet (*Setaria italica* L.). Plant Cell Rep 36:759–772

Park JH, Lee SY, Kim WY, Jung YJ, Chae HB, Jung HS et al (2011) Heat-induced chaperone activity of serine/threonine protein phosphatase 5 enhances thermotolerance in Arabidopsis *thaliana*. New Phytol 191:692–705

Park JH, Kim WY, Chae HB, Kim MG, Lee SY (2012) Serine/threonine protein phosphatase 5 (PP5) interacts with substrate under heat stress conditions and forms protein complex in Arabidopsis. Plant Signal Behav 7:535–538

Quan R, Lin H, Mendoza I, Zhang Y, Cao W, Yang Y, Shang M, Chen S, Pardo JM, Guo Y (2007) SCABP8/CBL10, a putative calcium sensor, interacts with the protein kinase SOS2 to protect Arabidopsis shoots from salt stress. Plant Cell 19:1415–1431

Quintero FJ, Martinez-Atienza J, Villalta I, Jiang X, Kim WY, Ali Z, Fujii H, Mendoza I, Yun DJ, Zhu JK (2011) Activation of the plasma membrane Na/H antiporter Salt-Overly Sensitive 1 (SOS1) by phosphorylation of an auto-inhibitory C-terminal domain. Proc Natl Acad Sci U S A 108:2611–2616

Roychoudhury A, Banerjee A (2017) Abscisic acid signaling and involvement of mitogen activated protein kinases and calcium-dependent protein kinases during plant abiotic stress. In: Pandey G (ed) Mechanism of plant hormone signaling under stress, vol 1. Wiley, Hoboken, NJ, pp 197–241

Schweighofer A, Meskiene I (2015) Phosphatases in plants. Methods Mol Biol 1306:25–46

Shankar A, Agrawal N, Sharma M, Pandey A, Pandey GK (2015) Role of protein tyrosine phosphatases in plants. Curr Genomics 16:224–236

Singh A, Giri J, Kapoor S, Tyagi AK, Pandey GK (2010) Protein phosphatase complement in rice: genome-wide identification and transcriptional analysis under abiotic stress conditions and reproductive development. BMC Genomics 11:435

Singh A, Jha SK, Bagri J, Pandey GK (2015) ABA inducible rice protein phosphatase 2C confers ABA insensitivity and abiotic stress tolerance in Arabidopsis. PLoS One 10:e0125168

Sopory SK, Munshi M (1998) Protein kinases and phosphatases and their role in cellular signaling in plants. Crit Rev Plant Sci 17:245–318

Sun L, Pehlivan N, Esmaeli N, Jiang W, Yang X et al (2018) Co-overexpression of *AVP1* and *PP2A-C5* in Arabidopsis makes plants tolerant to multiple abiotic stresses. Plant Sci 274:271–283

Tian H, Chen S, Yang W, Wang T, Zheng K et al (2017) A novel family of transcription factors conserved in angiosperms is required for ABA signaling. Plant Cell Environ 40:2958–2971

Uhrig RG, Labandera AM, Moorhead GB (2013) Arabidopsis PPP family of serine/threonine protein phosphatases: many targets but few engines. Trends Plant Sci 18:505–513

Van Bentem SDLF, Vossen JH, Vermeer JE, de Vroomen MJ, Gadella TW, Haring MA, Cornelissen BJ (2003) The subcellular localization of plant protein phosphatase 5 isoforms is determined by alternative splicing. Plant Physiol 133:702–712

Vollmer AH, Youssef NN, DeWald DB (2011) Unique cell wall abnormalities in the putative phosphoinositide phosphatase mutant AtSAC9. Planta 234:993–1005

Wang W, Vinocur B, Altman A (2003) Plant responses to drought, salinity and extreme temperatures: towards genetic engineering for stress tolerance. Planta 218:1–14

Wu Q, Bai X, Zhao W, Xiang D, Wan Y et al (2017) De novo assembly and analysis of tartary buckwheat (*Fagopyrum tataricum* Garetn.) transcriptome discloses key regulators involved in salt-stress response. Genes (Basel) 8:e255

Xu L, Zhang D, Xu Z, Huang Y, He X et al (2016) Comparative expression analysis of Calcineurin B-like family gene *CBL10A* between salt-tolerant and salt-sensitive cultivars in *B. oleracea*. Sci Total Environ 571:1–10

Zaidi I, Ebel C, Belgaroui N, Ghorbel M, Amara I, Hanin M (2016) The wheat MAP kinase phosphatase 1 alleviates salt stress and increases antioxidant activities in Arabidopsis. J Plant Physiol 193:12–21

Zhang J, Li X, He Z, Zhao X, Wang Q, Zhou B et al (2013) Molecular character of a phosphatase 2C (PP2C) gene relation to stress tolerance in Arabidopsis *thaliana*. Mol Biol Rep 40:2633–2644

Zhang Z, Li Y, Luo Z, Kong S, Zhao Y, Zhang C, Zhang W, Yuan H, Cheng L (2019) Expansion and functional divergence of inositol polyphosphate 5-phosphatases in angiosperms. Genes 10:393

Chapter 10
Phosphatases: The Critical Regulator of Abiotic Stress Tolerance in Plants

Sripati Abhiram Sahoo, Rishiraj Raghuvanshi, Ashish Kumar Srivastava, and Penna Suprasanna

10.1 Introduction

The major challenge for plant biologists in the twenty-first century is to sustain food security for growing population under the changing climatic conditions and increasing episodes of different abiotic stresses like salinity, drought, cold, submergence, and heavy metal toxicity. These abiotic stresses decrease the fertile agricultural yield and hence reduce crop productivity (Pandey et al. 2016; Zhu 2016). During the course of evolution, plants have evolved mechanisms to adapt under stress conditions. The perception of stress initiates a series of steps which leads to the expression of stress-responsive transcription factors and finally the downstream effector genes. The process by which plant cells sense stress signals and transmit them to activate adaptive responses is referred as "signal transduction" (Xiong and Zhu 2001). These signal transduction pathways are highly coordinated, and under stress conditions, they activate multiple genes/proteins/metabolites, which together functions to restore homeostasis at cellular [HSPs (heat shock proteins), LEA (late-embryogenic abundant)], hormonal [ABA (abscisic acid), GA (gibberellic acid), IAA (indoleacetic acid)], and redox [APX (ascorbate peroxidase), SOD (superoxide dismutase), CAT (catalase)], osmotic (proline), and ionic (NHX and HKT) levels (Fig. 10.1).

S. A. Sahoo · R. Raghuvanshi
Department of Plant Molecular Biology and Biotechnology, Indira Gandhi Krishi Vishwavidyalaya, Raipur, India

A. K. Srivastava (✉) · P. Suprasanna (✉)
Nuclear Agriculture and Biotechnology Division, Bhabha Atomic Research Centre, Mumbai, India

Homi Bhabha National Institute, Mumbai, India
e-mail: ashish@barc.gov.in; prasanna@barc.gov.in

G. K. Pandey (ed.), *Protein Phosphatases and Stress Management in Plants*,
https://doi.org/10.1007/978-3-030-48733-1_10

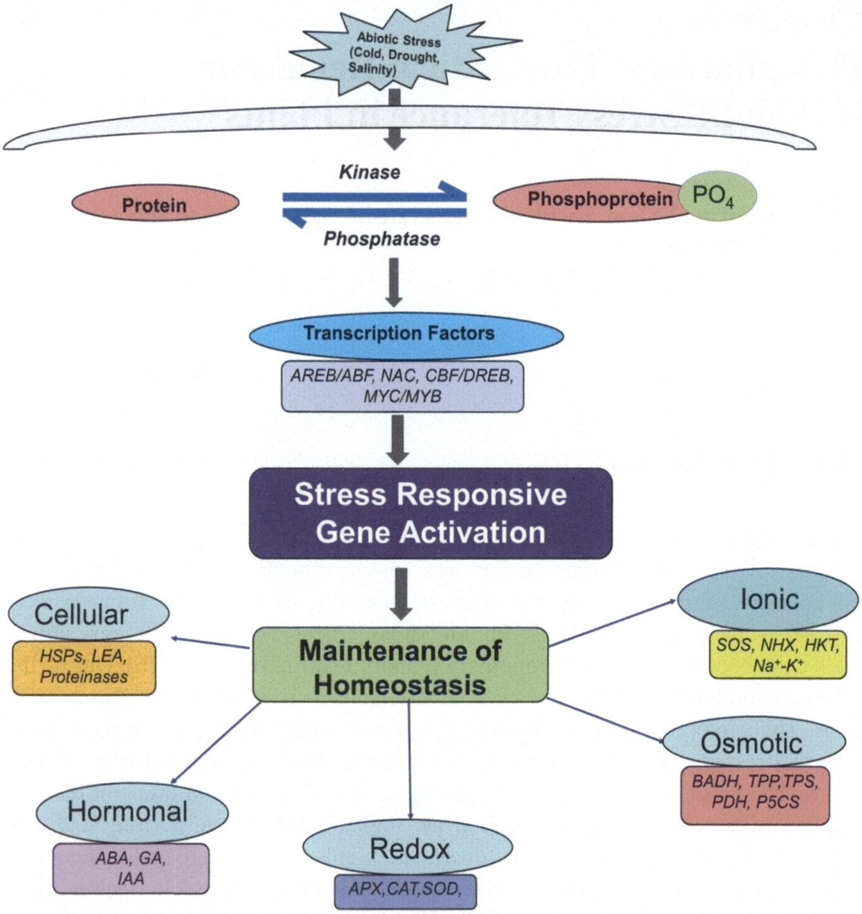

Fig. 10.1 Schematic representation of the Mechanism of stress tolerance in plants. In plants, stress is perceived through coordinated action of kinases and phosphatases. This in turn activates transcription factors, which upregulate the expression of stress-responsive genes (SRGs). SRGs function to restore the homeostasis at multiple levels including cellular, redox, hormones, ionic, and osmotic. Abbreviations: *AREB/ABF* ABA-Responsive Element Binding factor/ABA response element Binding Factor, *NAC* (NAM, ATAF1/2, CUC) [*NAM* No Apical Meristem, *ATAF* Arabidopsis Transcription Activation Factor, *CUC* CUp shaped Cotyledon], *CBF/DREB* C-repeat-Binding Factor/Dehydration Responsive Element-Binding factor, *MYC/MYB* Myelocytomatosis oncogene/Myeloblastosis oncogene, *SOS* Salt Overly Sensitive, *NHX* Sodium Hydrogen exchanger, *HKT* High-affinity K^+ Transporter, *BADH* Betaine-Aldehyde DeHydrogenase, *TPP* Trehalose 6-Phosphate Phosphatase, *TPS* Trehalose-6-Phosphate Synthase, *PDH* Proline DeHydrogenase, *P5CS* Pyrroline-5-Carboxylate Synthase, *APX* Ascorbate Peroxidase, *CAT* Catalase, *SOD* Superoxide Dismutase, *ABA* Abscisic Acid, *GA* Gibberelic Acid, *IAA* IndoleAcetic Acid, *HSPs* Heat Shock Proteins, *LEA* Late Embryogenesis Abundant protein

A large number of proteins are involved in the perception and signal transduction pathways (Pandey et al. 2016). Among these, kinases and phosphatases constitute an important family, which regulate signal transduction through "reversible protein phosphorylation and dephosphorylation." During protein phosphorylation, a protein kinase either activates or deactivates a substrate through the addition of a phosphate group, thereby propagating or regulating a signal. Protein phosphatases reverse the effect by removing the phosphate from the substrate (Chae et al. 2009). It can be well understood as protein phosphatases are the agents of the "OFF" state and the enzymes reverse the "ON" state encouraged by kinases, but several signaling systems feature phosphatases in positive dynamic and regulatory roles (DeLong 2006). In plants, protein phosphatases play important role in tuning cellular responses to physiological stimuli under varying developmental phases (Schweighofer and Meskiene 2015). The present review focuses on the role of phosphatases and different phosphatase gene families involved in stress signaling pathways. We discuss in-depth about the functional relevance of major phosphatases under the stress conditions and elaborate the integrated network in plants.

10.2 Major Protein Phosphatase Gene Families

Plant protein phosphatases are classified into three families based on the substrate of phosphorylation. These include serine/threonine-specific protein phosphatase (PPP), metal ion-dependent protein phosphatase (PPM), and protein tyrosine phosphatase (PTP). Advancements in molecular, biochemical, and genomics have enabled further classification of (1) PPP into PP1, PP2A, PP2B, and other distantly related phosphatases like PP4, PP5, PP6, and PP7 and (2) PPM for catalysis, including PP2C and other Mg^{2+}–dependent phosphatases (Chae et al. 2009; Singh et al. 2010, 2016). The protein tyrosine phosphatase (PTP) family consists of both tyrosine-specific phosphatases and dual-specificity phosphatases (dsPTPs) (Tonks and Neel 1996; Denu et al. 1996; Farkas et al. 2007; Chae et al. 2009). The phosphatase gene families are quite complex as most of the genes encode multiple transcripts with variable number of exon and introns. A comparable complexity is observed in both model plant Arabidopsis as well as crops like rice and soybean (Table 10.1). Although the occurrence of multiple genes in each subfamily of phosphatases makes genetic characterization difficult due to redundancy, however, by combining forward genetic analysis and biochemistry, valuable information has been generated on the biological functions of several members of plant phosphatases. Subsequently functional identification of phosphatase subunits and interacting factors has also highlighted the central roles of plant PPPs in cellular signaling.

Table 10.1 Gene model information of phosphatase gene families in plants

A. Arabidopsis

Protein tyrosine phosphatase (PTP) family (source: https://www.nsf.gov; https://www.arabidopsis.org)

Gene	Locus ID	CDS (bp)	Exons	Introns	No. of transcripts	Protein function/description
AtPTP1	At1g71860	1023	27	24	3	Protein tyrosine phosphatase
AtDSP1	At3g23610	687	18	15	3	Protein phosphatase dual-specificity phosphatase—may dephosphorylate MAPKs
AtDSP1-like	At5g23720	2790	32	29	3	Protein phosphatase dual-specificity function
AtDSP2	At3g06110	504	13	10	3	Protein phosphatase dual-specificity function
AtDSP3, IBR5	At2g04550	774	10	8	2	Mediating auxin response
AtDSP4, AtPTPKIS1, AtSEX4	At3g52180	1140	25	23	2	Protein phosphatase function, dephosporylation of starch; involved in starch metabolism regulation/kinase interaction sequence protein (PTPKIS1)
AtDSP5	At3g10940	849	4	3	1	Protein phosphatase dual-specificity function
AtDSP6	At3g01510	1776	10	9	1	Protein phosphatase dual-specificity function—putative/5′-AMP-activated protein kinase beta-1 subunit-related
AtDSP7	At2g32960	774	6	5	1	Protein phosphatase dual-specificity function
AtDSP8	At2g35680	1014	6	5	1	Protein phosphatase dual-specificity function
AtDSP9	At1g05000	744	11	9	2	Protein phosphatase dual-specificity function
AtDSP10	At5g56610	687	11	9	2	Protein phosphatase dual-specificity function
AtDSP11	At4g03960	597	5	4	1	Protein phosphatase dual-specificity function
AtDSP12	At5g16480	615	5	4	1	Protein phosphatase dual-specificity function
AtDSP13	At3g02800	612	5	4	1	Protein phosphatase dual-specificity function
AtDSP14, AtMKP1	At3g55270	2354	4	3	1	Protein phosphatase dual-specificity function—may dephosphorylate MAPK
CDC25-like	At5g03455	441	3	2	1	Rhodanese-like domain-containing protein

(continued)

Table 10.1 (continued)

A. Arabidopsis

Protein tyrosine phosphatase (PTP) family (source: https://www.nsf.gov; https://www.arabidopsis.org)

Gene	Locus ID	CDS (bp)	Exons	Introns	No. of transcripts	Protein function/description
AtPTEN1	At5g39400	1239	8	7	1	Protein phosphatase dual-specificity function, function in pollen development/ phosphatase and tensin
AtPTEN2	At3g19420	1836	12	11	1	Protein phosphatase dual-specificity function
AtPTEN3	At3g50110	1899	13	12	1	Protein phosphatase dual-specificity function
AtPTEN4	At5g58160	3975	15	14	1	Protein phosphatase dual-specificity function—putative/ FH2-domain containing
AtMTM1	At3g10550	2523	19	18	1	Protein phosphatase dual-specificity function
AtMTM2	At5g04540	2502	19	18	1	Protein phosphatase dual-specificity function
AtLMW	At3g44620	789	9	7	2	Protein phosphatase dual-specificity function—putative/ phosphotyrosine protein

Arabidopsis PP2A family genes (source: Durian et al. 2016; https://phytozome.jgi.doe.gov/pz/portal.html; https://www.arabidopsis.org)

Gene	Locus ID	CDS (bp)	Exons	Introns	No. of transcripts	Protein function/description
B alpha (Bα)	AT1g51690	1812	45	42	3	55 kDa B regulatory subunit of phosphatase 2A mRNA
B beta (Bβ)	AT1g17720	1506	28	26	2	Type 2A protein serine/ threonine phosphatase 55 kDa B
B′ alpha (B′α)	AT5g03470	1488	2	1	1	Encodes B′ regulatory subunit of PP2A (AtB′alpha), putative size of 57 kDa. Functions redundantly with the beta subunit do maintain sister chromatid cohesion during meiosis
B′ beta (B′β)	AT3g09880	1500	2	1	1	Encodes B′ regulatory subunit of PP2A (AtB′beta). Functions redundantly with the alpha subunit do maintain sister chromatid cohesion during meiosis
B′ gamma (B′γ)	AT4g15415	1569	5	3	2	B′ regulatory subunit of PP2A (AtB′gamma)

(continued)

Table 10.1 (continued)

Arabidopsis PP2A family genes (source: Durian et al. 2016; https://phytozome.jgi.doe.gov/pz/portal.html; https://www.arabidopsis.org)

Gene	Locus ID	CDS (bp)	Exons	Introns	No. of transcripts	Protein function/description
B′ delta (B′δ)	AT3g26030	1434	2	1	1	Protein phosphatase 2A regulatory subunit isoform B′ delta. the mRNA is cell-to-cell mobile
B′ epsilon (B′ε)	AT3g54930	1494	3	2	1	Protein phosphatase 2A regulatory B subunit family protein
B′ zeta (B′ζ)	AT3g21650	1641	3	2	1	Encodes protein phosphatase 2A (PP2A) B′zeta subunit; targeted to mitochondria
B′ eta (B′η)	AT3g26020	1590	21	17	4	Encodes protein phosphatase 2A (PP2A) B′eta subunit; targeted to nucleus and cytosol
B′ theta (B′θ)	AT1g13460	1479	6	4	2	Encodes protein phosphatase 2A (PP2A) B′theta subunit; targeted to peroxisomes
B′ kappa (B′κ)/B′ iota (B′ι) in gene bank	AT5g25510	1503	3	2	1	Protein phosphatase 2A regulatory B subunit family protein
B″ alpha (B″α)	AT5g44090	1617	13	12	1	Calcium-binding EF-hand family protein
B″ epsilon (B″ε) or B″ beta in GenBank	AT5g28850	1611	23	21	2	Calcium-binding EF-hand family protein
B″ delta (B″δ)	AT5g28900	1611	13	12	1	Calcium-binding EF-hand family protein
B″ gamma (B″γ)	AT1g54450	1608	11	10	1	Calcium-binding EF-hand family protein
B″ beta (B"β)/B″ epsilon in GenBank	AT1g03960	1590	26	24	2	Calcium-binding EF hand family protein
TON 2(FASS)	AT5g18580	1443	12	11	1	Fass mutants have aberrant cell shapes due to defects in arrangement of cortical microtubules

(continued)

Table 10.1 (continued)

Arabidopsis PP2C family genes (source: Schweighofer et al. 2004; https://www.arabidopsis.org)					
Locus id	CDS (bp)	Exons	Introns	No. of transcripts	Protein function/description
Group A					
AT3g11410	1200	4	3	1	Arabidopsis *thaliana* protein phosphatase 2Ca
AT1g17550	1536	4	3	1	ABA-hypersensitive germination 1
AT1g72770	1536	15	12	3	ABA-hypersensitive inhibition of seed germination
AT4g26080	1305	4	3	1	Regulates the activation of the Snf1-related kinase OST1 by abscisic acid
AT5g57050	1272	9	7	2	Involved in ABA signal transduction
Group B					
AT1g67820	1338	4	3	1	Protein phosphatase 2C family protein
AT2g40180	1173	3	2	1	PP2C5 acts as a MAPK phosphatase that positively regulates seed germination
AT1g07160	1143	3	2	1	Protein phosphatase 2C family protein
AT2g30020	1191	2	1	1	Magnesium-dependent protein serine/threonine phosphatase activity
AT3g27140	738	5	4	1	Acts as a MAPK phosphatase that negatively regulates MPK4 and MPK6
AT4g08260	639	4	3	1	Protein phosphatase 2C family protein (source: Araport11)
Group C					
AT2g46920	2571	9	7	2	Encodes a protein with similarity to the POL locus
AT2g35350	2352	4	3	1	Encodes a protein most similar to the POLTERGEIST locus
AT1g07630	1989	4	3	1	Encodes a protein phosphatase 2C-like gene, similar to POL
AT2g28890	1965	4	3	1	Encodes a protein phosphatase 2C-like gene
AT3g09400	1953	8	6	2	Similar to POLTERGEIST (POL) protein phosphatase 2C
AT5g02400	2025	4	3	1	Similarity to the POL locus which is a novel protein phosphatase 2C
Group D					
AT5g02760	1113	4	3	1	Encodes a phosphatase that functions in sustaining proper leaf longevity and preventing early senescence by suppressing or perturbing SARK-mediated senescence signal transduction
AT3g17090	1155	8	6	2	Protein phosphatase 2C family protein
AT3g12620	1158	9	7	2	Protein phosphatase 2C family protein
AT3g55050	1155	7	5	2	Protein phosphatase 2C family protein

(continued)

Table 10.1 (continued)

Arabidopsis PP2C family genes (source: Schweighofer et al. 2004; https://www.arabidopsis.org)					
Locus id	CDS (bp)	Exons	Introns	No. of transcripts	Protein function/description
AT4g38520	1203	9	7	2	Protein phosphatase 2C family protein
AT3g51370	1140	10	8	2	Protein phosphatase 2C family protein
AT5g66080	1158	4	3	1	Protein phosphatase 2C family protein
AT4g33920	1143	4	3	1	Protein phosphatase 2C family protein
AT5g06750	1182	13	10	3	Protein phosphatase 2C family protein
Group E					
AT1g03590	1389	6	5	1	Protein phosphatase 2C family protein
AT1g16220	1476	5	4	1	Protein phosphatase 2C family protein
AT1g79630	1515	16	13	3	Protein phosphatase 2C family protein
AT5g01700	1149	11	9	2	Protein phosphatase 2C family protein
AT3g02750	1584	17	14	3	Protein phosphatase 2C family protein
AT5g36250	1347	5	4	1	Encodes a myristoylated 2C-type protein phosphatase that interacts with the catalytic subunit of SnRK1. The mRNA is cell-to-cell mobile
AT5g26010	996	5	4	1	Protein phosphatase 2C family protein
AT4g32950	981	5	4	1	Protein phosphatase 2C family protein
AT3g16800	1056	17	14	3	EGR3 functions as a negative regulator of plant growth with prominent effect on plant growth during drought stress. EGR3 regulates microtubule organization and likely affects additional cytoskeleton and trafficking processes along the plasma membrane
AT3g05640	1077	12	10	2	EGR1 functions as a negative regulator of plant growth with prominent effect on plant growth during drought stress. EGR1 regulates microtubule organization and likely affects additional cytoskeleton and trafficking processes along the plasma membrane
AT5g27930	1122	11	9	2	EGR2 functions as a negative regulator of plant growth with prominent effect on plant growth during drought stress. EGR2 regulates microtubule organization and likely affects additional cytoskeleton and trafficking processes along the plasma membrane
AT2g20050	3285	28	26	2	Protein phosphatase 2C and cyclic nucleotide-binding/kinase domain-containing protein
AT3g06270	1047	2	1	1	Protein phosphatase 2C family protein
Group F					
AT3g23360	783	5	4	1	Protein phosphatase 2C family protein

(continued)

Table 10.1 (continued)

Arabidopsis PP2C family genes (source: Schweighofer et al. 2004; https://www.arabidopsis.org)

Locus id	CDS (bp)	Exons	Introns	No. of transcripts	Protein function/description
AT2g34740	1020	14	12	2	Protein phosphatase 2C family protein
AT1g78200	852	11	9	2	Protein phosphatase 2C family protein
AT1g22280	864	17	14	3	Encodes a phytochrome-associated protein, PAPP2C (phytochrome-associated protein phosphatase type 2C). PAPP2C interacts in the nucleus with phyA (phytochrome A) and phyB. Functions as a regulator of phytochrome-interacting factor PIF3 by dephosphorylating phytochromes in the nucleus
AT1g34750	849	6	5	1	Protein phosphatase 2C family protein
AT3g15260	870	11	9	2	Protein phosphatase 2C family protein
AT2g20630	873	11	9	2	PP2C induced by AVRRPM1
AT4g28400	852	5	4	1	Protein phosphatase 2C family protein
AT1g43900	1116	8	7	1	Protein phosphatase 2C family protein
AT5g53140	1263	9	8	1	Protein phosphatase 2C family protein
AT4g31750	936	9	8	1	Encodes HopW1-1-interacting protein 2 (WIN2). Interacts with the *P. syringae* effector HopW1-1. WIN2 has protein phosphatase activity. Modulates plant defenses against bacteria. Three WIN proteins are identified so far (WIN1: AT1G80600; WIN2: AT4G31750; WIN3: AT5G13320)
AT5g24940	1344	8	7	1	Protein phosphatase 2C family protein
AT5g10740	1065	9	8	1	Protein phosphatase 2C family protein
Group G					
AT3g62260	1155	8	6	2	Protein phosphatase 2C family protein
AT1g48040	1152	2	1	1	Protein phosphatase 2C family protein
AT3g17250	1269	2	1	1	Protein phosphatase 2C family protein
AT2g25620	1179	4	3	1	Encodes DBP1, a member of the DBP factors (DNA-binding protein phosphatases) featuring sequence-specific DNA-binding and protein phosphatase activity. DBP1 is involved in plant-potyvirus interactions. Loss of function of DBP1 renders resistance to potyviruses
AT2g33700	1143	4	3	1	Encodes a putative protein phosphatase 2C that positively regulates salt tolerance in abscisic acid-dependent manner
AT3g51470	1086	3	2	1	Protein phosphatase 2C family protein
Group H					
AT1g09160	1287	17	15	2	Protein phosphatase 2C family protein

(continued)

Table 10.1 (continued)

Arabidopsis PP2C family genes (source: Schweighofer et al. 2004; https://www.arabidopsis.org)					
Locus id	CDS (bp)	Exons	Introns	No. of transcripts	Protein function/description
AT1g47380	1287	10	9	1	Protein phosphatase 2C family protein
AT1g68410	1311	19	17	2	Protein phosphatase 2C family protein
Group I					
AT2g25070	1068	10	9	1	Protein phosphatase 2C family protein
AT4g31860	1074	21	19	2	Protein phosphatase 2C family protein
Group J					
AT3g63320	1272	6	5	1	Protein phosphatase 2C family protein
AT3g63340	3126	41	39	2	Kinase superfamily protein
Other PP2Cs					
AT4g11040	888	8	6	2	Encodes a nuclear localized protein with sequence similarity to PP2C phosphatases that is involved in seed dormancy. Loss-of-function mutations have reduced seed dormancy but do not act through ABA or DOG1 pathways
AT1g75010	2226	15	14	1	Encodes ARC3 (Accumulation and Replication of Chloroplast 3), a chloroplast division factor functioning in the initiation of chloroplast division. ARC3 is a chimera of the prokaryotic FtsZ and part of the eukaryotic phosphatidylinositol-4-phosphate 5-kinase (PIP5K)
AT1g18030	1056	20	18	2	Protein phosphatase 2C family protein
PPH1					
AT4g27800	1167	32	29	3	Choroplast protein phosphatase TAP38/PPH1 is required for efficient dephosphorylation of the LHCII anthena and state transition from state 2 to state 1
AT2g40860	1977	12	11	1	Protein kinase family protein/protein phosphatase 2C (PP2C) family protein
KAPP					
AT5g19280	1776	27	25	2	Kinase-associated protein phosphatase composed of three domains: an amino-terminal signal anchor, a kinase interaction (KI) domain, and a type 2C protein phosphatase catalytic region

B. Rice (source: https://phytozome.jgi.doe.gov/pz/portal.html)					
PP2A family genes					
MSU locus ID	CDS (bp)	Exons	Introns	No. of transcripts	Protein function/description
LOC_Os02g12580	924	6	5	1	Similar to isoform 2 of serine/threonine protein phosphatase PP2A-3 catalytic subunit

(continued)

Table 10.1 (continued)

B. Rice (source: https://phytozome.jgi.doe.gov/pz/portal.html)

PP2A family genes

MSU locus ID	CDS (bp)	Exons	Introns	No. of transcripts	Protein function/description
LOC_ Os03g07150	945	21	19	2	Similar to serine/threonine protein phosphatase PP2A-3 catalytic subunit
LOC_ Os03g62730	1554	17	14	3	Similar to protein phosphatase 2A B′ regulatory subunit
LOC_ Os04g40860	1530	2	1	1	Similar to protein phosphatase 2A B′ regulatory subunit
LOC_ Os05g48150	1362	5	3	2	Similar to protein phosphatase 2A B′kappa subunit
LOC_ Os06g11640	1191	2	1	1	Similar to protein phosphatase 2A, regulatory subunit B′ (PP2A, subunit B′, PR53 isoform) (phosphotyrosyl phosphatase activator)
LOC_ Os06g37660	921	6	5	1	Protein phosphatase 2A (Os06t0574500-01)
LOC_ Os08g02860	1551	3	2	1	Similar to protein phosphatase 2A B′ regulatory subunit. Similar to protein phosphatase 2A B′ regulatory subunit
LOC_ Os09g07510	1764	39	36	3	Protein phosphatase 2A A subunit (phosphatase 2A regulatory A subunit). Similar to phosphatase 2A regulatory A subunit
LOC_ Os10g33680	1626	13	12	1	Similar to protein phosphatase 2A 62 kDa B′ regulatory subunit (protein phosphatase 2A 62 kDa B regulatory subunit)
LOC_ Os11g04520	1257	28	25	3	Similar to PP2A regulatory subunit-like protein (Os11t0141000-01); similar to PP2A regulatory subunit-like protein

PP2C family genes

Gene name	MSU locus ID	CDS (bp)	Exons	Introns	No. of transcripts	Protein function/description
OsPP1	LOC_ Os01g07090	996	12	11	1	5-Azacytidine resistance protein azr1, putative, expressed
OsPP2	LOC_ Os01g19130	1143	4	3	1	PHS1, putative, expressed
OsPP3	LOC_ Os01g20940	2781	11	10	1	PHS1, putative, expressed
OsPP4	LOC_ Os01g24470	2616	19	17	2	Ser/Thr protein phosphatase family protein, putative, expressed
OsPP5	LOC_ Os01g24750	978	4	3	1	Dual-specificity protein phosphatase, putative, expressed

(continued)

Table 10.1 (continued)

PP2C family genes						
Gene name	MSU locus ID	CDS (bp)	Exons	Introns	No. of transcripts	Protein function/description
OsPP6	LOC_ Os01g29469	600	5	4	1	Protein phosphatase protein, putative, expressed
OsPP7	LOC_ Os01g32964	1005	8	7	1	Protein phosphatase 2C-containing protein, expressed
OsPP8	LOC_ Os01g36080	1974	45	41	4	Protein phosphatase 2C, putative, expressed
OsPP9	LOC_ Os01g37130	1170	22	20	2	Protein phosphatase 2C, putative, expressed
OsPP10	LOC_ Os01g40094	1404	4	3	1	Protein phosphatase 2C, putative, expressed
OsPP11	LOC_ Os01g43100	1176	8	7	1	Protein phosphatase 2C, putative, expressed
OsPP12	LOC_ Os01g46760	1212	3	2	1	Ser/Thr protein phosphatase family protein, putative, expressed
OsPP13	LOC_ Os01g49690	912	10	9	1	Dual-specificity protein phosphatase, putative, expressed
OsPP14	LOC_ Os01g53710	1026	6	5	1	Protein phosphatase 2C, putative, expressed
OsPP15	LOC_ Os01g62760	1245	4	3	1	Dual-specificity protein phosphatase, putative, expressed
OsPP16	LOC_ Os01g64010	780	9	7	2	Protein phosphatase 2C, putative, expressed
OsPP17	LOC_ Os01g74530	1131	5	4	1	Protein phosphatase 2C, putative, expressed
OsPP18	LOC_ Os02g05630	1047	30	26	4	Protein phosphatase 2C, putative, expressed
OsPP19	LOC_ Os02g08364	1089	13	12	1	OsPP2Ac-3—phosphatase 2A isoform 3 belonging to family 1
OsPP20	LOC_ Os02g12580	924	6	5	1	Protein phosphatase 2C, putative, expressed
OsPP21	LOC_ Os02g13100	1170	2	1	1	Protein phosphatase 2C, putative, expressed
OsPP22	LOC_ Os02g15594	1092	8	7	1	AGC_PKA/PKG_like.1—ACG kinases include homologs to PKA, P
OsPP23	LOC_ Os02g17970	3264	30	28	2	Protein phosphatase 2C, putative, expressed
OsPP24	LOC_ Os02g27220	1557	5	4	1	Protein phosphatase protein, putative, expressed
OsPP25	LOC_ Os02g35910	1329	9	8	1	Protein phosphatase 2C, putative, expressed

(continued)

Table 10.1 (continued)

PP2C family genes						
Gene name	MSU locus ID	CDS (bp)	Exons	Introns	No. of transcripts	Protein function/description
OsPP26	LOC_ Os02g38580	1566	11	10	1	Protein phosphatase 2C-containing protein, expressed
OsPP27	LOC_ Os02g38690	2292	13	12	1	Protein phosphatase 2C-containing protein, expressed
OsPP28	LOC_ Os02g38710	2415	13	12	1	Protein phosphatase 2C-containing protein, expressed
OsPP29	LOC_ Os02g38780	1959	40	37	3	Protein phosphatase 2C-containing protein, expressed
OsPP30	LOC_ Os02g38804	1554	12	11	1	Protein phosphatase 2C, putative, expressed
OsPP31	LOC_ Os02g39410	1467	29	26	3	Cyclin, N-terminal domain-containing protein, expressed
OsPP32	LOC_ Os02g39470	2916	15	14	1	Protein phosphatase 2C, putative, expressed
OsPP33	LOC_ Os02g39480	1746	13	12	1	5-Azacytidine resistance protein related, putative, expressed
OsPP34	LOC_ Os02g42250	960	1	0	1	Stage II sporulation protein E, putative, expressed
OsPP35	LOC_ Os02g42270	948	1	0	1	Protein phosphatase 2C, putative, expressed
OsPP36	LOC_ Os02g46080	1164	11	9	2	Protein phosphatase 2C, putative, expressed
OsPP37	LOC_ Os02g46490	1791	4	3	1	Dual-specificity protein phosphatase, putative, expressed
OsPP38	LOC_ Os02g48840	807	5	4	1	Tyrosine phosphatase family protein, putative, expressed
OsPP39	LOC_ Os02g53160	615	5	4	1	Protein phosphatase 2C, putative, expressed
OsPP40	LOC_ Os02g55560	1065	7	5	2	Ser/Thr protein phosphatase family protein, putative, expressed
OsPP41	LOC_ Os02g57450	951	9	7	2	Dual-specificity protein phosphatase, putative, expressed
OsPP42	LOC_ Os03g01750	1107	85	79	6	Protein phosphatase 2C, putative, expressed
OsPP43	LOC_ Os03g04430	1200	12	9	3	OsPP2Ac-5—phosphatase 2A isoform 5 belonging to family 2

(continued)

Table 10.1 (continued)

PP2C family genes						
Gene name	MSU locus ID	CDS (bp)	Exons	Introns	No. of transcripts	Protein function/description
OsPP44	LOC_ Os03g07150	945	21	19	2	Stage II sporulation protein E, putative, expressed
OsPP45	LOC_ Os03g09220	1797	55	49	6	Protein phosphatase 2C, putative, expressed
OsPP46	LOC_ Os03g10950	1179	3	2	1	Ser/Thr protein phosphatase family protein, putative, expressed
OsPP47	LOC_ Os03g16110	969	12	9	3	Protein phosphatase 2C, putative, expressed
OsPP48	LOC_ Os03g16170	1215	7	5	2	Protein phosphatase 2C, putative, expressed
OsPP49	LOC_ Os03g16760	1896	4	3	1	Protein phosphatase 2C, putative, expressed
OsPP50	LOC_ Os03g18150	1176	3	2	1	Protein phosphatase protein, putative, expressed
OsPP51	LOC_ Os03g18970	1299	7	6	1	Protein phosphatase 2C, putative, expressed
OsPP52	LOC_ Os03g25600	2934	11	9	2	Protein phosphotase protein, putative, expressed
OsPP53	LOC_ Os03g27780	1335	17	15	2	Serine/threonine protein phosphatase, putative, expressed
OsPP54	LOC_ Os03g44500	3012	21	20	1	Protein phosphatase 2C, putative, expressed
OsPP55	LOC_ Os03g55320	1143	12	9	3	OsPP2Ac-2—phosphatase 2A isoform 2 belonging to family 2
OsPP56	LOC_ Os03g59060	924	23	21	2	Stage II sporulation protein E, putative, expressed
OsPP57	LOC_ Os03g59470	1440	5	4	1	Protein kinase-associated protein phosphatase, putative, expressed
OsPP58	LOC_ Os03g59530	1686	13	12	1	Protein phosphatase 2C, putative, expressed
OsPP59	LOC_ Os03g60650	1920	4	3	1	Protein phosphatase 2C, putative, expressed
OsPP60	LOC_ Os03g61690	1161	7	5	2	Retrotransposon protein, putative, unclassified, expressed
OsPP61	LOC_ Os04g08560	1305	5	4	NA	Protein phosphatase 2C, putative, expressed
OsPP62	LOC_ Os04g17130	4533	16	15	1	Protein phosphatase 2C, putative, expressed
OsPP63	LOC_ Os04g25570	1383	5	4	1	Protein phosphatase protein, putative, expressed

(continued)

Table 10.1 (continued)

PP2C family genes						
Gene name	MSU locus ID	CDS (bp)	Exons	Introns	No. of transcripts	Protein function/description
OsPP64	LOC_ Os04g33080	1563	16	13	3	Protein phosphatase 2C, putative, expressed
OsPP65	LOC_ Os04g37660	1386	16	4	2	Protein phosphatase 2C, putative, expressed
OsPP66	LOC_ Os04g37904	855	12	10	2	Protein phosphatase 2C, putative, expressed
OsPP67	LOC_ Os04g42260	1059	12	11	1	Protein phosphatase 2C, putative, expressed
OsPP68	LOC_ Os04g49490	1167	9	7	2	Protein phosphatase 2C, putative, expressed
OsPP69	LOC_ Os04g52000	966	5	4	1	Protein phosphatase 2C, putative, expressed
OsPP70	LOC_ Os04g56450	849	15	13	2	Protein phosphatase 2C, putative, expressed
OsPP71	LOC_ Os05g02110	1782	4	3	1	Serine/threonine protein phosphatase, putative, expressed
OsPP72	LOC_ Os05g04360	1170	4	3	1	Serine/threonine protein phosphatase 5, putative, expressed
OsPP73	LOC_ Os05g05240	2676	42	40	2	Protein phosphatase 2C, putative, expressed
OsPP74	LOC_ Os05g11550	1452	25	23	2	Protein phosphatase 2C, putative, expressed
OsPP75	LOC_ Os05g29030	1176	8	6	2	Dual-specificity protein phosphatase, putative, expressed
OsPP76	LOC_ Os05g38290	1251	7	5	2	Protein phosphatase 2C, putative, expressed
OsPP77	LOC_ Os05g44910	1134	6	5	1	Protein phosphatase 2C, putative, expressed
OsPP78	LOC_ Os05g46040	1164	7	5	2	Protein phosphatase 2C, putative, expressed
OsPP79	LOC_ Os05g49730	1146	3	2	1	Protein phosphatase 2C, putative, expressed
OsPP80	LOC_ Os05g50970	1476	9	8	1	Dual-specificity protein phosphatase, putative, expressed
OsPP81	LOC_ Os05g51510	1338	8	6	2	Ser/Thr protein phosphatase family protein, putative, expressed
OsPP82	LOC_ Os06g05870	816	14	12	2	Protein phosphatase 2C, putative, expressed
OsPP83	LOC_ Os06g06880	969	3	2	1	Dual-specificity protein phosphatase, putative, expressed

(continued)

Table 10.1 (continued)

PP2C family genes						
Gene name	MSU locus ID	CDS (bp)	Exons	Introns	No. of transcripts	Protein function/description
OsPP84	LOC_ Os06g08140	1083	4	3	1	Protein phosphatase 2C, putative, expressed
OsPP85	LOC_ Os06g20340	846	9	7	2	Protein phosphatase 2C, putative, expressed
OsPP86	LOC_ Os06g33530	1062	7	6	1	OsPP2Ac-1—phosphatase 2A isoform 1 belonging to family 1
OsPP87	LOC_ Os06g33549	1059	8	7	1	Protein phosphatase 2C, putative, expressed
OsPP88	LOC_ Os06g37660	921	6	5	1	Protein phosphatase 2C, putative, expressed
OsPP89	LOC_ Os06g39600	1104	8	6	2	Protein phosphatase 2C, putative, expressed
OsPP90	LOC_ Os06g44210	1107	23	21	2	Protein phosphatase 2C, putative, expressed
OsPP91	LOC_ Os06g48300	984	8	7	1	Protein phosphatase 2C, putative, expressed
OsPP92	LOC_ Os06g50380	1179	16	13	3	Protein kinase-associated protein phosphatase, putative, expressed
OsPP93	LOC_ Os07g02330	1134	2	1	1	Protein phosphatase 2C, putative, expressed
OsPP94	LOC_ Os07g11010	1758	26	24	2	Expressed protein
OsPP95	LOC_ Os07g32380	873	5	4	1	Protein phosphatase 2C, putative, expressed
OsPP96	LOC_ Os07g33230	672	3	2	1	Protein phosphatase 2C, putative, expressed
OsPP97	LOC_ Os07g37890	1284	5	4	1	Protein phosphatase protein, putative, expressed
OsPP98	LOC_ Os07g45170	1338	3	2	1	5-AMP-activated protein kinase beta-1 subunit-related, putative
OsPP99	LOC_ Os07g49040	1296	7	6	1	Ser/Thr protein phosphatase family protein, putative, expressed
OsPP100	LOC_ Os08g29160	1782	10	9	1	Protein phosphatase 2C, putative, expressed
OsPP101	LOC_ Os08g35440	924	3	2	1	Ser/Thr protein phosphatase family protein, putative, expressed
OsPP102	LOC_ Os08g39100	1596	5	4	1	Low molecular weight protein-tyrosine-phosphatase slr0328

(continued)

Table 10.1 (continued)

PP2C family genes						
Gene name	MSU locus ID	CDS (bp)	Exons	Introns	No. of transcripts	Protein function/description
OsPP103	LOC_ Os08g40200	1287	5	4	1	Tyrosine phosphatase family protein, putative, expressed
OsPP104	LOC_ Os08g44320	807	6	5	1	Ser/Thr protein phosphatase family protein, putative, expressed
OsPP105	LOC_ Os09g05020	669	11	9	2	Protein phosphatase 2C, putative, expressed
OsPP106	LOC_ Os09g11230	924	8	7	1	Protein phosphatase 2C, putative, expressed
OsPP107	LOC_ Os09g14540	1311	11	10	1	Protein phosphatase protein, putative, expressed
OsPP108	LOC_ Os09g15670	1077	1	0	1	Protein phosphatase 2C, putative, expressed
OsPP109	LOC_ Os09g28560	1269	9	8	1	Expressed protein
OsPP110	LOC_ Os09g38550	1059	46	42	4	OsPP2Ac-4—phosphatase 2A isoform 4 belonging to family 2
OsPP111	LOC_ Os10g22460	1398	9	8	1	Protein phosphatase 2C, putative, expressed
OsPP112	LOC_ Os10g27050	945	11	10	1	Protein phosphatase 2C, putative, expressed
OsPP113	LOC_ Os10g39540	990	18	14	4	Dual-specificity protein phosphatase, putative, expressed
OsPP114	LOC_ Os10g39780	1182	18	14	4	Protein phosphatase 2C, putative, expressed
OsPP115	LOC_ Os10g41240	1089	11	9	2	Dual-specificity protein phosphatase, putative, expressed
OsPP116	LOC_ Os11g01790	1263	5	4	1	Dual-specificity protein phosphatase, putative, expressed
OsPP117	LOC_ Os11g02180	816	4	3	1	mRNA-capping enzyme, putative, expressed
OsPP118	LOC_ Os11g04180	1071	6	5	1	Protein phosphatase 2C, putative, expressed
OsPP119	LOC_ Os11g11070	2094	34	32	2	Protein phosphatase 2C, putative, expressed
OsPP120	LOC_ Os11g13820	1194	4	3	1	Protein phosphatase 2C, putative, expressed
OsPP121	LOC_ Os11g22404	1299	5	4	1	Protein phosphatase 2C, putative, expressed
OsPP122	LOC_ Os11g37540	3348	19	18	1	Dual-specificity protein phosphatase, putative, expressed

(continued)

Table 10.1 (continued)

PP2C family genes						
Gene name	MSU locus ID	CDS (bp)	Exons	Introns	No. of transcripts	Protein function/description
OsPP123	LOC_ Os12g01770	1431	5	4	1	Dual-specificity protein phosphatase, putative, expressed
OsPP124	LOC_ Os12g02120	816	10	8	2	Dual specificity protein phosphatase, putative, expressed
OsPP125	LOC_ Os12g03990	1071	6	5	1	Protein tyrosine phosphatase domain-containing protein, expressed
OsPP126	LOC_ Os12g05660	720	18	14	4	mRNA-capping enzyme, putative, expressed
OsPP127	LOC_ Os12g07590	987	24	21	3	Protein phosphatase 2C, putative, expressed
OsPP128	LOC_ Os12g09120	1998	18	17	1	PTEN, putative, expressed
OsPP129	LOC_ Os12g09640	1266	4	3	1	Protein phosphatase 2C, putative, expressed
OsPP130	LOC_ Os12g21890	1977	13	12	1	Serine/threonine protein phosphatase BSL2, putative, expressed
OsPP131	LOC_ Os12g39120	1179	4	3	1	Serine/threonine protein phosphatase BSL2, putative, expressed
OsPP132	LOC_ Os12g42310	3030	21	20	1	Serine/threonine protein phosphatase BSL2, putative, expressed

Dual-specificity phosphatase (DSP) genes					
MSU locus ID	CDS (bp)	Exons	Introns	No. of transcripts	Protein function/description
LOC_ Os01g29469	600	5	4	1	Similar to dual-specificity protein phosphatase-like protein (Os01t0390900-01); similar to dual-specificity protein phosphatase 9
LOC_ Os01g53710	1026	6	5	1	Protein tyrosine phosphatase, dual-specificity domain-containing protein (Os01t0739200-01)
LOC_ Os01g64010	780	9	7	2	Protein tyrosine phosphatase, dual-specificity domain-containing protein (Os01t0859400-01); protein tyrosine phosphatase
LOC_ Os02g48840	807	5	4	1	Protein tyrosine phosphatase, dual-specificity domain-containing protein (Os02t0720300-01)

(continued)

Table 10.1 (continued)

Dual-specificity phosphatase (DSP) genes					
MSU locus ID	CDS (bp)	Exons	Introns	No. of transcripts	Protein function/description
LOC_ Os02g53160	615	5	4	1	Protein tyrosine phosphatase, SIW14-like domain-containing protein (Os02t0771400-00)
LOC_ Os03g01770	393	3	2	1	Dual-specificity tyrosine phosphatase CDC25, arsenic metabolism (Os03t0108000-01)
LOC_ Os05g44910	1134	6	5	1	Protein tyrosine phosphatase, dual-specificity domain-containing protein (Os05t0524200-01)
LOC_ Os06g05870	816	14	12	2	Similar to dual-specificity protein phosphatase family protein (Os06t0152000-01)
LOC_ Os06g10650	681	5	4	1	Similar to dual-specificity phosphatase protein (Os06t0208700-01)
LOC_ Os06g20340	846	9	7	2	Protein tyrosine phosphatase, dual-specificity domain-containing protein (Os06t0308100-01)
LOC_ Os08g29160	1782	10	9	1	Protein tyrosine phosphatase, dual-specificity domain-containing protein (Os08t0379300-01); hypothetical conserved gene
LOC_ Os09g05020	669	11	9	2	Protein tyrosine phosphatase, negative regulation of drought stress response (Os09t0135700-01)
LOC_ Os10g39860	627	4	2	2	Similar to rhodanese-like protein (Os10t0545700-01); similar to dual-specificity phosphatase
LOC_ Os10g41240	1089	11	9	2	Protein tyrosine phosphatase, dual-specificity domain-containing protein
LOC_ Os11g02180	816	4	3	1	Protein tyrosine phosphatase, dual-specificity domain-containing protein (Os11t0113100-01)
LOC_ Os11g04180	1071	6	5	1	Similar to predicted protein (Os11t0136800-01); similar to dual-specificity protein phosphatase 19
LOC_ Os12g02120	816	10	8	2	Protein tyrosine phosphatase, dual-specificity domain-containing protein. (Os12t0112500-01)
LOC_ Os12g03990	1071	6	5	1	Similar to dual-specificity protein phosphatase 19 (EC 3.1.3.48) (EC 3.1.3.16) (stress-activated protein kinase pathway-regulating phosphatase 1)

(continued)

Table 10.1 (continued)

Dual-specificity phosphatase (DSP) genes					
MSU locus ID	CDS (bp)	Exons	Introns	No. of transcripts	Protein function/description
LOC_ Os12g19350	858	12	10	2	Similar to dual-specificity protein phosphatase 8 (EC 3.1.3.48) (EC 3.1.3.16) (neuronal tyrosine threonine phosphatase 1)
LOC_ Os12g23190	615	6	5	1	Similar to tyrosine-specific protein phosphatase family protein (Os12t0420300-01)

C. Soybean (source: https://phytozome.jgi.doe.gov/pz/portal.html)

PP2A family genes					
Locus ID	CDS (bp)	Exons	Introns	No. of transcripts	Protein function/description
Glyma.13G145700	1452	2	1	1	Serine/threonine protein phosphatase 2A 57 kDa regulatory subunit B′ alpha isoform
Glyma.08G195200	1548	28	26	2	Serine/threonine protein phosphatase 2A regulatory subunit B (PPP2R2)
Glyma.09G185700	1764	26	24	2	Serine/threonine protein phosphatase 2A regulatory subunit A (PPP2R1)
Glyma.19G077700	1611	3	2	1	Serine/threonine protein phosphatase 2A 55 kDa regulatory subunit B′ delta isoform
Glyma.19G204400	942	32	29	3	Serine/threonine protein phosphatase 2A catalytic subunit (PPP2C)
Glyma.02G097600	1743	26	24	2	Serine/threonine protein phosphatase 2A regulatory subunit A (PPP2R1)
Glyma.08G113200	936	21	19	2	Serine/threonine protein phosphatase 2A catalytic subunit (PPP2C)
Glyma.05G155300	936	21	19	2	Serine/threonine protein phosphatase 2A catalytic subunit (PPP2C)
Glyma.03G206900	942	21	19	2	Serine/threonine protein phosphatase 2A catalytic subunit (PPP2C)
Glyma.02G169200	945	11	10	1	Serine/threonine protein phosphatase 2A catalytic subunit (PPP2C)

(continued)

Table 10.1 (continued)

C. Soybean (source: https://phytozome.jgi.doe.gov/pz/portal.html)

PP2A family genes

Locus ID	CDS (bp)	Exons	Introns	No. of transcripts	Protein function/description
Glyma.20G114000	1758	13	12	1	Serine/threonine protein phosphatase 2A regulatory subunit A (PPP2R1)
Glyma.07G012000	1548	42	39	3	Serine/threonine protein phosphatase 2A regulatory subunit B (PPP2R2)

PP2C family genes

Locus ID	CDS (bp)	Exons	Introns	No. of transcripts	Protein function/description
Glyma.19G222700	2664	11	8	3	Protein phosphatase 2C 32
Glyma.11G042300	2220	4	3	1	Protein phosphatase 2C 29
Glyma.11G004000	1731	12	11	1	Protein phosphatase 2C 70
Glyma.11G042300	2220	4	3	1	Protein phosphatase 2C 29
Glyma.09G194200	3228	64	60	4	cGMP-dependent protein kinase/ PKG II
Glyma.01G199600	2262	9	7	2	Protein phosphatase 2C 29
Glyma.04G233900	1173	22	20	2	Protein phosphatase 2C 57
Glyma.01G146600	3291	33	31	2	cGMP-dependent protein kinase/ PKG II
Glyma.01G239700	1737	12	11	1	Protein phosphatase 2C 70
Glyma.06G130900	1179	30	27	3	Protein phosphatase 2C 57
Glyma.15G100900	1296	51	42	9	Protein phosphatase 1K, mitochondrial
Glyma.14G101400	1020	11	10	1	Protein phosphatase 1G [EC:3.1.3.16] (PPM1G, PP2CG)
Glyma.08G220100	1107	22	20	2	Protein phosphatase 1G [EC:3.1.3.16] (PPM1G, PP2CG)
Glyma.06G060800	1038	54	49	5	Protein phosphatase 1G [EC:3.1.3.16] (PPM1G, PP2CG)

The gene model (CDS length, number of exons, introns, and transcripts) information and annotation of different phosphatases are compiled in model plant Arabidopsis (A) and crops like rice (B) and soybean (C). [**Key note**: For most of the genes, more than one transcript and subsequently CDS were found in database. Hence, longest CDS (in bp) was chosen for compilation. The number of exons and the introns denotes cumulative total in different transcripts in a particular gene.]

10.2.1 Protein Phosphatase P (PPP)

The biochemical studies on animals classified serine/threonine phosphatases into type 1 (PP1) and type 2 (PP2) classes based on their substrate specificity and pharmacological properties. PP2 can be further divided into three classes PP2A, PP2B, and PP2C based on their dependency on divalent cations as cofactors. PP2A and PP2C require Ca^{2+} and Mg^{2+}, respectively, for their regulation. PP2A and PP1 phosphatases don't require divalent cations for their activity and regulation (Luan 2003; Chae et al. 2009). However, PP2B, a Ca^{2+}–dependent phosphatase, also known as calcineurin A (CNA), is not identified so far in plants (Uhrig et al. 2013; Singh et al. 2016). Sequence and structural analyses of these gene products demonstrate that PP1, PP2A, and PP2B are more closely related and defined as PPP family, whereas PP2C, pyruvate dehydrogenase phosphatase, and several other Mg^{2+}–dependent Ser/Thr phosphatases are similar to each other and hence are referred to as PPM family (Barford 1996; Cohen 1997; Luan 2003). For this reason, PP1, PP2A, and PP2B are now referred to as protein phosphatase P (PPP) family. PP2C and few other serine/threonine phosphatases that are magnesium (Mg^{2+}) dependent comprise a separate family of serine/threonine phosphatases known as protein phosphatase M (PPM) family (Chae et al. 2009). Besides the classical PPP enzymes included in PP1/2A/2B subgroups, other phosphatases with high homology to the PPP family have been identified that contain unique domains or play distinct functions. These more distantly related enzymes are referred to PP4, PP5, PP6, and PP7, which are sequence and structurally related (Cohen 1997; Luan 2003; Uhrig et al. 2013). Even within the same family, significant structural diversity can be generated by the presence of unique regulatory and targeting domains or attachment of regulatory subunits to the catalytic subunits. These regulatory domains or subunits may localize the protein complexes to a specific subcellular compartment, modulate the substrate specificity, or alter the catalytic activity (Luan 2000). Based on the advancement in bioinformatics and phylogenetic analysis tools, the function and the phylogenetic relationship between these protein phosphatases are now being understood (Uhrig et al. 2013). PPPs occur in various complexes with regulatory and targeting subunits, which modulate the activity and specificity of catalytic subunits, and target the enzymes to specific locations (Janssens and Goris 2001; Ceulemans and Bollen 2004; Farkas et al. 2007).

10.2.1.1 Protein Phosphatase 1

The unraveling of crystal structure of PP1 and PP2B has provided cues for different models for the structure and catalysis of PPP enzymes (Goldberg et al. 1995; Egloff et al. 1995; Luan 2003). Based on the primary sequence, it is confirmed that PP1c and PP2B share a common catalytic domain structure. The catalytic domain of PP1c and PP2B consists of a central "sandwich" which is formed by the two subdomains of a "helix-sheet" mixture (Luan 2003). The crystal structure of the two enzymes also reveals the importance of metal ions such as Mn^{2+} and Fe^{2+}/Fe^{3+} for PP1c (Egloff et al. 1995) and Zn^{2+} and Fe^{2+}/Fe^{3+} for PP2B (Yu et al. 1997) in the catalytic reaction of PPP family phosphatases (Luan 2003). Type1 phosphoprotein Ser/Thr phosphatases (PP1) are highly conserved enzymes across all the eukaryotes. In Arabidopsis, five distinct Arabidopsis cDNA clones (TOPP1, TOPP2, TOPP3, TOPP4, and TOPP5), which encode the catalytic subunit (PP1c) of type 1 protein phosphatases, were isolated, and later genomic southern blotting confirmed that these clones are the products of five distinct genes of type 1 protein phosphatase family (Smith and Walker 1993). Later it was found that plant TOPP and animal PP1 sequences show significant phylogenetic divergence (Lin et al. 1999). The protein phosphatases 1 are also found to be conserved across eukaryotes with nine genes in Arabidopsis (Kerk et al. 2008; Uhrig et al. 2013). Another protein phosphatase 1 gene isolated from *Phaseolus vulgaris* is believed to be involved in dephosphorylation events arising after the initial responses to biotic stress (Zimmerlin et al. 1995; Chae et al. 2009). Additionally, three genes (*NPP1*, *NPP2*, and *NPP3*) encoded type 1 PP (PP1) and two genes (*NPP4* and *NPP5*) encoded type 2A PP (PP2A) showing differential tissue-specific expression was isolated and cloned in *Nicotiana tabacum* (Suh et al. 1998). Other studies have also identified RVxF motif containing plant PP1 partner's inhibitor-3 (I3) (Takemiya et al. 2009; Uhrig et al. 2013) and the rice (*Oryza sativa*) protein RICE SALT SENSITIVE 1 (RSS1) (Ogawa et al. 2011; Uhrig et al. 2013). RVxF-containing PP1 regulatory protein RSS1 was identified through a combined approach of genetic screening for salt tolerance in rice and yeast two-hybrid screening; the loss of RSS1 resulted in short root and dwarf phenotypes under high salt stress (Ogawa et al. 2011; Uhrig et al. 2013). These effects of RSS1 are exerted by regulating the G1/S transition, possibly through an interaction of RSS1 with protein phosphatase 1, and are mediated by the phytohormone, cytokinin (Ogawa et al. 2011).

10.2.1.2 Protein Phosphatase 2A

The core enzyme of PP2A is a dimer ($PP2A_D$), consisting of a 36 kDa catalytic subunit ($PP2A_C$) and a regulatory subunit of molecular mass 65 kDa, termed PR65, or the A subunit along with a third regulatory B subunit can be associated with this core structure (Janssens and Goris 2001). The B subunits, which determine the substrate specificity and subcellular localization of PP2As, are classified into 55 kDa B, 54–74 kDa B′, and 72–130 kDa B′ subunit families (Farkas et al. 2007). Additionally, a number of genes encoding the catalytic subunit and the regulatory subunits of protein phosphatase 2A have already been found out. In Arabidopsis, isolation and sequencing of PP2A-1, PP2A-2, and PP2A-3 clones suggested that PP2A-1 and PP2A-2 encode almost identical proteins (97% identity). Both proteins are 306 amino acids in length and are very similar (79–80% identity) to the mammalian isotypes of the catalytic subunit of protein phosphatase 2A. The PP2A-3 protein is extremely similar (95% identity) to the predicted protein from a cDNA clone previously found in *Brassica napus*. It is also found that all the PP2As were differentially expressed in higher plants (Ariño et al. 1993). In rice, five genes (*OsPP2A-1*, *OsPP2A-2*, *OsPP2A-3*, *OsPP2A-4*, and *OsPP2A-5*) have been identified which encode the PP2A catalytic subunits (PP2Ac). It has been established that among the five genes, *OsPP2A-2*, *OsPP2A-4*, and *OsPP2A-5* are ubiquitously expressed in all rice tissues during plant development and differentially expressed in response to high salinity and the combined stresses of drought and heat (Yu et al. 2003, 2005). In Arabidopsis, TAP46 (a phosphatase 2A-associated protein of size 46 kDa), which is a homolog of *Saccharomyces cerevisiae* TAP42 and mammalian α4, was found to be involved in chilling response and suggested there could be existence of a target-of-rapamycin-like signaling pathway in plants (Harris et al. 1999). However, the involvement of TAP42 in the regulation of TOR- and SnRK-mediated energy homeostasis and abiotic stress tolerance mechanism has been established in recent times (Fu et al. 2020).

10.2.1.3 Protein Phosphatase 2B

Protein phosphatase 2B (PP2B; also known as the calcineurin A or CNA) requires divalent cation like Ca^{2+} for their activity and regulation. The activity and the role of PP2B has been reported in plants, but the genes encoding the catalytic subunit of PP2B are yet to be identified (Bethke and Jones 1997; Pardo et al. 1998; Chae et al. 2009). Despite extensive genome exploration of several plants, PP2B or calcineurin A (CNA) encoding gene has not been identified. Similar to animal PP2B regulatory subunits, i.e., calcineurin B, some genes have been found in Arabidopsis and rice (Kudla et al. 1999; Luan et al. 2002; Kolukisaoglu et al. 2004; Chae et al. 2009). In contrary to phosphatases, the calcineurin B-like proteins (CBLs) target a set of kinases known as CIPKs to form ABA and abiotic stress signaling (Kudla et al. 1999; Cheong et al. 2003; Kim et al. 2003; Kolukisaoglu et al. 2004; Pandey et al. 2004; Pandey 2008; Chae et al. 2009).

10.2.1.4 PP4, PP5, PP6, and PP7

The protein phosphatases, PP2A, PP4/PPX, and PP6 share close resemblance and a common ancestry among each other when compared phylogenetically with other members of the PPP family (Brewis et al. 1993; Moorhead et al. 2009; Uhrig et al. 2013). Despite its strong sequence conservation, PPX in plants have different functions than in animals as it has been demonstrated in *A. thaliana* by cloning two PPX-1 and PPX-2 genes, which were differentially expressed in all organs. PPX-1 shows co-localization with ferredoxin in root epidermal cells, and immunolocalization indicates that PPX-1 is a lumenal protein in root plastids (Pujol et al. 2000; Farkas et al. 2007). However, it is still unclear whether PPX-1 acts in the regulation of plastid-specific protein kinases and substrates (Farkas et al. 2007).

Protein serine/threonine phosphatase 5 (PP5) plays an important role in signal transduction in animal cells, but in plants, there is paucity of information about PP5. In tomato, LePP5, a PP5-encoding gene, has been cloned and found be a single-copy *LePP5* gene encoding two mRNA species that arise by alternative pre-mRNA splicing. Similarly, Arabidopsis was found to express two *PP5* transcripts, suggesting that alternative splicing of *PP5* pre-mRNA is not specific for tomato (Van Bentem et al. 2003). The larger isoform (62 kDa) maintains the additional exon and endoplasmic reticulum localization, whereas the smaller isoform (55 kDa) exhibits the well-conserved dual cytosolic-nuclear subcellular localization (Van Bentem et al. 2003; Uhrig et al. 2013). It has been demonstrated that disease resistance can be achieved by assembling cyclophilin 40 (CYP40), protein phosphatase 5 (PP5), and several other proteins with the tetratricopeptide repeat (TPR) domain associates with AGO1 in an HSP90-dependent manner (Iki et al. 2012). Another study indicated that overexpression of protein serine/threonine phosphatase 5 (AtPP5) gene enhanced heat shock tolerance in Arabidopsis *thaliana*. However, T-DNA-inserted *atpp5* mutant was sensitive to heat stress (Park et al. 2011). Apart from these, several other regulatory roles of PP5 have been unveiled so far which explains PP5 as a key component of many signaling pathways.

The function of *PP6* is limited to specific physiological functions like flowering. Arabidopsis *PP6* plays distinct roles in PhyA and PhyB regulation and possibly in flowering time control (Farkas et al. 2007). Pea PP6, named FyPP, interacts with oat phytochrome PhyA and Arabidopsis PhyB in yeast. *In vitro* pull-down assays indicate that FyPP preferentially binds the phosphorylated form of PhyA and far-red forms of PhyA and PhyB (Kim et al. 2002; Farkas et al. 2007). The phytochromes functionally interact with catalytic subunit of a Ser/Thr-specific protein phosphatase 2A-designated FyPP. The interactions were influenced by phosphorylation status and spectral conformation of the phytochromes. These results indicated that a self-regulatory phytochrome kinase-phosphatase coupling is a key signaling component in the photoperiodic control of flowering (Kim et al. 2002).

PP7 subfamily is very much new to the plant systems, and unlike the PPP family members the regulatory and catalytic subunits do not contain N-terminal and C-terminal adjunct. Although PP7 lacks EF-hand domains, Ca^{2+} still influences PP7 because the catalytic insert region has been believed to bind calmodulin *in vitro* in a Ca^{2+}-dependent manner (Kutuzov and Andreeva 2001; Uhrig et al. 2013). Like PP6 subfamily, PP7 is also involved in regulation of phytochrome, cryptochrome, and light-sensing mechanism (Møller et al. 2003; Genoud et al. 2008; Sun et al. 2012; Uhrig et al. 2013).

10.2.2 Protein Phosphatase M (PPM)

PPM family constitutes serine/threonine phosphatases including PP2C phosphatases and some other phosphatases that are dependent on the divalent cation Mg^{2+} for their activation and the catalysis. PP2Cs are monomeric enzymes present in both prokaryotes and the eukaryotes. Arabidopsis genome contains 76 PP2C genes, which are largest among the other eukaryotes (Kerk et al. 2002; Schweighofer et al. 2004). Plant PP2Cs are involved in many regulatory pathways and play an important role in signal transduction. Majority of Arabidopsis PP2Cs are classified into ten groups comprising A to J (Schweighofer et al. 2004). Several experiments have shown that group A contains ABA-associated PP2C genes, group B are homologues of alfalfa PP2Cs (Known as MP2Cs) believed to be involved in MAPK signaling, and group C contains the poltergeist phosphatase gene (POL) shown to be involved in flower development through CLAVATA pathway (Chae et al. 2009). PP2Cs are often considered as the negative regulators of the ABA signaling (Merlot et al. 2001; Schweighofer et al. 2004). In Arabidopsis, HAI-1 (a group A PP2C; AT5G59220) encodes a nuclear protein that has been mutated, and the loss-of-function mutant in water loss condition did not show any wilting and found to be having green rosette leaves. Similarly, the loss-of-function mutant showed higher insensitivity toward exogenous ABA and NaCl. HAI-1 overexpression did not improve the phenotype under stress conditions (Zhang et al. 2013). Arabidopsis plants overexpressing *OsPP108* (a group A PP2C in rice) are highly insensitive to ABA and tolerant to high salt and mannitol stresses during seed germination, root growth, and overall seedling growth, and at adult stage, *OsPP108* overexpression leads to high tolerance to salt, mannitol, and drought stresses in transgenic Arabidopsis plants (Singh et al. 2015). Another study identified interactors of ABI1 and ABI2 (type 2C protein phosphatases) as Regulatory Components of ABA Receptor (RCARs), which was shown to bind ABA, to mediate ABA-dependent inactivation of ABI1 or ABI2 *in vitro*, and to antagonize PP2C action in plants (Ma et al. 2009). *PP2Cs*, in both Arabidopsis and rice, have similar expression pattern when subjected to different stresses like drought, high salinity, and cold (Xue et al. 2008; Singh et al. 2016). This type of expression pattern showed that PP2C could be involved in many signal transduction pathways. Group-A PP2Cs regulate SnRK2 kinases including SnRK2.2, SnRK2.3, and SnRK2.6 and, hence, can be considered as positive regulators of ABA and abiotic stress signaling (Fujii et al. 2009; Fujita et al. 2011; Singh et al. 2016). Besides, PP2Cs regulate abiotic stress signaling through MAPK signaling pathways (Danquah et al. 2014).

10.2.3 Protein Tyrosine Phosphatases (PTP)

As their name suggests, protein tyrosine phosphatases (PTPs) can dephosphorylate the tyrosine residues. PTPs are divided in to two groups as (1) tyrosine-specific PTPs and (2) dual-specificity PTPs based on their specificity toward phosphoamino acid. The first group depends on the ability to dephosphorylate the tyrosine residues, whereas the second group can dephosphorylate the phosphotyrosine residues as well as the phosphoserine or threonine (Stone and Dixon 1994; Tonks and Neel 1996; Chae et al. 2009; Shankar et al. 2015). Despite having great difference in the amino acid sequences and the substrate specificity, both the classes of PTPs showed high similarity in their crystal structure (Barford 1999; Shankar et al. 2015). Further it has been classified into receptor-like PTPs and intracellular PTPs. The receptor-like PTPs have a ligand binding domain which is extracellular and one or two cytoplasmic PTP domains, whereas the intracellular PTPs like PTP1B and SHP1 contain a single catalytic domain and a number of amino or carboxyl terminal extensions. These include an SH2 domain that involves targeting and regulating mechanisms (Chae et al. 2009). The protein tyrosine kinases (PTKs) are well known to share sequence similarity with protein serine/threonine kinases; however, the enzymes belonging to PTPs share no sequence similarity with protein serine/threonine phosphatases (Walton and Dixon 1993; Shankar et al. 2015).

It has been well established that PTPs play crucial roles in signal transduction and various regulatory activities in number of organisms starting from protozoans to plant and animal kingdom. In the plant kingdom, in Arabidopsis and rice, the PTPs have been well characterized (Kerk et al. 2002; Singh et al. 2010). Using modern tools and techniques, the discoveries of plant PTPs and their genes were carried out a decade after the discovery of PTPs in animal system. In Arabidopsis AtPTP1 as the first plant PTP was identified and characterized by Luan and team (Xu et al. 1998; Gupta et al. 1998). Further rice (*Oryza sativa*) PTPs and dual-specific phosphatase (DSP) gene families have been identified by Singh et al. (2010) in crop plants (Shankar et al. 2015). In tomato and pine, the presence of tyrosine phosphatase-like genes has been identified (Fordham-Skelton et al. 2002; Gupta and Luan 2003). Subsequently the whole genome sequencing of Arabidopsis led to identification of a number of PTPs and DSPs like genes encoding proteins similar to catalytic domain of PTPs in other crops (Kerk et al. 2002). Out of 24 members of putative PTPs from Arabidopsis, one encodes a Tyr-specific PTP, 22 genes encode dual-specificity protein Tyr phosphatases (DSPs or DsPTPs), and a single gene of low molecular weight PTP (LMWPTP) (Shankar et al. 2015). Further genome-wide identification and transcriptional analysis revealed that rice is comprised of 23 DSPs, a single member of PTP, and LMWPTP (Kerk et al. 2008; Singh et al. 2010). Additionally, 29 PTPs have been identified from maize on the basis of sequence similarity of protein phosphatases from rice and Arabidopsis (Singh and Pandey 2012). Interestingly, genomes of Arabidopsis, rice, and maize contain much lesser PTPs than the human genome, comprised of more than 100 PTPs including 60 DSPs (Shankar et al. 2015).

10.3 Role of Phosphatases in Abiotic Stress Signaling in Crop Plants

The protein phosphorylation and dephosphorylation caused by protein kinases and the protein phosphatases, respectively, are crucial for signal transduction during abiotic stresses in plants (Hunter 2009; Pandey et al. 2014). Phosphatases regulate various plant processes including ABA signaling, stomatal opening/closing, and "K" sensing. Brief information about the regulatory role of different protein phosphatases for these processes is summarized below. In addition, an integrated network highlighting the interconnection is depicted in Fig. 10.2.

10.3.1 Salt Stress

Salinity stress is often considered as the major abiotic stress, which severely affects plant growth and development. The salt stress signaling is mainly regulated by the salt overly sensitive (SOS) pathway, which is primarily functioned by the protein kinase activities. However, protein phosphatases do regulate the salt stress mechanism by involving themselves in several other signaling pathways. In Arabidopsis, *AtPTP1*, a protein Tyr phosphatase, showed several-fold higher expressions under salt stress, while it remained downregulated under cold stress condition (Xu et al. 1998). *In vitro* biochemical analysis of AtPTP1 and AtDsPTP1 confirmed that they both dephosphorylate and inactivate MAPKs, which has an important role in regulation of abiotic stress signaling (Luan 2002). Besides signal transduction, the role of PTP has been found to be regulating ionic fluxes across the tonoplast during turgor regulations in guard cells, which is also an adaptive feature for managing ionic homeostasis (MacRobbie 2002). Another study indicated that an Arabidopsis protein phosphatase group 1 (AtPP2CG1) positively regulates salinity tolerance by upregulating several salt-responsive genes in an ABA-dependent manner (Liu et al. 2012b). However, overexpressing *OsPP108* (a rice protein phosphatase 2C, orthologous to Arabidopsis group A PP2C) in Arabidopsis was found to be highly insensitive to ABA and tolerant to high salt and mannitol stresses during seed germination, root growth, and overall seedling growth. Subsequently the overexpression of *OsPP108* in fully grown plants has also been found to impart high tolerance to salt, mannitol, and drought stresses (Singh et al. 2015). Also, AtDsPTP1 (a dual-specificity protein phosphatase 1) mutant led to improved phenotype under osmotic stress suggesting it was a negative regulator of osmotic stress (Shankar et al. 2015). In wheat, a protein phosphatase TaPP2C1 was found to negatively regulate ABA signaling and imparts salt tolerance positively by increasing the antioxidant defense mechanism and transcription of genes independent of ABA signaling which the overexpressed transgenic tobacco lines confirmed (Hu et al. 2015). Apart from that, in rice, the protein phosphatase 1a (*OsPP1a*) is found to be involved in salt stress tolerance. The transgenic rice overexpressing *OsPP1a* showed enhanced tolerance to high salt treatment and *SnRK1A,* and two other stress-responsive genes, *OsNAC5* and *OsNAC6*, were found to be upregulated in transgenic lines (Liao et al. 2016). Also in Arabidopsis, a protein phosphatase PP2A-C5 gene, which encodes the catalytic subunit 5 of PP2A, was found to be an important player for plant growth in salt stress, confirmed by the overexpressed phenotype, which showed better root and shoot growth than the loss-of-function phenotype under salt stress (Hu et al. 2017). Further, GhDsPTP3a (a cotton protein phosphatase), demonstrated to be a positive

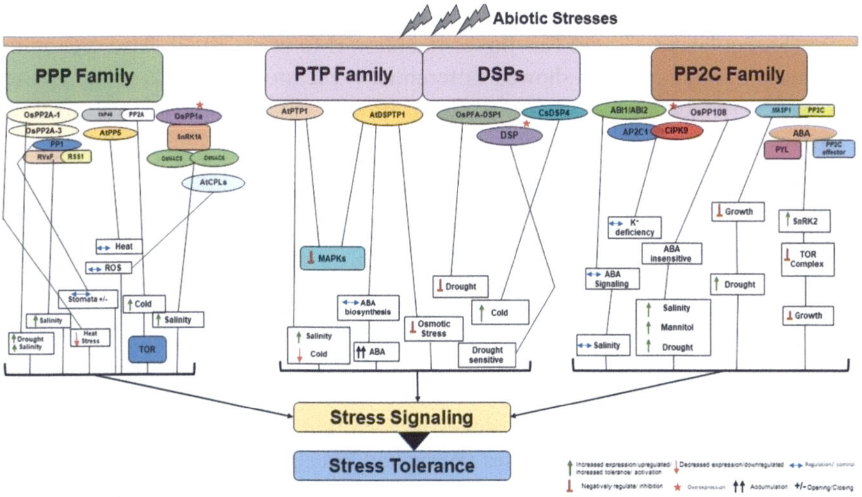

Fig. 10.2 Integrated view of different major phosphatases and their downstream targets. The regulatory roles of three families of plant phosphatase genes under abiotic stress conditions are summarized. **PPP family**: In Arabidopsis, AtCPLs are involved in the regulations and detoxification of ROS molecules bringing redox homeostasis to the plant. *AtPP5* is involved in the regulation of heat stress. AtTAP46 (a homolog of *Saccharomyces cerevisiae* TAP42 and mammalian α4) binds with catalytic subunit of PP2A and regulates target-of-rapamycin-like signaling pathway (TOR) in plants. In rice, *OsPP2A-1* and *OsPP2A-3* genes are upregulated under drought and salt stress, while *OsPP2A-1* remains downregulated under heat stress. *RSS1* (*RVxF*-containing PP1 regulatory protein) mutation results in short root and dwarf phenotypes under salt stress condition. **PTP/DSP family**: Arabidopsis *AtPTP1* gene shows a contrasting expression pattern as it is upregulated in salinity stress whereas downregulated in cold stress condition. Arabidopsis *AtDSPTP1* is shown to regulate the ABA biosynthesis and accumulate ABA in the cell and carry out the signal transduction mechanism for the abiotic stress tolerance mechanism. *AtDsPTP1* works as the negative regulator of osmotic stress as its mutant is shown to grow better than the wild type under osmotic stress. Both AtPTP1 and AtDsPTP1 dephosphorylate and inactivate MAPKs and, hence, act as a negative regulator of abiotic stress signaling. Similarly, rice *OsPFA-DSP1* (fungi atypical dual-specificity phosphatase) is also a negative regulator of drought stress. Chestnut *CsDSP4* is induced under low temperature conditions, and overexpression of DSP increased plant sensitivity toward drought stress. **PP2C family**: *ABI1/ABI2* functions as central components in abscisic acid (ABA) signal transduction and regulates salt tolerance. The role of *AP2C1* (a protein phosphatase 2C, PP2C) protein has been demonstrated to be associated with CBL-interacting kinase 9 (CIPK9) to regulate the K^+ deficiency in Arabidopsis. Arabidopsis plants overexpressing *OsPP108* (a group A PP2C in rice) are highly insensitive to ABA and tolerant to high salt and mannitol stresses during seed germination, root growth, and overall seedling growth and at adult stage, *OsPP108* overexpression leads to high tolerance to salt, mannitol, and drought stresses. Protein phosphatase 2Cs and *microtubule-associated stress protein1 (MASP1)* control microtubule stability, growth, and drought response in plants where three *clade E Growth-Regulating* (*EGR*) type 2C protein phosphatases are involved as a negative regulator of growth by restraining plant growth during drought. Target of rapamycin (TOR) kinase phosphorylates PYL ABA receptors at a conserved serine residue, which leads to inactivation of SnRK2 kinases by disrupting the PYL association with ABA and with PP2C phosphatase effectors. However, under stress conditions, ABA activates the SnRK2 kinases by TOR complex dissociation and inhibition along with growth, which enables plant to develop a strategy to combat the stress

regulator of salinity tolerance by interacting with an annexin protein GhANN8b, reversely regulates Ca^{2+} and Na^+ fluxes in cotton (Mu et al. 2019). Also, genomic and transcriptomic analysis showed differential expression of PP2C8 in the young seedling of rice under salt stress and found to be a target candidate gene for salt tolerance in the seedling stage (Sun et al. 2019).

10.3.2 Potassium (K^+) Deficiency

Potassium (K^+) is essentially required by the plants, and its deficiency results into retarded plant growth, inhibition in protein synthesis, impaired photosynthesis, as well as long distance transport (Chérel et al. 2014; Singh et al. 2016). The K^+ uptake via different transporters under K^+-deficient condition is coordinated through various protein kinases and phosphatases (Wang and Wu 2013). Besides, Ca^{2+} signaling components like calcineurin B-like protein (CBL)-interacting protein kinase9 (CIPK9) have been identified as critical regulator of low K^+ response (Pandey et al. 2007). In Arabidopsis, AIP1 (a protein phosphatase 2C) reverses the action of CIPK23 [CBL (calcineurin B-like)-interacting protein kinase23) by dephosphorylating potassium channel AKT1 and acts as a negative regulator of K^+ uptake under K^+-deficient condition (Lee et al. 2007). AKT1 channel inactivation could be directly related to the interaction of PP2C and CIPK23, which blocks the kinase activity of the later (Lan et al. 2011). AtPP2CA was shown to interact and dephosphorylate AKT2 to inhibit AKT2 channel activity (Wang and Wu 2013). It has already known that low K^+ condition activates several high affinity potassium (K^+) transporters like HAK5. However, some Ca^{2+} sensors like calcineurin B-like (CBL1), CBL8, CBL9, and CBL10, together with CBL-interacting protein kinase23 (CIPK23), demonstrated to activate HAK5 by phosphorylating N-terminus of HAK5 and facilitate the K^+ uptake (Ragel et al. 2015). The role of AP2C1 (a protein phosphatase 2C, PP2C) protein has been demonstrated to be associated with CIPK9 to regulate the K^+ deficiency in Arabidopsis by Singh et al. (2018). The null mutant alleles, *ap2c-1* and *ap2c-2*, and the overexpressed AP2C1 transgenic lines showed the contrasting phenotypes under low K^+ condition, which suggested CIPK9 as the positive regulator and AP2C1 as a negative regulator under low K^+ conditions in Arabidopsis (Singh et al. 2018). Thus, protein phosphatases are believed to be involved in the reversible phosphorylation to regulate the activity of kinases as well as the K^+ transporters in K^+-deprived conditions.

10.3.3 ABA and Stomatal Regulation

ABA is considered as the prime stress hormone, and it's role in stomatal closure mechanism is controlled by protein phosphorylation/dephosphorylation. It has been established that *AtDsPTP1* is associated with genes controlling ABA biosynthesis and ABA catabolism, which regulates the ABA accumulation in the cell (Luan 2002). Also a flux measurement experiment has demonstrated that a protein tyrosine dephosphorylation occurs due to Ca^{2+} signal which triggers ion efflux from the vacuole resulting in stomatal closure (MacRobbie 2002). Another experiment showed a contrasting result that protein phosphatase 1 acts as a positive regulator of blue light signaling between phototropins and the H^+-ATPase in guard cells of *Vicia faba*, enabling the opening of stomata (Takemiya et al. 2006). The protein tyrosine phosphatases (PTPs) are also found to be involved in stomatal movement as it regulates the ABA signaling through MAPKs (mitogen-activated protein kinases) which induces ABA-induced H_2O_2 generation in guard cells and stomatal closure (Shi et al. 2004). Similarly, protein phosphatase 2C (PP2C)-type phosphatases especially *ABI1/ABI2* function as central components in abscisic acid (ABA) signal transduction. In rice, *ABI1* is upregulated in IR29 (salinity-sensitive cultivar), while it is downregulated in FL478 (salinity-tolerant cultivar). This contributes to better growth of FL478, which possibly involves reduced PP2C activity and therefore limits ABA-dependent growth inhibition (Senadheera and Maathuis 2009). The regulatory A subunit of protein phosphatase 2A, RCN1 regulates the stomatal closure by cross talking between methyl jasmonate and ABA signaling where both methyl jasmonate and ABA failed to induce stomatal closure in Arabidopsis rcn1 knockout mutants unlike in wild-type plants (Saito et al. 2008; Munemasa et al. 2011).

In the absence of ABA, type A protein phosphatase 2C (PP2C) inhibits the kinase activity of OST1, while in the presence of ABA, ABA binds to its receptor PYRabactin resistance/PYrabactin-Like/Regulatory Components of ABA Receptor (PYR/PYL/RCAR) and releases the OST1 which culminates in closure of stomata (Zhang et al. 2014; Balmant et al. 2016). Similarly, ABI1 (a protein phosphatase 2C clade A) and okadaic acid-sensitive phosphatases of the PPP family were demonstrated as negative regulator of salt stress-activated SnRK2.4 (a SNF1-related protein kinase 2) in both ABA-dependent and ABA-independent signaling pathway confirming the dual role of protein phosphatases (Krzywińska et al. 2016). Another protein phosphatase PP2C has been shown to regulate two ion channels GORK and SLAC1 resulting in stomatal closure (Lefoulon et al. 2016). This stomatal opening and closing mechanism may have a significant role in governing abiotic stress tolerance in plants. A further study indicated that MAPK6 (a mitogen-activated protein kinase) acts as a positive regulator UV-B-induced stomatal closure by modulating H_2O_2-induced nitric oxide (NO) in guard cells (Li et al. 2017). The phosphatases along with protein kinases can also activate or deactivate stress-responsive genes to control growth defense trade-off mechanisms in plants. For example, target of rapamycin (TOR) kinase phosphorylates PYL ABA receptors at a conserved serine residue, which leads to inactivation of SnRK2 kinases due to the disruption of PYL association with ABA and with PP2C phosphatase effectors. But under stress, ABA activates SnRK2 kinases by TOR complex dissociation and inhibition. Thus, under control condition, TOR complex inhibits ABA signaling and stress response, whereas in stress condition, ABA signaling inhibits TOR signaling and growth to enable plant to develop a strategy to combat the stress (Wang et al. 2018). Similarly, in rice PP45 acts as negative regulator of ABA signaling. The PP2C domain of PP45 interacts with CaM binding domain of DMI3, *Oryza sativa* CCaMK (Ca^{2+}/calmodulin-dependent protein kinase). In the absence of ABA, PP45 directly inactivates DMI3 by dephosphorylating Thr-263 in DMI3, while in the presence of ABA, ABA-induced H_2O_2 production by the NADPH oxidases RbohB/E inhibits the activity of PP45 (Ni et al. 2019). Another study in Arabidopsis revealed the PR5-like protein kinase AtPR5K2 works in the ABA-dependent manner and by phosphorylating ABA-insensitive ABI1 and ABI2 (protein phosphatases 2C) confers the drought tolerance (Baek et al. 2019). Recently, OsPP2C50 and OsPP2C53 (two clade A protein phosphatase 2C) were found to be a negative regulator of ABA signaling which regulates OsSLAC1 (a slow-type anion channel) directly or indirectly through SAPK10 (a stress-activated serine/threonine protein kinase) and confers stomatal closure in rice (Min et al. 2019). These findings have resolved the key role of various protein phosphatases in ABA signaling-mediated stomatal regulation mechanism.

10.3.4 Other Abiotic Stresses

Apart from the salinity stress, protein phosphatases are also a major player in many other abiotic stresses like drought, UV, heat stress, low temperature by differentially regulating stress-responsive genes. In Arabidopsis, it was found that the C-terminal domain phosphatase-like family members (AtCPLs) differentially regulate abiotic stress signaling, growth, and development (Koiwa et al. 2002). Further *AtMKP1* (a *MAPK phosphatase 1*) has been demonstrated to be involved in UV resistance in Arabidopsis as any interference with this gene showed to develop sensitivity toward UV stress. Later MAP kinases, *MPK3* and *MPK6*, were found out to be the genes involved in UV resistance where *MKP1* was found to be a key regulator (Ulm et al. 2001, 2002). Thus, PTPs and DSPs were found to regulate abiotic stress tolerance through MAPK-dependent manner. Another study suggests that expression level of *OsPP2A-1* and *OsPP2A-3* was high in leaves subjected to drought and high salinity stress, whereas heat stress decreased the expression of *OsPP2A-1* in stems and induced expression of *OsPP2A-3* in all organs. This suggested that PP2Ac genes like *OsPP2A-1* and *OsPP2A-3* are under developmental and stress-related regulation (Singh et al. 2010). Exposure of chestnut plantlets at 4 °C was found to induce *CsDSP4* (a dual specificity protein phosphatase) gene expression which has a role in starch degradation and cold acclimation (Berrocal-Lobo et al. 2011). This *CsDSP4* has also been demonstrated to be highly induced in stems of chestnut seedlings under low temperature conditions (Liu et al. 2014). Other phosphatases also play a significant role in signal transduction as well as in the abiotic stress tolerance in crop plants.

Different regulatory and catalytic subunits of the phosphatases interact with different enzymes and signaling mediators resulting into activation/inactivation of stress-responsive genes. For instance, Arabidopsis PP2A-A1 subunit [ROOTS CURL IN NAPHTHYLPHTHALAMIC ACID 1 (RCN1)] and the B-subunits were reported to be directly involved in many signaling pathways (Trotta et al. 2011). In rice, *OsPFA-DSP1* (fungi atypical dual-specificity phosphatase) is known to be the negative regulator of drought stress (Liu et al. 2012a; Shankar et al. 2015). Similarly OsPP18, another rice protein phosphatase 2C that is regulated by STRESS-RESPONSIVE NAC1 (SNAC1), accounts for drought and oxidative stress tolerance by increasing the scavenging of ROS through antioxidant defense mechanism confirmed by the overexpressed transgenic lines where the mutant lines show hypersensitivity toward these stresses (You et al. 2014). Other studies have demonstrated the role of protein phosphatase 2A (PP2A) in ROS-dependent responses associated with light acclimation and pathogenesis in Arabidopsis *thaliana* (Rahikainen et al. 2016). In the same line, protein phosphatase 2Cs and *microtubule-associated stress protein1* (*MASP1*) are found to control microtubule stability, growth, and drought responses in plants. Three *clade E Growth-Regulating* (*EGR*) type 2C protein phosphatases act as negative regulator of growth by restraining plant growth during drought (Bhaskara et al. 2017). These protein phosphatases are also found to be involved in ROS signaling and maintain redox homeostasis through reversible oxidation of critical cysteine (Cys) residues which significantly contribute toward the development of abiotic and biotic stress tolerance mechanisms in plants (Bheri and Pandey 2019). Therefore, understanding the role of these protein phosphatases in signal transduction mechanism will help further research to highlight the key issues regarding development of abiotic stress-tolerant crops.

10.4 Conclusions

Protein phosphorylation and dephosphorylation are the key regulatory mechanisms of plant signal transduction. There has been considerable progress in understanding gene families that encode major protein phosphatases, their significance, and involvement in signal transduction process underlying plant growth, development, and stress tolerance. Physiological and genomic approaches have elucidated the pathways for utilizing phosphatases for modulating plant abiotic stress tolerance. Despite wealth of information on such interacting pathways, challenges await to unravel the interaction and cross talk between different phosphatases. Another daunting task is to address kinase-phosphatase regulation, especially in the context of single and combined stress conditions. Holistic and integrated research efforts are warranted to not only improve our basic understanding about phosphatase-dependent regulation of signal transduction and their multifaceted roles but also ensure their utilization as "genetic" conduits for developing climate-resilient crops.

References

Ariño J, Pérez-Callejón E, Cunillera N, Camps M, Posas F, Ferrer A (1993) Protein phosphatases in higher plants: multiplicity of type 2A phosphatases in Arabidopsis thaliana. Plant Mol Biol 21(3):475–485

Baek D, Kim MC, Kumar D, Park B, Cheong MS, Choi W et al (2019) AtPR5K2, a PR5-like receptor kinase, modulates plant responses to drought stress by phosphorylating protein phosphatase 2Cs. Front Plant Sci 10:1146

Balmant KM, Zhang T, Chen S (2016) Protein phosphorylation and redox modification in stomatal guard cells. Front Physiol 7:26

Barford D (1996) Molecular mechanisms of the protein serine/threonine phosphatases. Trends Biochem Sci 21(11):407–412

Barford D (1999) Structural studies of reversible protein phosphorylation and protein phosphatases. Biochem Soc Trans 27:751

Berrocal-Lobo M, Ibanez C, Acebo P, Ramos A, Perez-Solis E, Collada C et al (2011) Identification of a homolog of Arabidopsis DSP4 (SEX4) in chestnut: its induction and accumulation in stem amyloplasts during winter or in response to the cold. Plant Cell Environ 34(10):1693–1704

Bethke PC, Jones RL (1997) Reversible protein phosphorylation regulates the activity of the slow-vacuolar ion channel. Plant J 11(6):1227–1235

Bhaskara GB, Wen TN, Nguyen TT, Verslues PE (2017) Protein phosphatase 2Cs and microtubule-associated stress protein 1 control microtubule stability, plant growth, and drought response. Plant Cell 29(1):169–191

Bheri M, Pandey GK (2019) Protein phosphatases meet reactive oxygen species in plant signaling networks. Environ Exp Bot 161:26–40

Brewis ND, Street AJ, Prescott AR, Cohen PT (1993) PPX, a novel protein serine/threonine phosphatase localized to centrosomes. EMBO J 12(3):987–996

Ceulemans H, Bollen M (2004) Functional diversity of protein phosphatase-1, a cellular economizer and reset button. Physiol Rev 84:1–39

Chae L, Pandey GK, Luan S, Cheong YH, Kim KN (2009) Protein kinases and phosphatases for stress signal transduction in plants. In: Abiotic stress adaptation in plants. Springer, Dordrecht, pp 123–163

Cheong YH, Kim KN, Pandey GK, Gupta R, Grant JJ, Luan S (2003) CBL1, a calcium sensor that differentially regulates salt, drought, and cold responses in Arabidopsis. Plant Cell 15(8):1833–1845

Chérel I, Lefoulon C, Boeglin M, Sentenac H (2014) Molecular mechanisms involved in plant adaptation to low K+ availability. J Exp Bot 65(3):833–848

Cohen PT (1997) Novel protein serine/threonine phosphatases: variety is the spice of life. Trends Biochem Sci 22(7):245–251

Danquah A, de Zelicourt A, Colcombet J, Hirt H (2014) The role of ABA and MAPK signaling pathways in plant abiotic stress responses. Biotechnol Adv 32(1):40–52

DeLong A (2006) Switching the flip: protein phosphatase roles in signaling pathways. Curr Opin Plant Biol 9(5):470–477

Denu JM, Stuckey JA, Saper MA, Dixon JE (1996) Form and function in protein dephosphorylation. Cell 87(3):361–364

Durian G, Rahikainen M, Alegre S, Brosché M, Kangasjärvi S (2016) Protein phosphatase 2A in the regulatory network underlying biotic stress resistance in plants. Front Plant Sci 7:812

Egloff MP, Cohen PT, Reinemer P, Barford D (1995) Crystal structure of the catalytic subunit of human protein phosphatase 1 and its complex with tungstate. J Mol Biol 254(5):942–959

Farkas I, Dombradi V, Miskei M, Szabados L, Koncz C (2007) Arabidopsis PPP family of serine/threonine phosphatases. Trends Plant Sci 12(4):169–176

Fordham-Skelton AP, Chilley P, Lumbreras V, Reignoux S, Fenton TR, Dahm CC et al (2002) A novel higher plant protein tyrosine phosphatase interacts with SNF1-related protein kinases via a KIS (kinase interaction sequence) domain. Plant J 29(6):705–715

Fujii H, Chinnusamy V, Rodrigues A, Rubio S, Antoni R, Park SY et al (2009) *In vitro* reconstitution of an abscisic acid signaling pathway. Nature 462(7273):660

Fujita Y, Fujita M, Shinozaki K, Yamaguchi-Shinozaki K (2011) ABA-mediated transcriptional regulation in response to osmotic stress in plants. J Plant Res 124(4):509–525

Fu L, Wang P, Xiong Y (2020) Target of Rapamycin signaling in plant stress responses. Plant Physiol 182(4):1613–1623

Genoud T, Santa Cruz MT, Kulisic T, Sparla F, Fankhauser C, Métraux JP (2008) The protein phosphatase 7 regulates phytochrome signaling in Arabidopsis. PLoS One 3(7):e2699

Goldberg J, Huang HB, Kwon YG, Greengard P, Nairn AC, Kuriyan J (1995) Three-dimensional structure of the catalytic subunit of protein serine/threonine phosphatase-1. Nature 376(6543):745

Gupta R, Luan S (2003) Redox control of protein tyrosine phosphatases and mitogen-activated protein kinases in plants. Plant Physiol 132(3):1149–1152

Gupta R, Huang Y, Kieber J, Luan S (1998) Identification of a dual-specificity protein phosphatase that inactivates a MAP kinase from Arabidopsis. Plant J 16(5):581–589

Harris DM, Myrick TL, Rundle SJ (1999) The Arabidopsis homolog of yeast TAP42 and mammalian α4 binds to the catalytic subunit of protein phosphatase 2A and is induced by chilling. Plant Physiol 121(2):609–618

Hu W, Yan Y, Hou X, He Y, Wei Y, Yang G et al (2015) TaPP2C1, a group F2 protein phosphatase 2C gene, confers resistance to salt stress in transgenic tobacco. PLoS One 10(6):e0129589

Hu R, Zhu Y, Wei J, Chen J, Shi H, Shen G, Zhang H (2017) Overexpression of PP2A-C5 that encodes the catalytic subunit 5 of protein phosphatase 2A in Arabidopsis confers better root and shoot development under salt conditions. Plant Cell Environ 40(1):150–164

Hunter T (2009) Tyrosine phosphorylation: thirty years and counting. Curr Opin Cell Biol 21(2):140–146

Iki T, Yoshikawa M, Meshi T, Ishikawa M (2012) Cyclophilin 40 facilitates HSP90-mediated RISC assembly in plants. The EMBO J 31(2):267–278

Janssens V, Goris J (2001) Protein phosphatase 2A: a highly regulated family of serine/threonine phosphatases implicated in cell growth and signaling. Biochem J 353(3):417–439

Kerk D, Bulgrien J, Smith DW, Barsam B, Veretnik S, Gribskov M (2002) The complement of protein phosphatase catalytic subunits encoded in the genome of Arabidopsis. Plant Physiol 129(2):908–925

Kerk D, Templeton G, Moorhead GB (2008) Evolutionary radiation pattern of novel protein phosphatases revealed by analysis of protein data from the completely sequenced genomes of humans, green algae, and higher plants. Plant Physiol 146(2):351–367

Kim DH, Kang JG, Yang SS, Chung KS, Song PS, Park CM (2002) A phytochrome-associated protein phosphatase 2A modulates light signals in flowering time control in Arabidopsis. Plant Cell 14(12):3043–3056

Kim KN, Cheong YH, Grant JJ, Pandey GK, Luan S (2003) CIPK3, a calcium sensor–associated protein kinase that regulates abscisic acid and cold signal transduction in Arabidopsis. Plant Cell 15(2):411–423

Koiwa H, Barb AW, Xiong L, Li F, McCully MG, Lee BH et al (2002) C-terminal domain phosphatase-like family members (AtCPLs) differentially regulate Arabidopsis thaliana abiotic stress signaling, growth, and development. Proc Natl Acad Sci 99(16):10893–10898

Kolukisaoglu Ü, Weinl S, Blazevic D, Batistic O, Kudla J (2004) Calcium sensors and their interacting protein kinases: genomics of the Arabidopsis and rice CBL-CIPK signaling networks. Plant Physiol 134(1):43–58

Krzywińska E, Bucholc M, Kulik A, Ciesielski A, Lichocka M, Dębski J et al (2016) Phosphatase ABI1 and okadaic acid-sensitive phosphoprotein phosphatases inhibit salt stress-activated SnRK2.4 kinase. BMC Plant Biol 16(1):136

Kudla J, Xu Q, Harter K, Gruissem W, Luan S (1999) Genes for calcineurin B-like proteins in Arabidopsis are differentially regulated by stress signals. Proc Natl Acad Sci 96(8):4718–4723

Kutuzov MA, Andreeva AV (2001) Noncompetitive inhibition of plant protein Ser/Thr phosphatase PP7 by phosphate. Biochem Biophys Res Commun 283(1):93–96

Lan WZ, Lee SC, Che YF, Jiang YQ, Luan S (2011) Mechanistic analysis of AKT1 regulation by the CBL–CIPK–PP2CA interactions. Mol Plant 4(3):527–536

Lee SC, Lan WZ, Kim BG, Li L, Cheong YH, Pandey GK et al (2007) A protein phosphorylation/dephosphorylation network regulates a plant potassium channel. Proc Natl Acad Sci 104(40):15959–15964

Lefoulon C, Boeglin M, Moreau B, Véry AA, Szponarski W, Dauzat M et al (2016) The Arabidopsis AtPP2CA protein phosphatase inhibits the GORK K+ efflux channel and exerts a dominant suppressive effect on phosphomimetic-activating mutations. J Biol Chem 291(12):6521–6533

Li FC, Wang J, Wu MM, Fan CM, Li X, He JM (2017) Mitogen-activated protein kinase phosphatases affect UV-B-induced stomatal closure via controlling NO in guard cells. Plant Physiol 173(1):760–770

Liao YD, Lin KH, Chen CC, Chiang CM (2016) Oryza sativa protein phosphatase 1a (OsPP1a) involved in salt stress tolerance in transgenic rice. Mol Breed 36(3):22

Lin Q, Buckler ES IV, Muse SV, Walker JC (1999) Molecular evolution of type 1 serine/threonine protein phosphatases. Mol Phylogenet Evol 12(1):57–66

Liu B, Fan J, Zhang Y, Mu P, Wang P, Su J, Wang H (2012a) OsPFA-DSP1, a rice protein tyrosine phosphatase, negatively regulates drought stress responses in transgenic tobacco and rice plants. Plant Cell Rep 31(6):1021–1032

Liu X, Zhu Y, Zhai H, Cai H, Ji W, Luo X et al (2012b) AtPP2CG1, a protein phosphatase 2C, positively regulates salt tolerance of Arabidopsis in abscisic acid-dependent manner. Biochem Biophys Res Commun 422(4):710–715

Liu R, Liu Y, Ye N, Zhu G, Chen M, Jia L et al (2014) AtDsPTP1 acts as a negative regulator in osmotic stress signaling during Arabidopsis seed germination and seedling establishment. J Exp Bot 66(5):1339–1353

Luan S (2000) Protein phosphatases: structure, regulation and function. Adv Bot Res 32:67–107

Luan S (2002) Tyrosine phosphorylation in plant cell signaling. Proc Natl Acad Sci 99(18):11567–11569

Luan S (2003) Protein phosphatases in plants. Annu Rev Plant Biol 54(1):63–92

Luan S, Kudla J, Rodriguez-Concepcion M, Yalovsky S, Gruissem W (2002) Calmodulins and calcineurin B–like proteins: calcium sensors for specific signal response coupling in plants. Plant Cell 14(Suppl 1):S389–S400

Ma Y, Szostkiewicz I, Korte A, Moes D, Yang Y, Christmann A, Grill E (2009) Regulators of PP2C phosphatase activity function as abscisic acid sensors. Science 324(5930):1064–1068

MacRobbie EA (2002) Evidence for a role for protein tyrosine phosphatase in the control of ion release from the guard cell vacuole in stomatal closure. Proc Natl Acad Sci 99(18):11963–11968

Merlot S, Gosti F, Guerrier D, Vavasseur A, Giraudat J (2001) The ABI1 and ABI2 protein phosphatases 2C act in a negative feedback regulatory loop of the abscisic acid signaling pathway. Plant J 25(3):295–303

Min MK, Choi EH, Kim JA, Yoon IS, Han S, Lee Y et al (2019) Two clade a phosphatase 2Cs expressed in guard cells physically interact with abscisic acid signaling components to induce stomatal closure in rice. Rice 12(1):1–13

Møller SG, Kim YS, Kunkel T, Chua NH (2003) PP7 is a positive regulator of blue light signaling in Arabidopsis. Plant Cell 15(5):1111–1119

Moorhead GB, De Wever V, Templeton G, Kerk D (2009) Evolution of protein phosphatases in plants and animals. Biochem J 417(2):401–409

Mu C, Zhou L, Shan L, Li F, Li Z (2019) Phosphatase GhDs PTP 3a interacts with annexin protein Gh ANN 8b to reversely regulate salt tolerance in cotton (Gossypium spp.). New Phytol 223(4):1856–1872

Munemasa S, Mori IC, Murata Y (2011) Methyl jasmonate signaling and signal crosstalk between methyl jasmonate and abscisic acid in guard cells. Plant Signal Behav 6(7):939–941

Ni L, Fu X, Zhang H, Li X, Cai X, Zhang P et al (2019) Abscisic acid inhibits Rice protein phosphatase PP45 via H2O2 and relieves repression of the Ca2+/CaM-dependent protein kinase DMI3. Plant Cell 31(1):128–152

Ogawa D, Abe K, Miyao A, Kojima M, Sakakibara H, Mizutani M et al (2011) RSS1 regulates the cell cycle and maintains meristematic activity under stress conditions in rice. Nat Commun 2:278
Pandey GK, Cheong,YH, Kim BG, Grant JJ, Li L, Luan S (2007) CIPK9: a calcium sensor-interacting protein kinase required for low-potassium tolerance in Arabidopsis. Cell research 17(5):411–421
Pandey GK (2008) Emergence of a novel calcium signaling pathway in plants: CBL-CIPK signaling network. Physiol Mol Biol Plants 14(1–2):51
Pandey GK, Cheong YH, Kim KN, Grant JJ, Li L, Hung W et al (2004) The calcium sensor calcineurin B-like 9 modulates abscisic acid sensitivity and biosynthesis in Arabidopsis. Plant Cell 16(7):1912–1924
Pandey GK, Kanwar P, Pandey A (2014) Global comparative analysis of CBL-CIPK gene families in plants. Springer, Heidelberg
Pandey GK, Pandey A, Prasad M, Böhmer M (2016) Editorial: abiotic stress signaling in plants: functional genomic intervention. Front Plant Sci 7:681. https://doi.org/10.3389/fpls.2016.00681
Pardo JM, Reddy MP, Yang S, Maggio A, Huh GH, Matsumoto T et al (1998) Stress signaling through Ca2+/calmodulin-dependent protein phosphatase calcineurin mediates salt adaptation in plants. Proc Natl Acad Sci 95(16):9681–9686
Park JH, Lee SY, Kim WY, Jung YJ, Chae HB, Jung HS et al (2011) Heat-induced chaperone activity of serine/threonine protein phosphatase 5 enhances thermotolerance in Arabidopsis thaliana. New Phytol 191(3):692–705
Pujol G, Baskin TI, Casamayor A, Cortadellas N, Ferrer A, Ariño J (2000) The Arabidopsis thaliana PPX/PP4 phosphatases: molecular cloning and structural organization of the genes and immunolocalization of the proteins to plastids. Plant Mol Biol 44(4):499–511
Ragel P, Ródenas R, García-Martín E, Andrés Z, Villalta I, Nieves-Cordones M et al (2015) The CBL-interacting protein kinase CIPK23 regulates HAK5-mediated high-affinity K+ uptake in Arabidopsis roots. Plant Physiol 169(4):2863–2873
Rahikainen M, Pascual J, Alegre S, Durian G, Kangasjärvi S (2016) PP2A phosphatase as a regulator of ROS signaling in plants. Antioxidants 5(1):8
Saito N, Munemasa S, Nakamura Y, Shimoishi Y, Mori IC, Murata Y (2008) Roles of RCN1, regulatory A subunit of protein phosphatase 2A, in methyl jasmonate signaling and signal crosstalk between methyl jasmonate and abscisic acid. Plant Cell Physiol 49(9):1396–1401
Schweighofer A, Hirt H, Meskiene I (2004) Plant PP2C phosphatases: emerging functions in stress signaling. Trends Plant Sci 9(5):236–243
Senadheera P, Maathuis FJ (2009) Differentially regulated kinases and phosphatases in roots may contribute to inter-cultivar difference in rice salinity tolerance. Plant Signal Behav 4(12):1163–1165
Shankar A, Agrawal N, Sharma M, Pandey A, Pandey GK (2015) Role of protein tyrosine phosphatases in plants. Curr Genomics 16(4):224–236
Shi W, Jia W, Liu X, Zhang S (2004) Protein tyrosine phosphatases involved in signaling of the ABA-induced H_2O_2 generation in guard cells of Vicia faba L. Chin Sci Bull 49(17):1841–1846
Singh A, Pandey GK (2012) Protein phosphatases: a genomic outlook to understand their function in plants. J Plant Biochem Biotechnol 21(1):100–107
Singh A, Giri J, Kapoor S, Tyagi AK, Pandey GK (2010) Protein phosphatase complement in rice: genome-wide identification and transcriptional analysis under abiotic stress conditions and reproductive development. BMC Genomics 11(1):435
Singh A, Jha SK, Bagri J, Pandey GK (2015) ABA inducible rice protein phosphatase 2C confers ABA insensitivity and abiotic stress tolerance in Arabidopsis. PLoS One 10(4):e0125168
Singh A, Pandey A, Srivastava AK, Tran LSP, Pandey GK (2016) Plant protein phosphatases 2C: from genomic diversity to functional multiplicity and importance in stress management. Crit Rev Biotechnol 36(6):1023–1035
Singh A, Yadav AK, Kaur K, Sanyal SK, Jha SK, Fernandes JL et al (2018) A protein phosphatase 2C, AP2C1, interacts with and negatively regulates the function of CIPK9 under potassium-deficient conditions in Arabidopsis. J Exp Bot 69(16):4003–4015

Smith RD, Walker JC (1993) Expression of multiple type 1 phosphoprotein phosphatases in Arabidopsis thaliana. Plant Mol Biol 21(2):307–316
Stone RL, Dixon JE (1994) Protein-tyrosine phosphatases. J Biol Chem 269(50):31323–31326
Suh MC, Cho HS, Kim YS, Liu JR, Lee HS (1998) Multiple genes encoding serine/threonine protein phosphatases and their differential expression in Nicotiana tabacum. Plant Mol Biol 36(2):315–322
Sun X, Kang X, Ni M (2012) Hypersensitive to red and blue 1 and its modification by protein phosphatase 7 are implicated in the control of Arabidopsis stomatal aperture. PLoS Genet 8(5):e1002674
Sun BR, Fu CY, Fan ZL, Chen Y, Chen WF, Zhang J et al (2019) Genomic and transcriptomic analysis reveal molecular basis of salinity tolerance in a novel strong salt-tolerant rice landrace Changmaogu. Rice 12(1):99
Takemiya A, Kinoshita T, Asanuma M, Shimazaki KI (2006) Protein phosphatase 1 positively regulates stomatal opening in response to blue light in Vicia faba. Proc Natl Acad Sci 103(36):13549–13554
Takemiya A, Ariyoshi C, Shimazaki KI (2009) Identification and functional characterization of inhibitor-3, a regulatory subunit of protein phosphatase 1 in plants. Plant Physiol 150(1):144–156
Tonks NK, Neel BG (1996) From form to function: signaling by protein tyrosine phosphatases. Cell 87(3):365–368
Trotta A, Wrzaczek M, Scharte J, Tikkanen M, Konert G, Rahikainen M et al (2011) Regulatory subunit B′ γ of protein phosphatase 2A prevents unnecessary defense reactions under low light in Arabidopsis. Plant Physiol 156(3):1464–1480
Uhrig RG, Labandera AM, Moorhead GB (2013) Arabidopsis PPP family of serine/threonine protein phosphatases: many targets but few engines. Trends Plant Sci 18(9):505–513
Ulm R, Revenkova E, di Sansebastiano GP, Bechtold N, Paszkowski J (2001) Mitogen-activated protein kinase phosphatase is required for genotoxic stress relief in Arabidopsis. Genes Dev 15(6):699–709
Ulm R, Ichimura K, Mizoguchi T, Peck SC, Zhu T, Wang X et al (2002) Distinct regulation of salinity and genotoxic stress responses by Arabidopsis MAP kinase phosphatase 1. EMBO J 21(23):6483–6493
Van Bentem SDLF, Vossen JH, Vermeer JE, de Vroomen MJ, Gadella TW, Haring MA, Cornelissen BJ (2003) The subcellular localization of plant protein phosphatase 5 isoforms is determined by alternative splicing. Plant Physiol 133(2):702–712
Walton KM, Dixon JE (1993) Protein tyrosine phosphatases. Annu Rev Biochem 62(1):101–120
Wang Y, Wu WH (2013) Potassium transport and signaling in higher plants. Annu Rev Plant Biol 64:451–476
Wang P, Zhao Y, Li Z, Hsu CC, Liu X, Fu L et al (2018) Reciprocal regulation of the TOR kinase and ABA receptor balances plant growth and stress response. Mol Cell 69(1):100–112
Xiong L, Zhu JK (2001) Abiotic stress signal transduction in plants: molecular and genetic perspectives. Physiol Plant 112(2):152–166
Xu Q, Fu HH, Gupta R, Luan S (1998) Molecular characterization of a tyrosine-specific protein phosphatase encoded by a stress-responsive gene in Arabidopsis. Plant Cell 10(5):849–857
Xue T, Wang D, Zhang S, Ehlting J, Ni F, Jakab S et al (2008) Genome-wide and expression analysis of protein phosphatase 2C in rice and Arabidopsis. BMC Genomics 9(1):550
You J, Zong W, Hu H, Li X, Xiao J, Xiong L (2014) A STRESS-RESPONSIVE NAC1-regulated protein phosphatase gene rice protein phosphatase18 modulates drought and oxidative stress tolerance through abscisic acid-independent reactive oxygen species scavenging in rice. Plant Physiol 166(4):2100–2114
Yu L, Golbeck J, Yao J, Rusnak F (1997) Spectroscopic and enzymatic characterization of the active site dinuclear metal center of calcineurin: implications for a mechanistic role. Biochemistry 36(35):10727–10734

Yu RMK, Zhou Y, Xu ZF, Chye ML, Kong RYC (2003) Two genes encoding protein phosphatase 2A catalytic subunits are differentially expressed in rice. Plant Mol Biol 51:295–311
Yu RMK, Wong MML, Jack RW, Kong RYC (2005) Structure, evolution and expression of a second subfamily of protein phosphatase 2A catalytic subunit genes in the rice plant (Oryza sativa L.). Planta 222(5):757–768
Zhang J, Li X, He Z, Zhao X, Wang Q, Zhou B et al (2013) Molecular character of a phosphatase 2C (PP2C) gene relation to stress tolerance in Arabidopsis thaliana. Mol Biol Rep 40(3):2633–2644
Zhang T, Chen S, Harmon AC (2014) Protein phosphorylation in stomatal movement. Plant Signal Behav 9(11):e972845
Zhu J-K (2016) Abiotic stress signaling and responses in plants. Cell 167(2):313–324
Zimmerlin A, Jupe SC, Bolwell GP (1995) Molecular cloning of the cDNA encoding a stress-inducible protein phosphatase 1 (PP1) catalytic subunit from French bean (Phaseolus vulgaris L.). Plant Mol Biol 28(3):363–368

Chapter 11
Role of Protein Phosphatases in Signaling, Potassium Transport, and Abiotic Stress Responses

Sibaji K. Sanyal, G. Rajasheker, P. B. Kavi Kishor, S. Anil Kumar, P. Hima Kumari, K. V. Saritha, P. Rathnagiri, and Girdhar K. Pandey

11.1 Introduction

Plants experience different types of biotic and abiotic stresses and nutrient deficiency during their life cycle. In a given environment, multiple stress factors are exerting their effect on plant growth and development simultaneously. Enhancing crop resilience in response to these abiotic stresses or mineral deprivation is a challenge in the endeavor to improve crop productivity (Mittler and Blumwald 2010). Abiotic stresses, in general, hamper plant growth and productivity by affecting several physiological, biochemical, and molecular responses (Zhu 2016). The role of potassium (K^+) during plant stress tolerance especially salt and drought as well as in plant growth and development is well known. In comparison, it has been reported that K^+ retains the photosynthetic electron transport activity by decreasing the activity of nicotinamide adenine dinucleotide phosphate (NADPH) oxidases (Waraich et al. 2012). This phenomenon, in turn, decreases the production of ROS under stress conditions. On the other hand, deficiencies in K^+ can reduce the dark CO_2

[SC]Sibaji K. Sanyal and G. Rajasheker contributed equally to this chapter.

S. K. Sanyal · G. K. Pandey (✉)
Department of Plant Molecular Biology, University of Delhi South Campus, New Delhi, India
e-mail: gkpandey@south.du.ac.in

G. Rajasheker · P. B. K. Kishor (✉) · P. H. Kumari
Department of Genetics, Osmania University, Hyderabad, India

S. A. Kumar
Department of Biotechnology, Vignan's Foundation for Science, Technology and Research, Guntur, Andhra Pradesh, India

K. V. Saritha
Department of Biotechnology, S.V. University, Tirupati, India

P. Rathnagiri
Genomix CARL Pvt. Ltd., Pulivendula, Andhra Pradesh, India

G. K. Pandey (ed.), *Protein Phosphatases and Stress Management in Plants*,
https://doi.org/10.1007/978-3-030-48733-1_11

fixation and productivity (Waraich et al. 2012). K^+ reduces the effects of drought, cold, and high light intensities (Waraich et al. 2012). Its accumulation is triggered under a combination of stresses like drought and high temperature (Halford 2009). K^+ homeostasis in the cytoplasm and the ability of plant tissues to retain K^+ under stress conditions is an important mechanism of salinity stress tolerance (Shabala et al. 2016). A clear positive correlation between tissue retention of K^+ and salt stress tolerance was noticed in cotton, pepper, and other crops (Shabala and Cuin 2008). Further, it appears that the differential K^+ retention ability imparts diverse salinity stress tolerance between halophytes and glycophytes (Assaha et al. 2017). Thus, K^+ appears to play crucial roles during abiotic stress tolerance. It is not clear how K^+ transporters are regulated by the phytohormonal network, reactive oxygen species (ROS) signaling, and other factors especially under K^+-deficient or high salt/drought stress conditions. Fifty-six K^+ transporters have been identified in higher plants like *Pyrus bretschneideri*, *Fragaria vesca*, *Vitis vinifera*, and wheat (Cheng et al. 2018; Li et al. 2018), but only few of them (e.g., Arabidopsis K^+ transporter 1 (AKT1), high-affinity K^+ transporter 5 (HAK5), K^+ channel in Arabidopsis *thaliana* (KAT), gated outwardly rectifying K^+ channel (GORK), K^+ efflux antiporter (KEA), tandem-pore K^+ channel (TPK/KCO)) have been characterized functionally under K^+ nutrition/transport (Dreyer and Uozumi 2011; Hedrich 2012; Ragel et al. 2019). These transporters play crucial roles in K^+ deprivation conditions and help the plants to survive under harsh environments.

Kinases and protein phosphatases (PPs) are the two enzymes involved in protein phosphorylation and dephosphorylation, respectively (Luan 2003). PP is an enzyme, which removes the phosphate group from the phosphorylated amino acid residue by counteracting the action of the protein kinases and provides modulation and reversibility of the phosphoregulatory mechanism (Luan 2003). In plant and animal, phosphorylation occurs on nine different amino acids, but the most common are Ser, Thr, and Tyr (Moorhead et al. 2009). Phosphorylation of His has been reported in fungi (Ota and Varshavsky 1993) and higher plant (Chang et al. 1993; Nongpiur et al. 2012). PPs occur in chloroplast, cytosol, mitochondria, and nuclei (MacKintosh et al. 1991; Huber et al. 1994). Based on the function, sequence, and structure, plant and animal PPs are evolutionarily conserved and are classified into four groups: (1) the phosphoprotein phosphatases (PPP), (2) the Mg^{2+}/Mn^{2+}-dependent enzymes (PP2C/PPM), (3) the protein tyrosine phosphatases (PTP), and (4) the Asp-based enzymes (Uhrig et al. 2013b). The PPPs are responsible for ~80–90% protein dephosphorylation in animal. In a plant cell, the PP2C/PPMs have a higher number of PP2Cs (compared to animal), and in the plant, the above ratio (of PPPs performing 80–90% dephosphorylation) may be altered (Uhrig et al. 2013a). Many studies point out that protein phosphatases participate in signaling cascades including abscisic acid (ABA) (Luan 1998), auxin (Xu and Zhang 2015), brassinosteroid (Tang et al. 2011), and biotic stress responses (Durian et al. 2016). Phosphatases play an important role in plant growth and development (Luan 2002) and are also involved in the regulation of mineral deprivation conditions (Pandey et al. 2014). This chapter summarizes some of the recent advances made on the regulation and functional aspects of phosphatases and their role in K^+ homeostasis under low K^+

availability and abiotic stress. Elucidation of such a mechanism(s) would help in designing crop plants that can cope with high salinity stress conditions and yet produce better yields.

11.2 Protein Phosphatases (PPs): Their Classes and Structure

Post-transcriptional regulation of proteins by reversible activities of kinases (phosphorylation) and phosphatases (dephosphorylation) is an important aspect in the life cycle of a plant and involves in diverse functions such as metabolism, ion transport, developmental control, and stress responses (Luan 2003). Attachment of a phosphate group to a protein regulates the enzyme activities by the initiation of allosteric conformational changes. Such an attachment blocks the access to the active site of an enzyme (Johnson et al. 1993; Johnson and O'Reilly 1996) and also the interaction among protein partners for proper functioning at the destination sites (Pawson 1995). Therefore, it is interesting to study the key roles being played by both phosphatases and kinases during plant development. Both diversity and functions of plant protein phosphatases and kinases have been well recognized.

In the PPP group, the catalytic subunit is the core enzyme and is more specific for Ser/Thr residues than Tyr (Moorhead et al. 2009). The PPP can be divided into seven members, protein phosphatase PP1, PP2A, PP2B (aka PP3), PP4, PP5, PP6, and PP7 (Moorhead et al. 2009). All the four groups are conserved in plants except PP2B/PP3 (aka calcineurin), which is absent in plants (Singh et al. 2015). Probably as compensation, plants have Kelch-repeat domain-containing protein phosphatases (associated with brassinosteroid signaling) (Mora-Garcia et al. 2004) and *Shewanella*-like phosphatases (SLPs) (Andreeva and Kutuzov 2004). These two also fall under the PPP and are unique to plants although they may also show Tyr specificity (Uhrig et al. 2013b). The PP2Cs in Arabidopsis and rice are divided into 11 subclades (A-K) and, like PPPs, are Ser/Thr dephosphorylating enzymes (Singh et al. 2015). The PTPs and Asp-based enzymes can be phosphatases dephosphorylating Tyr, Ser/Thr, and both (aka dual-specificity phosphatase) (Uhrig et al. 2013b). The Like-SEX4-1 and Like-SEX4-2 (LSF1 and LSF2) are the best-characterized examples of PTP (Silver et al. 2014; Gentry et al. 2016). Both PPP and PPM members share similar catalytic mechanisms but differ in many ways (Uhrig and Moorhead 2011). PPMs are Mg^{2+}/Mn^{2+} dependent, without any associated regulatory subunits, and are inhibited by okadaic acid and microcystin (Shi 2009). On the other hand, members of the PPP family lack accessory domains in their catalytic subunits but are associated with regulatory subunits (Moorhead et al. 2009). Members of the PPP family have a catalytic domain of nearly 280 amino acids, and the sequence is mostly conserved. On the other hand, members of the PP2C family are similar to animal homologs but diversified (Luan 2003). Though the primary sequence of PP2C and other members of PPM family do not share homology with

PPP enzymes, the structural folds are mostly similar (Das et al. 1996). The N-terminal catalytic region forms a central β-sandwich that binds to Mn^{2+} surrounded by α-helices, and the C-terminal region is an antiparallel helix structure attached to the catalytic domain (Das et al. 1996). In Arabidopsis PPP and PPM families comprise 102 of the nearly 150 protein phosphatases, while there are only 31 members (PPP and PPM proteins) out of 148 protein phosphatases in humans (Kerk et al. 2008; Uhrig and Moorhead 2011). In Arabidopsis, 80 PP2C members have been annotated so far (Xue et al. 2008; Singh et al. 2015). Identification of such a large number of PP2C members indicates that they may have diverse functions to perform during the growth and development of a plant.

The PTPs and Asp-based enzymes are both Tyr phosphatases, the difference between them arising from the residue (Cys in case of PTP and Asp in the latter case) that mediates dephosphorylation (Hobiger and Friedrich 2015). PTPs can be divided into three classes (I to III). The Class I comprises classical and dual-specificity phosphatases (DSPs). The classical PTPs show specificity toward Tyr, but the DSPs toward both Ser/Thr and Tyr. The DSPs can also choose non-protein substrates (mRNA, phosphoinositol, myotubularins, inositol-4-phosphatases, and carbohydrate). The Class II phosphatases often work at low pH values (≤ 6) and hence are also called acid phosphatases. These are smaller than Class I substrates (and hence also termed as low molecular weight phosphatases) and are also Tyr specific. The Class III phosphatases are involved in cell division and are also known as cell division control (CDC) phosphatases. Some of the members of the CDC can target Thr as well as Tyr (Hobiger and Friedrich 2015). Plants lack typical Tyr kinase and consequently typical classical PTPs. Class II and Class III PTPs are absent in plants (Moorhead et al. 2009; Hobiger and Friedrich 2015). The Asp-based enzymes consist of the FCP/SCP, eyes absent phosphatases and haloacid dehalogenase (HAD) family enzymes (Moorhead et al. 2009; Shankar et al. 2015). The PTPs and DSPs share a common catalytic domain of around 250 amino acids. At the N-terminal extremity of the catalytic domain has a conserved Lys residue for ATP binding. The catalytic signature of the PTP family is CX_5R and the Asp-based enzyme is DXDXT/V (D is Asp) (Moorhead et al. 2009; Shankar et al. 2015). The 24 members of PTP identified in Arabidopsis are one Tyr-specific PTP, 22 DSPs, and a low molecular weight PTP (probably a homolog of Class II PTPs) (Shankar et al. 2015). Rice genome sequence revealed 23 DSPs, one PTP, and one low molecular weight PTP (Shankar et al. 2015). Till date, no Asp-based phosphatase has been identified in plants. All information on the classification of phosphatases has been summarized in Fig. 11.1.

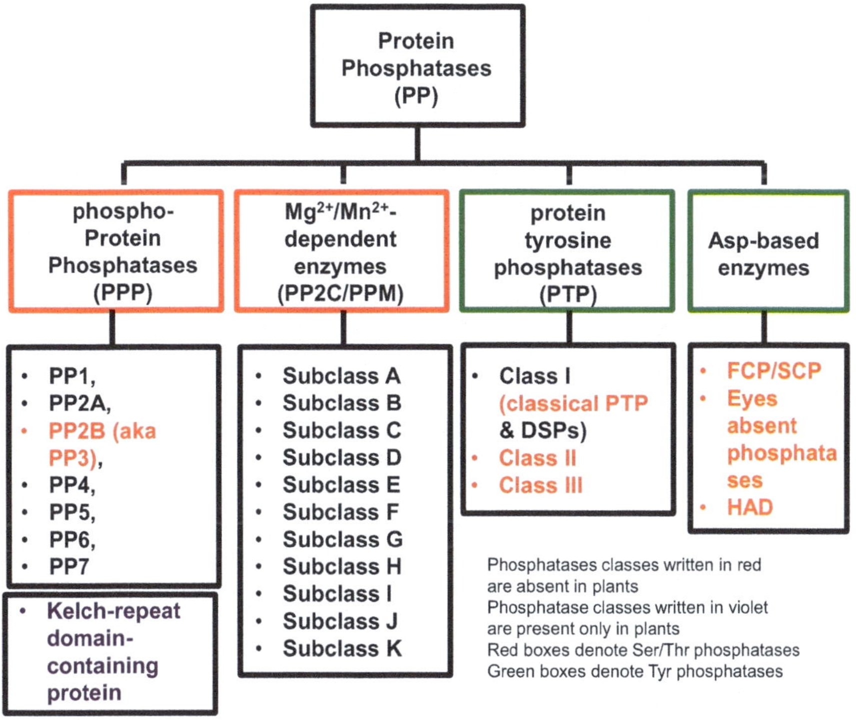

Fig. 11.1 Classification of protein phosphatases. The classification of protein phosphatases present in animal and plant is presented. The PPP and the PP2C/PPM group are the Ser/Thr phosphatases. Except for PP2B/PP3, all others are present in the plant. The plant additionally has some more phosphatases of this class. The PTP- and ASP-based enzymes are the Tyr phosphatases. Till date, more DSPs have been identified in the plant, which serves as the plant PTPs

11.3 The CBL-CIPK Family, Plants Modified Version of PP2B Family Phosphatases

As mentioned in the previous paragraphs, the PP2B family members are absent in plants. But animals have calcineurin, which consists of two subunits, i.e., CNA (PP2B phosphatase) and CNB (Ca^{2+} sensor), which is one of the important signaling molecules involved in Ca^{2+}-mediated signal transduction in animal and fungi. Search for a similar molecule in plant led to the identification of calcineurin B-like proteins (CBLs), a reminiscent of CNB (Luan 2009). CBL interacts and targets a specific group of kinases called CBL-interacting protein kinases (CIPKs) and mediates the Ca^{2+} signaling in plants (Sanyal et al. 2015). Studies on the CBL-CIPK module have led to the identification of multiple CBLs and CIPKs in diverse plant systems. A total of ten CBLs and CIPKs (26 and 33) have been identified in Arabidopsis and rice, respectively (Kolukisaoglu et al. 2004; Kanwar et al. 2014). Over the years, the CBL-CIPK module has been identified in several plant species

(Pandey et al. 2014). But lower plants like *Selaginella* and charophytes have lesser number of CBLs and CIPKs (Weinl and Kudla 2009; Kleist et al. 2014).

The CBL proteins contain calcium-binding four EF-hand (helix-loop-helix) domains (Sanyal et al. 2015). The N-terminal sequence of CBL protein contains the typical signal sequence that leads to the subcellular localization of a CBL and its interacting CIPK. Majorly, the myristoylation and acetylation sites (type I), the tonoplast targeting sequence (type II), and the transmembrane helix (type III) are present in the CBLs (Kleist et al. 2014). There is also a C-terminal phosphorylation motif present, phosphorylated by an interacting CIPK to enhance the activity of the module (Sanyal et al. 2016). On the other hand, CIPKs contain an N-terminal catalytic and C-terminal regulatory domains [having the NAF (CBL binding) and PPI (protein phosphatase binding domain)] (Sanyal et al. 2015). Similar to the SNF1 protein kinase, the CIPKs are Ser/Thr kinases with ATP binding site and a catalytic domain with an activation loop (Sanyal et al. 2015). CBLs are the Ca^{2+} sensors that bind to Ca^{2+} and change their conformation thus allowing CIPKs to bind to them through NAF domain. This binding relieves the autoinhibition of CIPK making the kinase active and ready to phosphorylate the targets (Sanyal et al. 2015). It has also been reported that CBL can interact with other proteins (than CIPK) indicating that the sensor module might also act independently (Sanyal et al. 2015).

11.4 Phosphatases and Signal Transduction

In eukaryotes, PP regulates all signaling pathways at different levels such as regulation of gene expression and phosphoregulation of a substrate(s). Here, we discuss the role of phosphatases in some well-known signal transduction pathways.

11.4.1 Phosphatases and Abscisic Acid (ABA) Signaling

PPs have an elaborate role in ABA signaling. ABA is a key plant hormone involved in abiotic stress, signal transduction, as well as during development (Zhu 2016). Accumulation of ABA under varied environmental stress conditions is common. It's signaling modulates many responses to the stress conditions. ABA regulates several of the downstream genes associated with stress (Verslues and Zhu 2005). Five components such as ABA receptors, positive regulators, negative regulators, ABA-responsive genes, and ABA-responsive transcription factors are associated with its signaling (Yang et al. 2017). Members of ABA receptors like pyrabactin resistance 1 (PYR)/PYR1-like (PYLs)/regulatory components of ABA receptors (RCAR), type 2C protein phosphatases (negative regulators), SnRK2 protein kinases (positive regulators), transcription factors, and ion channels play pivotal roles in ABA signaling (Wang et al. 2013). Dephosphorylation by PP2Cs and phosphorylation by SnRK2s are essential components in ABA signaling events (Fujii et al. 2009). The interaction of PYL and ABA is vital for the PP2C activity and downstream components like sucrose non-fermenting 1 related kinase group 2 (SnRK2s) to be phosphorylated (Fujii et al. 2009). Under normal condition, PP2Cs bind to SnRKs and keep them dephosphorylated. Stress-induced ABA binds to PYR/PYL; it interacts with PP2Cs repressing the phosphatase activity. Thus, the inhibition of SnRK2s is relieved by the above complex components (Fujii et al. 2009). SnRK2s are then autophosphorylated, and activated SnRKs phosphorylate the downstream targets like transcription factors and other targets (Wang et al. 2013). Such phosphorylations of SnRK2s bring about ABA-associated responses including stomatal closure (Wang et al. 2013). Both protein phosphorylation and redox modification play key roles in stomatal regulations.

Out of nine clade A protein phosphatase 2Cs in *A. thaliana*, functions are known for six of them [ABA Insensitive1 (ABI1), ABI2, ABA Hypersensitive Germination1 (AHG1), AHG3/AtPP2CA, Hypersensitive to ABA1 (HAB1), and HAB2] as negative regulators of ABA signaling (Sheen 1998; Yoshida et al. 2006; Nishimura et al. 2007; Xue et al. 2008). It has been demonstrated that PP2Cs have overlapping and redundant functions in ABA signaling (Rubio et al. 2009). It has been shown that ABA receptors such as ABI1, ABI2, HAB1, and AHG3 interact with RCAR/PYR/PYL (Singh et al. 2015).

11.4.2 Protein Phosphatases in Defense Signaling and the Regulation of Primary and Secondary Metabolism and the Regulation of Mitogen-Activated Protein Kinases (MAPKs)

It is vital to understand how pathogen infection produces signals that can succor the plants to prepare for combat mechanisms. We also need to find out how plants deploy converging signaling pathways to decipher stimuli generated at the membrane (in response to pathogen attack) and get it translated into varied physiological and biochemical changes to bring about disease resistance in the plants in addition to other stresses. Dephosphorylation of proteins is slowly becoming prominent as an important regulatory mechanism. Activities of both protein kinases and phosphatases appear to be involved in such combat mechanisms. Receptor-like kinases (RLKs) or proteins (RLPs) sense the presence of pathogen-associated molecular patterns (PAMPs) and elicit the downstream response in plants (Gomez-Gomez and Boller 2000; Gust and Felix 2014; Prince et al. 2014). PP2A is a crucial component and performs many functions like the regulation of signaling (receptor and organellar), metabolic pathways, and gene expression during such biotic stresses or plant immunity (Durian et al. 2016). PP2A is a trimeric protein with catalytic subunit C, scaffold subunit A, and regulatory subunit B (Trotta et al. 2011). Subunit B is generally referred to as "specificity unit" since it determines the target specificity of PP2A enzyme (Rahikainen et al. 2016). They are highly conserved both in animal and plant and modulate stress signaling (Rahikainen et al. 2016). It appears that different subunits of PP2A conciliate posttranslational regulation of metabolic enzymes besides signaling events (Durian et al. 2016). It has also been noticed that transcript levels of some of the catalytic subunits of PP2A like PP2A-C_5 and PP2A-C_2 enhance in response to plant pathogen attack (Durian et al. 2016). Similarly, transcript levels of subunits of PP2A-B (regulatory subunit) like B′ζ, B′η, and B′θ are also upregulated under biotic stress conditions (Durian et al. 2016). This infers that both the catalytic and regulatory subunits are involved in perceiving biotic cues. The information accumulated till date reveals that PP2A is functionally controlled at multiple levels perhaps to safeguard specificity in cellular signaling cascades with different heterotrimeric compositions (Bheri and Pandey 2019a, b).

PP2A also acts as a modulator of cell death (Durian et al. 2016). PP2A is negatively regulated by bacterial type III effectors (AvrEs) (Durian et al. 2016). They bind to PP2A and inhibit molecular pattern-triggered immunity (Degrave et al. 2015). Further, it has been reported that a mutant deficient in B′θ of PP2A exhibits resistance to pathogens (Kataya et al. 2015a). The B′θ subunit of PP2A co-localizes with PP2A-C2, PP2A-C5, and PP2A-A2 subunits in peroxisomes and influences β-oxidation of fatty acids (Kataya et al. 2015b). Overall it is inferred that PP2A is involved in complex and tightly interconnected molecular events in plant biotic stress such as perception, signaling, and response (Durian et al. 2016). The appearance of several primary metabolites like carbohydrates and amino acids is central to the plant immunity (Bolton 2009). In an interesting discovery, it was found that PP2A-B′γ subunit regulates the phosphorylation level of the cytosolic ACONITASE 3 (Konert et al. 2015). Metabolic reprogramming in plants appears partially mediated by PAMP-induced genes involved in the biosynthesis of carbohydrates, amino acids, lignins, flavonols, and phytoalexins (Truman et al. 2007). PP2A is also involved in the control of metabolic responses (Durian et al. 2016). Several proteins like MPK3 and MPK6 are constitutively active and trigger secondary metabolic activities and biosynthesis of camalexin and agmantine derivatives (both are defense-related metabolites) (Lassowskat et al. 2014). PP2A also appears to negatively regulate ethylene biosynthesis under such stress conditions (Durian et al. 2016). Overall PP2A is well connected with primary and secondary metabolic pathways, metabolic adjustments, and cellular signaling cascades indicating that PPs are key central components in plant immunity.

A chloroplast-localized DSP has been discovered which contains an ancient carbohydrate-binding domain and binds the starch (Kerk et al. 2006). The data suggest that DSPs in plant systems could be part of a protein assemblage at the starch granule. Thus, some DSPs bind to starch and modulate starch metabolism through reversible phosphorylations (Shankar et al. 2015). Dual-specificity plant tyrosine-specific phosphatases (PTPases) can modulate the mitogen-activated protein kinases (MAPK) (He et al. 2012). The Arabidopsis dual-specificity PTPase (AtDsPTp1) can inactivate MAPK by dephosphorylation (Gupta et al. 1998).

11.4.3 Role of Auxin and Brassinosteroid and Protein Phosphatases

Both auxin and brassinosteroid are two important phytohormones implicated in plant growth and development. They act synergistically to usher in multiple physiological and biochemical changes like hypocotyl elongation, vascular bundle and root development, and abiotic stress tolerance (Tian et al. 2017). They share the same target genes and regulate multiple processes on multiple levels. Auxin perception and degradation of AUX/IAA proteins and consequent release of auxin response factors (ARFs) in higher plants mediate auxin response (Tian et al. 2017). The PP2A protein is vital for the maintenance of auxin transport streams and hence plant growth and responses to salt stress (Han et al. 2017). Further, PP2A can bind to phosphatidic acid (a product of phospholipase D) to regulate auxin transport (Gao et al. 2013). Decreased PP2A activity enhances the level of phosphorylation of PIN1, PIN2, and PIN4 (Gao et al. 2013). The phosphatases thus play a major role in plant signaling by cross talking with hormone (ABA, brassinosteroid, auxin) and in biotic stress pathway.

11.5 Conditions that Necessitate K^+ Uptake Systems (AKT1, HAK5, KUP7) in Plants and K^+ Deficiency Sensing

Conditions such as low soil K^+ concentrations and salt and drought stresses can necessitate K^+ uptake systems in higher plants. It was reported that plant could sense the change of external K^+ levels and initiate many physiological responses (Schachtman and Shin 2007). The root epidermal cells and root hairs help the plant systems to absorb soil K^+ (Aleman et al. 2011). But, what is vital for its transport is the membrane potential of plant cells that depend mostly on the level of externally available K^+ (Cheeseman and Hanson 1979; Maathuis and Sanders 1994). It appears that plant cells behave similarly to K^+-specific electrodes with respect to external K^+ concentrations (Wang and Wu 2010). Hyperpolarization of root cell membrane potential is the first symptom that appears due to K^+ deficiency in plants (Maathuis and Sanders 1993), but depolarization occurs if the external K^+ concentration increases (Spalding et al. 1999). Uptake of K^+ is coupled to H^+ extrusion (Behl and Raschke 1987). The ATPases also help in K^+ uptake from the soil by establishing an electrical gradient and/or proton motive force (Dreyer and Uozumi 2011). Importantly, the activities of these channels are increased by protons (Fuchs et al. 2005). Thus, low external K^+-induced membrane hyperpolarization, coupled with extracellular acidification, is responsible to absorb K^+ under conditions of K^+ deprivation/salt stress (Wang and Wu 2010). The K^+ channels actually responsible for K^+ uptake from soil are HAK5 (at external K^+ concentrations below 0.01 mM), HAK5 and AKT1 (at external K^+ concentrations between 0.01 mM and 0.05 mM), and AKT1 and other unknown low-affinity K^+ uptake systems (at further higher K^+

concentration) (Sharma et al. 2013). Thus, the uptake of K^+ from the soil is carried out by transport proteins in an organized and well-coordinated fashion (Aleman et al. 2011). In addition to the AKT1 channel and HAK5 transporter proteins, KUP7 (a member of the HAK family) helps in K^+ uptake in the roots (Han et al. 2016). KUP7 mediates K^+ transport into the xylem as well as into shoots (Han et al. 2016). AKT1, HAK5, and KUP7 are localized to the plasma membrane of the Arabidopsis roots (Lagarde et al. 1996; Qi et al. 2008; Han et al. 2016). Once K^+ is absorbed by peripheral root cells, it is transported into root stelar tissues and then to shoots by gated outwardly rectifying K^+ channel (GORK) and stelar K^+ outward rectifier (SKOR) (Han et al. 2016). While GORK regulates K^+ efflux from roots (Ivashikina et al. 2001), SKOR is expressed in stelar tissues and releases K^+ into xylem sap (Gaymard et al. 1998).

AKT1 is a member of the plant *shaker* family of voltage-gated K^+ channels. In Arabidopsis, the family has nine members divided into four subfamilies depending on their response to the membrane voltage (Lebaudy et al. 2007; Dreyer and Blatt 2009). Five members (including AKT1, AKT5, AKT6/SPIK, KAT1, and KAT2) get activated upon membrane hyperpolarization and are closed when the driving force for K^+ is outwardly directed (and elicit only inward K^+ currents (K_{in})) (Lebaudy et al. 2007). GORK and SKOR get activated upon membrane depolarization and function opposite to K_{in} (i.e., they are closed when the driving force for K^+ is inwardly directed and elicit only outward K^+ currents (K_{out})) (Lebaudy et al. 2007; Sharma et al. 2013). Both of them regulate the release of K^+ (leakage pathways) from the cells (Demidchik et al. 2014). It was demonstrated that *SKOR*-encoded channels catalyze the K^+ leakage from stelar cells of roots into the xylem (Gaymard et al. 1998). They are activated at voltages more positive than resting potential (Demidchik et al. 2014). The GORK-encoded channels are expressed in the root atrichoblasts and trichoblasts (epidermal cells) and guard cells (Demidchik et al. 2014). GORK has the functions of ABA and ethylene-controlled K^+ efflux (Demidchik et al. 2014). Once K^+ is released through GORK, stomata are closed, which is regulated by Ca^{2+} and phytohormones (Demidchik et al. 2014). AKT2 exhibits weak voltage dependence and can mediate both, K^+ efflux and K^+ influx (K_{weak}) (Dreyer and Uozumi 2011; Sharma et al. 2013). The KC1 (aka KAT3) can modify the property of some K_{in} channels and is classified in K_{silent} category (Dreyer and Uozumi 2011). However, K_{out} channels are not regulated by the KAT3 protein (Ragel et al. 2019). But the interaction of KAT3 with K_{in} channels is important for negative regulation of the K_{in} channels (Ragel et al. 2019). Also, heteromerization of different subunits of K_{in} channels plays a crucial role in the functional diversity (Ragel et al. 2019). The initial identification of the HAK family of transporters in different systems by different groups led to the HAK, KUP, or KT nomenclature, but presently they are recognized as one group *KT/HAK/KUP* (Grabov 2007). The phylogenetic tree prepared for the plant *KT/HAK/KUP* transporters revealed that it contains four clades (Rubio et al. 2009). All plants have transporters homologous to members of clades I or clades II (Grabov 2007). Clade III members are majorly found in Arabidopsis and rice (Grabov 2007). The smallest is cluster IV, comprising only four rice genes (Grabov 2007). Recent work has increased this classification to six clusters, with the sixth cluster only present in bryophytes (Nieves-Cordones et al. 2016; Santa-Maria et al. 2018). The current information states that the family has members in Arabidopsis, peach, grapevine, *Medicago*, *Cassava*, rice, maize, *Brachypodium,* and *Panicum* (Ragel et al. 2019). Important members from Arabidopsis fall in clade Ia (HAK5), clade IIa (KUP3, KUP4/TRH1), clade IIb (KUP1), clade IIc (KUP2, KUP6, KUP8), clade III (KUP9, KUP10, KUP11, KUP12), and clade V (KUP7) (Nieves-Cordones et al. 2016; Ragel et al. 2019). In general terms, transporters involved in root high-affinity K^+ uptake fall into clade Ia, and transporters associated with developmental processes, especially those which demand turgor-driven cell expansion, fall into clade II (in Arabidopsis only) (Nieves-Cordones et al. 2016). However, some members of the clade I show different functions. For example, the *Cq*HAK5-like transporters from quinoa are involved in K^+ influx into the cells of the leaf salt bladders to contribute to the osmotic balance of the cytosol (Bohm et al. 2018). The other clade members have not been characterized sufficiently to affix a general function with these clades.

11.5.1 Plant Non-voltage-Gated K^+ Channels (TPK /KCO) and KEA K^+ Transporters

Other than the voltage-gated class, plants also possess the non-voltage-gated class of K^+ transporters. In Arabidopsis, five tandem-pore channels (TPK1–TPK5) (aka KCO) have been reported along with a sixth duplicated (and partially deleted) KCO_3 (Dreyer and Uozumi 2011). A functional TPK channel would exist as a dimer and is more relaxed in their cellular localization (they can be found in the tonoplast, or other organelle membranes, and the plasma membrane) (Dreyer and Uozumi 2011). The KEA group of efflux proteins are the final group of K^+ transporters in plants having homology to bacterial K^+/H^+ transporters (Grabov 2007). Arabidopsis has six KEA members in its genome (Tsujii et al. 2019). KEAs belong to the monovalent cation/proton antiporter (CPA) superfamily. The CPA family consists of two subfamilies, CPA1 (Na^+-H^+ exchangers (NHX) are members) and CPA2 (KEA are members) (Tsujii et al. 2019). The KEAs themselves can be divided into two clades, KEAI (having KEA1-KEA3 as members) and KEAII (having KEA4-KEA6 as members). The KEAI group can be further divided into two groups, KEA-Ia (with KEA1 and KEA2) having a longer N-terminal sequence and KEA-Ib (with KEA3) (Tsujii et al. 2019). KEAI members possess a putative chloroplast transit peptide in their N-terminal regions. KEA1 and KEA2 are localized in the inner envelope of the chloroplast, and that KEA3 is present in the thylakoid membrane (Tsujii et al. 2019). The role of KEA proteins has been implicated in photosynthetic regulation, chloroplast development, and stress tolerance (Tsujii et al. 2019). Using a bacterial system, it has been demonstrated that the Arabidopsis KEAs show K^+ transport activity (Tsujii et al. 2019).

11.5.2 K^+ Deprivation and Calcium Signaling

Other intracellular events help in the activation of the K^+ channels once low K^+ is sensed by the plant. The decrease in external K^+ concentration elevates cytoplasmic Ca^{2+} in root cells, which is mediated by hyperpolarization and ROS-activated Ca^{2+} channels (Shin et al. 2005; Demidchik and Maathuis 2007). These Ca^{2+} signatures generated in the cytoplasm due to K^+ concentration are decoded by members of several Ca^{2+} sensors such as CDPKs and CBL-CIPK module (Wang and Wu 2017). Further, it has been observed that K^+ deficiency causes the accumulation of ROS and activation of low K^+-responsive genes like *HAK5* (Wang and Wu 2010). So in the following sections, we discuss our current understanding of the K^+ sensing mechanism of plants and the role of phosphatases in it.

11.5.3 *Regulation of* Arabidopsis *K^+ Transporter 1 (AKT1) K^+ Selective Channel*

As already mentioned, several of the *Shaker*-like K^+ channels function in K^+ nutrition in plants (Very and Sentenac 2003). AKT1 was the first K^+ transporter identified in plants (Hirsch et al. 1998). As already mentioned, AKT1 majorly contributes to K^+ uptake along with HAK5 in plants. Therefore, AKT1 is one of the main targets of the regulatory network in higher plants. The search for this regulatory network resulted in the identification of the CBL-CIPK module that could modulate AKT1 activity (Li et al. 2006; Xu et al. 2006). CIPK23 phosphorylates AKT1 and increases the uptake of K^+ under conditions of low-K^+ stress (Li et al. 2006; Xu et al. 2006). On further probing, they identified that CBL1 and CBL9 are necessary to activate AKT1 and also act as the upstream regulators of CIPK23 (Li et al. 2006; Xu et al. 2006). On sensing the Ca^{2+} signature generated due to low K^+, the CBLs bind to Ca^{2+} and get activated (Behera et al. 2017). They then bind to the CIPK23 and direct its localization to the plasma membrane (where AKT1 is located). Here the module CBL1/CBL9-CIPK23 phosphorylates AKT1 and activates it (Li et al. 2006; Xu et al. 2006). This paradigm that the CBL-CIPK module leads to phosphorylation of AKT1 for its modulation has been conclusively proved in the oocytes of *Xenopus laevis* (Lee et al. 2007). However, a regulatory network is not complete until the antagonist can be identified that can reverse the action of the agonist. So, this search resulted in the identification of a protein phosphatase2C (AIP1) that binds to the ankyrin domain of AKT1 and keeps it deactivated by dephosphorylating it (Lee et al. 2007) (Fig. 11.2). Further investigation on the CBL-CIPK module and the PP2C for their role in AKT1 regulation has proved that the PP2C can also bind directly to the CIPK (CIPK6) (that is already bound to ankyrin domain of AKT1). The bound PP2C can inhibit AKT1 activation by probably inhibiting the kinase activity of the CIPK6 (Lan et al. 2011). This is supposed to be another mechanism that allows the regulation of CIPK kinase activity (besides the CBL binding and

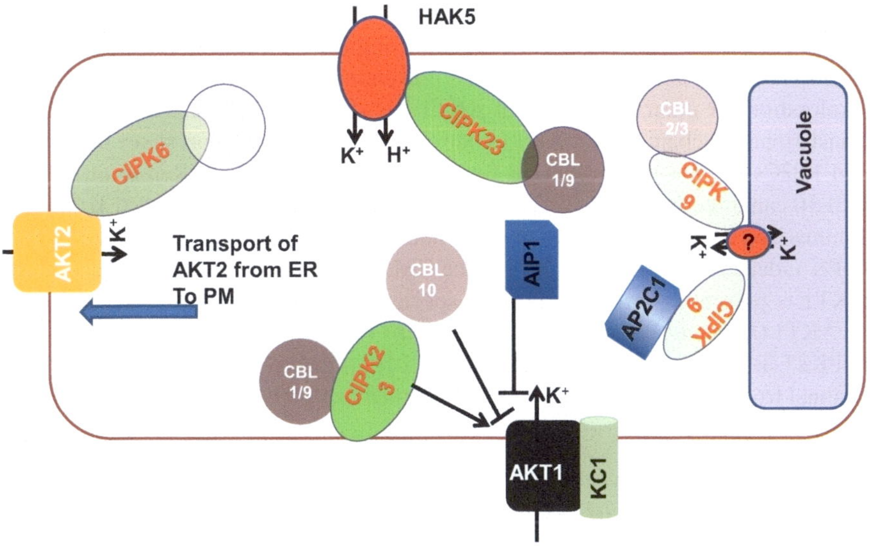

Fig. 11.2 The role of phosphatases in kinases in plant K^+ uptake. HAK5 transporters are modulated by the CBL1/9-CIPK23 module. The CBL1/9 can take the CIPK23 to the plasma membrane where it phosphorylates and modulates HAK5 for K^+ uptake. At further higher K^+ concentrations, the AKT1 channel is activated. It is under the control of AIP1 and CBL10 (and others) that are its negative regulators. CIPK23 and KC1 help in modulating AKT1 positively. The CBL4-CIPK6 module helps in targeting of AKT2 channel from endoplasmic reticulum (ER) to plasma membrane. The CBL2/3-CIPK9 module is targeted to vacuole where it can regulate a yet unknown transporter/channel for K^+ movement to or from the vacuole. AP2C1 can directly interact with CIPK9 and can directly inhibit its activity, acting as a dephosphorylation switch

relieving autoinhibition of CIPK) (Lan et al. 2011). Another CBL (one different from the CBL present in the module) can interact with the PP2C and relieves its inhibitory effect on CIPK6 (Lan et al. 2011). The entire mechanism points to very complex control of the AKT1 by the CBL-CIPK module and the PP2Cs (Lan et al. 2011). In the same work, it was also shown that CIPKs could directly interact with PP2Cs belonging to ABA signaling pathway in addition to the ones involved in K^+ signaling pathway (Lan et al. 2011). So this clearly shows that PP2Cs and CIPKs form major phosphorylation/dephosphorylation pair controlling important physiological processes (Lan et al. 2011).

The importance of CIPK23 and CBLs in the regulation of AKT1 is unambiguous. As already mentioned, there is another very robust mechanism that plants further employ to regulate the K^+ uptake through AKT1. The KC1 inhibits the AKT1 channel and is thought to control any leakage of K^+ from AKT1 (Reintanz et al. 2002; Duby et al. 2008). The effects of KC1 on inward and outward currents of AKT1 and also the K^+-dependent stability of the pore alteration in AKT1-KC1 heteromers under varying K^+ concentrations were reported (Geiger et al. 2009). The heteromerization of AKT1/KC1 makes the channel highly efficient at blocking K^+ permeation in the outward direction (Geiger et al. 2009). The CIPK23 (along with

CBLs) and KC1 are thought to work synergistically to control plant K^+ uptake (Wang et al. 2016). The CBL(1/9)-CIPK23 complex would then activate AKT1 to take up more K^+, and KC1 should lock it preventing K^+ leakage (Wang et al. 2016). Exploration for other controllers of AKT1 has shown that there are several post-translational mechanisms for AKT1 control. One interesting report states that CIPK and PP2Cs are not required at all for the regulation of AKT1 (Ren et al. 2013). CBL10 can directly bind to AKT1 and reduces its activity (Ren et al. 2013). This is summarized in Fig. 11.2. This activity is interestingly concentration-dependent, but CIPK-independent (Ren et al. 2013). The CBL(s)-CIPK23-mediated modulation of AKT1 is present in orthologous plant species as well. AKT1-like channels from *Hv*AKT1 (*Hordeum vulgare*) and *Vv*K1.1 (*Vitis vinifera*) are activated by CBL1 and CIPK23 (Boscari et al. 2009; Cuellar et al. 2010). Further, voltage-gated inward K^+ channel from grapevine was detected and named as *Vv*K1.2 (Cuellar et al. 2013).

11.5.4 Regulation of **Arabidopsis** *K^+ Transporter 2 (AKT2) K^+ Selective Channels*

Interaction of PP2CA with the plasma membrane K^+ transporters like Arabidopsis K^+ transporter 2 (AKT2) and Slowly Activating Anion Channel 1 (SLAC1) has been reported (Chérel et al. 2002; Lee et al. 2009). The coexpression studies of the *AKT2* and *AtPP2CA* genes were performed in mesophyll cells and observed that *AtPP2CA* gene expression levels were overlapping with that of *AKT2* and were upregulated by ABA in root and shoot tissues (Chérel et al. 2002). Dephosphorylation events mediate this protein-protein interaction because the changes in AKT2 activity are inhibited by vanadate (Chérel et al. 2002). Further, it is also suppressed by a point mutation in the catalytic site of the *At*PP2CA protein, resulting in the loss of its phosphatase activity (Chérel et al. 2002). The Ca^{2+} sensor CBL4 interacts with CIPK6 and mediates the translocation of AKT2 to the plasma membrane in Arabidopsis (Held et al. 2011). The Ca^{2+} sensor CBL4 interacts with CIPK6 and forms CBL4-CIPK6 module, which enables translocation of AKT2 K^+ channel to the plasma membrane and hence regulates its activity in phosphorylation-independent manner (Held et al. 2011). This is described in Fig. 11.2.

11.5.5 Role of Other CBL-CIPK Modules in Ion Homeostasis Pathways

Ca^{2+} sensors such as CBL2 and CBL3 are located in the vacuolar membrane (Tang et al. 2012). Therefore, they can interact with tonoplast-bound H^+-ATPase and modulate intracellular ion homeostasis (Tang et al. 2012). It has been shown that *cbl2cbl3* double mutants are sensitive to several metal ions like Ca^{2+}, Cu^{2+}, K^+, Zn^{2+}, and

Mg^{2+}, but not to Na^+ (Tang et al. 2012). Later, it was demonstrated that CBL2 and CBL3 function in the sequestration of vacuolar Mg^{2+} along with a quartet of CIPKs (CIPK3/9/23/26) (Tang et al. 2015). Another CBL-CIPK module, CBL3, and CIPK9 are involved in the transport of K^+ (Liu et al. 2013). CBL2 and CBL3 share 92% similarity at the level of amino acid sequence (Liu et al. 2013). Accordingly, overexpression of *CBL2* and *CBL3* displayed phenotypes that are sensitive to low K^+ conditions (Liu et al. 2013). Therefore, it is inferred that CBL2 and CBL3 are paralogous gene pair that mostly have similar physiological functions. The role of CIPK23 has already been discussed in K^+ homeostasis. Besides K^+, CIPK23 also plays a role in nitrate and iron homeostasis (Ho et al. 2009; Dubeaux et al. 2018). Two other CBL-CIPK modules involved in Na^+ homeostasis are CBL4-CIPK24 (SOS pathway members) and CBL10-CIPK24 (Kim et al. 2007; Zhu 2016). Further, it has been shown that CBL10 functions in the shoots and increases the tonoplast-bound Na^+/H^+ exchanger (NHX) activity by interacting with CIPK24 protein and brings about ion homeostasis to impart salt stress tolerance (Kim et al. 2007).

11.5.6 The GORK and SKOR Channels in K^+ Homeostasis and Other Functions in Plant Cell

It is not only the uptake of K^+ that is pivotal, but also its internal distribution equally within the plant and its homeostasis under abiotic stress conditions. The gating of GORK depends on the extracellular concentrations of K^+, and GORK can sense the changes in K^+ levels and enable the roots to react accordingly (Sharma et al. 2013). The transport of K^+ from roots to the leaf is vital and takes place via the xylem tissues (Gaymard et al. 1998). The SKOR channel has been found to express strongly in root tissues like pericycle and the xylem parenchyma (Gaymard et al. 1998). SKOR can be modulated by external K^+ concentrations which increases the voltage required by the channel to open to a higher threshold (Johansson et al. 2006). This is achieved by a complex interplay between the pore region of SKOR and last transmembrane domains of the channel responsible for opening and closing (Johansson et al. 2006). At high external K^+ concentration, the pore region strongly interacts with the last transmembrane domain of the channel and keeps the channel in a closed state (Johansson et al. 2006). At low external K^+ concentration, the pore region has fewer K^+, and as a result, the pore region of SKOR does not interact with the transmembrane domain (Johansson et al. 2006). This brings down the channel opening voltages at a lower threshold. Following the molecular rearrangement during low-sK^+ state, the transmembrane domain arranges itself in such a way that a K^+ outward current is allowed so that the SKOR can now allow K^+ to the stellar apoplast for K^+ transport within the cell (Johansson et al. 2006; Sharma et al. 2013).

It has been shown that the expression of SKOR is negatively regulated by ABA (Gaymard et al. 1998). Further, acidification (pH) also inhibits the SKOR currents, thus preventing the loss of K^+ from roots to shoots (Lacombe et al. 2000). Since reactive oxygen species (ROS) (of which H_2O_2 is a part) acts as a signal and impacts plant growth and development under abiotic stress conditions (Gapper and Dolan 2006), exogenous application of H_2O_2 leads in the enhancement of SKOR outward currents (Garcia-Mata et al. 2010). Thus, K^+ uptake and its redistribution/partitioning within the root and shoot systems appear to be part of the complex stress response preeminently played by the ROS. Investigations have shown that the regulation of the GORK channel, which mediates K^+ efflux, is carried out by PP2CA in *A. thaliana* (Lefoulon et al. 2016). It has also been recently shown that Ca^{2+}-dependent protein kinase 21 (CPK21) activates the GORK channel (van Kleeff et al. 2018). So like the AKT1-PP2C and CBL-CIPK functioning, we can imagine a similar fine-tuning of GORK activity by the kinase-phosphatase pair.

11.5.7 The Regulation of AtHAK5

Since the availability of soil K^+ varies, plants utilize multiple transporters and mechanisms for acquiring it (Maathuis 2009). Members of the family KT/HAK/KUP are ubiquitous, distributed in different subcellular compartments with variable numbers and involved in high-affinity K^+ uptake from the soil across membranes and K^+ supply, besides K^+ translocation, control of water movement at the plant level, and salt and drought tolerance also (Li et al. 2018; Santa-Maria et al. 2018). As already mentioned, members of the clade Ia of KT/HAK/KUP transporters are associated with K^+ uptake from the soil (e.g., *At*HAK5, *Os*HAK1, and *Os*HAK5) (Ragel et al. 2019). The HAK5 (and other HAK-like transporters) is generally upregulated under K^+ deprivation and downregulated when K^+ is resupplied (Ragel et al. 2019). It is generally accepted that under any stress conditions that affect K^+ acquisition (including salinity), high-affinity K^+ uptake systems should be transcriptionally (or post-translationally) activated to maintain the K^+ supply and K^+/Na^+ homeostasis (Ragel et al. 2019). It must also be mentioned that most of clade II to V members of KT/HAK/KUP do not exhibit transcriptional regulation in response to K^+ deficiency (Ragel et al. 2019).

As HAK5 is the most investigated protein of the KT/HAK/KUP class, we focus more on it to understand the regulation of this class. Both K^+ deficiency and salt stress regulate *HAK5* in Arabidopsis and tomato and are associated with hyperpolarization of the plasma membrane of root cells (Nieves-Cordones et al. 2008; Rubio et al. 2014). Roots deprived of K^+ induce the expression of ethylene (biosynthesis and signaling) and ROS metabolism genes promoting higher levels of ethylene and the increase in hydrogen peroxide (H_2O_2) (Ragel et al. 2019). Both, as a result, enhance transcription of *HAK5* (Ragel et al. 2019). Salt stress also results in modifications of *AtHAK5* expression or the low-K^+ response. As only high salt stress induces *AtHAK5* expression, it may be inferred that plants may recognize high Na^+ levels as K^+ under K^+ deprivation (Ragel et al. 2019). It appears that the expression of *HAK5* gene is also dependent on several other factors like the hormones like auxin, cytokinin, jasmonic acid, ABA (K^+ deficiency also induces ethylene biosynthesis), and many transcription factors in plants (Ragel et al. 2019). Taking a cue from the above findings, if we look into the transcription factors (TFs) that have been identified in HAK5, we find that HAK5 is under a strict regulation depending on the availability of external K^+. Under normal condition (external K^+ sufficient), ARF2 (Auxin Response Factor 2) binds to the auxin-responsive elements (AuxREs) within the *AtHAK5* promoter and represses transcription, so that the K^+ uptake is only through channels and not through more energy requiring HAK5 (which is a symport system) (Zhao et al. 2016). Under low external K^+ condition, ARF2 is phosphorylated by a kinase (yet unknown) and loses DNA binding activity and is removed from the *AtHAK5* promoter, which relieves the repression on *AtHAK5* transcription (Zhao et al. 2016; Ragel et al. 2019). Now another set of TFs (including RAP2.11, DDF2, JLO, bHLH121, and TFII_A) binds to the promoter of HAK5 to initiate its transcription (Kim et al. 2012; Hong et al. 2013). The RAP2.11 binds to the ethylene-responsive element (ERE) and the GCC-box of the *AtHAK5* promoter (Kim et al. 2012). The information on the binding sites of others is still unavailable (Ragel et al. 2019).

Moving from transcriptional regulation of HAK5 to posttranslational regulation, it has been reported that the CBL1/9-CIPK23 module can also regulate the HAK5 protein (through phosphorylation) (Ragel et al. 2015; Scherzer et al. 2015; Bohm et al. 2018). The phosphorylation of the cytosolic N-terminal of HAK5 by CBL-CIPK23 complex is similar to the AKT1 activation by the same module (Li et al. 2006; Xu et al. 2006). The regulation could also be like CBL4-CIPK6 and AKT2, i.e., the CBL-CIPK23 module could traffic HAK5 to the plasma membrane (as under K^+ starvation HAK5 enriches in the plasma membrane) (Qi et al. 2008; Ragel et al. 2019). HAK5 can also be regulated by raf-like integrin-linked kinase1 (ILK1) (probably undergoing phosphorylation). The calmodulin-like protein 9 (CML9) and ILK1 are together required for regulation of HAK5 (Brauer et al. 2016). However, it is unknown at this point of time if this happens in coordination with CBL1/9-CIPK23 pathway or independent of it. The direct phosphatase modulator of HAK5 is yet to be identified. ABI2 has been identified as direct interactors of CBL1 and CIPK23 (although during nitrate sensing) (Leran et al. 2015). The same phosphatase may modulate the HAK5 activity by controlling the CBL1 and CIPK23.

11.5.8 K^+ Uptake Regulation by a Novel CIPK and PP2C Pair

Pandey and colleagues first identified *CIPK9*, a calcium sensor-interacting protein kinase which is required for low K^+ tolerance in *A. thaliana* (Pandey et al. 2007). Its transcript levels both in roots and shoots are upregulated by K^+ deprivation conditions (Pandey et al. 2007). Based on the T-DNA inserted loss-of-function mutant analysis, they have suggested the functional role of CIPK9 in K^+ utilization or sensing mechanism (Pandey et al. 2007). Another Ser/Thr phosphatase type 2C1 (AP2C1), a stress signal regulator in *A. thaliana*, negatively regulates both MAPK4 and MAPK6 (Schweighofer et al. 2007). It also modulates innate immunity, jasmonic acid, and ethylene levels in *A. thaliana*. The authors further showed that *ap2c1* mutant plants produce higher amounts of jasmonate upon wounding. Their experiments revealed an important role for AP2C1 phosphatase in moderating defense responses and activities of MAPK. While searching for up- and downstream components of CIPK9-mediated K^+ deficiency signaling, *AP2C1* was identified as the interactor of CIPK9 in a yeast two-hybrid screen (Singh et al. 2018). AP2C1 physically interacts with CIPK9 *in vitro* and *in planta* (Singh et al. 2018). Just like *CIPK23*, *CIPK9* is known to be associated with the modulation of K^+ signaling in Arabidopsis (Pandey et al. 2007). As mentioned above, AP2C1 was recognized earlier as a MAPK phosphatase, but the discovery that it interacts with CIPK9 was a new finding. *AP2C1* was characterized as a negative regulator of K^+ signaling under low K^+ availability (Singh et al. 2018). Genetic analysis of null mutants of *AP2C1* and *CIPK9* and *AP2C1*-overexpressing transgenic *A. thaliana* lines revealed that they indeed modulate K^+ deprivation conditions (Singh et al. 2018). Further, AP2C1 has KIM domain (a conserved structural feature) necessary for interaction with different kinases including CIPK9 (Singh et al. 2018). Though CIPK proteins are known to be cytoplasmic, their cellular location and action depend on the interaction with CBLs (Batistic et al. 2010). CIPK9 also interacts with a Ca^{2+} pump, ACA8, in Arabidopsis and brings about changes in cytosolic Ca^{2+} levels (Costa et al. 2017). Although the target/substrate of CIPK9 has not been identified, authors speculate that CIPK9 might be regulating some of the transporters/channels present on the tonoplast under K^+ deficiency condition (Singh et al. 2018). Authors have shown that CIPK9 and AP2C1 act as an important phosphorylation-dephosphorylation switch where CIPK9 might act as a positive regulator while AP2C1 might be acting as a negative regulator of K^+ deficiency response (explained in Fig. 11.2). The phosphatases thus play a very crucial and critical role in plant K^+ homeostasis.

11.6 Ser/Thr Protein Phosphatases in Stress Adaptation

Protein phosphatase 2Cs are the negative modulators of protein kinase pathways and are associated with many environmental stress responses (Xue et al. 2008). Since diverse gene family members are present, the tissue-specific expressions under diverse environmental stresses are critical for the functional understanding of the genes. So in the following section, we look into the role of protein phosphatases in abiotic stress.

11.6.1 Protein Phosphatase Expression Profile Under Stress Conditions

Several PP2C members in the subfamily A displayed their expressions under stress conditions and also during ABA responses (Yang et al. 2018). Expression analysis in Arabidopsis, rice, maize, and tomato has revealed that group-A PP2C genes are highly inducible in response to different abiotic stresses (Singh et al. 2015). The PP2C genes, both from Arabidopsis and rice, exhibit specific as well as an overlapping expression during drought and high salinity (Singh et al. 2015). One hundred and thirty-two *PP2Cs* were identified in the rice genome (Singh et al. 2010). Of the 132 identified, 128 genes were differentially expressed under environmental stresses such as salinity, cold, and drought and 11 under reproductive developmental stages (Singh et al. 2010). The catalytic subunit genes of *OsPp2A-1-5* are upregulated in leaves under salt stress conditions (Yu et al. 2003, 2005). The transgenic tobacco plants overexpressing *TaPP2Ac-1* (a catalytic subunit) could tolerate salt and drought stress conditions (Xu et al. 2007). In potato, during salinity stress, high transcript levels were recorded for *StPp2Ac1*, *StPP2Ac2b*, and *StPP2Ac3* (Pais et al. 2009). In *Medicago truncatula,* the expression profiles revealed differential expression patterns under cold, drought, and ABA stress conditions (Yang et al. 2018). In wheat, the *TaPP2C* genes regulate developmental processes as well as stress responses (Yu et al. 2019). Taken together, it appears that *PP2Cs* are implicated in regulating both stresses and in developmental conditions in plants.

11.6.2 Phosphatases Are Involved in Modulating Kinases During Salt Stress

As already mentioned some of the clade A PP2C members are involved in ABA signaling. The remaining three members of Highly ABA-Induced1 (HAI1), AKT1-Interacting PP2C1/HAI2, and HAI3 (collectively HAI) exhibited a more ABA-independent role in plants (Bhaskara et al. 2012). It was speculated that this HAI group may cross talk with the ABA-dependent and ABA-independent pathway during abiotic stress (Bhaskara et al. 2012). Clade A PP2Cs (ABI1 and PP2CA) inhibit ABA-activated SnRK2s during salt stress (Krzywinska et al. 2016; Krzywińska et al. 2016). Thus, PP2CA along with ABI1 inhibits the activity of SnRK2.4 and regulates root growth and salt stress tolerance (Krzywinska et al. 2016; Krzywińska et al. 2016). Further, they showed that salt-induced SnRK2.4/SnRK2.10 activity is better in the double-mutant *abi1-2 pp2ca-1* in comparison with controls and also single-mutant plants (*abi1* or *pp2ca*) (Krzywinska et al. 2016; Krzywińska et al. 2016). This points out that these phosphatases are inhibitors of SnRK2.4 activity in plants under stress conditions. Another member of PP2C-type protein phosphatase *ABI2* was identified as a CIPK24 [aka SOS2 (salt overly sensitive)]-interacting phosphatase (Ohta et al. 2003). ABI2 plays a critical role in dephosphorylating

CIPK24. The *abi2* mutant plants are salt-tolerant unlike the *sos2* mutant (and other SOS pathway mutants). The results inferred that ABI2 is a negative regulator of *CIPK24* in *SOS* pathway (Ohta et al. 2003).

11.6.3 Phosphatases in Regulating Guard Cell: The Best-Characterized ABA Signaling Pathway

In Arabidopsis, ABI1 and ABI2 encode PP2Cs that negatively regulate ABA signaling including stomatal closure. In guard cells, open stomata 1 (OST1/SRK2E) has been identified as an important component of ABA signal transduction, and mutations in the gene *SRK2E/OST1/SnRK2.6* impair stomatal closure (Yoshida et al. 2002). The *ost1* mutants have ABA-insensitive stomata (Mustilli et al. 2002). It is known that ABI1 and ABI2 encode PP2C-type protein phosphatases and negatively regulate ABA signaling events including stomatal closure (Mustilli et al. 2002). It was demonstrated that ABI1 interacted with OST1 and plays a critical role in the stomatal opening (Yoshida et al. 2006).

The classical ABA signaling is followed in this case. The absence of ABA keeps the PP2C bound to SnRK2 and inactivated (Umezawa et al. 2010). Upon exposure of plants to drought stress, ABA accumulates leading to the closure of stomata. In response to dehydration stress, ABA binds to PYR/PYL receptor and forms a complex with PP2Cs (ABI1 and others) (Umezawa et al. 2010). As soon as there is ABA, PP2C releases OST1 (an SnRK), and the activated OST1 can phosphorylate channels [slow anion channel1 (SLAC1), quickly activating anion channel 1 (QUAC1), GORK] and respiratory burst oxidase homolog F (RBOH F) (Balmant et al. 2016). The channels mediate anion release from stomata causing a depolarization of the membrane of stomata. The depolarization (of the membrane) and KAT1 phosphorylation block K^+ influx into the stomata (Balmant et al. 2016). The phosphorylation of RBOHF produces ROS in the cell, and it functions as second messengers for ABA signaling and activates Ca^{2+} channels and also OST1 (aka SnRK2E) acts upstream of ROS production (Pei et al. 2000; Mustilli et al. 2002). The ROS in the cell leads to Ca^{2+} spike in the cell. The Ca^{2+} spike is picked up by Ca^{2+} sensors, which further helps in stomatal closure. Thus the PP2C serves as a switch to serve as a control of stomatal physiology (Balmant et al. 2016) (summarized in Fig. 11.3). This model gives us a complete picture of the events involved in the modulation of stomata under stress. ABA signaling, K^+ transport, ROS, and Ca^{2+} signaling all play an important role to modulate stomata, and phosphatase (PP2C) stands at the top to control this pathway. By keeping the SnRK2 (OST1) dephosphorylated, phosphatase acts as a key controller of the pathway.

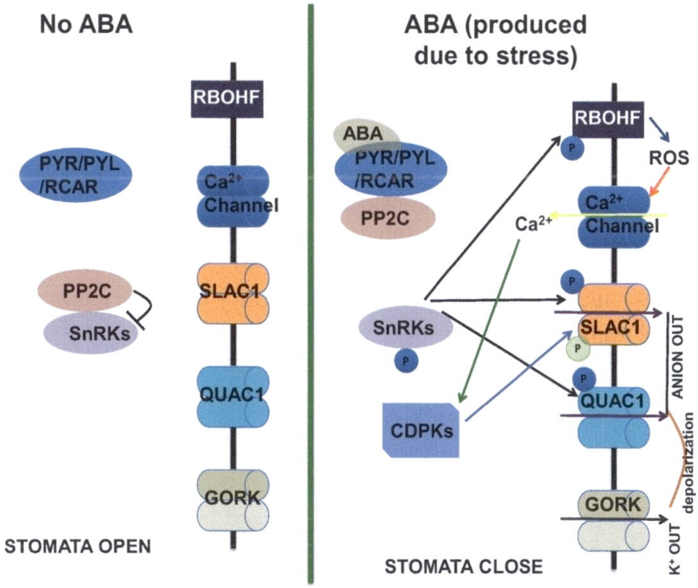

Fig. 11.3 The PP2C phosphatases play a major role in modulating stomatal physiology. Under normal condition, PP2C binds to SnRK (OST1) and keeps the stomata open. Under stress (drought), ABA binds to the receptors and removes PP2C from SnRK. SnRK then through a series of phosphorylation events activates the RBOHF, SLAC1, and QUAC1. Activation of RBOHF produces ROS, which in turn activates Ca^{2+} channels, thus resulting in activation of CDPKs. CDPKs can further activate SLAC1 through phosphorylation. The anions being pushed out of stomata by SLAC1 and QUAC1 result in depolarization of membrane and activation of GORK which pushes out K^+ from stomata causing the stomatal closure

11.7 Conclusions

The different classes of plant phosphatases are yet to be classified fully. The future research should focus on characterizing the neglected classes more so that our understanding of the plant phosphatases is increased holistically. What we do know for sure is that phosphatases play a key role in fine-tuning the physiological pathways at the molecular level. There are innumerable examples that are testimony to this fact. The role of phosphatases in the core ABA signaling pathway is well established. Several of the protein phosphatases like PP2Cs are Ser/Thr phosphatases working as vital components of ABA signal transduction. This is the best-characterized pathway for plant phosphatases to date. We have discussed in the chapter some major K^+ signaling pathways where the role of phosphatases has been proven, and the AKT1 pathway is an example of this. In all probability, the GORK channel may be modulated by the CPK21 and PP2CA pair. Some of the other important K^+ transport elements still need their phosphatase pair to be identified. We know that CBL1/9-CIPK23 module can regulate HAK5, but the phosphatase to counter the module's action *in vivo* is still unknown. The new kinase and phosphatase pair of CIPK9 and AP2C1 awaits the identification of their target transport element. Taking into account that both are involved in plant K^+ response, it is possible that the target will be involved in K^+ transport. We need to think where the phosphatase binds to when regulating a target. We know from the AIP and AKT1 model that the phosphatase binds to ankyrin domain of AKT1. But there is an alternate complex model proposed by Lan and colleagues (Lan et al. 2011) as discussed earlier in the chapter which should be kept into consideration for future explanations.

Besides, we have clues that phosphatases are involved in modulating development, ion accumulation, biotic stress pathway, and others. For the SOS pathway, we have the kinase and phosphatase pair identified. But overall very few actual pairs (kinase and phosphatase) are known. We are certain that abiotic stress stimuli perturb the expression of phosphatase genes. So in the next step, high-throughput interactome studies should be designed that enable us to find out the physical targets of phosphatases.

Acknowledgments P.B.K. is thankful for the CSIR Emeritus Scientist Fellowship through Grant No. 38(1325)/12/EMR-II) from the Council of Scientific and Industrial Research, New Delhi. Research in GKP's lab is funded by Department of Biotechnology, Science and Engineering Research Board, University Grant Commission (UGC), and Delhi University (R&D grant) India. Financial support from the DBT-RA Program in Biotechnology and Life Sciences is gratefully acknowledged by S.K.S.

References

Aleman F, Nieves-Cordones M, Martinez V, Rubio F (2011) Root K(+) acquisition in plants: the Arabidopsis thaliana model. Plant Cell Physiol 52:1603–1612

Andreeva AV, Kutuzov MA (2004) Widespread presence of "bacterial-like" PPP phosphatases in eukaryotes. BMC Evol Biol 4:47

Assaha DVM, Ueda A, Saneoka H, Al-Yahyai R, Yaish MW (2017) The role of Na+ and K+ transporters in salt stress adaptation in glycophytes. Front Physiol 8:509

Balmant KM, Zhang T, Chen S (2016) Protein phosphorylation and redox modification in stomatal guard cells. Front Physiol 7:26

Batistic O, Waadt R, Steinhorst L, Held K, Kudla J (2010) CBL-mediated targeting of CIPKs facilitates the decoding of calcium signals emanating from distinct cellular stores. Plant J 61:211–222

Behera S, Long Y, Schmitz-Thom I, Wang XP, Zhang C, Li H, Steinhorst L, Manishankar P, Ren XL, Offenborn JN, Wu WH, Kudla J, Wang Y (2017) Two spatially and temporally distinct Ca(2+) signals convey Arabidopsis thaliana responses to K(+) deficiency. New Phytol 213:739–750

Behl R, Raschke K (1987) Close coupling between extrusion of H(+) and uptake of K (+) by barley roots. Planta 172:531–538

Bhaskara GB, Nguyen TT, Verslues PE (2012) Unique drought resistance functions of the highly ABA-induced clade A protein phosphatase 2Cs. Plant Physiol 160:379–395

Bheri M, Pandey GK (2019a) PP2A phosphatases take a giant leap in the post-genomics era. Curr Genomics 20:154–171

Bheri M, Pandey GK (2019b) Protein phosphatases meet reactive oxygen species in plant signaling networks. Env Exp Bot 161:26–40

Bohm J, Messerer M, Muller HM, Scholz-Starke J, Gradogna A, Scherzer S, Maierhofer T, Bazihizina N, Zhang H, Stigloher C, Ache P, Al-Rasheid KAS, Mayer KFX, Shabala S, Carpaneto A, Haberer G, Zhu JK, Hedrich R (2018) Understanding the molecular basis of salt sequestration in epidermal bladder cells of Chenopodium quinoa. Curr Biol 28:3075–3085.e3077

Bolton MD (2009) Primary metabolism and plant defense—fuel for the fire. Mol Plant-Microbe Interact 22:487–497

Boscari A, Clement M, Volkov V, Golldack D, Hybiak J, Miller AJ, Amtmann A, Fricke W (2009) Potassium channels in barley: cloning, functional characterization and expression analyses in relation to leaf growth and development. Plant Cell Environ 32:1761–1777

Brauer EK, Ahsan N, Dale R, Kato N, Coluccio AE, Pineros MA, Kochian LV, Thelen JJ, Popescu SC (2016) The Raf-like kinase ILK1 and the high affinity K+ transporter HAK5 are required for innate immunity and abiotic stress response. Plant Physiol 171:1470–1484

Chang C, Kwok SF, Bleecker AB, Meyerowitz EM (1993) Arabidopsis ethylene-response gene ETR1: similarity of product to two-component regulators. Science 262:539–544

Cheeseman JM, Hanson JB (1979) Energy-linked potassium influx as related to cell potential in corn roots. Plant Physiol 64:842–845

Cheng X, Liu X, Mao W, Zhang X, Chen S, Zhan K, Bi H, Xu H (2018) Genome-wide identification and analysis of HAK/KUP/KT potassium transporters gene family in wheat (Triticum aestivum L.). Int J Mol Sci 19:3969

Chérel I, Michard E, Platet N, Mouline K, Alcon C, Sentenac H, Thibaud JB (2002) Physical and functional interaction of the Arabidopsis K+ channel AKT2 and phosphatase AtPP2CA. Plant Cell 14:1133–1146

Costa A, Luoni L, Marrano CA, Hashimoto K, Koster P, Giacometti S, De Michelis MI, Kudla J, Bonza MC (2017) Ca2+-dependent phosphoregulation of the plasma membrane Ca2+-ATPase ACA8 modulates stimulus-induced calcium signatures. J Exp Bot 68:3215–3230

Cuellar T, Azeem F, Andrianteranagna M, Pascaud F, Verdeil JL, Sentenac H, Zimmermann S, Gaillard I (2013) Potassium transport in developing fleshy fruits: the grapevine inward K^+ channel VvK1.2 is activated by CIPK-CBL complexes and induced in ripening berry flesh cells. Plant J 73:1006–1018

Cuellar T, Pascaud F, Verdeil JL, Torregrosa L, Adam-Blondon AF, Thibaud JB, Sentenac H, Gaillard I (2010) A grapevine shaker inward K^+ channel activated by the calcineurin B-like calcium sensor 1-protein kinase CIPK23 network is expressed in grape berries under drought stress conditions. Plant J 61:58–69

Das AK, Helps NR, Cohen PT, Barford D (1996) Crystal structure of the protein serine/threonine phosphatase 2C at 2.0 A resolution. EMBO J 15:6798–6809

Degrave A, Siamer S, Boureau T, Barny MA (2015) The AvrE superfamily: ancestral type III effectors involved in suppression of pathogen-associated molecular pattern-triggered immunity. Mol Plant Pathol 16:899–905

Demidchik V, Maathuis FJ (2007) Physiological roles of nonselective cation channels in plants: from salt stress to signaling and development. New Phytol 175:387–404

Demidchik V, Straltsova D, Medvedev SS, Pozhvanov GA, Sokolik A, Yurin V (2014) Stress-induced electrolyte leakage: the role of K+-permeable channels and involvement in programmed cell death and metabolic adjustment. J Exp Bot 65:1259–1270

Dreyer I, Blatt MR (2009) What makes a gate? The ins and outs of Kv-like K+ channels in plants. Trends Plant Sci 14:383–390

Dreyer I, Uozumi N (2011) Potassium channels in plant cells. FEBS J 278:4293–4303

Dubeaux G, Neveu J, Zelazny E, Vert G (2018) Metal sensing by the IRT1 transporter-receptor orchestrates its own degradation and plant metal nutrition. Mol Cell 69:953–964.e955

Duby G, Hosy E, Fizames C, Alcon C, Costa A, Sentenac H, Thibaud JB (2008) AtKC1, a conditionally targeted shaker-type subunit, regulates the activity of plant K+ channels. Plant J 53:115–123

Durian G, Rahikainen M, Alegre S, Brosche M, Kangasjarvi S (2016) Protein phosphatase 2A in the regulatory network underlying biotic stress resistance in plants. Front Plant Sci 7:812

Fuchs I, Stolzle S, Ivashikina N, Hedrich R (2005) Rice K+ uptake channel OsAKT1 is sensitive to salt stress. Planta 221:212–221

Fujii H, Chinnusamy V, Rodrigues A, Rubio S, Antoni R, Park SY, Cutler SR, Sheen J, Rodriguez PL, Zhu JK (2009) *In vitro* reconstitution of an abscisic acid signaling pathway. Nature 462:660–664

Gao HB, Chu YJ, Xue HW (2013) Phosphatidic acid (PA) binds PP2AA1 to regulate PP2A activity and PIN1 polar localization. Mol Plant 6:1692–1702

Gapper C, Dolan L (2006) Control of plant development by reactive oxygen species. Plant Physiol 141:341–345

Garcia-Mata C, Wang J, Gajdanowicz P, Gonzalez W, Hills A, Donald N, Riedelsberger J, Amtmann A, Dreyer I, Blatt MR (2010) A minimal cysteine motif required to activate the SKOR K+ channel of Arabidopsis by the reactive oxygen species H2O2. J Biol Chem 285:29286–29294

Gaymard F, Pilot G, Lacombe B, Bouchez D, Bruneau D, Boucherez J, Michaux-Ferriere N, Thibaud JB, Sentenac H (1998) Identification and disruption of a plant shaker-like outward channel involved in K+ release into the xylem sap. Cell 94:647–655

Geiger D, Becker D, Vosloh D, Gambale F, Palme K, Rehers M, Anschuetz U, Dreyer I, Kudla J, Hedrich R (2009) Heteromeric AtKC1-AKT1 channels in Arabidopsis roots facilitate growth under K+-limiting conditions. J Biol Chem 284:21288–21295

Gentry MS, Brewer MK, Vander Kooi CW (2016) Structural biology of glucan phosphatases from humans to plants. Curr Opin Struct Biol 40:62–69

Gomez-Gomez L, Boller T (2000) FLS2: an LRR receptor-like kinase involved in the perception of the bacterial elicitor flagellin in Arabidopsis. Mol Cell 5:1003–1011

Grabov A (2007) Plant KT/KUP/HAK potassium transporters: single family—multiple functions. Ann Bot 99:1035–1041

Gupta R, Huang Y, Kieber J, Luan S (1998) Identification of a dual-specificity protein phosphatase that inactivates a MAP kinase from Arabidopsis. Plant J 16:581–589

Gust AA, Felix G (2014) Receptor like proteins associate with SOBIR1-type of adaptors to form bimolecular receptor kinases. Curr Opin Plant Biol 21:104–111

Halford NG (2009) New insights on the effects of heat stress on crops. J Exp Bot 60:4215–4216

Han EH, Petrella DP, Blakeslee JJ (2017) 'Bending' models of halotropism: incorporating protein phosphatase 2A, ABCB transporters, and auxin metabolism. J Exp Bot 68:3071–3089

Han M, Wu W, Wu WH, Wang Y (2016) Potassium transporter KUP7 is involved in K(+) acquisition and translocation in Arabidopsis root under K(+)-limited conditions. Mol Plant 9:437–446

He H, Su J, Shu S, Zhang Y, Ao Y, Liu B, Feng D, Wang J, Wang H (2012) Two homologous putative protein tyrosine phosphatases, OsPFA-DSP2 and AtPFA-DSP4, negatively regulate the pathogen response in transgenic plants. PLoS One 7:e34995

Hedrich R (2012) Ion channels in plants. Physiol Rev 92:1777–1811

Held K, Pascaud F, Eckert C, Gajdanowicz P, Hashimoto K, Corratge-Faillie C, Offenborn JN, Lacombe B, Dreyer I, Thibaud JB, Kudla J (2011) Calcium-dependent modulation and plasma membrane targeting of the AKT2 potassium channel by the CBL4/CIPK6 calcium sensor/protein kinase complex. Cell Res 21:1116–1130

Hirsch RE, Lewis BD, Spalding EP, Sussman MR (1998) A role for the AKT1 potassium channel in plant nutrition. Science 280:918–921

Ho CH, Lin SH, Hu HC, Tsay YF (2009) CHL1 functions as a nitrate sensor in plants. Cell 138:1184–1194

Hobiger K, Friedrich T (2015) Voltage sensitive phosphatases: emerging kinship to protein tyrosine phosphatases from structure-function research. Front Pharmacol 6:20

Hong JP, Takeshi Y, Kondou Y, Schachtman DP, Matsui M, Shin R (2013) Identification and characterization of transcription factors regulating Arabidopsis HAK5. Plant Cell Physiol 54:1478–1490

Huber SC, Huber JL, McMichael RW (1994) Control of plant enzyme activity by reversible protein phosphorylation. Int Rev Cytol 149:47–98

Ivashikina N, Becker D, Ache P, Meyerhoff O, Felle HH, Hedrich R (2001) K(+) channel profile and electrical properties of Arabidopsis root hairs. FEBS Lett 508:463–469

Johansson I, Wulfetange K, Poree F, Michard E, Gajdanowicz P, Lacombe B, Sentenac H, Thibaud JB, Mueller-Roeber B, Blatt MR, Dreyer I (2006) External K+ modulates the activity of the Arabidopsis potassium channel SKOR via an unusual mechanism. Plant J 46:269–281

Johnson LN, O'Reilly M (1996) Control by phosphorylation. Curr Opin Struct Biol 6:762–769

Johnson LN, Snape P, Martin JL, Acharya KR, Barford D, Oikonomakos NG (1993) Crystallographic binding studies on the allosteric inhibitor glucose-6-phosphate to T state glycogen phosphorylase b. J Mol Biol 232:253–267

Kanwar P, Sanyal SK, Tokas I, Yadav AK, Pandey A, Kapoor S, Pandey GK (2014) Comprehensive structural, interaction and expression analysis of CBL and CIPK complement during abiotic stresses and development in rice. Cell Calcium 56:81–95

Kataya AR, Heidari B, Hagen L, Kommedal R, Slupphaug G, Lillo C (2015b) Protein phosphatase 2A holoenzyme is targeted to peroxisomes by piggybacking and positively affects peroxisomal beta-oxidation. Plant Physiol 167:493–506

Kataya AR, Heidari B, Lillo C (2015a) Protein phosphatase 2A regulatory subunits affecting plant innate immunity, energy metabolism, and flowering time—joint functions among B'eta subfamily members. Plant Signal Behav 10:e1026024

Kerk D, Conley TR, Rodriguez FA, Tran HT, Nimick M, Muench DG, Moorhead GB (2006) A chloroplast-localized dual-specificity protein phosphatase in Arabidopsis contains a phylogenetically dispersed and ancient carbohydrate-binding domain, which binds the polysaccharide starch. Plant J 46:400–413

Kerk D, Templeton G, Moorhead GB (2008) Evolutionary radiation pattern of novel protein phosphatases revealed by analysis of protein data from the completely sequenced genomes of humans, green algae, and higher plants. Plant Physiol 146:351–367

Kim BG, Waadt R, Cheong YH, Pandey GK, Dominguez-Solis JR, Schultke S, Lee SC, Kudla J, Luan S (2007) The calcium sensor CBL10 mediates salt tolerance by regulating ion homeostasis in Arabidopsis. Plant J 52:473–484

Kim MJ, Ruzicka D, Shin R, Schachtman DP (2012) The Arabidopsis AP2/ERF transcription factor RAP2.11 modulates plant response to low-potassium conditions. Mol Plant 5:1042–1057

Kleist TJ, Spencley AL, Luan S (2014) Comparative phylogenomics of the CBL-CIPK calcium-decoding network in the moss Physcomitrella, Arabidopsis, and other green lineages. Front Plant Sci 5:187

Kolukisaoglu U, Weinl S, Blazevic D, Batistic O, Kudla J (2004) Calcium sensors and their interacting protein kinases: genomics of the Arabidopsis and rice CBL-CIPK signaling networks. Plant Physiol 134:43–58

Konert G, Trotta A, Kouvonen P, Rahikainen M, Durian G, Blokhina O, Fagerstedt K, Muth D, Corthals GL, Kangasjarvi S (2015) Protein phosphatase 2A (PP2A) regulatory subunit B'gamma interacts with cytoplasmic ACONITASE 3 and modulates the abundance of AOX1A and AOX1D in Arabidopsis thaliana. New Phytol 205:1250–1263

Krzywinska E, Bucholc M, Kulik A, Ciesielski A, Lichocka M, Debski J, Ludwikow A, Dadlez M, Rodriguez PL, Dobrowolska G (2016) Phosphatase ABI1 and okadaic acid-sensitive phosphoprotein phosphatases inhibit salt stress-activated SnRK2.4 kinase. BMC Plant Biol 16:136

Krzywińska E, Kulik A, Bucholc M, Fernandez MA, Rodriguez PL, Dobrowolska G (2016) Protein phosphatase type 2C PP2CA together with ABI1 inhibits SnRK2.4 activity and regulates plant responses to salinity. Plant Signal Behav 11:e1253647

Lacombe B, Pilot G, Gaymard F, Sentenac H, Thibaud JB (2000) pH control of the plant outwardly-rectifying potassium channel SKOR. FEBS Lett 466:351–354

Lagarde D, Basset M, Lepetit M, Conejero G, Gaymard F, Astruc S, Grignon C (1996) Tissue-specific expression of Arabidopsis AKT1 gene is consistent with a role in K+ nutrition. Plant J 9:195–203

Lan WZ, Lee SC, Che YF, Jiang YQ, Luan S (2011) Mechanistic analysis of AKT1 regulation by the CBL-CIPK-PP2CA interactions. Mol Plant 4:527–536

Lassowskat I, Bottcher C, Eschen-Lippold L, Scheel D, Lee J (2014) Sustained mitogen-activated protein kinase activation reprograms defense metabolism and phosphoprotein profile in Arabidopsis thaliana. Front Plant Sci 5:554

Lebaudy A, Very AA, Sentenac H (2007) K+ channel activity in plants: genes, regulations and functions. FEBS Lett 581:2357–2366

Lee SC, Lan W, Buchanan BB, Luan S (2009) A protein kinase-phosphatase pair interacts with an ion channel to regulate ABA signaling in plant guard cells. Proc Natl Acad Sci U S A 106:21419–21424

Lee SC, Lan WZ, Kim BG, Li L, Cheong YH, Pandey GK, Lu G, Buchanan BB, Luan S (2007) A protein phosphorylation/dephosphorylation network regulates a plant potassium channel. Proc Natl Acad Sci U S A 104:15959–15964

Lefoulon C, Boeglin M, Moreau B, Very AA, Szponarski W, Dauzat M, Michard E, Gaillard I, Cherel I (2016) The Arabidopsis AtPP2CA protein phosphatase inhibits the GORK K+ efflux channel and exerts a dominant suppressive effect on Pphosphomimetic-activating mutations. J Biol Chem 291:6521–6533

Leran S, Edel KH, Pervent M, Hashimoto K, Corratge-Faillie C, Offenborn JN, Tillard P, Gojon A, Kudla J, Lacombe B (2015) Nitrate sensing and uptake in Arabidopsis are enhanced by ABI2, a phosphatase inactivated by the stress hormone abscisic acid. Sci Signal 8:ra43

Li L, Kim BG, Cheong YH, Pandey GK, Luan S (2006) A Ca^{2+} signaling pathway regulates a K^+ channel for low-K response in Arabidopsis. Proc Natl Acad Sci U S A 103:12625–12630

Li W, Xu G, Alli A, Yu L (2018) Plant HAK/KUP/KT K(+) transporters: function and regulation. Semin Cell Dev Biol 74:133–141

Liu LL, Ren HM, Chen LQ, Wang Y, Wu WH (2013) A protein kinase, calcineurin B-like protein-interacting protein Kinase9, interacts with calcium sensor calcineurin B-like Protein3 and regulates potassium homeostasis under low-potassium stress in Arabidopsis. Plant Physiol 161:266–277

Luan S (1998) Protein phosphatases and signaling cascades in higher plants. Trends Plant Sci 3:271–275

Luan S (2002) Tyrosine phosphorylation in plant cell signaling. Proc Natl Acad Sci U S A 99:11567–11569

Luan S (2003) Protein phosphatases in plants. Annu Rev Plant Biol 54:63–92

Luan S (2009) The CBL-CIPK network in plant calcium signaling. Trends Plant Sci 14:37–42

Maathuis FJ (2009) Physiological functions of mineral macronutrients. Curr Opin Plant Biol 12:250–258

Maathuis FJ, Sanders D (1994) Mechanism of high-affinity potassium uptake in roots of Arabidopsis thaliana. Proc Natl Acad Sci U S A 91:9272–9276

Maathuis FJM, Sanders D (1993) Energization of potassium uptake in Arabidopsis thaliana. Planta 191:302–307

MacKintosh C, Coggins J, Cohen P (1991) Plant protein phosphatases. Subcellular distribution, detection of protein phosphatase 2C and identification of protein phosphatase 2A as the major quinate dehydrogenase phosphatase. Biochem J 273(Pt 3):733–738

Mittler R, Blumwald E (2010) Genetic engineering for modern agriculture: challenges and perspectives. Annu Rev Plant Biol 61:443–462

Moorhead GB, De Wever V, Templeton G, Kerk D (2009) Evolution of protein phosphatases in plants and animals. Biochem J 417:401–409

Mora-Garcia S, Vert G, Yin Y, Cano-Delgado A, Cheong H, Chory J (2004) Nuclear protein phosphatases with Kelch-repeat domains modulate the response to brassinosteroids in Arabidopsis. Genes Dev 18:448–460

Mustilli AC, Merlot S, Vavasseur A, Fenzi F, Giraudat J (2002) Arabidopsis OST1 protein kinase mediates the regulation of stomatal aperture by Abscisic acid and acts upstream of reactive oxygen species production. Plant Cell 14:3089–3099

Nieves-Cordones M, Miller AJ, Aleman F, Martinez V, Rubio F (2008) A putative role for the plasma membrane potential in the control of the expression of the gene encoding the tomato high-affinity potassium transporter HAK5. Plant Mol Biol 68:521–532

Nieves-Cordones M, Rodenas R, Chavanieu A, Rivero RM, Martinez V, Gaillard I, Rubio F (2016) Uneven HAK/KUP/KT protein diversity among angiosperms: species distribution and perspectives. Front Plant Sci 7:127

Nishimura N, Yoshida T, Kitahata N, Asami T, Shinozaki K, Hirayama T (2007) ABA-Hypersensitive Germination1 encodes a protein phosphatase 2C, an essential component of abscisic acid signaling in Arabidopsis seed. Plant J 50:935–949

Nongpiur R, Soni P, Karan R, Singla-Pareek SL, Pareek A (2012) Histidine kinases in plants: cross talk between hormone and stress responses. Plant Signal Behav 7:1230–1237

Ohta M, Guo Y, Halfter U, Zhu JK (2003) A novel domain in the protein kinase SOS2 mediates interaction with the protein phosphatase 2C ABI2. Proc Natl Acad Sci U S A 100:11771–11776

Ota IM, Varshavsky A (1993) A yeast protein similar to bacterial two-component regulators. Science 262:566–569

Pais SM, Gonzalez MA, Tellez-Inon MT, Capiati DA (2009) Characterization of potato (Solanum tuberosum) and tomato (Solanum lycopersicum) protein phosphatases type 2A catalytic subunits and their involvement in stress responses. Planta 230:13–25

Pandey GK, Cheong YH, Kim BG, Grant JJ, Li L, Luan S (2007) CIPK9: a calcium sensor-interacting protein kinase required for low-potassium tolerance in Arabidopsis. Cell Res 17:411–421

Pandey GK, Kanwar P, Pandey A (2014) Global comparative analysis of CBL-CIPK gene families in plants. Springer, New York

Pawson T (1995) Protein modules and signaling networks. Nature 373:573–580

Pei ZM, Murata Y, Benning G, Thomine S, Klusener B, Allen GJ, Grill E, Schroeder JI (2000) Calcium channels activated by hydrogen peroxide mediate abscisic acid signaling in guard cells. Nature 406:731–734

Prince DC, Drurey C, Zipfel C, Hogenhout SA (2014) The leucine-rich repeat receptor-like kinase BRASSINOSTEROID INSENSITIVE1-ASSOCIATED KINASE1 and the cytochrome P450 PHYTOALEXIN DEFICIENT3 contribute to innate immunity to aphids in Arabidopsis. Plant Physiol 164:2207–2219

Qi Z, Hampton CR, Shin R, Barkla BJ, White PJ, Schachtman DP (2008) The high affinity K+ transporter AtHAK5 plays a physiological role in planta at very low K+ concentrations and provides a caesium uptake pathway in Arabidopsis. J Exp Bot 59:595–607

Ragel P, Raddatz N, Leidi EO, Quintero FJ, Pardo JM (2019) Regulation of K+ nutrition in plants. Front Plant Sci 10:281

Ragel P, Rodenas R, Garcia-Martin E, Andres Z, Villalta I, Nieves-Cordones M, Rivero RM, Martinez V, Pardo JM, Quintero FJ, Rubio F (2015) The CBL-interacting protein kinase CIPK23 regulates HAK5-mediated high-affinity K^+ uptake in Arabidopsis roots. Plant Physiol 169:2863–2873

Rahikainen M, Pascual J, Alegre S, Durian G, Kangasjärvi S (2016) PP2A phosphatase as a regulator of ROS signaling in plants. Antioxidants (Basel) 5:E8

Reintanz B, Szyroki A, Ivashikina N, Ache P, Godde M, Becker D, Palme K, Hedrich R (2002) AtKC1, a silent Arabidopsis potassium channel alpha -subunit modulates root hair K+ influx. Proc Natl Acad Sci U S A 99:4079–4084

Ren XL, Qi GN, Feng HQ, Zhao S, Zhao SS, Wang Y, Wu WH (2013) Calcineurin B-like protein CBL10 directly interacts with AKT1 and modulates K^+ homeostasis in Arabidopsis. Plant J 74:258–266

Rubio F, Fon M, Rodenas R, Nieves-Cordones M, Aleman F, Rivero RM, Martinez V (2014) A low K+ signal is required for functional high-affinity K+ uptake through HAK5 transporters. Physiol Plant 152:558–570

Rubio S, Rodrigues A, Saez A, Dizon MB, Galle A, Kim TH, Santiago J, Flexas J, Schroeder JI, Rodriguez PL (2009) Triple loss of function of protein phosphatases type 2C leads to partial constitutive response to endogenous abscisic acid. Plant Physiol 150:1345–1355

Santa-Maria GE, Oliferuk S, Moriconi JI (2018) KT-HAK-KUP transporters in major terrestrial photosynthetic organisms: a twenty years tale. J Plant Physiol 226:77–90

Sanyal SK, Pandey A, Pandey GK (2015) The CBL-CIPK signaling module in plants: a mechanistic perspective. Physiol Plant 155:89–108

Sanyal SK, Rao S, Mishra LK, Sharma M, Pandey GK (2016) Plant stress responses mediated by CBL-CIPK phosphorylation network. Enzymes 40:31–64

Schachtman DP, Shin R (2007) Nutrient sensing and signaling: NPKS. Annu Rev Plant Biol 58:47–69

Scherzer S, Bohm J, Krol E, Shabala L, Kreuzer I, Larisch C, Bemm F, Al-Rasheid KA, Shabala S, Rennenberg H, Neher E, Hedrich R (2015) Calcium sensor kinase activates potassium uptake systems in gland cells of Venus flytraps. Proc Natl Acad Sci U S A 112:7309–7314

Schweighofer A, Kazanaviciute V, Scheikl E, Teige M, Doczi R, Hirt H, Schwanninger M, Kant M, Schuurink R, Mauch F, Buchala A, Cardinale F, Meskiene I (2007) The PP2C-type phosphatase AP2C1, which negatively regulates MPK4 and MPK6, modulates innate immunity, jasmonic acid, and ethylene levels in Arabidopsis. Plant Cell 19:2213–2224

Shabala L, Zhang J, Pottosin I, Bose J, Zhu M, Fuglsang AT, Velarde-Buendia A, Massart A, Hill CB, Roessner U, Bacic A, Wu H, Azzarello E, Pandolfi C, Zhou M, Poschenrieder C, Mancuso S, Shabala S (2016) Cell-type-specific H+-ATPase activity in root tissues enables K+ retention and mediates acclimation of barley (Hordeum vulgare) to salinity stress. Plant Physiol 172:2445–2458

Shabala S, Cuin TA (2008) Potassium transport and plant salt tolerance. Physiol Plant 133:651–669

Shankar A, Agrawal N, Sharma M, Pandey A, Girdhar KPM (2015) Role of protein tyrosine phosphatases in plants. Curr Genomics 16:224–236

Sharma T, Dreyer I, Riedelsberger J (2013) The role of K+ channels in uptake and redistribution of potassium in the model plant Arabidopsis thaliana. Front Plant Sci 4:224

Sheen J (1998) Mutational analysis of protein phosphatase 2C involved in abscisic acid signal transduction in higher plants. Proc Natl Acad Sci U S A 95:975–980

Shi Y (2009) Serine/threonine phosphatases: mechanism through structure. Cell 139:468–484

Shin R, Berg RH, Schachtman DP (2005) Reactive oxygen species and root hairs in Arabidopsis root response to nitrogen, phosphorus and potassium deficiency. Plant Cell Physiol 46:1350–1357

Silver DM, Kotting O, Moorhead GB (2014) Phosphoglucan phosphatase function sheds light on starch degradation. Trends Plant Sci 19:471–478

Singh A, Giri J, Kapoor S, Tyagi AK, Pandey GK (2010) Protein phosphatase complement in rice: genome-wide identification and transcriptional analysis under abiotic stress conditions and reproductive development. BMC Genomics 11:435

Singh A, Pandey A, Srivastava AK, Tran LP, Pandey GK (2015) Plant protein phosphatases 2C: from genomic diversity to functional multiplicity and importance in stress management. Crit Rev Biotechnol 36:1023–1035

Singh A, Yadav AK, Kaur K, Sanyal SK, Jha SK, Fernandes JL, Sharma P, Tokas I, Pandey A, Luan S, Pandey GK (2018) Protein phosphatase 2C, AP2C1 interacts with and negatively regulates the function of CIPK9 under potassium deficient conditions in Arabidopsis. J Exp Bot 69:4003–4015

Spalding EP, Hirsch RE, Lewis DR, Qi Z, Sussman MR, Lewis BD (1999) Potassium uptake supporting plant growth in the absence of AKT1 channel activity: inhibition by ammonium and stimulation by sodium. J Gen Physiol 113:909–918

Tang RJ, Liu H, Yang Y, Yang L, Gao XS, Garcia VJ, Luan S, Zhang HX (2012) Tonoplast calcium sensors CBL2 and CBL3 control plant growth and ion homeostasis through regulating V-ATPase activity in Arabidopsis. Cell Res 22:1650–1665

Tang RJ, Zhao FG, Garcia VJ, Kleist TJ, Yang L, Zhang HX, Luan S (2015) Tonoplast CBL-CIPK calcium signaling network regulates magnesium homeostasis in Arabidopsis. Proc Natl Acad Sci U S A 112:3134–3139

Tang W, Yuan M, Wang R, Yang Y, Wang C, Oses-Prieto JA, Kim TW, Zhou HW, Deng Z, Gampala SS, Gendron JM, Jonassen EM, Lillo C, DeLong A, Burlingame AL, Sun Y, Wang ZY (2011) PP2A activates brassinosteroid-responsive gene expression and plant growth by dephosphorylating BZR1. Nat Cell Biol 13:124–131

Tian H, Lv B, Ding T, Bai M, Ding Z (2017) Auxin-BR interaction regulates plant growth and development. Front Plant Sci 8:2256

Trotta A, Wrzaczek M, Scharte J, Tikkanen M, Konert G, Rahikainen M, Holmstrom M, Hiltunen HM, Rips S, Sipari N, Mulo P, Weis E, von Schaewen A, Aro EM, Kangasjarvi S (2011) Regulatory subunit B'gamma of protein phosphatase 2A prevents unnecessary defense reactions under low light in Arabidopsis. Plant Physiol 156:1464–1480

Truman W, Bennett MH, Kubigsteltig I, Turnbull C, Grant M (2007) Arabidopsis systemic immunity uses conserved defense signaling pathways and is mediated by jasmonates. Proc Natl Acad Sci U S A 104:1075–1080

Tsujii M, Kera K, Hamamoto S, Kuromori T, Shikanai T, Uozumi N (2019) Evidence for potassium transport activity of Arabidopsis KEA1-KEA6. Sci Rep 9:10040

Uhrig RG, Kerk D, Moorhead GB (2013b) Evolution of bacterial-like phosphoprotein phosphatases in photosynthetic eukaryotes features ancestral mitochondrial or archaeal origin and possible lateral gene transfer. Plant Physiol 163:1829–1843

Uhrig RG, Labandera AM, Moorhead GB (2013a) Arabidopsis PPP family of serine/threonine protein phosphatases: many targets but few engines. Trends Plant Sci 18:505–513

Uhrig RG, Moorhead GB (2011) Two ancient bacterial-like PPP family phosphatases from Arabidopsis are highly conserved plant proteins that possess unique properties. Plant Physiol 157:1778–1792

Umezawa T, Nakashima K, Miyakawa T, Kuromori T, Tanokura M, Shinozaki K, Yamaguchi-Shinozaki K (2010) Molecular basis of the Core regulatory network in ABA responses: sensing, signaling and transport. Plant Cell Physiol 51:1821–1839

van Kleeff PJM, Gao J, Mol S, Zwart N, Zhang H, Li KW, de Boer AH (2018) The Arabidopsis GORK K(+)-channel is phosphorylated by calcium-dependent protein kinase 21 (CPK21), which in turn is activated by 14-3-3 proteins. Plant Physiol Biochem 125:219–231

Verslues PE, Zhu JK (2005) Before and beyond ABA: upstream sensing and internal signals that determine ABA accumulation and response under abiotic stress. Biochem Soc Trans 33:375–379

Very AA, Sentenac H (2003) Molecular mechanisms and regulation of K+ transport in higher plants. Annu Rev Plant Biol 54:575–603

Wang P, Xue L, Batelli G, Lee S, Hou YJ, Van Oosten MJ, Zhang H, Tao WA, Zhu JK (2013) Quantitative phosphoproteomics identifies SnRK2 protein kinase substrates and reveals the effectors of abscisic acid action. Proc Natl Acad Sci U S A 110:11205–11210

Wang XP, Chen LM, Liu WX, Shen L, Wang FL, Zhou Y, Zhang Z, Wu WH, Wang Y (2016) AtKC1 and CIPK23 synergistically modulate AKT1-mediated low potassium stress responses in Arabidopsis. Plant Physiol 170:2264–2277

Wang Y, Wu WH (2010) Plant sensing and signaling in response to K+-deficiency. Mol Plant 3:280–287

Wang Y, Wu WH (2017) Regulation of potassium transport and signaling in plants. Curr Opin Plant Biol 39:123–128

Waraich EA, Ahmad R, Halim A, Aziz T (2012) Alleviation of temperature stress by nutrient management in crop plants: a review. J Soil Sci Plant Nutr 12:221–244

Weinl S, Kudla J (2009) The CBL-CIPK Ca^{2+}-decoding signaling network: function and perspectives. New Phytol 184:517–528

Xu C, Jing R, Mao X, Jia X, Chang X (2007) A wheat (Triticum aestivum) protein phosphatase 2A catalytic subunit gene provides enhanced drought tolerance in tobacco. Ann Bot 99:439–450

Xu J, Li HD, Chen LQ, Wang Y, Liu LL, He L, Wu WH (2006) A protein kinase, interacting with two calcineurin B-like proteins, regulates K^+ transporter AKT1 in Arabidopsis. Cell 125:1347–1360

Xu J, Zhang S (2015) Mitogen-activated protein kinase cascades in signaling plant growth and development. Trends Plant Sci 20:56–64

Xue T, Wang D, Zhang S, Ehlting J, Ni F, Jakab S, Zheng C, Zhong Y (2008) Genome-wide and expression analysis of protein phosphatase 2C in rice and Arabidopsis. BMC Genomics 9:550

Yang Q, Liu K, Niu X, Wang Q, Wan Y, Yang F, Li G, Wang Y, Wang R (2018) Genome-wide identification of PP2C genes and their expression profiling in response to drought and cold stresses in Medicago truncatula. Sci Rep 8:12841

Yang W, Zhang W, Wang X (2017) Post-translational control of ABA signaling: the roles of protein phosphorylation and ubiquitination. Plant Biotechnol J 15:4–14

Yoshida R, Hobo T, Ichimura K, Mizoguchi T, Takahashi F, Aronso J, Ecker JR, Shinozaki K (2002) ABA-activated SnRK2 protein kinase is required for dehydration stress signaling in Arabidopsis. Plant Cell Physiol 43:1473–1483

Yoshida R, Umezawa T, Mizoguchi T, Takahashi S, Takahashi F, Shinozaki K (2006) The regulatory domain of SRK2E/OST1/SnRK2.6 interacts with ABI1 and integrates abscisic acid (ABA) and osmotic stress signals controlling stomatal closure in Arabidopsis. J Biol Chem 281:5310–5318

Yu RM, Wong MM, Jack RW, Kong RY (2005) Structure, evolution and expression of a second subfamily of protein phosphatase 2A catalytic subunit genes in the rice plant (Oryza sativa L.). Planta 222:757–768

Yu RM, Zhou Y, Xu ZF, Chye ML, Kong RY (2003) Two genes encoding protein phosphatase 2A catalytic subunits are differentially expressed in rice. Plant Mol Biol 51:295–311

Yu X, Han J, Wang E, Xiao J, Hu R, Yang G, He G (2019) Genome-wide identification and homoeologous expression analysis of PP2C genes in wheat (Triticum aestivum L.). Front Genet 10:561

Zhao S, Zhang ML, Ma TL, Wang Y (2016) Phosphorylation of ARF2 relieves its repression of transcription of the K+ transporter gene HAK5 in response to low potassium stress. Plant Cell 28:3005–3019

Zhu JK (2016) Abiotic stress signaling and responses in plants. Cell 167:313–324

Chapter 12
Protein Phosphatases in N Response and NUE in Crops

Supriya Kumari and Nandula Raghuram

12.1 Introduction

Nitrogen is an essential macronutrient for plant growth and crop yield, as most crop plants barring N-fixing legumes depend on reactive N compounds such as urea, ammonium salts, or nitrates from manures or fertilizers as their main nitrogen source. Fertilizers are the world's largest anthropogenic source of reactive nitrogen, with annual global consumption at approximately 119.40 million tons and annual growth of 1.4% (FAO 2018). Due to the poor N use efficiency (NUE) of most crops at a global average of <30%, unused reactive N compounds from fertilizers and other sources are a major source of pollution of land, water, and air, affecting human and animal health, biodiversity, and climate change (Sutton and Bleeker 2013; Sutton et al. 2019).The fourth UN Environment Assembly held in March 2019 has adopted a resolution to work toward sustainable nitrogen management (https://sdg.iisd.org/events/fourth-session-of-the-un-environment-assembly-unea-4/). Therefore, the development of N use efficient crops is an extremely important goal, not only for sustainable agriculture but also for environmental sustainability and mitigation of climate change (Raghuram et al. 2007; Raghuram and Sharma 2019).

NUE can be defined in several ways (Raghuram and Sharma 2019), but it is best understood as yield per unit N, or the maximal output with minimal N input. NUE is a complex genetic trait involving several genes and their interactions with the environment, which are yet to be fully characterized (Mandal et al. 2018). NUE involves a combination of processes such as N uptake, retention, assimilation, and remobilization of internal N reserves. Plants possess families of transporters to take up urea, ammonium, and nitrate ions (NO_3^-), of which NO_3^- ions not only acts as nutrient but also play a very important role in signal transduction and regulation of

S. Kumari · N. Raghuram (✉)
School of Biotechnology, Guru Gobind Singh Indraprastha University, New Delhi, India
e-mail: raghuram@ipu.ac.in

G. K. Pandey (ed.), *Protein Phosphatases and Stress Management in Plants*,
https://doi.org/10.1007/978-3-030-48733-1_12

one or more of the above processes (Raghuram et al. 2006; Pathak et al. 2008, 2011; Chakraborty and Raghuram 2011; Mandal et al. 2018; Raghuram and Sharma 2019).

The progress on the precise characterization of the biological avenues for improvement for NUE has been painfully slow, despite some occasional advances on phenotype (Sun et al. 2014; Sharma et al. 2018). Indeed, even though low nutrient stress is known to be an important abiotic stress for crops in developing countries, the interface of N and stress has not been explored in great detail (Jangam et al. 2016). While this has hampered crop improvement by forward genetics, recent advances in genomics and functional genomics of several crops have opened the opportunity for using reverse genetics to address NUE using candidate gene approach. To understand the physiological and molecular basis of N response under normal and low supply of nitrogen, transcriptomic and proteomic studies have identified a large number of differentially expressed genes in crops like rice (Yang et al. 2015;Waqas et al. 2018; Sinha et al. 2018), sorghum (Gelli et al. 2014), maize (Jiang et al. 2018), and barley (Quan et al. 2016, 2019).

Among the multitude of pathways and processes that constitute N response, protein kinases and phosphatases as well as transcription factors and other signaling intermediates regulated by phosphorylation emerged significant. Protein kinases as well as protein phosphatases have been widely studied in rice, Arabidopsis, wheat, maize, etc. (Singh et al. 2010; Xue et al. 2008; Bradai et al. 2018; Trevisan et al. 2011). PP2Cs are a major class of phosphatases and play an important role in stress in plants (Moorhead et al. 2009; Singh et al. 2010, 2015). In this chapter, we discuss the role of protein phosphatases in N-sensing and signaling and their role in carbon and nitrogen metabolism and phosphatases involved in N-response and NUE with enhancement potential for further validation of phosphatases.

12.2 Phosphatases in N Uptake and Primary Nitrate Response

For optimal growth as well as ion balance and homeostasis of plant, adjustment of uptake of ammonium (NH_4^+) and NO_3^- is essential as compared to other major ions (Haynes 1990). Phosphorylation and dephosphorylation play major roles in regulating transport of these ions by transporter protein through plasma membrane (Straub et al. 2017). An important target of such regulation is NRT1.1/CHL1/NPF6.3, a dual-affinity nitrate transporter. It works under high as well as low NO_3^- condition. PP2C plays a major role in dephosphorylation of NRT1.1 by binding to the calcineurin B-like-interacting protein kinase (CIPK). CIPK consists of two domains; one is the conserved N-terminal kinase catalytic domain, and the other is the highly variant C-terminal regulatory domain having NAF/FISL motif and a phosphatase interaction motif (Guo et al. 2001). In response to low NO_3^- condition, NO_3^- binds to a high affinity site, which activates CIPK23 to phosphorylate CHL1 at T101. The phosphorylated CHL1 prevents higher primary nitrate response to low

NO_3^- concentration. At high NO_3^- concentrations, NO_3^- binds to a low nitrate affinity site, which dephosphorylates CHL1 at T101 and generates a high primary nitrate response (Ho et al. 2009; Vert and Chory 2009; Straub et al. 2017). In low-affinity phase of (NO_3^-), CIPK8 is activated by an unknown CBL, which phosphorylates CHL1 at a residue different from T101 (Hu et al. 2009). CIPK23 acts as a negative regulator of high-affinity NO_3^- response, while it acts as a positive regulator of primary nitrate response in low-affinity phase of NO_3^- response.

12.3 PP2Cs Are Negative Regulators of ABA Signaling in NO_3^- Sensing

ABA signaling pathway includes a type 2C protein phosphatase (PP2C; a negative regulator) and a SNF1-related protein kinase 2 (SnRK2; a positive regulator) (Umezawa et al. 2009). ABA can inhibit or stimulate the functioning of the root meristem and modulate root growth depending on its concentration (Cheng et al. 2002). Development of root architecture is strongly regulated by the concentration of NO_3^- in the root environment, as well as its distribution. If the environment surrounding the root system is uniformly high in NO_3^-, lateral root growth is generally inhibited (Zhang and Forde 2000; Walch-Liu et al. 2005). NO_3^- locally stimulates lateral root elongation and in some species initiation, when NO_3^- presents only in patch (Zhang and Forde 1998). In Arabidopsis, this local stimulation of lateral root elongation by patches of NO_3^- requires ABA signaling (Signora et al. 2001). In Arabidopsis, PP2C proteins such as ABA-insensitive 1 (ABI1), ABI2, and Hypersensitive to ABA 1 (HAB1) have been found to function in regulation of root development. ABA-induced signaling functions under both biotic and abiotic stresses by interacting with SnRK2s and PYR/PYL/RCARs.

12.4 Phosphatases: Key Players in Carbon and Nitrogen Balance

In addition to the independent utilization of carbon and nitrogen metabolites, their ratio (known as C/N balance) is more important for the regulation of plant growth. The signaling mechanism underlying C/N balance is not clear till date (Sulpice et al. 2013; Lu et al. 2015). In leaves, NO_3^- and nitrite are reduced to ammonia and then to glutamate (Glu) via the glutamine synthetase (GS)-glutamine-2-oxoglutarate aminotransferase (GOGAT) pathway. Glutamate is a source of C and N for the biosynthesis of most other amino acids (Forde and Lea 2007). GOGAT is found in two isoforms (FdGOGAT form and NADH-GOGAT) and located in the chloroplast in higher plants. Among these, Fd-GOGAT was found to be very active in chloroplast of photosynthetic tissues (Nigro et al. 2014). It plays a very important role in re-assimilation of ammonia released during photorespiration and is potential target to improve NUE (Zeng et al. 2017). Phosphatases and kinases both play very important role in posttranslational modification of chloroplast protein in regulating distribution of light energy between the photosystem I and photosystem II (Michelet et al. 2013; Rochaix 2013; Grabsztunowicz et al. 2017). The PSII core proteins D1 and D2, inner antenna protein CP43, and a minor PSII subunit PsbH are targets for light-dependent Thr phosphorylation catalyzed mainly by the STN8 kinase (Fristedt and Vener 2011), while the PSII core phosphatase is responsible for the dephosphorylation (Samol et al. 2012). Dephosphorylation of light harvesting complex (LHC) by the PPH1/TAP38 (chloroplast protein phosphatase/thylakoid-associated phosphatase of 38 kDa) protein phosphatase results in redistribution of excitation energy toward PSII (Pribil et al. 2010; Shapiguzov et al. 2010).

Ribulose-1,5-bisphosphate carboxylase-oxygenase (Rubisco) plays a central role in CO_2 assimilation of all photosynthetic organisms. It catalyzes the photosynthetic CO_2 fixation by the carboxylation of ribulose-1,5-bisphosphate (RuBP) in C3 photosynthesis. Rubisco is a slow enzyme, and its large amount (up to 25% of leaf N and 50% of leaf-soluble protein) is needed for photosynthesis (Parry et al. 2013; Whitney et al. 2015). In a study, it has been shown that in antisense plants by decreasing Rubisco content up to 15–20%, nitrogen demand reduced by 10% without negatively affecting photosynthetic carbon fixation (Sitt and Schulze 1994; Parry et al. 2013). Thus, increasing CO_2 fixing capacity by changing Rubisco and/or Rubisco activase function, i.e., more carbon fixed for same amount of leaf protein, would increase nitrogen use efficiency. Rubisco has been reported as a target of reversible phosphorylation in many plant species (Wang et al. 2014; Roitinger et al. 2015). Two subunits of Rubisco, RBCL (Rubisco large chain gene) and RBCS (Rubisco small chain gene), contain multiple phosphorylation sites (Cao et al. 2011; Wang et al. 2014). Phosphorylation of the highly conserved RBCL residues Ser208, Thr246, Tyr239, and Thr330, located in the close proximity to RuBP binding site, might affect the catalytic activity of the enzyme (Hodges et al. 2013). Dephosphorylation of RBCL has been found to dissociate Rubisco holocomplex and decreased activity of the enzyme by affecting the interaction between Rubisco and Rubisco activase (RA) (Chen et al. 2011; Hodges et al. 2013).

12.5 PP2A-TOR in Regulation of Nitrate Metabolism

Target of rapamycin (TOR) is a serine/threonine kinase and is highly conserved among all eukaryotes. TOR is activated by both nitrogen and carbon metabolites and promotes cell division, mRNA translation, and repressing nutrient remobilization through autophagy (Dobrenel et al. 2016). It is a master regulator of growth and development including transcription, ribosome biogenesis, autophagy, nutrition recycling, and cellular metabolism in plants (Bakshi et al. 2019). The plant TOR complex consists of TOR, Raptor, and LST8. The TOR complex promotes cell growth and translation through the regulation of S6K and its substrate ribosomal

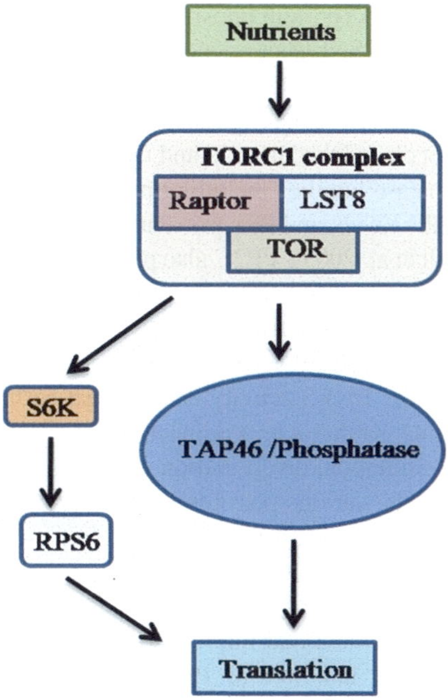

Fig. 12.1 Schematic representation of TOR signaling in plants in response to nutrients (Modified from Ahn et al. 2011)

protein S6(RPS6). It has been reported that TAP46 and its associated phosphatases (PP2Ac and PP2Ac-like subunits) play a critical role in mediating TOR signaling, leading to the promotion of protein translation, the repression of autophagy, and nitrate metabolism (Fig. 12.1). TAP42-phosphatase complex associates with TOR complex 1 (TORC1), under normal condition, while in nutrient deprivation condition or rapamycin treatment, TAP42-phosphatase complex dissociates and is released to cytosol. Interaction of TAP42 with PP2A or PP2A-like phosphatases modulates the activity of downstream effectors and regulates transcription and translation of starvation-responsive genes (Ahn et al. 2011). In Arabidopsis transgenic plants, overexpression of TAP46 leads to upregulation of genes related to nitrogen metabolism and nitrate assimilation and also increased TOR activity, whereas TAP46-silenced plants exhibited a reduction in N-assimilating enzymes (Ahn et al. 2015).

12.6 Protein Phosphatases: Fine-Tuning of Nitrate Reductase

Phosphatases are not only passive partners of kinases, rather they play a dynamic regulatory role in several metabolic processes. They activate or deactivate themselves through posttranslational modification (Heidari et al. 2011). Posttranslational regulation of nitrate reductase (NR) activity through phosphorylation/dephosphorylation by PP2A (protein phosphatase 2A), CDPK (calcium-dependent protein kinase), and SNRK (SNF/sucrose nonfermenting-related kinase) is light dependent (Raghuram and Sopory 1995; Huber et al. 1996). Activation of NR is triggered by photosynthesis, but the signaling cascades from chloroplasts to cytosol, where NR is located, are not clear. Okadaic acid and microcystin-LR are known inhibitors of the protein phosphatase families including protein phosphatase 1 (PP1), PP2A, PP4, PP5, PP6, and PPP-Kelch, and they prevent light activation of NR in plant system (Lillo et al. 1996; Ali et al. 2007). PP2C also plays a major regulatory role in different stress signalings in plants (Moorhead et al. 2007; Singh et al. 2015). Dephosphorylation of NR is essential for activation of NR.14-3-3 which is highly conserved family of proteins found in all eukaryotes. Depending on the developmental and environmental requirements, 14-3-3 activity could direct carbon either into sucrose and storage carbohydrate synthesis or, via inactivation of carbon metabolism and activation of nitrogen assimilation, divert carbon skeletons into the synthesis of amino acids. Plants could use the various possible combinations of 14-3-3 homo- and heterodimers for fine-tuning (Comparot et al. 2003). In a study in rice, dephosphorylation of SnRK and 14-3-3 by PP2C led to the activation of nitrate reductase for increasing NUE and assimilating efficiency of plants (Waqas et al. 2018). In extract of spinach leaf, okadaic acid and microcystin were found to inactivate NR activity, but inhibitor 2 (known inhibitor of PP1 family) showed no effect on NR. Mammalian PP2A activates NR *in vitro*, and it further suggested the role of other phosphatases including PP2A in this process (Mackintosh 1992). BSL1, BSL2, and BSL3 are expressed in leaf and can be considered as potential candidate genes for dephosphorylating NR (Mora-García et al. 2004). Heidari et al. (2011) showed that by knocking down of all three subunits of PP2A in Arabidopsis, NR activity was negatively regulated. This provides a confirmation of involvement of PP2A in NR activation in light-dark transition in plants.

12.7 Phosphatases Identified in N Response/NUE

Global gene expression and comparative analysis of genotypes contrasting for NUE allow the enrichment of candidate genes for NUE. Considerable literatures are now available on transcriptomics, proteomics, as well as on quantitative trait loci (QTL) affecting N response and NUE (Sinha et al. 2018). However, only limited studies explored the regulatory role of different classes of phosphatases in N responses and NUE (Waqas et al. 2018; Sinha et al. 2018; Xiong et al. 2019).

A compilation of the different phosphatases reported to be involved in plant N response and/or NUE is provided in Table 12.1. In a study conducted by Sinha et al. (2018), two contrasting rice genotypes IR64 and Nagina22 were used for genome-wide transcriptome analysis under optimal and chronic starvation of nitrogen from 15-day-old root and shoot tissues. Two genes, *LOC_Os01g71420* and *LOC_Os10g25430* belonging to Ser/Thr phosphatase family, are reported as N-responsive genes.In transgenic line, over expression of PP2C9TL has been reported to improve

Table 12.1 List of genes encoding phosphatase for N response/NUE

Organism	Gene symbol/locus id	Description	References
Rice	LOC_Os01g47580	Lipid phosphate phosphatase 2	Hsieh et al. (2018)
Rice	LOC_Os09g15670	Protein phosphatase 2 C 68 (PP2C68)	Hsieh et al. (2018)
Rice	LOC_Os05g02110	Protein phosphatase 2 C 46 (PP2C46)	Hsieh et al. (2018)
Rice	LOC_Os04g33080	Protein phosphatase 2 C 39 (PP2C39)	Hsieh et al. (2018)
Rice	LOC_Os01g71420	Ser/Thr protein phosphatase family protein, putative, expressed	Sinha et al. (2018)
Rice	LOC_Os10g25430	Ser/Thr protein phosphatase family protein, putative, expressed	Sinha et al. (2018)
Rice	LOC_Os08g33370.2	14-3-3 protein	Waqas et al. (2018)
Rice	LOC_Os05g11550.1	Ser/Thr protein phosphatase5	Waqas et al. (2018)
Rice	LOC_Os07g32380.1	Protein phosphatase 2C	Waqas et al. (2018)
Rice	LOC_Os09g06230.1	Ser/Thr protein phosphatase 16	Waqas et al. (2018)
Rice	LOC_Os04g56450.1	Protein phosphatase 2C	Waqas et al. (2018)
Barley	hv_10051	T6P phosphatase (TPP)	Fataftah et al. (2018)
Maize	GRMZM2G010855	Protein phosphatase 2c (pp2c)	Liu et al. (2011)
Maize	GRMZM2G152447	Purple acid phosphatase 1	Jiang et al. ((2018)
Maize	GRMZM2G134054	Ser/Thr protein phosphatase	Jiang et al. (2018)
Wheat	Traes_6BS_143FEF476	Bifunctional protein-serine/ threonine kinase/phosphatase	Xiong et al. (2019)
Sorghum	Sb08g019110	Phosphatases	Gelli et al. (2014)
Poplar	POPTR_0007s05670	Haloacid dehalogenase-like hydrolase (HAD) superfamilyprotein/trehalose-phosphatase family protein	Plavcova et al. (2013)
Poplar	POPTR_0010s11510	UDP-glycosyltransferase/ trehalose-phosphatase familyprotein	Plavcova et al. (2013)
Poplar	POPTR_0008s13590	Glycosyl hydrolase 9C2	Plavcova et al. (2013)
Poplar	POPTR_0001s19180	HAD superfamily, subfamily IIIB acid phosphatase	Plavcova et al. (2013)
Poplar	POPTR_0004s16720(Pt-PAP.2)	Purple acid phoshatase 10	Plavcova et al. (2013)
Arabidopsis	TAP46	2A phosphatase-associated protein of 46 kDa	Ahn et al. (2015)
Arabidopsis	ABI2	ABA-insensitive 2	Leran et al. (2015)
Arabidopsis	BSL1	BSU-like phosphatase 1	Heidari et al. (2011)
Arabidopsis	BSL2	BSU-like phosphatase 2	Heidari et al. (2011)
Arabidopsis	BSL3	BSU-like phosphatase 3	Heidari et al. (2011)

(continued)

Table 12.1 (continued)

Organism	Gene symbol/locus id	Description	References
Arabidopsis	PBCP	PSII core phosphatase	Samol et al. (2012)
Arabidopsis	PPH1/TAP38	Chloroplast protein phosphatase/ thylakoid-associated phosphatase of 38 kDa	Samol et al. (2012), Pribil et al. (2010), Shapiguzov et al. (2010)

NUE due to higher activity of NR by downregulation of SnRK and 14-3-3 (Waqas et al. 2018). The higher activity of NR leads to more production of nitric oxide, which then increases N uptake by enhancing the lateral root development. Besides NR, phosphoenolpyruvate carboxylase (PEPC) was also found to be downregulated due to overexpression of PP2C, which dephosphorylates PEPC. Upregulation of PSII and Rubisco increases photosynthetic rate under low N (Waqas et al. 2018). PP2C68 and lipid phosphatase 2 have been reported as upregulated N-responsive phosphatases, while PP2C39 and PP2C46 have been reported to be downregulated in root under N starvation condition (Hsieh et al. 2018). Purple acid phosphatase 1 and serine threonine phosphatase have been found to be nitrate-responsive gene under low nitrate condition in maize (Jiang et al. 2018). Two nitrogen-responsive genes (trehalose-6-phosphate, T6P) in barley have been found to be downregulated under low N condition (Fataftah et al. 2018). These genes along with other genes reverse reaction from pyruvate back to phosphoenolpyruvate during N starvation as an efficient pathway for the remobilization of N sources. In wheat, Traes_6BS_143FEF476 encoding bifunctional protein-serine/threonine kinase/ phosphatase was found to be a candidate gene providing tolerance to low N (Xiong et al. 2019).

12.8 Conclusions and Future Prospects

Understanding the genetic basis of tolerance to low N in crops is important for the development of NUE improvement strategies. Protein phosphatases are emerging as an important gene family among the candidate genes/families identified from different plants, as they regulate N uptake, assimilation, and remobilization. While functional genomic approaches have provided several leads in this regard, their potential in the improvement of NUE needs further validation through mutant/transgenic lines, before using them as targets in crop breeding programs for NUE. In the meantime, integration of all the available information at the molecular, physiological, and genetic level will be useful to mine SNPs through genome-wide association studies (GWAS) and marker development using candidate gene approach. We can expect to see some developments in this direction in the years ahead.

Acknowledgments This work was supported in part by research grant to N.R. and fellowship to S.K. from the Indo-UK Virtual Centre on Nitrogen Efficiency of Whole cropping Systems (NEWS) BT/IN/UK-VNC/44/NR/ 2015-16 and UKRI GCRF South Asian Nitrogen Hub (NE/S009019/1).

References

Ahn CS, Ahn HK, Pai HS (2015) Overexpression of the PP2A regulatory subunit Tap46 leads to enhanced plant growth through stimulation of the TOR signaling pathway. J Exp Bot 66:827–840

Ahn CS, Han JA, Lee HS, Lee S, Pai HS (2011) The PP2A regulatory subunit Tap46, a component of the TOR signaling pathway, modulates growth and metabolism in plants. Plant Cell 23:185–209

Ali A, Sivakami S, Raghuram N (2007) Regulation of activity and transcript levels of NR in rice (*Oryza sativa*): roles of protein kinase and G-proteins. Plant Sci 172:406–413

Bakshi A, Moin M, Madhav MS, Kirti PB (2019) Target of rapamycin, a master regulator of multiple signaling pathways and a potential candidate gene for crop improvement. Plant Biol 21:190–205

Bradai M, Mahjoubi H, Chini A, ChabouteÂ M-E, Hanin M, Ebel C (2018) Genome wide identification of wheat and Brachypodium type one protein phosphatases and functional characterization of durum wheat TdPP1a. PLoS One 13:e0191272

Cao X, Gao Y, Wang Y, Li CM, Zhao YB, Han ZH, Zhang XZ (2011) Differential expression and modification of proteins during ontogenesis in *Malus domestica*. Proteomics 11(24):4688–4701

Chakraborty N, Raghuram N (2011) Nitrate sensing and signaling in genomewide plant N response. In: Jain V, Anandakumar P (eds) Nitrogen use efficiency in plants. New India Publishing Agency, New Delhi, pp 45–62

Chen X, Zhang W, Zhang B, Zhou J, Wang Y, Yang Q, Ke Y, He H (2011) Phosphoproteins regulated by heat stress in rice leaves. Proteome Sci 9:37

Cheng WH, Endo A, Zhou L, Penney J, Chen HC, Arroyo A, Leon P, Nambara E, Asami T, Seo M, Koshiba T, Sheen J (2002) A unique short-chain dehydrogenase/reductase in Arabidopsis glucose signaling and abscisic acid biosynthesis and functions. Plant Cell 14:2723–2743

Comparot S, Lingiah G, Martin T (2003) Function and specificity of 14-3-3 proteins in the regulation of carbohydrate and nitrogen metabolism. J Exp Bot 54:595–604

Dobrenel T, Caldana C, Hanson J, Robaglia C, Vincentz M, Veit B, Meyer C (2016) TOR signaling and nutrient sensing. Annu Rev Plant Biol 67:261–285

FAO (2018) World Fertilizer Trends and Outlook to 2018. FAO, Rome, p 66

Fataftah N, Mohr C, Hajirezaei MR, Wirén NV, Humbeck K (2018) Changes in nitrogen availability lead to a reprogramming of pyruvate metabolism. BMC Plant Biol 18:77

Forde BG, Lea PJ (2007) Glutamate in plants: metabolism, regulation, and signaling. J Exp Bot 58:2339–2358

Fristedt R, Vener AV (2011) High light-induced disassembly of photosystem II super complexes in Arabidopsis requires STN7-dependent phosphorylation of CP29. PLoS One 6(9):e24565

Gelli M, Duo Y, Konda AR, Zhang C, Holding D, Dweikat I (2014) Identification of differentially expressed genes between sorghum genotypes with contrasting nitrogen stress tolerance by genome-wide transcriptional profiling. BMC Genom 15(1):179

Grabsztunowicz M, Koskela MM, Mulo P (2017) Post-translational modifications in regulation of chloroplast function: recent advances. Front Plant Sci 8:240

Guo Y, Halfter U, Ishitani M, Zhu JK (2001) Molecular characterization of functional domains in the protein kinase SOS2 that is required for plant salt tolerance. Plant Cell 13:1383–1400

Haynes RJ (1990) Active ion uptake and maintenance of cation-anion balance: a critical examination of their role in regulating rhizosphere pH. Plant Soil 126:247–264

Heidari B, Matre P, Nemie-Feyissa D, Meyer C, Rognli OA, Moller SG, Lillo C (2011) Protein phosphatase 2A B55 and a regulatory subunits interact with nitrate reductase and are essential for nitrate reductase activation. Plant Physiol 156:165–172

Ho CH, Lin SH, Hu HC, Tsay YF (2009) CHL1 functions as a nitrate sensor in plants. Cell 138:1184–1194

Hodges M, Jossier M, Boex-Fontvieille E, Tcherkez G (2013) Protein phosphorylation and photorespiration. Plant Biol 15:694–706

Hsieh PH, Kan CC, Wu HY, Yang HC, Hsieh MH (2018) Early molecular events associated with nitrogen deficiency in rice seedling roots. Sci Rep 8:12207

Hu HC, Wang YY, Tsay YF (2009) AtCIPK8, a CBL-interacting protein kinase, regulates the low-affinity phase of the primary nitrate response. Plant J 57:264–278

Huber SC, Bachmann M, Huber JL (1996) Post-translation regulation of nitrate reductase activity: a role for Ca^{2+} and 14-3-3 proteins. Trends Plant Sci 1:432–438

Jangam AP, Pathak RR, Raghuram N (2016) Microarray analysis of Rice d1 (RGA1) mutant reveals the potential role of G-protein alpha subunit in regulating multiple abiotic stresses such as drought, salinity, heat, and old. Front Plant Sci 7:11

Jiang L, Ball G, Hodgman C, Coules A, Zhao H, Lu C (2018) Analysis of gene regulatory networks of maize in response to nitrogen. Genes 9:1–20

Leran S, Edel KH, Pervent M, Hashimoto K, Corratge-Faillie C, Offenborn JN, Tillard P, Gojon A, Kudla J, Lacombe B (2015) Nitrate sensing and uptake in Arabidopsis are enhanced by ABI2, a phosphatase inactivated by the stress hormone abscisic acid. Sci Signal 8:ra43

Lillo C, Smith LH, Nimmo HG, Wilkins MB (1996) Regulation of nitrate reductase and phosphoenolpyruvate carboxylase activities in barley leaf protoplasts. Planta 200:181–185

Liu R, Zhang H, Zhao P, Zhang Z, Liang W, Tian Z, Zheng Y (2011) Mining of candidate maize genes for nitrogen use efficiency by integrating gene expression and QTL data. Plant Mol Biol Rep 30:297–308

Lu Y, Sasaki Y, Li X, Mori IC, Matsuura T, Hirayama T, Sato T, Yamaguchi J (2015) ABI1 regulates carbon/nitrogen-nutrient signal transduction independent of ABA biosynthesis and canonical ABA signaling pathways in Arabidopsis. J Exp Bot 66:2763–2771

Mackintosh C (1992) Regulation of spinach-leaf nitrate reductase by reversible phosphorylation. Biochim Biophys Acta 1137:121–126

Mandal VK, Sharma N, Raghuram N (2018) Molecular targets for improvement of crop nitrogen use efficiency: current and emerging options. In: Shrawat A, Zayed A, Lightfoot DA (eds) Engineering nitrogen utilization in crop plants. Elsevier, New York, NY

Michelet L, Zaffagnini M, Morisse S, Sparla F, Pérez-Pérez ME, Francia F, Danon A, Marchand CH, Fermani S, Trost P, Lemaire SD (2013) Redox regulation of the calvin-benson cycle: something old, something new. Front Plant Sci 4:470

Moorhead GB, Trinkle-Mulcahy L, Ulke-Lemee A (2007) Emerging roles of nuclear protein phosphatases. Nat Rev Mol Cell Biol 8:234–244

Moorhead GBG, De Wever V, Templeton G, Kerk D (2009) Evolution of protein phosphatases in plants and animals. Biochem J 417:401–409

Mora-García S, Vert G, Yin Y, Caño-Delgado A, Cheong H, Chory J (2004) Nuclear protein phosphatases with Kelch-repeat domains modulate the response to brassinosteroids in Arabidopsis. Genes Dev 18:448–460

Nigro D, Blanco A, Anderson OD, Gadaleta A (2014) Characterization of ferredoxin-dependent glutamine-oxoglutarate amidotransferase (Fd-GOGAT) genes and their relationship with grain protein content QTL in wheat. PLoS One 9:e103869

Parry MAJ, Andralojc PJ, Scales JC, Salvucci ME, Carmo-Silva AE, Alonso H, Whitney SM (2013) Rubisco activity and regulation as targets for crop improvement. J Exp Bot 64:717–730

Pathak RR, Ahmad A, Lochab S, Raghuram N (2008) Molecular physiology of plant N-use efficiency and biotechnological options for its enhancement. Curr Sci 94(11):1394–1403
Pathak RR, Lochab S, Raghuram N (2011) Improving nitrogen use efficiency. In: Comprehensive biotechnology, vol 4, 2nd edn. Elsevier, Oxford, pp 209–218
Plavcova L, Hacke UG, Almeida-Rodriguez AM, Li E, Douglas CJ (2013) Gene expression patterns underlying changes in xylem structure and function in response to increased nitrogen availability in hybrid poplar. Plant Cell Environ 36:186–199
Pribil M, Pesaresi P, Hertle A, Barbato R, Leister D (2010) Role of plastid protein phosphatase TAP38 in LHCII dephosphorylation and thylakoid electron flow. PLoS Biol 8:e100028810
Quan X, Zeng J, Chen G, Zhang G (2019) Transcriptomic analysis reveals adaptive strategies to chronic low nitrogen in Tibetan wild barley. BMC Plant Biol 19:68
Quan XY, Zeng JB, Ye LZ, Chen G, Han ZG, Shah JM, Zhang G (2016) Transcriptome profiling analysis for two Tibetan wild barley genotypes in responses to low nitrogen. BMC Plant Biol 16:30
Raghuram N, Pathak RR, Sharma P (2006) Signaling and the molecular aspects of N-use efficiency in higher plants. In: Singh RP, Jaiwal PK (eds) Biotechnological approaches to improve nitrogen use efficiency in plants. Studium Press LLC, Houston, TX, pp 19–40
Raghuram N, Sachdev MS, Abrol YP (2007) Towards an integrative understanding of reactive nitrogen. In: Abrol YP, Raghuram N, Sachdev MS (eds) Agricultural nitrogen use and its environmental implications. IK International, Delhi, pp 1–6
Raghuram N, Sharma N (2019) Improving crop nitrogen use efficiency. In: Moo-Young M (ed) Comprehensive biotechnology, Vol. 4. Elsevier, Pergamon, pp 211–220
Raghuram N, Sopory SK (1995) Evidence for some common signal transduction events for opposite regulation of nitrate reductase and phytochrome-I gene expression by light. Plant Mol Biol 29:25–35
Rochaix JD (2013) Redox regulation of thylakoid protein kinases and photosynthetic gene expression. Antioxid Redox Signal 18:2184–2201
Roitinger E, Hofer M, Kocher T, Pichler P, Novatchkova M, Yang J, Schlogelhofer P, Mechtler K (2015) Quantitative phosphoproteomics of the ataxia telangiectasia-mutated (ATM) and ataxia telangiectasia-mutated and rad3-related (ATR) dependent DNA damage response in Arabidopsis *thaliana*. Mol Cell Proteomics 14:556–571
Samol I, Shapiguzov A, Ingelsson B, Fucile G, Crevecoeur M, Vener AV, Rochaix JD, Goldschmidt-Clermont M (2012) Identification of a photosystem II phosphatase involved in light acclimation in Arabidopsis. Plant Cell 24:2596–2609
Shapiguzov A, Ingelsson B, Samol I, Andres C, Kessler F, Rochaix JD, Vener AV, Goldschmidt-Clermont M (2010) The PPH1 phophatase is specifically involved in LHCII dephosphorylation and state transitions in Arabidopsis. Proc Natl Acad Sci U S A 107(10):4782–4787
Sharma N, Sinha VB, Gupta N, Rajpal S, Kuchi S, Sitaramam V, Parsad R, Raghuram N (2018) Phenotyping for nitrogen use efficiency (NUE): Rice genotypes differ in N-responsive germination, oxygen consumption, seed urease activities, root growth, crop duration and yield at low N. Front Plant Sci 9:1452
Signora L, De Smet I, Foyer CH, Zhang H (2001) ABA plays a central role in mediating the regulatory effects of nitrate on root branching in Arabidopsis. Plant J 28:655–662
Singh A, Giri J, Kapoor S, Tyagi AK, Pandey GK (2010) Protein phosphatase complement in rice: genome-wide identification and transcriptional analysis under abiotic stress conditions and reproductive development. BMC Genomics 11:435
Singh A, Jha SK, Bagri J, Pandey GK (2015) ABA inducible rice protein phosphatase 2 C confers ABA insensitivity and abiotic stress tolerance in Arabidopsis. PLoS One 10:e0125168
Sinha SK, Sevanthi VAM, Chaudhary S, Tyagi P, Venkadesan S, Rani M, Mandal PK (2018) Transcriptome analysis of two rice varieties contrasting for nitrogen use efficiency under chronic N starvation reveals differences in chloroplast and starch metabolism-related genes. Genes 11:E206

Sitt M, Schulze ED (1994) Does Rubisco control the rate of photosynthesis and plant growth? An exercise in molecular ecophysiology. Plant Cell Environ 17:465–487

Straub T, Ludewig U, Neuhauser B (2017) The kinase CIPK23 inhibits ammonium transport in Arabidopsis *thaliana*. Plant Cell 29:409–422

Sulpice R, Nikoloski Z, Tschoep H, Antonio C, Kleessen S, Larhlimi A, Selbig J, Ishihara H, Gibon Y, Fernie AR, Stitt M (2013) Impact of the carbon and nitrogen supply on relationships and connectivity between metabolism and biomass in a broad panel of Arabidopsis accessions. Plant Physiol 162:347–363

Sun H, Qian Q, Wu K, Luo J, Wang S, Zhang C, Ma Y, Liu Q, Huang X, Yuan Q, Han R, Zhao M, Dong G, Guo L, Zhu X, Gou Z, Wang W, Wu Y, Lin H, Fu X (2014) Heterotrimeric G proteins regulate nitrogen-use efficiency in rice. Nat Genet 46(6):652–656

Sutton M, Raghuram N, Adhya TK (2019) The nitrogen fix: from nitrogen cycle pollution to nitrogen circular economy in emerging issues of environmental concern, Frontiers 2018/19, pp 52–64

Sutton MA, Bleeker A (2013) The shape of nitrogen to come. Nature 494:435–437

Trevisan S, Manoli A, Begheldo M, Nonis A, Enna M, Vaccaro S, Caporale G, Ruperti B, Quaggiotti S (2011) Transcriptome analysis reveals coordinated spatiotemporal regulation of hemoglobin and nitrate reductase in response to nitrate in maize roots. New Phytol 192:338–352

Umezawa T, Sugiyama N, Mizoguchi M, Hayashi S, Myouga F, Yamaguchi-Shinozaki K, Ishihama Y, Hirayama T, Shinozaki K (2009) Type 2C protein phosphatases directly regulate abscisic acid-activated protein kinases in Arabidopsis. Proc Natl Acad Sci U S A 106:17588–17593

Vert G, Chory J (2009) A toggle switch in plant nitrate uptake. Cell 138:1064–1066

Walch-Liu P, Filleur S, Gan Y, Forde BG (2005) Signaling mechanisms integrating root and shoot responses to changes in the nitrogen supply. Photosynth Res 83:239–250

Wang Y, Wang Y, Zhao YB, Chen DM, Han ZH, Zhang XZ (2014) Protein phosphorylation differs significantly among ontogenetic phases in malus seedlings. Proteome Sci 12:31

Waqas M, Feng S, Amjad H, Letuma P, Zhan W, Li Z, Fang C, Arafat Y, Khan MU, Tayyab M, Lin W (2018) Protein phosphatase (PP2C9) induces protein expression differentially to mediate nitrogen utilization efficiency in rice under nitrogen-deficient condition. Int J Mol Sci 19:2827

Whitney SM, Birch R, Kelso C, Beck JL, Kapralov MV (2015) Improving recombinant rubisco biogenesis, plant photosynthesis and growth by co-expressing its ancillary *RAF1* chaperone. Proc Natl Acad Sci U S A 112:3564–3569

Xiong H, Guo H, Zhou C, Guo X, Xie Y, Zhao L, Gu J, Zhao S, Ding Y, Liu L (2019) A combined association mapping and t-test analysis of SNP loci and candidate genes involving in resistance to low nitrogen traits by a wheat mutant population. PLoS One 14(1):e0211492

Xue T, Wang D, Zhang S, Ehlting J, Ni F, Jakab S, ZhengC Zhong Y (2008) Genome-wide and expression analysis of protein phosphatase 2C in rice and Arabidopsis. BMC Genom 9:550

Yang W, Yoon J, Choi H, Fan Y, Chen R, An G (2015) Transcriptome analysis of nitrogen-starvation-responsive genes in rice. BMC Plant Biol 15:31

Zeng DD, Qin R, Li M, Alamin M, Jin XL, Liu Y, Shi CH (2017) The ferredoxin-dependent glutamate synthase (OsFd-GOGAT) participates in leaf senescence and the nitrogen remobilization in rice. Mol Genet Genomics 292(2):385–395

Zhang H, Forde B (2000) Regulation of *Arabidopsis* root development by nitrate availability. J Exp Bot 51:51–59

Zhang H, Forde BG (1998) An Arabidopsis MADS box gene that controls nutrient-induced changes in root architecture. Science 279:407–409

Chapter 13
Protein Phosphatases of Cereals and Millets: Identification, Structural Organization, and Their Involvement in the Regulation of Abiotic Stresses

Roshan Kumar Singh, Ashish Prasad, and Manoj Prasad

13.1 Introduction

Posttranslational modification of cellular proteins is a universal mechanism to determine their function and regulation. Two such modifications are phosphorylation and dephosphorylation of proteins. Protein kinases and phosphatases play an essential role in maintaining the balance of proteome phosphorylation status of the cell. Kinases belong to phosphotransferase, whereas phosphatases belong to a phosphohydrolase group of enzymes and are involved in the conjugation of γ-phosphate groups to their substrate proteins or their removal, respectively (Fig. 13.1). The occurrence of these two kinds of regulatory protein is universal, which is found in prokaryotic organisms to higher plants and animals. The study on the identification, structure, function, and regulation of protein kinases has been predominated for a while, whereas limited reports on PPases are available (Brautigan 2013). The dephosphorylation reactions are of immense importance in biological catalysis owing to the fact that about 30% of proteins in eukaryotic cells have the potential to be phosphorylated (Ubersax and Ferrell 2007); modification happens predominantly on serine and threonine residues (Virshup and Shenolikar 2009). The other significant phosphorylation site is tyrosine residue, which accounts for 10–15% of total phosphorylated amino acids. Other than these amino acids, histidine, arginine, lysine, aspartate, and glutamate are also reported to be phosphorylated in specific cellular processes; they contribute lesser extent of total phosphorylation of the cellular protein.

PPases are broadly classified into serine/threonine (Ser/Thr) phosphatase and tyrosine (Tyr) phosphatase based on amino acid residues dephosphorylated by them. Ser/Thr phosphatases are further grouped into phosphoprotein Ser/Thr phosphatases (PPPs) and metal-dependent Ser/Thr phosphatases (PPMs). The PPP

R. K. Singh · A. Prasad · M. Prasad (✉)
National Institute of Plant Genome Research, New Delhi, India
e-mail: manoj_prasad@nipgr.ac.in

G. K. Pandey (ed.), *Protein Phosphatases and Stress Management in Plants*,
https://doi.org/10.1007/978-3-030-48733-1_13

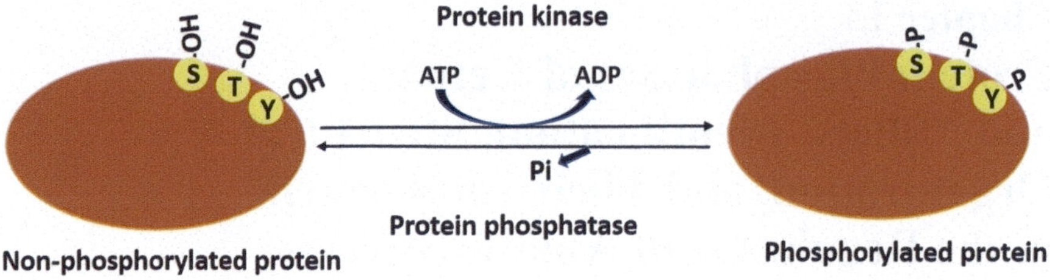

Fig. 13.1 Maintenance of balance between phosphorylated and non-phosphorylated proteins by protein kinase and protein phosphatase. Protein kinases catalyze the addition of phosphate group to either through ATP hydrolysis and removal of phosphate group is resulted by the activity of protein phosphatase. The Protein kinases and phosphatase activity mostly happens at serine (S), threonine (T), and tyrosine (Y) residues

family comprises PP1, PP2A, PP2B (generally known as calcineurin), and other phosphatases such as PP4, PP5, PP6, and PP7, whereas PPM is represented by PP2C and other Mg^{2+}- or Mn^{2+}-dependent protein phosphatases (País et al. 2009). Interestingly, PPPs and PPMs are diverse in sequence similarity, perhaps evolved from different ancestral genes, but share highly structural features at the catalytic center (Das et al. 1996). PPP phosphatase types are structurally multimeric in nature, containing a small number of catalytic subunits united with different combinations of a diverse array of regulatory subunits (Smith and Walker 1996). The regulatory subunit plays an essential role in modulating substrate specificity, catalytic activity, and spatial localization of protein phosphatase within the cell. This subunit is encoded by a relatively more significant number of genes with lesser sequence conservation within them (Smith and Walker 1996). Unlike the PPP subfamily, PPM phosphatases lack a regulatory subunit but consist of an additional domain and conserve motif to determine substrate specificity. The most abundantly found and extensively studied Ser/Thr phosphatases are PP1 and PP2A subfamilies. Plant Tyr phosphatases are also classified into two classes, Tyr-specific phosphatases and dual-specificity phosphatases (DSP), which can hydrolyze phosphate group from tyrosine as well as serine and threonine residues (Bentem and Hirt 2009). Physiological and biochemical studies have shown evidence on the essential role of both the classes of protein phosphatases in the diverse cellular process (signal transduction, cell cycle, hormonal regulation, cellular metabolism) and stress regulation (biotic and abiotic).

The genome-wide study of PPase has become feasible in the post-genomic era where a whole genome sequence of several plants is available (Singh et al. 2016). However, global protein phosphatase has explored only in a small number of plant species like Arabidopsis, *Medicago truncatula*, rice, maize, *Brachypodium*, and wheat. In this chapter, we are drawing an overview of genome-wide identification, characterization, and transcript profiling of protein phosphatase-encoding genes, structural organization of proteins, and the role they play in abiotic stress responses in cereal crops.

13.2 Ser/Thr Phosphatases

13.2.1 Protein Phosphatase 1 (PP1)

PP1 is a highly conserved Ser/Thr phosphatase and universally present in all higher organisms. A large number of cellular processes are regulated by them, including cell division, cytoskeleton association, translation, sugar/protein metabolism, and regulation of membrane-bound receptors and channel proteins (Cohen 2002; Ceulemans and Bollen 2004). The holoenzyme is a complex of conserved catalytic subunits with one or many highly distinct regulatory subunits essential for substrate specificity and subcellular location. More than 100 regulatory subunits have already been identified in different organisms and postulated that their numbers have increased sharply with the evolution to complex organism types (Ceulemans et al. 2002). The catalytic subunits might possess a sequence similarity of up to 70–80% with conserved tertiary fold and similar placed catalytic site for all members (Shi 2009). It comprises a condensed α/β-fold, with a β-sheet compactly packed between two α-helices. The metal ions, Mn^{2+} and Fe^{2+}, associate with the catalytic subunits. These metal ions catalyze dephosphorylation reaction by binding to the water molecules which initiate a nucleophilic attack on the phosphorous atom of the substrate protein (Goldberg et al. 1995).

13.2.2 Protein Phosphatase 2A (PP2A)

PP2A is highly conserved and one of the most occurred protein phosphatases of the cellular system estimated for up to 1% of total proteome fraction in some tissues (Shi 2009). Like PP1, it also plays an imperative role in diverse cellular functions such as growth, cell division and death, cytoskeleton structure, cell cycle regulation, metabolic activities, cellular mobility, and hormonal and stress signaling pathways (Janssens and Goris 2001). The PP2A holoenzyme exists in a heterotrimeric form—consists of a heterodimeric core subunit and regulatory protein. The core subunit consists of a catalytic center and a variable scaffold subunit. Both the catalytic center and scaffold subunit are comprised of two isoforms, α and β, where isoform is highly abundant. The regulatory subunit consists of four families each with 2–5 isoforms, where all are encoded by different genes. These regulatory subunits are highly diverse in sequence similarities and expression in different tissue types. The scaffold protein consists of 15 tandem HEAT (huntingtin-elongation-A subunit-TOR) repeats; each contains a pair of α-helices in an antiparallel orientation with a highly conserved interhelical loop. Fifteen interhelical loop sequences form a contiguous ridge which is recognized by a catalytic subunit (Xing et al. 2006). The regulatory protein part of PP2A holoenzyme consists of seven WD40 repeat elements and a β-hairpin handle with many other secondary elements on the top of β-propeller-like structure (Xu et al. 2008).

13.2.3 Protein Phosphatase 2B (PP2B)/Calcineurin

PP2B is involved in various calcium-dependent cellular processes and stress-responsive signal transductions. The catalytic subunit of calcineurin is composed of an N-terminal phosphatase domain and a regulatory protein binding motif, followed by a calcium-calmodulin (Ca^{2+}-cal) binding domain and an autoinhibition domain. The phosphatase activity of calcineurin is highly dependent on the association of calcium-calmodulin to its binding domain. PP2B is inactive in the absence of calcium-calmodulin binding. The autoinhibitory element prevents the PP2B activity by forming an α-helix that hinders access to the catalytic center (Kissinger et al. 1995). Here, the regulatory subunit contains two calcium-binding domains, each consisting of two EF-hand elements which consist of a helix-loop-helix topology, more like a spread thumb and forefinger of the human hand. The active regulatory subunit exists in the calcium-bound state at all four calcium-binding domains. Recent studies indicate that PP2B may dimerize upon calcium-calmodulin activation (Ye et al. 2008). Calcineurin (CN) is made up of two subunits named CAN (calcineurin-A, which is a PP2B phosphatase) and the regulatory Ca^{2+} sensor (CNB). CN also requires calmodulin (CaM) for its activation in the presence of transient Ca2+, which activates both CNB and CaM, and these in turn activate CAN, i.e., PP2B phosphatase.

13.2.4 Protein Phosphatase 2C (PP2C)

PP2C is highly conserved and one of the most abundant and large families of a well-studied Mn^{2+}- and Mg^{2+}-dependent member of the Ser/Thr phosphatase. Compared to animals, plants possess a large number of PP2Cs; 22 distinct isoforms of PP2C encoded by 16 genes have been identified in humans, whereas Arabidopsis alone contains 80 PP2C-encoding genes (Xue et al. 2008). The different isoforms of PP2C are distinct in sequence similarity, domain organization, function, and subcellular localization. PP2Cs are primarily involved in the regulation of the biotic and abiotic stress signals; they also govern plant development and metabolism and potassium deficiency signals (Singh et al. 2016). The domain architecture of higher organism PP2C uncovered the variable location of the conserved catalytic domain either at N- or C-terminal region (Schweighofer et al. 2004). The catalytic core of PP2C consists of a central β-pleated sheet flanked by a pair of α-helices. This organization leads to the formation of a cleft between two β-sheets where a couple of metal ions associate at the basal region. The additional three α-helices of the core domain attribute to substrate specificity. PP2Cs share a similar dephosphorylation mechanism as the PPP family, by nucleophilic attack of the phosphate group of the substrate protein by a divalent metal-induced water nucleophile (Shi 2009).

13.3 Protein Tyrosine (Tyr) Phosphatase

The molecular structure and function of protein Tyr phosphatases are well established in animals, but little studied in the plant system which suggests its active role in biotic and abiotic stress signaling, hormonal signaling, plant development, and starch metabolism. There is no evidence of sequence similarity of protein Tyr phosphatase with protein Ser/Thr phosphatases. All the proteins which belong to tyrosine phosphatase contain a signature sequence (V/I)HCXAGXGR(S/T)G in the conserved catalytic domain (Luan 2003). This domain comprises an essential cysteinyl molecule involved in the construction of a phosphoenzyme intermediate. An animal model-based crystal structure suggests the presence of four-stranded central parallel β-sheets followed by the presence of α-helices on both sides (Zhang 2002). Crystal structure of both Tyr-specific and dual-specificity classes of protein Tyr phosphatase shows high similarities despite divergence in sequence and substrate specificity (Luan 2003). The signature element of the catalytic site is located within a single loop inside the crevices on the protein surface. The specific cysteine residue is in the site for a nucleophilic attack on phosphoryl molecule to form a thiophosphate intermediate. The arginyl residue (Arg221) of the signature element at the catalytic domain catalyzes the hydrolysis of a thiophosphate intermediate.

13.4 Global Identification of Protein Phosphatase in Cereals and Millets

The initial genome-wide identification of PPase-encoding genes in Arabidopsis and rice has exposed the way to study global identification, distribution, expression analysis, and evolution in other crops. A total of 78 and 80 PP2C genes have been identified in rice and Arabidopsis genome, respectively, which are much larger than in yeasts and humans (Xue et al. 2008). The OsPP2C- and AtPP2C-encoding genes were further phylogenetically classified into 1 and 13 subfamilies, respectively, where each gene follows a monophyletic in origin and shares a common structure and protein motif. The expansion of PP2C genes in Arabidopsis and rice is mainly due to duplication events that have occurred in the genome; whole genome and chromosomal segmental duplication had a significant role compared to the local or tandem duplication. Subsequently, a second genome-wide investigation which was performed in rice leads to the identification of 132 protein phosphatase-encoding genes belonging to different subfamilies following domain architecture and phylogenetic analysis (Singh et al. 2010). With 90 PP2C genes represented the largest class of protein phosphatase followed by 17 PP2A, 23 dual specific phosphatases and one each of PTP and LMWP. The study could not find any genes representing the PP2B class. The investigation also revealed that the number of protein Tyr phosphatase in plants is much lower than that of humans, where more than 100 putative protein Tyr phosphatases have been identified (Alonso et al. 2004).

Interestingly, plants have a much higher number of protein kinases than in humans; there could be a possibility that either lower tyrosine phosphorylation component in plants or plant Tyr phosphatase/DSP could target a broad spectrum of the substrate in signaling cascade (Singh et al. 2010). Five isoforms of PP1 have also been found in rice genome named as *OsPP1a*, *OsPP1b*, *OsPP1c*, *OsPP1d*, and *OsPP1e* (Ogawa et al. 2012). All the *OsPP1s* share highly conserved amino acids essential for catalytic activity and variable N- and C-terminal domains. Phylogenetic investigations revealed their independent diversification in plants and animals.

The genome sequence of maize (*Zea mays*) enabled the identification of 159 protein phosphatase-encoding genes (Wei and Pan 2014). With 104 encoding genes, PP2Cs were the most abundant class; protein Tyr phosphatase and PP2A were represented by 29 and 26 encoding genes, respectively. Fifteen genes showed tandem duplication with a maximum of ten intervening genes in between. These tandemly duplicated genes form nine different tandemly duplicated gene clusters, one cluster consisting of three genes and the rest possessing two genes each. Similarly, in *Brachypodium distachyon*, genome-wide analysis resulted in the identification of 86 PP2C genes (Cao et al. 2016). Chromosomal distribution of these genes revealed that most of the BdPP2C genes were found within the low CpG density region. Phylogenetic analysis suggests that BdPP2Cs are evolutionarily closer to the rice than of Arabidopsis and this gene family has evolved from a common ancestor in these three plants. There were no tandemly duplicated gene pairs found in *Brachypodium*, but 22 pairs of genes were reported to be involved in segmental duplication event. In another study, 18 PP1-encoding genes were reported from common wheat, and its five and eight orthologues were identified from rice and *Brachypodium*, respectively (Bradai et al. 2018). PP1s from these crops were conserved in size ranging from 296 to 354 amino acids in length and molecular weight (from 32.4 to 37.9 kDa), differing in the length of N-terminal domain. There is an abundance of negatively charged amino acids in most of the PP1s of rice, wheat, and *Brachypodium* which confer the lower isoelectric point near to 5 (5.03–5.48).

The cereals and millets include a large number of crop plants, although the number of plants in which genome-wide identification of PPase has carried out is limited. The available genome sequence of barley (Milner et al. 2019), sorghum (McCormick et al. 2018), and millets including foxtail millet (Bennetzen et al. 2012; Zhang et al. 2012), finger millet (Hittalmani et al. 2017), pearl millet (Varshney et al. 2017), and broomcorn millet (Zou et al. 2019) is providing an excellent opportunity of a genome-wide analysis of PPase genes and their evolution. Transcriptomics resources may also serve as a useful resource for the comprehensive study of gene function where the genomic sequence is unavailable. Among millets, transcriptome resources are available in finger millet (Hittalmani et al. 2017), broomcorn millet (Yue et al. 2016), and pearl millet (Jaiswal et al. 2018; Dudhate et al. 2018).

13.5 Expression Pattern of Protein Phosphatase in Cereals

The expression pattern of protein phosphatase-encoding genes can provide an important intimation regarding their function. The techniques involved in the measurement of gene expression pattern have evolved from microarray, massively parallel signature sequence (MPSS), EST profiling, and semiquantitative PCR to the quantitative real-time PCR (qRT-PCR) and recently advanced RNA sequencing (RNA-seq). The comparative transcript profiling of the entire PPase gene family or specific subfamilies has been performed in several cereal crops including rice, maize, *Brachypodium*, and common wheat. In rice, the expression profiling of 128 *OsPPase* genes under different abiotic stresses (salinity, drought, and cold) was analyzed using microarray expression data (Singh et al. 2010). A total of 46 genes were expressed differentially in response to an abiotic stress; 31 genes are found to be upregulated and 15 genes were downregulated during any of the abiotic stress conditions. *OsPP2*, *OsPP40*, *OsPP46*, *OsPP48*, *OsPP50*, and *OsPP55* showed higher expression in all three abiotic stresses, whereas none of any *OsPPase* genes are found to be downregulated during all three stress situations. Thirteen *OsPP* genes were upregulated in drought and salinity together, but none of the genes were upregulated together in cold and drought or salt and cold together. The overlapping expression pattern revealed the involvement of the same protein phosphatase in the distinct signaling cascade. The authors have shown an interesting finding that all the genes that were upregulated belong to the PP2C subclass while downregulated members were represented by a subset of PP2A, PP2C, and DSP groups. This finding suggests the frontier role of PP2C in the regulation of abiotic stress response in plants (Singh et al. 2010).

In maize, among 159 reported protein phosphatase-encoding genes, the expression of 152 genes was detected in different tissues under different developmental stages and abiotic stress conditions using microarray and RNA-seq (Wei and Pan 2014). Maize is very sensitive to salt stress, and significant transcriptional distinction has been noticed in response to salinity. The differential expression was observed in four protein phosphatase-encoding genes, namely, *ZmPP66*, *ZmPP107*, *ZmPP127*, and *ZmPP149*, under high salt concentration in roots, and surprisingly, all four genes represented the PP2C class of protein phosphatase. In response to drought conditions, a twofold upregulation in the expression of *ZmPP67*, a member of protein Tyr phosphatase, was reported in roots of drought-tolerant maize line Han21. Upregulation in the expression of *ZmPP54* and *ZmPP59* was also reported from RNA-seq data and further validated through a real-time PCR. *ZmPP68*, *ZmPP101*, *ZmPP125*, and *ZmPP133* were found to be downregulated in water deficit conditions. Similarly in leaves, ten genes (*ZmPP6*, *ZmPP21*, *ZmPP29*, *ZmPP31*, *ZmPP113*, *ZmPP124*, *ZmPP127*, *ZmPP130*, *ZmPP146*, and *ZmPP*154) were shown to be upregulated more than twofold. Except for *ZmPP124*, all nine genes were members of the PP2C class. Transcriptome profiling under cold stress has reported the upregulation of 13 *ZmPP* genes (*ZmPP6*, *ZmPP24*, *ZmPP29*, *ZmPP66*, *ZmPP77*, *ZmPP82*, *ZmPP92*, *ZmPP112*, *ZmPP116*, *ZmPP127*, *ZmPP149*, *ZmPP154*, and

ZmPP155) and downregulation of 3 *ZmPP* genes (*ZmPP69, ZmPP87*, and *ZmPP101*). The expression pattern studies of *ZmPP* genes explained their probable role during abiotic stresses, and their functional validation in stress signaling needs to be investigated in the near future.

The expression profile of all identified 86 PP2C-encoding genes from *Brachypodium* was investigated through qRT-PCR under controlled conditions and abiotic stress treatments (Cao et al. 2016). The expression of 75 genes was observed, whereas no transcript for 9 genes was detected. More than half of the *BdPP2C* genes showed a differential expression pattern by the application of abiotic stress condition or exogenous ABA. The expression of *BdPP2C44*, *BdPP2C37*, and *BdPP2C36* was upregulated during the heat, drought, cold, or H_2O_2 treatment. These genes seem to have a broad-spectrum abiotic stress-responsive role. The expression of *BdPP2C46* and *BdPP2C47* was enhanced upon drought, cold, or oxidative stress, but repressed during heat induction. Similarly, *BdPP2C75* was positively regulated by salt or drought stress and negatively regulated by heat, cold, and oxidative stress. There are some *BdPP2C* genes whose expression got elevated during abiotic stress as well as upon exogenous treatment of ABA to the plants. These genes including *BdPP2C13*, *BdPP2C32*, and *BdPP2C70* are suggested to play a significant role in a ABA-dependent stress signaling pathway. It has been established from different studies that the same *PP2C* gene can be activated by different stress signals and dephosphorylate distinct or common substrate molecules (Singh et al. 2010; Wei and Pan 2014; Cao et al. 2016).

The expression data of PP1 from rice, wheat, and *Brachypodium* provide the insight into its regulatory role in abiotic stress response (Bradai et al. 2018). The expression of *BdPP1a* is enhanced by heat, salt, and drought, while *BdPP1e1* and *BdPP1e2* are differentially downregulated in cold, salt, and drought stresses. *BdPP1d* has been shown to be strongly induced during a cold condition only. In common wheat, the expression of *TaPP1e2* and *TaPP1g3* could not be observed in RNA-seq data, while drought stress induces the expression of *TaPP1e4*. Downregulation in the expression of *TaPP1b-A* is observed under the combined effect of heat and drought. In rice *OsPP1b* expression is retarded by salt and drought, and *OsPP1a* is induced by cold and drought. These observations revealed that although the number of PP1 is less in plants, their importance in abiotic stress response is significant.

13.6 Role of Protein Phosphatases in Abiotic Stress Signaling

Fine-tuning between the phosphorylation and dephosphorylation status of proteins regulates processes in plants ranging from development to stress adaptation, and any defect affecting this balance, if not lethal, may have severe implications in a plant's life cycle. Studies on cell signaling mechanism associated with the different phosphatase classes have mainly been done in response to abiotic stress (Fig. 13.2).

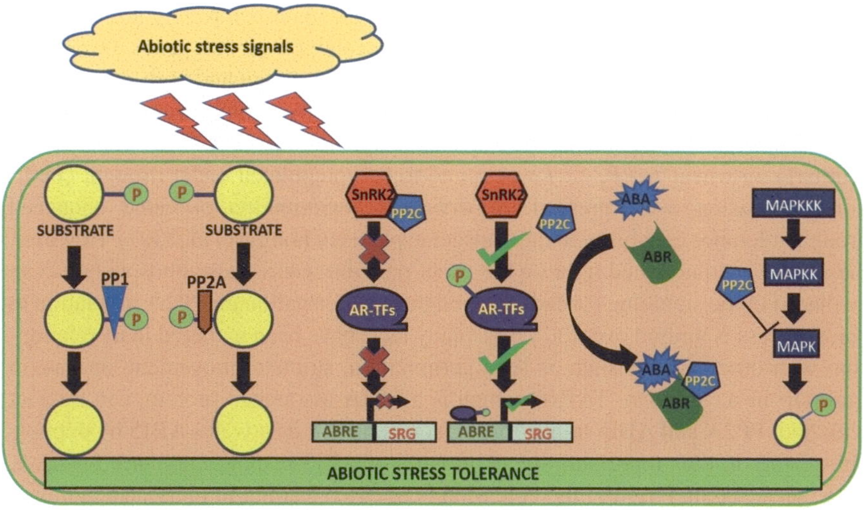

Fig. 13.2 Signaling mechanism associated with various classes of plant protein phosphatases. *P* phosphate, *PP1* protein phosphatase 1, *PP2A* protein phosphatase 2A, *PP2C* protein phosphatase 2C, *SnRK2* SNF1-related Protein kinases 2, *AR-TFs* abscisic acid-responsive transcription factors, *ABRE* abscisic acid-responsive elements, *SRGs* stress-related abscisic acid-responsive genes, *ABA* abscisic acid, *ABR* abscisic acid receptor, *MAPK* mitogen-activated Protein kinases, *MAPKK* mitogen-activated Protein kinases kinase, *MAPKKK* mitogen-activated Protein kinases kinase kinase. A balance between the phosphorylation and dephosphorylation states of proteins regulated by PP1 and PP2A may serve as an essential regulatory mechanism to withstand adverse conditions. PP2C inhibits SnRK2 activity under normal conditions, but under nutrient starvation, ABA binds to its receptor and this ABA-receptor complex binds to PP2C, releasing SnRK2 from inhibition. SnRK2 phosphorylates AR-TFs, which bind to ABRE and mediate the transcription of various SRGs. PP2C also regulates MAPK signaling cascades by regulating the activity of MAPK

The PP1 class of phosphatases, which have been so well characterized in animal systems, still lags far behind in the context of plants. There are still no reports on how PP1s modulate abiotic stress tolerance in cereals and millets; however in *A. thaliana,* it was recently shown that a PP1 along with its regulatory protein inhibitor 2 (Inh2) negatively regulates abscisic acid (ABA) signaling. Association of the PP1 with Inh2 leads to the suppression of SNF1-related protein kinases 2.6 (SnRK2.6), which is a kinase and an essential component of ABA signaling and is generally regulated by type 2C phosphatases. Mutant plants for both PP1 and Inh2 showed hypersensitivity to both salt and ABA treatments showcasing the importance of this novel regulation of the ABA signal transduction pathway (Hou et al. 2016). Other studies have identified the function of Inh3 as another regulatory protein of PP1 and an essential component of early embryogenesis and another regulatory protein—PP1 regulatory subunit2-like protein1 (PRSL1)—as an important regulator of a blue light-mediated stomatal opening (Takemiya et al. 2006, 2009, 2013). Our knowledge in respect to PP1 and its role in abiotic stress response in cereals and millets is still lacking, and extensive studies are still required to get a picture of the signaling cascade involving this class of phosphatases.

Some studies have shown the involvement of PP2A in controlling abiotic stress responses in cereals such as rice and wheat. Two catalytic subunit genes, *OsPP2A-1* and *OsPP2A-3*, were found to be upregulated in water deficit and high salinity conditions implying that they were subject to regulation by abiotic stresses and might be an important component of the associated signaling cascade (Man et al. 2003). Similarly, in wheat *TaPP2Ac-1* was found to be upregulated under drought conditions and when overexpressed in *Nicotiana benthamiana* provided enhanced drought tolerance capability to the transgenic plants (Chongyi et al. 2007). However, these studies do not shed light on the exact position where these phosphatases can be placed in the signaling cascade. Although the information on PP2A in relation to abiotic stress is limited over the years, many facts have been gathered in its relationship with other aspects such as seed germination, stomatal movement, and auxin transport in *A. thaliana*. TAP46, which is a PP2A-associated protein, can interact with both PP2A and ABI5 transcription factors. PP2A inactivates ABI5 by dephosphorylating it, thus repressing the expression of ABI5-responsive genes; however, in the presence of TAP46, this repression is released leading to more accumulation of the active phosphorylated form of ABI5 and expression of its target genes, which inhibit seed germination and promote seed maturation (Hu et al. 2014). A fascinating study showed that PIN1 apical-basal polarity is governed by its phosphorylation status and PINOID kinase and PP2A act in an antagonistic manner to direct auxin flux. Phosphorylated PIN1 is targeted to the apical region of cells, and dephosphorylated form (in the presence of higher concentration of PP2A) is targeted to the basal portion of the cells (Michniewicz et al. 2007). Again, as in the case of PP1s, the role of PP2As in respect to abiotic stress in cereals and millets still needs to be characterized, and current knowledge in this regard is still in its infancy.

PP2B and calcineurin (CN) are calcium-calmodulin complex-dependent serine/threonine phosphatases that are activated at high Ca^{2+} levels. Calcineurin is composed of two subunits: calcineurin-A (CNA) which is the catalytic subunit and calcineurin-B (CNB), which is the regulatory subunit. However, CNA has not been reported in plants till date, but genes similar to the regulatory subunit called Calcineurin-B like genes (*CBL*) have been found (Pandey 2008; Sanyal et al. 2015). CBLs have been found to interact with CBL-Interacting Protein Kinases (CIPKs) and activate their kinase activity after sensing Ca^{2+} signals during stress responses in plants (Kim 2013). Genes encoding for CNA, which is the catalytic counterpart of CN and has phosphatase activity, have not been identified in plants till date suggesting that PP2B in plants is just a sensor for calcium and activator of CIPK rather than controlling the signaling pathway by dephosphorylation.

Extensive studies in the last few years have placed PP2Cs as important components of abiotic stress signaling, and this class of phosphatases has been found to play a vital role in the abscisic acid and mitogen-activated protein kinase (MAPK) signaling cascade in response to abiotic stress (Singh et al. 2016). SnRK1 and SnRK2s are positive regulators of ABA signaling, and studies have shown that PP2Cs negatively regulate the signaling cascade by dephosphorylating the SnRKs (Umezawa et al. 2009; Rodrigues et al. 2013). SnRK1 is triggered during energy-deprived conditions, and two type-1 PP2Cs, PP2CA and ABI1, are employed by plants to reset SnRK1 when conditions are back to normal (Rodrigues et al. 2013). SnRK2s have been shown to positively regulate ABA signaling and their deactivation by clad1 PP2Cs leads to the dampening of ABA signaling. ABA receptors when bound to ABA sequester the PP2Cs, thus releasing the SnRK2s from inhibition and activate their downstream targets (Umezawa et al. 2009). MAPK stress signaling cascades also function in a similar way with PP2Cs acting as negative regulators of the signaling cascade by dephosphorylating MAPKs (Danquah et al. 2014). A fascinating study identified a rice *OsPP18* that provides tolerance to drought and oxidative stress not through ABA pathway but ROS scavenging. *OsPP18* did not interact with SnRK2, and ABA-responsive genes were not affected in *ospp18* mutant. However, the mutant had altered reactive oxygen species (ROS) scavenging enzyme expression suggesting that *OsPP18* regulated ROS homeostasis through an ABA-independent mechanism (You et al. 2014). Ca^{2+}-/calmodulin-dependent Protein kinases (CCaMK) has been shown to be a positive regulator of ABA responses and provide tolerance to drought and oxidative stress. Very recently, its molecular mechanism was deciphered, and results indicated that DMI3, which is a rice CCaMK, was controlled by PP45's phosphatase activity. At the basal state, PP45 inactivates DMI3 by dephosphorylation, and production of H_2O_2, which is induced by ABA, leads to the inhibition of PP45 activity, thus releasing DMI3 from inhibition (Ni et al. 2019). A study on *Brachypodium distachyon* revealed that PP2Cs were upregulated during a variety of abiotic stresses like heat, cold, drought, and salinity (Cao et al. 2016). A similar study on *Medicago truncatula* showed a differential expression of several *PP2C* genes under drought and cold treatments (Qi et al. 2018). However, there is an urgent need to characterize the exact point of action where these phosphatases function and control the signaling mechanism.

13.7 Concluding Remarks

Numerous studies have shown that all classes of phosphatases are differentially expressed during abiotic stress and may play important roles in regulating the abiotic stress signaling pathways (see Table 13.1). This also indicates the potential to utilize these proteins for providing tolerance against the wide range of environmental pressure that exists and ultimately leads to the inferior performance of plants under such conditions. A study showed that transgenic rice plants overexpressing *OsPP1a* were more tolerant to saline conditions and exhibited greater plant height and survivability highlighting the potential of PP1s in plant genetic improvements. SnRK1A and two stress-related transcription factors OsNAC5 and OsNAC6 were found to be upregulated in the transgenic lines (Liao et al. 2016). Similarly, *Triticum aestivum TaPP2Ac-1* when overexpressed in *N. benthamiana* showed enhanced drought tolerance. Transgenic plants had better water use efficiency, relative water content, and improved membrane integrity under water deficit conditions (Chongyi et al. 2007). *A. thaliana* plants overexpressing rice *OsPP108* have been shown to be highly insensitive to ABA treatments and tolerant to salinity, drought, and mannitol

Table 13.1 Transgenic approaches in cereals that have demonstrated the overexpression of protein phosphatases in conferring tolerance to several abiotic stresses

Gene name	Type	Source	Transgenic	Phenotype	Reference
OsBIPP2C1	PP2C	*O. sativa*	*N. benthamiana*	Biotic as well as abiotic stress tolerance	Hu et al. (2006)
OsPP108	PP2C	*O. sativa*	*A. thaliana*	Better salinity, drought and mannitol stress tolerance, and also enhanced physiological parameters	Singh et al. (2015)
OsPP18	PP2C	*O. sativa*	*O. sativa*	Drought tolerance	You et al. (2014)
OsPP1a	PP1	*O. sativa*	*O. sativa*	Salinity tolerance, better height and survivability	Liao et al. (2016)
OsPP1a	PP1	*O. sativa*	*O. sativa*	Salinity tolerance, better height and survivability	Liao et al. (2016)
TaPP2Ac-1	PP2A	*T. aestivum*	*N. benthamiana*	Drought tolerance	Chongyi et al. (2007)
ZmPP2AA1	PP2A regulator	*Z. mays*	*Z. mays*	Better phosphate uptake	Wang et al. (2017)
ZmPP2C2	PP2C	*Z. mays*	*N. benthamiana*	Cold tolerance	Hu et al. (2010)
ZmPP2C2	PP2C	*Z. mays*	*A. thaliana*	Drought and salinity susceptibility	Liu et al. (2009)

stresses (Singh et al. 2015). Transgenic plants have far better photosynthetic efficiency, chlorophyll content, and fresh weight when compared to non-transgenic plants; thus, the study highlights the potential of PPs for equipping plants not only with improved tolerance features but also with boosted physiological parameters (Singh et al. 2015). In a study, it was shown that rice plants overexpressing *OsPP18* were tolerant to drought and oxidative stress and plants with artificial miRNA against the gene were hypersensitive to drought stress (You et al. 2014). The gene was regulated by a stress-responsive *NAC* transcription factor as transgenic plants overexpressing *NAC1* accumulated more *OsPP18* and artificial miRNA transgenic plants against *NAC1* displayed a reduction in the level of *OsPP18* transcripts (You et al. 2014). *OsBIPP2C1,* which is a PP2C, has shown to be an excellent candidate for plant genetic improvement programs (Hu et al. 2006). Overexpression of this gene in *N. benthamiana* conferred tolerance to a variety of biotic and abiotic stresses. Transgenic plants were tolerant not only to fungal and viral infections but also to drought, cold, and salinity stresses. Specific studies in *Zea mays* have also indicated the potential of PP2C genes for their genetic improvement. ZmPP2C2, when overexpressed in *N. benthamiana*, showed tolerance to cold stress, but the same gene provided susceptibility to drought and salinity in *A. thaliana* (Liu et al. 2009; Hu et al. 2010).

Another study revealed the potential of regulatory subunit 2A in modifying the root system architecture (RSA) in response to low phosphate availability (Wang et al. 2017). *Z. mays* overexpressing *ZmPP2AA1* exhibited a profusely branched

RSA with an increase in lateral branching, and this coincided with better phosphate uptake under low Pi availability. Overexpression lines also displayed an increase in grain production compared to the non-transgenic plants under low Pi (Wang et al. 2017). A transgenic approach utilizing protein phosphatases to modify cereals and millets for abiotic stress tolerance has been minimal, and there is vast potential to use these genes for future genetic improvement programs. In fact, there is no study on millets in this zone, and considering the hardiness of millets to abiotic stresses, it will be exciting to know the role played by phosphatases for providing such adaptability to these stress-resistant crops and utilize the information gathered for their enhancement. All the studies reported above have shown the superiority of transgenic plants over non-transgenic plants and the potential to utilize these genes for future genetic improvements. This is the need of the hour with the human population exceeding its threshold and land fertility deteriorating at an unprecedented pace.

Acknowledgments The authors' work in this area is supported by the Core Grant of National Institute of Plant Genome Research, New Delhi, India. R.K.S. and A.P. acknowledge the Council of Scientific and Industrial Research, Govt. of India, India, for the award of research fellowships. The authors are thankful to the DBT-e-Library Consortium (DeLCON) for providing access to e-resources.

References

Alonso A, Sasin J, Bottini N, Friedberg I, Friedberg I, Osterman A, Godzik A, Hunter T, Dixon J, Mustelin T (2004) Protein tyrosine phosphatases in the human genome. Cell 117:699–711

Bennetzen JL, Schmutz J, Wang H, Percifield R, Hawkins J et al (2012) Reference genome sequence of the model plant *Setaria*. Nat Biotechnol 30:555–561

Bradai M, Mahjoubi H, Chini A, Chabouté ME, Hanin M, Ebel C (2018) Genome wide identification of wheat and Brachypodium type one protein phosphatases and functional characterization of durum wheat TdPP1a. PLoS One 13:e0191272. https://doi.org/10.1371/journal.pone.0191272

Brautigan DL (2013) Protein Ser/Thr phosphatases-the ugly ducklings of cell signaling. FEBS J 280:324–345

Cao J, Jiang M, Li P, Chu Z (2016) Genome-wide identification and evolutionary analyses of the PP2C gene family with their expression profiling in response to multiple stresses in *Brachypodium distachyon*. BMC Genomics 17:175

Ceulemans H, Bollen M (2004) Functional diversity of protein phosphatase-1, a cellular economizer and reset button. Physiol Rev 84:1–39

Ceulemans H, Stalmans W, Bollen M (2002) Regulator-driven functional diversification of protein phosphatase-1 in eukaryotic evolution. BioEssays 24:371–381

Chongyi X, Jing R, Mao X, Jia X, Chang X (2007) A wheat (*Triticum aestivum*) protein phosphatase 2A catalytic subunit gene provides enhanced drought tolerance in tobacco. Ann Bot 99:439–450

Cohen PT (2002) Protein phosphatase 1-targeted in many directions. J Cell Sci 115:241–256

Danquah A, De Zelicourt A, Colcombet J, Hirt H (2014) The role of ABA and MAPK signaling pathways in plant abiotic stress responses. Biotechnol Adv 32:40–52

Das AK, Helps NR, Cohen PT, Barford D (1996) Crystal structure of the protein serine/threonine phosphatase 2C at 2.0 A resolution. EMBO J 15:6798–6809

de la Fuente van Bentem S, Hirt H (2009) Protein tyrosine phosphorylation in plants: more abundant than expected? Trends Plant Sci 14:71–76

Dudhate A, Shinde H, Tsugama D, Liu S, Takano T (2018) Transcriptomic analysis reveals the differentially expressed genes and pathways involved in drought tolerance in pearl millet [*Pennisetum glaucum* (L.) R. Br]. PLoS One 13:e0195908. https://doi.org/10.1371/journal.pone.0195908

Goldberg J, Huang HB, Kwon YG, Greengard P, Nairn AC, Kuriyan J (1995) Three-dimensional structure of the catalytic subunit of protein serine/threonine phosphatase-1. Nature 376:745–753

Hittalmani S, Mahesh HB, Shirke MD, Biradar H, Uday G, Aruna YR, Lohithaswa HC, Mohanrao A (2017) Genome and Transcriptome sequence of Finger millet (*Eleusine coracana* (L.) Gaertn.) provides insights into drought tolerance and nutraceutical properties. BMC Genomics 18:465

Hou Y, Zhu Y, Wang P, Zhao Y, Xie S et al (2016) Type one protein phosphatase 1 and its regulatory protein inhibitor 2 negatively regulate ABA signaling. PLoS Genet 12:e1005835

Hu R, Zhu Y, Shen G, Zhang H (2014) TAP46 plays a positive role in the ABSCISIC ACID INSENSITIVE5-regulated gene. Plant Physiol 164:721–734

Hu X, Liu L, Xiao B, Li D, Xing X, Kong X, Li D (2010) Enhanced tolerance to low temperature in tobacco by over-expression of a new maize protein phosphatase 2C, ZmPP2C2. J Plant Physiol 167:1307–1315

Hu X, Song F, Zheng Z (2006) Molecular characterization and expression analysis of a rice protein phosphatase 2C gene, OsBIPP2C1, and overexpression in transgenic tobacco conferred enhanced disease resistance and abiotic tolerance. Physiol Plant 127:225–236

Jaiswal S, Antala TH, Mandavia MK, Chopra M et al (2018) Transcriptomic signature of drought response in pearl millet (*Pennisetum glaucum* (L.)) and development of web-genomic resources. Sci Rep 8:3382. https://doi.org/10.1038/s41598-018-21560-1

Janssens V, Goris J (2001) Protein phosphatase 2A: a highly regulated family of serine/threonine phosphatases implicated in cell growth and signaling. Biochem J 353:417–439

Kim K (2013) Stress responses mediated by the CBL calcium sensors in plants. Plant Biotechnol Rep 7:1–8

Kissinger CR, Parge HE, Knighton DR, Lewis CT et al (1995) Crystal structures of human calcineurin and the human FKBP12-FK506-calcineurin complex. Nature 378:641–644

Liao Y, Lin K, Chiang C, Chih-ming C (2016) Oryza sativa protein phosphatase 1a (*OsPP1a*) involved in salt stress tolerance in transgenic rice. Mol Breed 36:22

Liu L, Hu X, Song J, Zong X, Li D, Li D (2009) Over-expression of a *Zea mays* L. protein phosphatase 2C gene (*ZmPP2C*) in Arabidopsis *thaliana* decreases tolerance to salt and drought. J Plant Physiol 166:531–542

Luan S (2003) Protein phosphatases in plants. Annu Rev Plant Biol 54:63–92

Man R, Yu K, Zhou Y, Xu Z, Chye M, Kong RYC (2003) Two genes encoding protein phosphatase 2A catalytic subunits are differentially expressed in rice. Plant Mol Biol 51:295–311

McCormick RF, Truong SK, Sreedasyam A, Jenkins J, Shu S et al (2018) The *Sorghum bicolor* reference genome: improved assembly, gene annotations, a transcriptome atlas, and signatures of genome organization. Plant J 93:338–354

Michniewicz M, Zago MK, Abas L, Weijers D, Schweighofer A et al (2007) Antagonistic regulation of PIN phosphorylation by PP2A and PINOID directs auxin flux. Cell 130:1044–1056

Milner SG, Jost M, Taketa S, Mazón ER, Himmelbach A (2019) Genebank genomics highlights the diversity of a global barley collection. Nat Genet 51:319–326

Ni L, Fu X, Zhang H, Li X, Cai X et al (2019) Abscisic acid inhibits rice protein phosphatase PP45 via H_2O_2 and relieves repression of the Ca^{2+}/CaM-dependent protein kinase DMI3. Plant Cell 31:128–152

Ogawa D, Morita H, Hattori T, Takeda S (2012) Molecular characterization of the rice protein RSS1 required for meristematic activity under stressful conditions. Plant Physiol Biochem 61:54–60

País SM, Téllez-Iñón MT, Capiati DA (2009) Serine/threonine protein phosphatases type 2A and their roles in stress signaling. Plant Signal Behav 4:1013–1015

Pandey GK (2008) Emergence of a novel calcium signaling pathway in plants: CBL-CIPK signaling network. Physiol Mol Biol Plants 14:51–68

Qi Y, Li K, Ni X, Qi W, Yongqing W et al (2018) Genome-wide identification of PP2C genes and their expression profiling in response to drought and cold stresses in *Medicago truncatula*. Sci Rep 8:12841

Rodrigues A, Adamo M, Crozet P, Margalha L, Confraria A et al (2013) ABI1 and PP2CA phosphatases are negative regulators of Snf1-related protein kinase1 signaling in arabidopsis. Plant Cell 25:3871–3884

Sanyal SK, Pandey A, Pandey GK (2015) The CBL–CIPK signaling module in plants: a mechanistic perspective. Physiol Plant 155:89–108

Schweighofer A, Hirt H, Meskiene I (2004) Plant PP2C phosphatases: emerging functions in stress signaling. Trends Plant Sci 9:236–243

Shi Y (2009) Serine/threonine phosphatases: mechanism through structure. Cell 139:468–484

Singh A, Giri J, Kapoor S, Tyagi AK, Pandey GK (2010) Protein phosphatase complement in rice: genome-wide identification and transcriptional analysis under abiotic stress conditions and reproductive development. BMC Genomics 11:435

Singh A, Jha SK, Bagri J, Pandey GK (2015) ABA inducible rice protein phosphatase 2C confers ABA insensitivity and abiotic stress tolerance in Arabidopsis. PLoS One 10:e0125168

Singh A, Pandey A, Srivastava AK, Tran LS, Pandey GK (2016) Plant protein phosphatases 2C: from genomic diversity to functional multiplicity and importance in stress management. Crit Rev Biotechnol 36:1023–1035

Smith RD, Walker JC (1996) Plant protein phosphatases. Annu Rev Plant Physiol Plant Mol Biol 47:101–125

Takemiya A, Ariyoshi C, Shimazaki K (2009) Identification and functional characterization of inhibitor-3, a regulatory subunit of protein. Plant Physiol 150:144–156

Takemiya A, Kinoshita T, Asanuma M, Shimazaki K (2006) Protein phosphatase 1 positively regulates stomatal opening in response to blue light in *Vicia faba*. Proc Natl Acad Sci U S A 103:13549–13554

Takemiya A, Yamauchi S, Yano T, Ariyoshi C, Shimazaki K (2013) Identification of a regulatory subunit of protein phosphatase 1 which mediates blue light signaling for stomatal opening special focus issue-regular paper. Plant Cell Physiol 54:24–35

Ubersax JA, Ferrell JE Jr (2007) Mechanisms of specificity in protein phosphorylation. Nat Rev Mol Cell Biol 8:530–542

Umezawa T, Sugiyama N, Mizoguchi M, Hayashi S, Myouga F et al (2009) Type 2C protein phosphatases directly regulate abscisic acid-activated protein kinases in Arabidopsis. Proc Natl Acad Sci U S A 106:17588–17593

Varshney RK, Shi C, Thudi M, Mariac C, Wallace J (2017) Pearl millet genome sequence provides a resource to improve agronomic traits in arid environments. Nat Biotechnol 35:969–976

Virshup DM, Shenolikar S (2009) From promiscuity to precision: protein phosphatases get a makeover. Mol Cell 33:537–545

Wang J, Pei L, Jin Z, Zhang K, Zhang J (2017) Overexpression of the protein phosphatase 2A regulatory subunit a gene *ZmPP2AA1* improves low phosphate tolerance by remodeling the root system architecture of maize. PLoS One 1212:e0176538

Wei K, Pan S (2014) Maize protein phosphatase gene family: identification and molecular characterization. BMC Genomics 15:773

Xing Y, Xu Y, Chen Y, Jeffrey PD, Chao Y et al (2006) Structure of protein phosphatase 2A core enzyme bound to tumor-inducing toxins. Cell 127:341–353

Xu Y, Chen Y, Zhang P, Jeffrey PD, Shi Y (2008) Structure of a protein phosphatase 2A holoenzyme: insights into B55-mediated Tau dephosphorylation. Mol Cell 31:873–885

Xue T, Wang D, Zhang S, Ehlting J, Ni F, Jakab S, Zheng C, Zhong Y (2008) Genome-wide and expression analysis of protein phosphatase 2C in rice and Arabidopsis. BMC Genomics 9:550

Ye Q, Wang H, Zheng J, Wei Q, Jia Z (2008) The complex structure of calmodulin bound to a calcineurin peptide. Proteins 73:19–27

You J, Zong W, Hu H, Li X, Xiao J, Xiong L (2014) A STRESS-RESPONSIVE NAC1-regulated protein phosphatase gene rice protein phosphatase18 modulates drought and oxidative stress tolerance through abscisic acid-independent reactive oxygen species scavenging in rice 1 [W]. Plant Physiol 166:2100–2114

Yue H, Wang L, Liu H, Yue W, Du X, Song W, Nie X (2016) De novo assembly and characterization of the transcriptome of broomcorn millet (*Panicum miliaceum* L.) for gene discovery and marker development. Front Plant Sci 7:1083

Zhang G, Liu X, Quan Z, Cheng S, Xu X et al (2012) Genome sequence of foxtail millet (*Setaria italica*) provides insights into grass evolution and biofuel potential. Nat Biotechnol 30:549–554

Zhang ZY (2002) Protein tyrosine phosphatases: structure and function, substrate specificity, and inhibitor development. Annu Rev Pharmacol Toxicol 42:209–234

Zou C, Li L, Miki D, Li D, Tang Q, Xiao L et al (2019) The genome of broomcorn millet. Nat Commun 10:436

Chapter 14
Interplay of Protein Phosphatases with Cytoskeleton Signaling in Response to Stress Factors in Plants

Dariia O. Samofalova, Pavlo A. Karpov, Alexey V. Raevsky, and Yaroslav B. Blume

14.1 Introduction

Plant stress implies a series of processes and states where growth and development conditions are extremely different from the optimal ones. Biotic and abiotic stresses can lead to growth defects, drop in yield or even death. Canonical abiotic stress factors include low or high soil moisture, extremely low or high temperatures, abnormal light conditions, salinity, etc. At the same time, biotic factors are the result of unwarranted interference of vast group of organisms, such as different types of pathogens, pests, and weeds (Máthé et al. 2019).

The numerous data indicate that plant responses to stress involve a wide range of molecular mechanisms, such as changes in regulatory networks and gene expression or in work of their reaction products (Lichtenthaler 1996; Kranner et al. 2010; Mosa et al. 2017). In the end of the past century, it was demonstrated that changes in protein phosphorylation state are strongly associated with all the abovementioned factors and extracellular signals (Ho 2015; Máthé et al. 2019). Self-incompatibility (Goring et al. 1993), initiation of mitosis (Li and Roux 1992; Duerr et al. 1993), isoprenoid biosynthesis (MacKintosh et al. 1992), cytoplasmic streaming (Tominga et al. 1987; McCurdy and Harmon 1992), sucrose-phosphate synthase activity (Huber and Huber 1991), MSERK1 (Duerr et al. 1993), and phosphoenolpyruvate carboxylase activity (Bakrim et al. 1992) are the examples of regulation of cellular responses involving different protein kinases and protein phosphatases (Ling 2015; Máthé et al. 2019).

Now there is no doubt that, along with gene regulation, the role of reversible protein phosphorylation in the stress response is indispensable (Máthé et al. 2019). Nonetheless, the study of plant protein phosphatase features and functions is overdue, as opposed to their functional antagonists—protein kinases. Perhaps the reason

D. O. Samofalova · P. A. Karpov · A. V. Raevsky · Y. B. Blume (✉)
Institute of Food Biotechnology and Genomics, National Academy of Sciences of Ukraine, Kyiv, Ukraine
e-mail: blume.yaroslav@nas.gov.ua

G. K. Pandey (ed.), *Protein Phosphatases and Stress Management in Plants*,
https://doi.org/10.1007/978-3-030-48733-1_14

lies in more pronounced structural differences between plant and animal phosphatases (Samofalova et al. 2015, 2019). Historically, mammalian protein phosphatases were classified by their substrate specificity as serine/threonine (Ser/Thr), tyrosine (PTPs), and dual-specificity phosphatases (DSPs) (Máthé et al. 2019; Samofalova et al. 2015, 2019). Further, based on a sequence homology, spatial structure, and substrate specificity, protein phosphatases were divided into these three groups. Now, plant protein phosphatases can be grouped into four evolutionarily distant families: PPP (phosphoprotein phosphatase), PPM/PP2C (Mg^{2+}- or Mn^{2+}-dependent protein phosphatase/protein phosphatase 2C), PTP (phosphotyrosine phosphatase), and aspartate-dependent phosphatase families (Uhrig et al. 2013). PPP and PPM/PP2C families are responsible for dephosphorylation of serine and threonine residues, and it is considered that the evolutionarily highly conserved PPPs engaged in about 80% of total protein dephosphorylation in eukaryotes (Moorhead et al. 2009; Lillo et al. 2014). Plant members of the PPP family can be divided into eight subfamilies: PP1 (protein phosphatase type 1), PP2A (protein phosphatase 2A), PP4, PP5, PP6, PP7, SLP (Shewanella-like protein) phosphatase and PPKL (protein phosphatase with kelch-like repeat domains) (Uhrig et al. 2013). The third superfamily comprises the group of tyrosine phosphatases and includes three classes: class 1 (classical PTPs: receptor, non-receptor and dual tyrosine phosphatases (DSPs)—DSPsI, MAPKP, PTEN, myotubularins, mRNA capping, etc.), class 2 (represented by yeast like dual-specificity phosphatases CDC25(s)), and class 3 (represented by low-molecular-weight protein tyrosine phosphatases (LMWPTPs)). The fourth and the smallest group of protein phosphatases is represented by two families of aspartate-specific protein phosphatases: FCP/SCP-like (F-cell production/small CTD carboxy-terminal domain phosphatases) and HADs (haloacid dehalogenases). The catalytic subunits of these protein phosphatases are characterized by a conserved DXDXT/V motif located in the N-terminal part of the molecule (Table 14.1, Fig. 14.1).

Such heterogeneity of protein phosphatases increasingly leads to the conclusion that, in contrast to protein kinases with a common evolutionary origin, the groups of plant protein phosphatases originate from different ancestral sequences. In particular, this is confirmed by the significant differences in their spatial structures and mechanisms of action (Samofalova et al. 2015).

14.2 The Role of Plant Protein Phosphatases in Stress

Mammalian and plant stress signal transduction pathways start with a signal perception, followed by the activation of protein phosphorylation cascades involving protein kinases and phosphatases, that finally targets proteins involved in cellular protection or transcription factors controlling sets of stress-regulated genes (Tamura et al. 2002; Fujita et al. 2006; País et al. 2009). Essentially, protein phosphatases are a complex family represented by hundreds of enzymes acting as a multicomponent system of several quite conserved catalytic subunits combined with different types

Table 14.1 Plant protein phosphatases from Arabidopsis *thaliana* (Samofalova et al. 2015)

Class	Number of type/isoform	Swiss-Prot status	Number of closest human homologue	Indicators (min-max) of sequence similarity		
				Ident. (%)	Simil. (%)	Gap (%)
Ser/Thr protein phosphatases of the phosphoprotein phosphatase family						
PP1	12	±	2	45–84	62–93	0–4.9
PP2A	7	±	2	81–82	89–94	0
PP4/PPX	3	±	1	39–84	45–92	0.4–48
PP5	2	±	1	38–63	49–78	0–36
PP6	2	+	1	73	86	0.4
PP7	3	±	2	21–31	40–46	6–11
BSL	3	+	2	46–47	62–64	4.9
BSU	1	+	1	44	64	5.9
Mg^{2+}/Mn^{2+}-dependent protein phosphatase family						
PP2C	80	+	24	12–45	27–59	4–38
Protein tyrosine phosphatase superfamily						
PTPs I (classic)	2	+	2	17–37	31–52	5–20
DSPs I (DSPs)	9	+	6	16–44	33–65	0.6–14
PTEN	10	±	5	19–44	30–63	0–31
Myotub-PP	2	–	1	35–36	44–45	28
mRNA capping	3	–	1	32–35	51–53	8–9
PTPs II (Cdc25)	1	+	1	22.2	37.6	8.5
PTPs III (LMWPTP)	9	–	1	12–39	19–52	5–73
Asp-specific protein phosphatases						
FCP/FCP-like	23	±	4	23–58	37–79	0–17
HAD3	3	–	1	33–35	49–53	14–18

of regulating subunits (Peti et al. 2013). Mainly, variability of regulatory subunits determines specificity, selectivity, and cell localization of protein phosphatases (Virshup and Shenolikar 2009). Most of them participate in sensing and signaling mechanisms replying pathogen infections, stress signaling, plasma membrane sensor systems drawn into intracellular signaling networks, cell death, metabolic responses, primary metabolism, stress induction of secondary metabolism, and hormonal signaling (Durian et al. 2016). In contrast to stress-activated protein kinases, protein phosphatases are acting separately and usually not combined in joint groups due to the nature of stress factor. According to experimental data, the role of plant protein phosphatases (Figs. 14.2 and 14.3) in stress is commonly associated with tyrosine protein phosphatases—DSPsI (ABA regulation, abiotic stress, MKPs, gene expression and microtubule regulation, abiotic and biotic stress, necrotrophic fungal

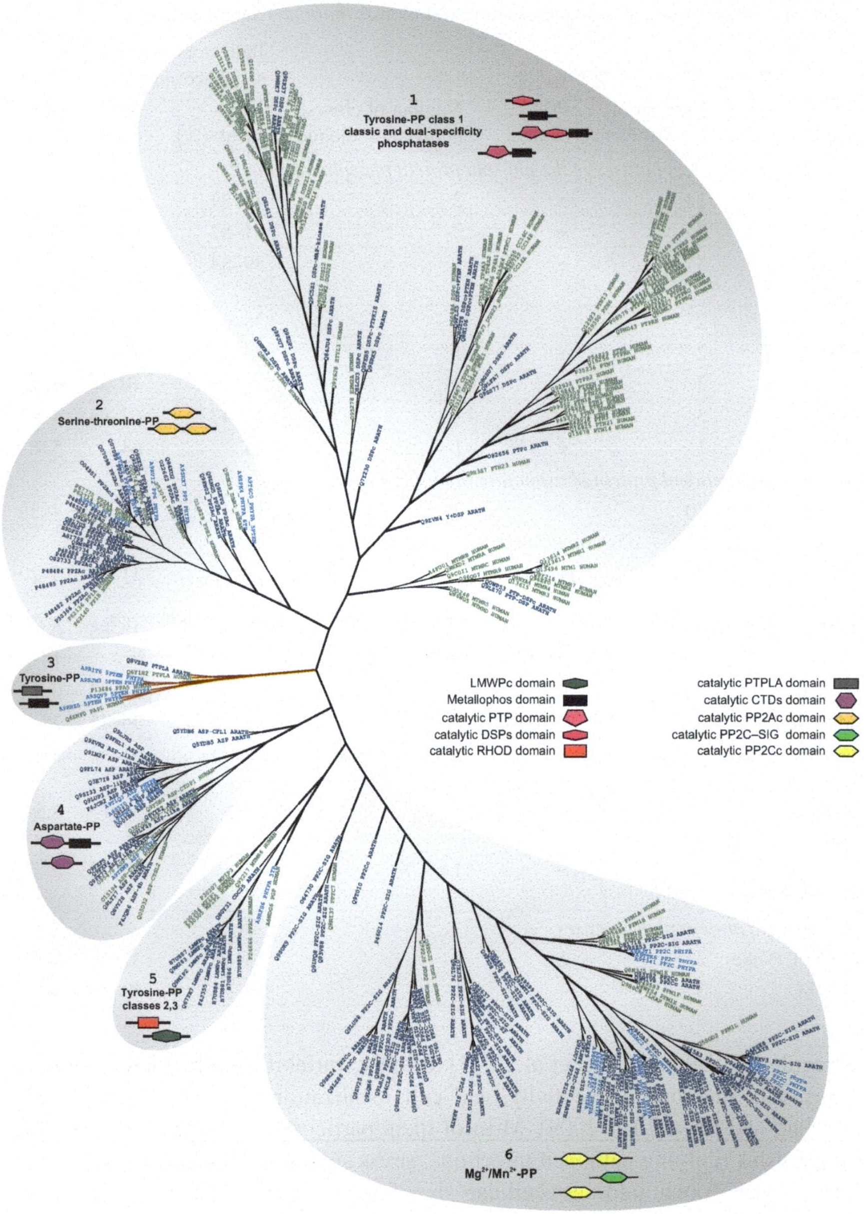

Fig. 14.1 Joint NJ-clustering analyses of human and plant (including predicted) protein phosphatases from *Physcomitrella patens* and *A. thaliana* based on similarity of catalytic domain amino acid sequences (Samofalova et al. 2015). The phylogram indicates clades common for these protein phosphatases: (1) Clade 1 consolidates tyrosine protein phosphatases of class 1; (2) clade 2 consolidates serine/threonine protein phosphatases; (3) clade 3 consolidates tyrosine protein phosphatases (according Dendroscope software prediction, this clade is the probable area of the root); (4) clade 4 joints aspartate protein phosphatases; (5) clade 5 consolidates tyrosine protein phosphatases of classes 2 and 3; (6) clade 6 includes Mg^{2+}/Mn^{2+}-dependent serine/threonine protein phosphatases

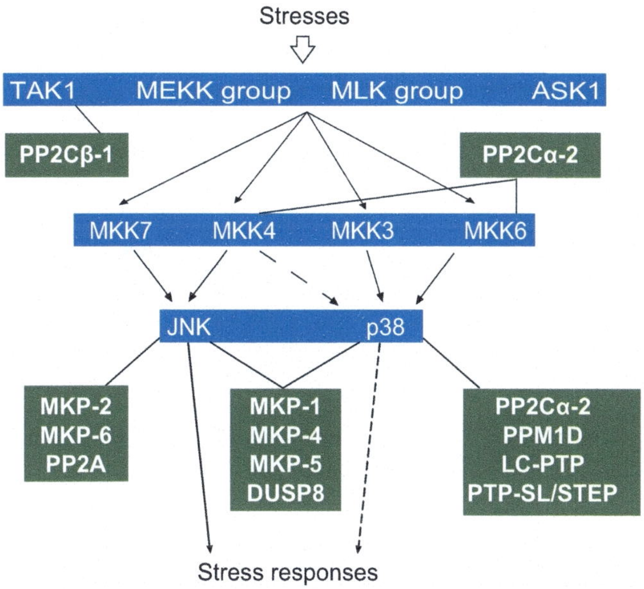

Fig. 14.2 Protein kinases and protein phosphatases of stress-activated protein kinase (SAPK) signaling pathways in mammals (based on Tamura et al. 2002). Abbreviated names of protein kinases: TAK1, mitogen-activated protein kinase 7 (MAP 3K7); MEKK, MAP kinase (or MAP 3K); MLK, mixed lineage kinase; ASK1, apoptosis signal-regulating kinase 1; MKK7, dual-specificity mitogen-activated protein kinase 7; MKK4, mitogen-activated protein kinase 4; MKK3, mitogen-activated protein kinase 3; MKK6, mitogen-activated protein kinase 6; JNK, c-Jun N-terminal kinase; p38, mitogen-activated protein kinases. Abbreviated names of protein phosphatases: PPC2β-1, protein phosphatase 2Cβ; PP2Cα-2, protein phosphatase 2Cα; MKP-2, mitogen-activated protein kinase phosphatase 2; MKP-6, dual-specificity phosphatase 14 (DUSP14); PP2A, Ser/Thr protein phosphatase 2A; MKP-1, mitogen-activated protein kinase phosphatase 1; MKP-4, dual-specificity protein phosphatase (DUSP9); MKP-5, dual-specificity phosphatase 10 (DUSP10); DUSP8, dual-specificity phosphatase 8 (DUSP8); PP2Cα-2, protein phosphatase 2Cα-2; Wip1/PPM1D, 2C family serine/threonine phosphatase; HePTP/LC-PTP, leukocyte tyrosine phosphatase; PTP-SL/STEP, protein tyrosine phosphatases.

or biotrophic bacterial pathogen)—and Ser/Thr protein phosphatases of types 1, 2A and 2C combined with different regulating subunits. The type of regulating subunit is usually determined by the type of biotic stress factor (Durian et al. 2016; Rahikainen et al. 2016).

During the infectious cycle of *Leishmania*, its alternative morphological forms are represented with either extracellular flagellated promastigote or intracellular pathogenic amastigote. In general, *Leishmania* differentiation is triggered by changes in environmental cues, mainly pH and temperature; after that, extracellular signals are translated into stage-specific gene expression by a cascade of reversible protein phosphorylation regulated by protein kinases and phosphatases. Though protein kinases have been actively studied as potential anti-parasitic drug targets, some results indicate the importance of PP5 in regulation of parasite-stress and

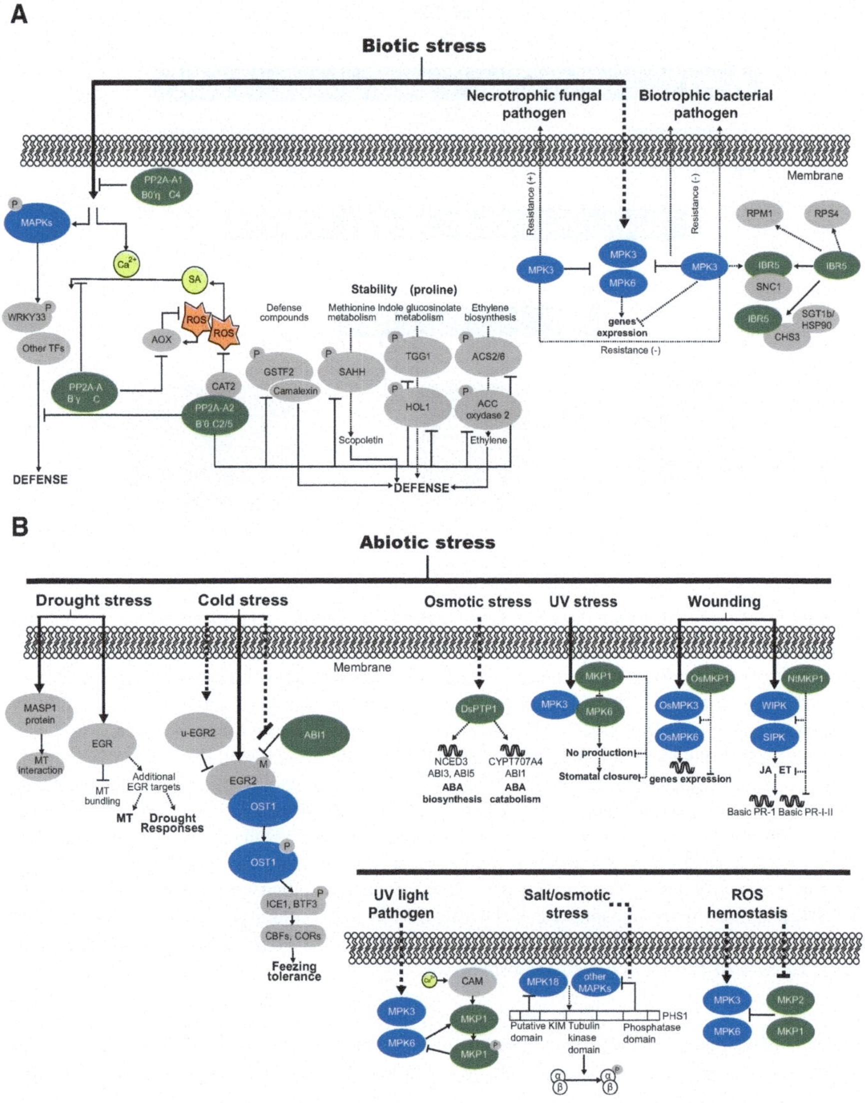

Fig. 14.3 Protein kinases and phosphatases of plant cells' response on biotic (**a**) and abiotic (**b**) stress (based on fragments of the figures from Jiang et al. 2018 and Durian et al. 2016). Part A: PP2A, Ser/Thr protein phosphatase 2A; MAPKs, mitogen-activated protein kinases; WRKY33, WRKY DNA-binding protein 33; TFs, transcription factors; ROS, reactive oxygen species; SA, ribosomal protein (RPSA, LBP, LRP, p40, 67LR, ICAS, lamR, 37LRP, LAMBR, LAMR1, LRP/LR, LBP/p40, and NEM/1CHD4); AOX, alternative oxidase; CAT2, catalase 2; GSTF2, glutathione *S*-transferase PHI 2; SAHH, *S*-adenosylhomocysteine hydrolase; TGG1, thioglucoside glucohydrolase 1; HOL1, harmless to ozone layer protein 1; ACS2/6, 1-aminocyclopropane-1-carboxylate synthase 2; ACC, ox1-aminocyclopropane-1-carboxylic oxidase 2; MPK3 and MPK6, mitogen-activated protein kinases; RPM1, NB-ARC domain-containing disease resistance protein; RPS4, disease resistance protein; IBR5, indole-3-butyric acid-response 5; SNC1, TIR-NBS-LRR class disease

adaptation during differentiation, making this protein an attractive potential target for therapeutic intervention (Norris-Mullins et al. 2018).

High expression of PP2C (SjPtc1) was identified in *Schistosoma japonicum*, and SjPTC1 expression reversed the sensitivity of yeast ptc1 null mutants towards H_2O_2, $ZnCl_2$, cisplatin, and rapamycin. SjPtc1 in *S. japonicum* may take part in the regulation of cellular responses to oxidative stress, DNA damage, and the TOR (target of rapamycin) signaling pathway (Lu et al. 2018). Eukaryotic PP2Cs are involved in manifold cellular processes. At the same time, their functions in filamentous fungi remain almost unknown. *Botrytis cinerea* have four PP2C-like genes, named BcPTC1, BcPTC3, BcPTC5, and BcPTC6. BcPtc3, but not BcPtc1, negatively regulates phosphorylation of BcSak1 (the homologue of *S. cerevisiae* Hog1) in *B. cinerea*, although both BcPTC1 and BcPTC3 were able to rescue the growth defects in yeast PTC1 deletion mutant under various stress conditions (Yang et al. 2013a). Also, expression contol at the promoter level is crucial for adaptive responses to salt stress in the yeast genome (Mizuno et al. 2018; Fu et al. 2018).

It was demonstrated that calcineurin (PP2B) and CchA are required to modulate cellular calcium levels and synergistically coordinate calcium influx under salt stress. The member of TRPC family of vacuolar Ca^{2+} channels, YvcA, was proven to compensate for calcineurin-CchA in fungal salt stress adaption (Wang et al. 2016; Park et al. 2016; Chow et al. 2017; Lahiri et al. 2014). Parasite-specific eIF2α kinases and phosphatases are also required for proper developmental transitions and adaptation to cellular stresses during life cycle. Apicomplexan-induced inhibition of eIF2α kinases may interfere in parasite translational control and lay the basis for development of new therapies against malaria and toxoplasmosis (Zhang et al. 2013).

The causative agent of African trypanosomiasis in human and domestic animals, *Trypanosoma brucei*, adapts in various environments during their digenetic life cycle. Heat shock protein 90 (Hsp90) is an important participant for survival of the parasite. Artificial induction of TbPP5 overproduction reduces the growth inhibitory effect of GA (geldanamycin), while knockdown of TbPP5 reduces cell growth more in the presence of GA, as compared to parental control. However, depletion of TbPP5 did not prevent the increase in Hsp90 level during GA treatment. Together,

Fig. 14.3 (continued) resistance protein; CHS3, TIR-NB-LRR resistance protein; SGT1b, phosphatase-like protein; HSP90, heat shock protein 90. Part B: MASP1, microtubule-associated stress protein 1; EGR, clade E growth-regulating protein; ABI1, protein phosphatase 2C family protein 1; OST1, protein kinase superfamily protein; ICE1, inducer of CBF expression 1, an upstream transcription factor; BTF3, basal transcription factor 3; CBFs, transcriptional activators; CORs, cold-regulated polypeptides; DsPTP1, dual-specificity protein phosphatase 1; NCED3, nine-cis-epoxycarotenoid dioxygenase 3 gene; ABI1, ABI3, ABI5, abscisic acid insensitive 1, 3 and 5 genes; MPK1, mitogen-activated protein kinase 1; MPK3 and MPK6, mitogen-activated protein kinases; OsMPK1, OsMPK3 and OsMPK6, mitogen-activated protein kinases of *Oryza sativa*; NtMPK1, mitogen-activated protein kinase 1 of *Nicotiana tabacum*; WIPK, wound-induced protein kinase; SIPK, salicylic acid-induced protein kinase; JA, jasmonic acid; ET, ethylene; PR1, pathogenesis-related protein 1; PI-II, probable transcription factors; CaM, calmodulin; MKP1, mitogen-activated protein kinase phosphatase 1; MPK18, mitogen-activated protein kinase 18

this suggests maintaining TbPP5 as positive regulator of trypanosomal Hsp90 under proteotoxic stresses (Jones et al. 2008).

In *Candida albicans*, the main stress protection is closely associated with trehalose biosynthetic pathway. In particular, it concerns the cellular response to oxidative stress and resistance to phagocytosis. TPS2-dependent defense mechanism encodes trehalose-6P phosphatase and causes a pleiotropic defective phenotype of *C. albicans*, maintaining cell wall integrity and the ability to form chlamydospores (Martínez-Esparza et al. 2009). At the same time, yeast homologue ScPFA-DSP1 (plant and fungi atypical DSPs) from *S. cerevisiae* is involved in response to caffeine and rapamycin stresses (Romá-Mateo et al. 2011).

Another mechanism, which is critical for the regulation of stress responses in fungi, is associated with calcium signaling via calmodulin and calcineurin. These functions of calmodulin and calcineurin are conserved among pathogenic fungi and model saprophytic fungi, but the mechanisms of action have diverged (Kraus and Heitman 2003).

It is noteworthy that PP2A are also required for nuclear accumulation of Msn2p and response to heat, osmotic, nitrogen (not glucose), and starvation stresses in *Saccharomyces cerevisiae*. PP2A and Tor kinase pathway transduce stress and nitrogen starvation signals to Msn2p. Finally, Msn2p localization is unaffected by conditional loss of 14-3-3 protein function, ruling out the possibility that 14-3-3 proteins act as a scaffold to sequester Msn2p in the cytoplasm (Santhanam et al. 2004). It was demonstrated that complete activation of PP2A-Cdc55 complex under environmental stress response is mediated by the transcription factors Msn2 and Msn4. It was established that PP2A-Cdc55 complex is not involved in cytosolic stress signal transduction. At the same time, PP2A-Cdc55 participates in a specific intranuclear mechanism regulating nuclear accumulation of Msn2 and Msn4, as well as chromatin association in response to stress (Reiter et al. 2013). At the same time, in *C. albicans*, the phosphatases of another homological type, PP4, act as a negative regulator of DNA damage-induced filamentation and virulence (Feng et al. 2017).

It was demonstrated that functions of PTPs in filamentous fungi *B. cinerea* are different from those in *S. cerevisiae*. So, BcPtpA and BcPtpB play important roles in the regulation of vegetative development, virulence and adaptation to oxidative, osmotic, and cell wall damage stresses in *B. cinerea* (Yang et al. 2013a, b).

It is known that fungal mitogen-activated protein kinases (MAPKs) are directly involved in stresses response and developmental processes. Since the precise regulation of MAPKs is fundamental for cell physiology, fungi bear dual-specificity phosphatases (DUSPs) that act as MAP kinase phosphatases (MKPs). Fungi offer a model to gain insight into the regulatory mechanisms that control MKPs and couple oxidative stress with substrate recognition (González-Rubio et al. 2019).

One more unique fungal "stress phosphatases" is protein phosphatases Z. This group is closely associated with resistance to high salt concentrations, cell wall integrity, cell cycle regulation, and oxidative stress in fungi. In *Aspergillus fumigatus*, it was shown that PHZA is under control of the transcription factor Skn7 and is involved in the control of the oxidative stress. Accordingly, the Δ*phzA* mutant showed a defect in virulence in an experimental model of corneal infection in immunocompetent animals and impact on susceptibility of cell wall for drugs (Muszkieta et al. 2014).

14.3 Dephosphorylation of Serine and Threonine Residues in Plant Stress Response

Biotic stress factors are the major threat to the plant health and can significantly decrease crop productivity through impairing the physiological functions. To resist the wide range of pathogens and insect herbivores, plants deploy converging signaling pathways. The counteracting activities of protein kinases and phosphatases form basic mechanisms that recognize the stress factor and determine appropriate defensive measures. Epistasis analysis of nine PP1 members from *A. thaliana* revealed their role in defense activation of topp4-1-dependent (Ser/Thr-protein phosphatase PP1 isozyme 4) non-race-specific disease resistance1, phytoalexin deficient 4, and the salicylic acid pathways. Mutations on PP1 orthologues in tomato *TOPP1*, *TOPP4*, *TOPP5*, *TOPP6*, *TOPP7*, *TOPP8*, and *TOPP9* confirmed their role in plant defense response against *Pseudomonas syringae* pv. tomato (Pst) DC3000 and expression of defense genes. Also, it was demonstrated that TOPPs interact with mitogen-activated protein kinases MPK3, MPK4, and MPK6 and affect MAPK-mediated downstream defense pathway (Qin et al. 2014; Liu et al. 2020).

Recent studies identified protein phosphatase 2A (PP2A) as a central and universal component of plant response to stress (Rahikainen et al. 2016). Genetic, proteomic, and metabolomic approaches revealed a versatile nature of PP2A. It was demonstrated that PP2A essentially impacts on the plant immunity via receptor and organelle signaling, regulates gene expression, participates in metabolic pathways, and contributes to cell death signaling. In turn, different subunits of this complex enzyme mediate post-translational regulation of numerous metabolic pathways and signaling components (Durian et al. 2016). Awotunde et al. (2003) demonstrated the interaction between intact holoenzyme of PP2A and tubulin in etiolated maize seedlings at the stage of intensive cell elongation. They also have proved the association of PP2A and HDA14 (histone deacetylase 14), which in turn can deacetylate α-tubulin and promote interplay between protein phosphorylation and acetylation. It was found that DA14, ELP3 (contractor of HDA14), and the PP2AA-subunits A1, A2 and A3 are all located both in nucleus and cytosol. For example, in *A. thaliana*, HDA14 is associated with PP2A and enriches in the microtubule fraction along with the putative histone acetyltransferase ELP3 (Tran et al. 2012).

Also, it was shown that protein phosphatase 2A (PP2A) regulates the dynamics of the cortical microtubules in Arabidopsis, through interaction with TONNEAU2 (TON2)/FASS and direct dephosphorylation of α-tubulin (Zhang 2008; Kirik et al. 2012). It was demonstrated that *in vivo* TON2 mainly interacts with the PP2AC subfamily II isoforms, PP2AC-3 and PP2AC-4. Virus-induced gene silencing of PP2AC subfamily II causes several defects in plant development, whereas silencing of subfamily I causes no visible phenotype changes. Silencing of *PP2AC* subfamily II causes cell shape defects in the leaf epidermis and decreases the density of cortical MT arrays in Arabidopsis. It was demonstrated that upon removal of inorganic impurities, PP2AC subfamily II dephosphorylates α-tubulin. That suggests involvement of plant PP2AC in regulation of cortical MT under normal and salt stress conditions (Yoon et al. 2018). At the same time, a transverse microtubule array in TON 2–15 mutants did not reorient in response to light exposure. This indicates the role of TON2 in array reorientations and its effect on dynamics of such reorganizations (Kirik et al. 2012).

Bhaskara et al. (2017) found that three *Growth-Regulating* (*EGR*) type 2C protein phosphatases target the cytoskeleton, plasma membrane-associated proteins, and act as negative growth regulators during drought. Phosphoproteomic analysis identified putative targets of these PP2Cs, including a new MT binding protein that accumulated during low ψ_w stress and promoted MT stability and growth in a phosphorylation-specific manner. Namely, microtubule-associated stress protein 1 (MASP1) overexpression enhanced growth, *in vivo* MT stability, and recovery of microtubule organization during drought acclimation. It was detected *in vivo* that MASP1 functions depend on phosphorylation of a single serine residue. The EGR-MASP1 system selectively regulates microtubule recovery and stability to adjusting cell and plant growth. PP2C protein phosphatases were identified as important regulators of MT organization and stabilizing factors of continued plant growth during drought stress (Bhaskara et al. 2017).

It was found out that reversible protein phosphorylation is extremely important for drought and abscisic acid (ABA) signaling (Jenks and Wood 2009). Abscisic acid is an essential phytohormone that contributes to growth, development, and responses to drought stress, salinity, or pathogens (Yoshida et al. 2002; Hirayama and Shinozaki 2007). In *Populus trichocarpa* and Arabidopsis, *HAB1* and *HAB15* were identified as PP2C protein phosphatases, regulating ABA signaling (Umezawa et al. 2010; Wang et al. 2015; Bhaskara et al. 2017). It was found that the interacting ensemble of *HAB1*, *HAB3*, *HAB12*, and *HAB15* in *Populus* is the most highly expressed genes under non-stress conditions. At the same time, these plant PP2C genes demonstrate different expression under different drought treatments. The expression of *HAB1* through *HAB3* was unchanged during downregulated response to drought, while all other *HAB* genes were weakly to strongly upregulated. It was reported that HAB2, HAB12, HAB13, and HAB14 interact with the mitogen-activated protein kinase 7 (PtrMPK7). Beyond the canonical ABA signaling, the role of PP2Cs in coordination of plant growth and development under the influence of environment conditions was predicted (Jenks and Wood 2009; Rigoulot et al. 2019).

Also, it was detected that activity of numerous plant PP2Cs is regulated by interaction with pyrabactin resistant-like/regulatory component of abscisic acid receptors (PYL/RCAR) (Cutler et al. 2010; Melcher et al. 2010; Fuchs et al. 2014). Later, it was established that interaction with PYLs is specific only for 9 of all 80 PP2Cs encoded in Arabidopsis genome (Fuchs et al. 2014; Sugimoto et al. 2014). For the major part of plant PP2Cs, such interactions have not been detected and their physiological function remains unknown (Bhaskara et al. 2017).

It was demonstrated recently that Highly ABA-Induced 1 (HAI1) protein phosphatase 2C plays an important role in restriction of plant growth under low-water potential stress. Protein phosphatase HAI1 controls expression of jasmonic acid, auxin-related genes, *SHOOTMERISTEMLESS* genes, and some other regulators of development. It was demonstrated that HAI1-dependent phosphorylation of AT-Hook-Like10 (AHL10) on Ser314 determinate its function and localization. It indicates that HAI1-AHL10 signaling coordinates plant growth in stress and defense responses (Wong et al. 2019).

14.4 Dephosphorylation of Tyrosine Residues and Its Role in Plant Stress Response

Initially it was reported that phosphorylation more often occurs on Ser and Thr residues whereas Tyr phosphorylation accounts for only 0.05% (Hunter and Sefton 1980). Later, phosphoproteomics studies based on more accurate detection of phosphopeptides have shown that phosphorylation on Ser, Thr, and Tyr residues occurs at a ratio of 88:11:1 (Olsen et al. 2006). Thus, it appears that Tyr phosphorylation is minor as compared to Ser and Thr. Nevertheless, Tyr phosphorylation plays a crucial role in regulation of many cellular processes in eukaryotic cells, such as cell division, growth, and differentiation (Yemets et al. 2008; Hunter 2009; Sheremet et al. 2012). Very few studies have tried to elucidate the involvement of Tyr phosphorylation in plant cells. This is due to the lack of typical *PTK* genes in plants and also to the fact that *PTP* genes have been identified in the genome of Arabidopsis (Luan 2002). Sensitive proteomic approaches have, however, confirmed the existence of protein Tyr phosphorylation in plants (Ghelis 2011).

During the first decade of the century, several studies demonstrated that phosphorylation of tyrosine (Tyr) in plants considerably contributes to microtubule depolymerization and stress response (Blume et al. 2008a; Yemets et al. 2008; Sheremet et al. 2012). Furthermore, inhibition of PTKs in Arabidopsis altered root hair growth and development, probably as a result of their significant influences on MT organization in root hairs (Yemets et al. 2008; Sheremet et al. 2012). Inhibition of PTPs resulted in intense induction of root hair development and growth and caused a significant shortening of the root elongation zone. In Arabidopsis, it also led to changes in MT orientation from transverse to longitudinal in epidermis and cortex cells of the elongation and differentiation zones of the root (Yemets et al. 2008).

It was reported that Tyr phosphorylation is closely related to the abiotic stress responses (Ghelis 2011). In *Arachis hypogaea*, a *PTK* gene expression and its activity are induced by *cold and salt* stress tolerance mechanisms (Rudrabhatla and

Rajasekharan 2002). Cold, heat, and salt stresses induce high levels of gene expression of 14 *DsPTK* genes and downregulate three *DsPTK* genes (Rudrabhatla et al. 2006). The sole member of canonical PTP in Arabidopsis, AtPTP1, was upregulated by salt and downregulated by cold treatment (Xu et al. 1998; Fordham-Skelton et al. 1999). Finally, it has been shown that MAPK phosphatase 1 (MKP1) is a DsPTP that participates to the response to salt stress. It was predicted based on expression profiling of wild-type vs. *mkp1* mutant lines and increased resistance to salinity of *mkp1* in mutant plants (Ulm et al. 2001).

In a number of investigations, it was found that during a water stress, transpirational loss of water is reduced as stomata were closing in response to ABA. The involvement of PTKs and PTPs in the signaling pathway was illustrated on the relation between treatment with specific inhibitors and subsequent stomatal closure (Shi et al. 2005; MacRobbie 2002; Ghelis et al. 2008). The *phs1-3* mutation in a *DsPTP* gene causes a deregulation of the ABA-dependent stomatal closure (Quettier et al. 2006). Perception of water deficit gives rise to increases in internal ABA concentrations. Involvement of Tyr phosphorylation has been observed in ABA signaling as the DsPTP PHS1 is a negative regulator of ABA signal transduction pathway whereas IBR5, another DsPTP, has been shown to regulate positively this pathway (Monroe-Augustus et al. 2003; Quettier et al. 2006).

On the other hand, Tyr phosphorylation controls both the oxidative stress tolerance and the response to genotoxic stress. Reactive oxygen species are produced during the normal operation of respiratory and photosynthetic electron transport. They are toxic for the plant as they induce the production of highly destructive species. Plants have elaborated mechanisms to minimize the action of these compounds. AtDsPTP2, also called MKP2, has been shown to regulate positively the physiological responses to oxidative stress generated during ozone treatment (Lee and Ellis 2007). The *mkp2* mutant plants inhibit hypersensitivity to oxidative stress induced by methyl viologen during germination, confirming the role of this DsPTP as a positive regulator (Lumbreras et al. 2010). DsPTP1 was the first dual-specificity protein phosphatase from higher plants shown to inactivate MAPK (MPK4) *in vitro* (Gupta et al. 1998; Opdenakker et al. 2012; Jiang et al. 2018). Besides the stress of increased light intensity, plants are subjected to stress from the ultraviolet (UV) wavelengths in incident irradiation. UV-C is the most damaging factor as it modifies DNA and proteins. Plants use both restorative and repair mechanisms to counter this stress. Screens for UV-sensitive mutants in Arabidopsis led to the identification of the DsPTP MKP1 as essential for UV resistance that interacts with a group of stress-activated MAPK3, MAPK4, and MAPK6 (Ulm et al. 2002; Seo et al. 2007). The Arabidopsis *mkp1* mutant, which is resistant to elevated salinity, is also hypersensitive to UV radiation (Ulm et al. 2001, 2002; Ghelis 2011).

In general, MAPK phosphatases play an integrative role in plants responding to diverse environmental stimuli. *In vitro* and/or *in vivo* studies have shown that MKPs physically interact with MAPKs and/or regulate their activation. Besides, it has been indicated that DsPTP1 inactivates a MPK4 *in vitro* (Gupta et al. 1998). MKP1 interacts with MPK3, MPK4, and MPK6 both *in vitro* and *in vivo* and deactivates MPK6 in protoplast-based assays (Ulm et al. 2002; Bartels et al. 2009). MKP2, which interacts with MPK3 and MPK6 both *in vitro* and *in vivo*, is able to dephosphorylate phospho-MPK3 and phospho-MPK6 *in vitro* (Lee and Ellis 2007). IBR5, a MPK12 interacting partner, has been shown to dephosphorylate and thus deactivate *in vitro* and *in vivo* MPK12 (Lee et al. 2009). Walia et al. (2009) demonstrated that PHS1 interacts with Arabidopsis MPK12 and MPK18 in a yeast two-hybrid interaction assay, and recombinant PHS1 dephosphorylates the activated MPK18 *in vitro* (Walia et al. 2009). In general, all these studies proved the central role of MAPK signaling cascades in regulation of multiple cellular responses in eukaryotes and the role of protein phosphatases as the main negative regulators of MAPK signaling. Among these, dual-specificity (Ser/Thr and Tyr) phosphatases (DSPs), which belong to a subfamily of the tyrosine phosphatases (PTPs), catalyzed the dephosphorylation both on pSer/pThr and on pTyr residues, located in the activation loops of different MAPK members (Keyse and Emslie 1992; Alessiet al. 1993; Sun et al. 1993; Ward et al. 1994; Jiang et al. 2018). In general, from 22 plant DSPs identified in Arabidopsis, experimental evidence of direct interaction and dephosphorylation of different MAPKs has been outlined only for 5 of them (Gupta et al. 1998; Ulm et al. 2001, 2002; Lee and Ellis 2007; Kerk et al. 2008; Lee et al. 2009; Walia et al. 2009). All DSPs are closely associated with regulation of plant growth and development. MAP kinase phosphatase IBR5 is a positive regulator of auxin responses, indicating a role for MKPs as negative coordinators of plant growth and development. Another vitally important area where MKPs have been shown to affect plant growth and development is the control of the dynamics and organization of microtubules, in which PHS1 has been implicated to play an essential role (Jiang et al. 2018). Also, it was demonstrated that PHS1 is involved in controlling of flowering time in Arabidopsis (Jaeger et al. 2006). MKP1 has also been shown to be involved in controlling the cell fate transition during stomata development (Tamnanloo et al. 2018).

In general, DSP MAPK phosphatases are central hubs integrating biotic and abiotic stress (Jiang et al. 2018). In Arabidopsis, several DSP-type phosphatases have been implicated in regulating pathogen-associated responses and resistance. MAPK PHOSPHATASE 1 (MKP1) is an important negative regulator of plant immunity. In short, MAPK PHOSPHATASE 2 (MKP2) dephosphorylates phospho-MPK3 and phospho-MPK6 *in vitro*, possessing distinct functions in regulation of different pathogen impacts (Lee and Ellis 2007; Lumbreras et al. 2010). MKPs also contribute to the regulation of several resistance (R) proteins. IBR5 plays a positive role in regulating R protein CHS3 involved in controlling disease resistance mediated by R proteins RPM1 and RPS4 (Liu et al. 2015a, b). According to Bartels et al. (2010), MKP1 plays an important role in plant growth homeostasis by repressing SNC1-mediated stress signaling (Bartels et al. 2010). In addition to the resistance against pathogen infections, MKPs also constitute important components regulating multiple abiotic stresses, genotoxic stress, osmotic/drought stress, and salinity stresses (Jiang et al. 2018). Finally, MKP1, as a positive regulator of genotoxic stress survival, has been identified to be a negative regulator of salinity resistance, as demonstrated by the fact that loss of MKP1 increased resistance to salt stress (Ulm et al. 2002). Despite little sequence homology (49% identity), MKP1 and TMKP1 seem to act in an antagonistic manner to regulate salt stress responses, which might be explained by distinct subcellular localization and differential catalytic regulation by Ca^{2+} (Lee et al. 2008; Bartels et al. 2009; Zaïdi et al. 2010; Ghorbel et al. 2015). In response to tissue wounding, plant MKPs have been reported to be general negative regulators (Katou et al. 2007). There is also increasing evidence highlighting the importance of MKPs in osmotic stress signaling pathways (Kültz and Burg 1998; Jiang et al. 2018). Liu et al. (2015a, b) demonstrated that Arabidopsis PTP1 function as a negative regulator in response to osmotic stress during seed germination and seedling establishment (Liu et al. 2015a, b). Fujita et al. (2013) demonstrated that PHS1 is closely associated with a salt/osmotic stress-induced depolymerization of cortical microtubules (Fujita et al. 2013). Their research illustrated that tubulin kinase activity was suppressed by the phosphatase activity of PHS1 under normal growth conditions. At the same time, upon osmotic stress, such suppression was relieved, leading to the phosphorylation on Thr349 residue of α-tubulin, contributing to the formation of polymerization-inefficient tubulins (Fujita et al. 2013).

Zhou et al. (2017) demonstrated that Arabidopsis homologues of E3-histone monoubiquitination 1 (HUB1/H2Bub1) and 2 (HUB2/H2Bub1) play an important regulatory role in response to salt stress. At the same time, it was demonstrated that H2Bub1 regulates salt stress-induced MT depolymerization and the PTP-MPK3/MPK6 signaling. In this way, PTPs modulate integrating signaling by regulation of MT stability and plant salt stress tolerance (Zhou et al. 2017).

Recently, the effect of high temperatures on plant immunity was linked with dependence between high-temperature resistance and PTP signaling (Zhu et al. 2010; Hua 2013). SUPPRESSOR OF npr1-1, CONSTITUTIVE 1 (SNC1) is the first identified R gene mediating high temperature inhibition of resistance (Yang and Hua 2004), which is negatively regulated by BONZAI1 (BON1) (Zhu et al. 2010). At 22 °C, the *bon1-1* loss-of-function mutation activates SNC1, which induces constitutive salicylic acid (SA)-mediated defense responses and inhibits plant growth. At the same time, at 28 °C the nuclear accumulation of SNC1 is reduced by high temperature, which may inhibit the activity of this protein and suppress the defense responses (Zhu et al. 2010). Besides BON1, other negative regulators of SNC1 have been identified, such as MAP KINASE PHOSPHATASE 1 (MKP1) and others (Zou et al. 2014; Gou et al. 2012; Yaish 2017).

Another example of the role of tyrosine-dependent dephosphorylation is presented by the plasma membrane-localized clade E growth-regulating 2 phosphatase. EGR2 interacts with OST1 (open stomata 1/SNF1-Related protein kinase 2.6) and inhibits its activity under normal conditions. Normally, EGR2 is N-myristoylated by N-myristoyltransferase NMT1 at 22 °C, predetermining its interaction with OST1. Moreover, myristoylation is required for EGR2 function in plant freezing tolerance. Under cold stress, the interaction of EGR2 and NMT1 is attenuated, leading to the suppression of EGR2 myristoylation in plants. Therefore, mutations of EGRs cause plant tolerance to freezing, whereas overexpression of EGR2 exhibits decrease in freezing tolerance (Ding et al. 2015, 2019).

14.5 The Role of Protein Phosphatases in Cytoskeleton Regulation

Cytoskeleton coordinates basic physiological processes such as cell division, growth and differentiation, cell inner/outer motility, vesicle transport, cell support and scaffolding, polymer crosslinking, and membrane anchorage (Baluška et al. 2003; Foster et al. 2003; Klyachko 2005; Gonzalez-Quevedo et al. 2005; Soda et al. 2016). These integrated signaling networks undergo structural changes in plant responses to the internal and external cues and play a role of a trigger for stress tolerance/resistance to various abiotic (photoperiod, mechanical stimuli, temperature changes, etc.) (Nick 2008, 2013) and biotic (infection with viral, bacterial, or fungal pathogens) stresses (Schmelzer 2002; Takemoto and Hardham 2004; Kobayashi and Kobayashi 2007; Soda et al. 2016). The orientation of microtubules (MTs) and actin filaments (AFs) along the shoot/root main axis (transverse, oblique, longitudinal, randomized) and their organization (bundled, relaxed, loose, stabilized, fragmented, depolymerized) vary during cell division, elongation, and differentiation. Tissue "growth status," developmental stage, and progress in stress responses are reflected in MT orientation (Duckett and Lloyd 1994; Blume et al. 2016).

The expression of different isoforms and various post-translational modifications of microtubular and microfilament proteins are the basis of functional specialization and adaptation of these cytoskeletal structures (Janke 2014; Soda et al. 2016; Blume et al. 2016). These two factors underlie the functional heterogeneity in tubulin and actin cytoskeletons and were named "tubulin code" and "actin code," respectively (Janke 2014; Gadadhar et al. 2017; Blume et al. 2016; Vedula and Kashina 2018). Along with a number of other types of post-translational modifications, reversible phosphorylation is one of the predetermining factors of specialization and adaptation of these cytoskeleton structures, including the responses to the different types of stress (Gimona 2008; Blume et al. 2016; Soda et al. 2016). Thus, among the huge variety of plant protein kinases and plant protein phosphatases, there is a small and coordinated group of enzymes directly related to MT and AF functionalization (Karpov et al. 2010a, b; Karpov et al. 2014; Samofalova et al. 2019).

There is a certain relation between microtubule functioning and activity of Ser/Thr-specific protein phosphatases PP1, PP2A/PP2B, PP4 (PPX) and PP6 PP7 (Awotunde et al. 2003; Farkas et al. 2007; Moorhead et al. 2007; Blume et al. 2008a, b). Some results of *in vitro* testing demonstrate that protein phosphatases PP1, PP2A, and PP4 are able to directly dephosphorylate α-, β-, and γ-tubulins. At the same time, PP6 and PP7 regulate mitosis and cell cycle, but their role in direct interaction with tubulin is not clear (Kumar et al. 2004; Bollen et al. 2009; De Wulf et al. 2009; Zeng et al. 2010). Particular attention must be paid to a group of classical and dual-specificity tyrosine phosphatases (PTP1B, CDC25, PTPH1, PTRN11, PTRN13, PTP14, DSP-DEP1, DSP7, DSP14B) (Yang and Tonks 1991; Alonso et al. 2004; Cho et al. 2004, 2005; (Lindqvist et al. 2005; Liu et al. 2012; Trush et al. 2014). It has been shown that CDC25a and CDC25b activate cyclin-dependent protein kinases, which in turn control progression of mitosis. PTP1B promotes cell proliferation, and PTPN11 (SHP2) is supporting stability of chromosome organization (Baldin et al. 1997; Lindqvist et al. 2005; Liu et al. 2012; Trush et al. 2014). However, despite the obvious colocalization of the abovementioned protein phosphatases with tubulins, their individual functions, as well as their direct contribution in plant "tubulin code," remain controversial (Tournebize et al. 1997; Cho et al. 2005; Sines et al. 2007; Wang et al. 2010,).

After all, the *A. thaliana* phosphatome was studied analyzing the literature data with subsequent profile search. From this data, the Ser/Thr-specific group (PP1, PP2A, PP4, PP6, PP7), classical non-receptor tyrosine (PTPN1, PTPN3, PTPN11, PTPN13, PTPRJ), and dual protein phosphatases (CDC25, DUSP7, DSP14) are associated with dephosphorylation of the MTs of higher plants (Samofalova et al. 2015) (Table 14.2). In particular, this set of protein phosphatases was predicted as potentially capable of dephosphorylating of α-, β-, and γ-tubulin.

Also, the respective groups were determined in dicotyledons (*Nicotiana tabacum*, *Medicago sativa*) and monocotyledons (*Oryza sativa*, *Zea mays*, *Triticum aestivum*). In total, 151 plant protein phosphatases that potentially participate in the MT regulation were analyzed (Samofalova et al. 2019). Our results confirmed earlier conclusions about the differences in phosphatomes of different families of *Magnoliophyta* (Samofalova et al. 2019). For example, search results in databases of sequence homology and keywords revealed the absence of PP4-type phosphatase homologues from *Medicago sativa* (Samofalova et al. 2019).

For the reconstruction of the spatial structure, some plant MT protein phosphatases, closest homologues with the experimentally resolved spatial structure, were studied, and thus the significant conservation of the Ser/Thr-specific PP of plant and animal origin was confirmed. It was determined that for these types of plant PPs, the

Table 14.2 Plant MT protein phosphatases

	Organism (UniProtKB numbers of PPs)					
	A. thaliana	*N. tabacum*	*O. sativa*	*M. sativa*	*Z. mays*	*T. aestivum*
	Ident. Homologues PPs (%)					
Type of PPs		95–63	94–81	80–78	94–65	94–76
PP1	+	+	+	+	+	+
PP2A	+	+	+	+	+	+
PP4/PPX	+	+	+	–	+	+
PP6	+	+	+	–	+	+
PP7	+	–	+	–	+	+
PTPN1/PTP1B	+	–	+	–	+	+
PTPN3/PTPH1	+	–	+	–	+	+
PTPN11/PTP2C	+	–	+	–	+	+
PTPN13/PTP14	+	–	+	–	+	+
PTPRJ/D-DEP1	+	–	+	–	+	+
CDC25	+	–	+	–	+	+
DUSP7/DSP7	+	+	+	–	+	+
DSP14	+	+	+	–	+	+

type of folding (alpha+beta) was identical to their homologues from animals. It has been shown that despite the high level of heterogeneity of the primary structure, the physical and chemical properties of the protein phosphatases PP1, PP2, and PP4 coincide by the main indicators at the level of the entire group (Samofalova et al. 2019). At the same time, for all studied catalytic subunits PP1, PP2, and PP4, the characteristic conservation of the C-terminal region was confirmed, and the unique α-structure of the catalytic domain was found, typical for the entire group of serine/threonine-specific PPs. After the reconstruction of the spatial structure of plant protein phosphatases PP1, PP2A, and PP4, the comparison of the topology of their functional motifs, such as the binding sites of ATP (for ATP-Mg-dependent form), the site of interaction with the cofactors (metal ions) as well as the regions responsible for binding of inhibitors of the corresponding PPs, was carried out. Difference in globular patterns of plant protein phosphatases was observed only for individual amino acids of loop domains. In particular, histidine, which forms the active site of PP and is responsible for ATP binding in most plant PP1, has the conformation distinct from the marker PP1 from *A. thaliana* (Samofalova et al. 2011). However, it should be noted that this does not cause a significant violation of the binding site structure. The analysis of amino acids directly responsible for inhibitors binding revealed only replacements of individual residues: For example, Val instead of Ile in 143 position in PP1 from *A. thaliana* and *Z. mays*. In the case of PP4 from *Nicotiana tabacum*, there is tryptophan instead of Phe279, and in PP4 from *T. aestivum*, the residue of Arg109 is replaced with Ser (Samofalova et al. 2015, 2019).

The comparison results of the complete amino acid sequences of potential serine/threonine PPs (PP1, PP2A, PP4, PP6, PP7) imply a sufficiently high level of sequence identity—within the range of 76–95%. At the same time, an analysis of a group of non-receptor tyrosine phosphatases (PTPN1, PTPN3, PTPN11, PTPN13, PTPRJ) and dual-specificity PP homologues (CDC25, DUSP7, DSP14) revealed a much lower level of identity of plant homologues, ranging from 10 to 35% when compared only to their catalytic domains, high frequency of gaps and lack of experimental evidence of enzyme activity (Fig. 14.4). However, due to the low level of sequence similarity and insufficient structural data, the analysis and reconstruction of the spatial structures was impossible for PP6, PP7, PTPN1, PTPN3, PTPN11, PTPN13, PTPRJ, and CDC25.

The next stage of the study of binding mechanisms of inhibitors with plant protein phosphatases included the analysis of chemical databases. Finally, 231 inhibitors of Ser/Thr-specific protein phosphatases with proven biological activity were selected. The aforementioned sample was used as a control. Based on the 2D fingerprints and Tanimoto and Tversky indices (with the threshold of similarity 85%), 1105 compounds from PubChem, ChEMBL, and ZINK databases were selected for the test set. Among the test set, only 11 substances had biochemical confirmation of activity and the described mechanism of ligand-protein interaction, namely, the interaction of PP1 and PP2A protein phosphatases with okadaic acid and microcystin-LR; PP1 with microcystin-LA, calyculin A, nodularin, and tautomycin; PP2A with dinophysistoxins; and PP5 with cantharidin and endothall.

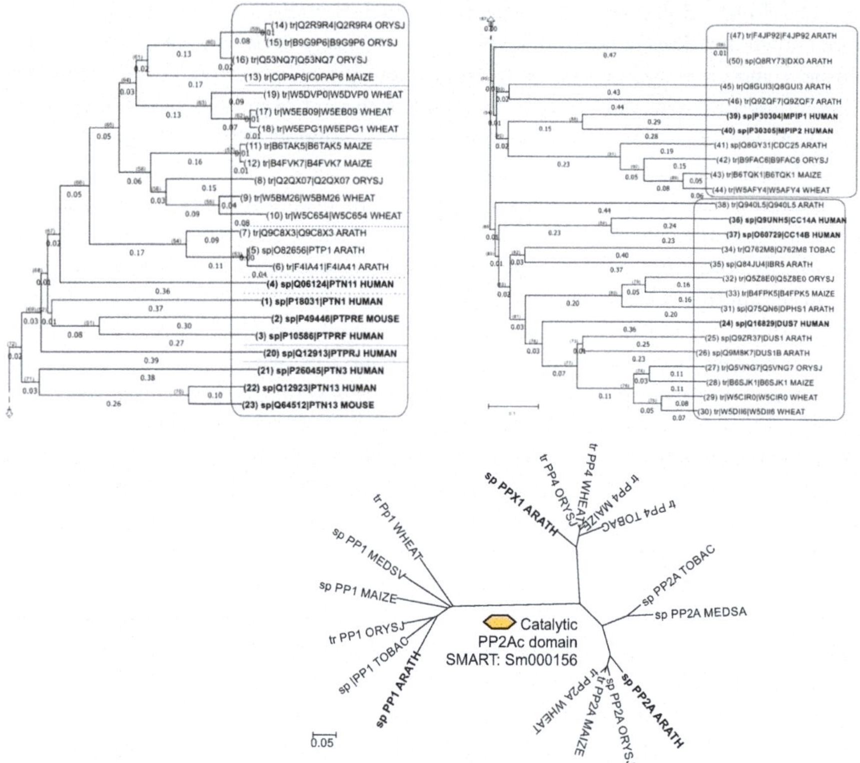

Fig. 14.4 Results of an NJ-clustering of potential plant homologues and annotated sequences of protein phosphatases associated with cytoskeleton regulation based on catalytic domain sequences of Tyr (**a**) and Ser/Thr (**b**) protein phosphatases

Subsequently, this information was used in the reconstruction and analysis of the interaction of these inhibitors with plant PPs. This stage included the profiling of the physical and chemical properties of complexes, the analysis of the ligand and protein interaction, to assess the existence of alternative targets for inhibitors in plants. It has been shown that the interaction of inhibitors with protein phosphatases occurs with the participation of 1–18 functionally significant water molecules. The exceptions were microcystins (LR and LA) that interact covalently with PP2A, as well as cantharidin and endothall that form coordination bonds with the active site metal ion in PP5, whereas in case of their absence during docking, they intercalate into molecules of protein phosphatases PP1, PP2A, and PP4. Based on the results of the comparison of PDB complexes of protein phosphatase PP1 with okadaic acid, microcystin-LR and LA, calyculin A, tautomycin, nodularin, and motuporin, such residues as Arg96, Tyr221, Tyr272, Val223, Asn224, and His225 were identified as the key amino acids for the binding site formation. In the case of protein phosphatase PP2A, which interacts with okadaic acid, microcystin-LR and dinophysistoxin of the types 1 and 2, the respective residues Arg89, Tyr272, Arg214, Ala216, Asn217, and His218 were determined. While for complexes of protein phosphatase type 5 with cantharidin and endothall, the conservative residues are Arg275, Tyr451, Arg400, Val402, Asn303, and His304.

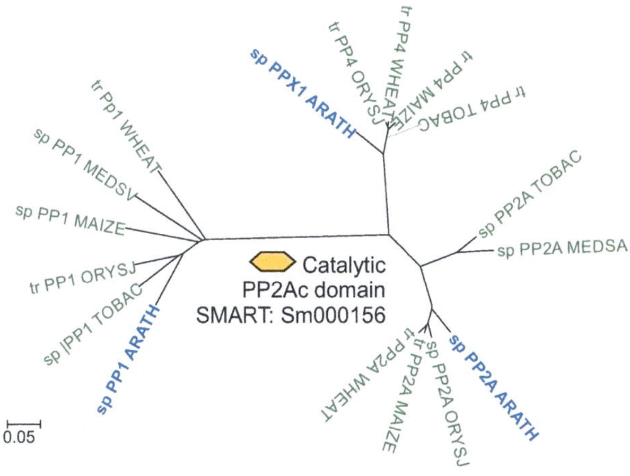

Fig. 14.4 (continued)

The implementation of the mentioned stages allowed us to work out the most convenient algorithm for the reconstruction of plant protein phosphatase complexes with potential inhibitors. Thus, by analyzing the results of molecular docking in the PP site and verification of the resulting complexes, we produced several assumptions. Unlike the derivatives of dinophysistoxins type 1 and 2, okadaic acid has a higher overall affinity for plant PPs. Regarding the different mode of action, the more significant indicators of binding affinity are typical to protein phosphatase PP2A than to protein phosphatases PP1 and PP4. Calyculin, which is similar to okadaic acid, ranks second in the range of evaluation functions of docking and also has a greater selectivity against protein phosphatase PP2A than to protein phosphatase PP1. Thus, compared to the standard protocol and molecular docking of calyculin and protein phosphatase PP4, the docking on the surface of protein phosphatase PP1 and further tests using the molecular dynamics method required parameter adjustment. Tautomycin has a little bit lower score of the docking evaluation function. However, it, in contrast to the first two leaders, has a greater selectivity to protein phosphatase PP1 than to protein phosphatases PP2A and PP4, which is completely in line with the literature data. It should be noted that, according to the profile analysis, okadaic acid, calyculin and tautomycin, which are selective inhibitors of protein phosphatases PP1 and PP2A, are also capable of inhibiting the activity of plant protein phosphatases PP3, PP5, and PP6. This is also confirmed by analysis of docking results and available literature data. Nodularin and motuporin, being conservative PP1 protein phosphatase inhibitors in plant cells, may decrease the activity of the protein phosphatase PP2A, to which they have greater selectivity than to protein phosphatases PP1 and PP4. Moreover, it has been found, by comparing the binding sites of protein phosphatase inhibitors, using profile techniques, that the inhibitors can suppress the activity of protein phosphatases PP5 and PP6. Concerning cantharidin and endothall, according to the profile analysis, it was assumed that they have low specificity and may act on almost all plant serine/threonine-specific protein phosphatases except the protein phosphatase PP7. However, these observations were derived and formulated by means of molecular modeling methods. We assume that derivatives of cantharidin reduce the activity of the protein phosphatase PP2A, which is in line with the literature data.

Thus, a group of plant protein phosphatases associated with the regulation of the structure and functions of microtubules was selected, and structural features of their interaction with specific inhibitors were determined. The obtained data improve the understanding of mechanisms that determine the selectivity of existing inhibitors to plant protein phosphatases and are important for the subsequent rational design of compounds with high affinity for these molecular targets and allow optimizing the process of their development.

14.6 Conclusion

As a result, many studies clearly demonstrate the key role of protein phosphatases in the response to stress. Despite the importance of protein phosphatases in stress responses that remain unknown, recent advances in their functional analysis revealed that they are key components of stress signal transduction pathways, balancing the action of protein kinases. In but unlike antagonists, in plant cell, protein phosphatases are acting separately and usually not combined in broad groups divided by the type of a stress factor (biotic or abiotic). According to experimental data, the role of plant phosphatases in stress is commonly associated with tyrosine protein phosphatases—PTP1, DSPsI, MKPs (Alonso et al. 2004; Awotunde et al. 2003; Bakrim et al. 1992; Bartels et al. 2009)—and Ser/Thr protein phosphatases of types PP1 (all isoforms), PP2 (A1, A2, A3) and PP2C (all isoforms) combined with different regulating subunits. The participation of the certain subunit is usually determined by the type of biotic stress factor. All these types of protein phosphatases were present in most of the studied plants, such as *A. thaliana*, *N. tabacum*, *M. sativa*, *O. sativa*, *Z. mays*, and *T. aestivum*, and had unique structures and functions, which is a consequence of the data from molecular phylogenetics and structural bioinformatics. First and foremost, it relates to the protein phosphatases involved in mitotic processes included in most signaling pathways and is associated with cytoskeleton regulation.

References

Alonso A, Sasin J, Bottini N et al (2004) Protein tyrosine phosphatases in the human genome. Cell 117(6):699–711

Awotunde OS, Lechward K, Krajewska K et al (2003) Interaction of maize (*Zea mays*) protein phosphatase 2A with tubulin. Acta Biochim Pol 50(1):131–138

Bakrim N, Echevarria C, Cretin C et al (1992) Regulatory phosphorylation of *Sorghum* leaf phosphoenolpyruvate carboxylase. Identification of the protein-serine kinase and some elements of the signal-transduction cascade. Eur J Biochem 204(2):821–830

Baldin V, Cans C, Superti-Furga G, Ducommun B (1997) Alternative splicing of the human CDC25B tyrosine phosphatase. Possible implications for growth control? Oncogene 14(20):2485–2495

Baluška F, Wojtaszek P, Volkmann D, Barlow P (2003) The architecture of polarized cell growth: the unique status of elongating plant cells. Bioessays 25(6):569–76. https://doi.org/10.1002/bies.10282

Bartels S, Anderson JC, González Besteiro MA et al (2009) MAP kinase phosphatase1 and protein tyrosine phosphatase1 are repressors of salicylic acid synthesis and SNC1-mediated responses in Arabidopsis. Plant Cell 21(9):2884–2897

Bartels S, González Besteiro MA, Lang D, Ulm R (2010) Emerging functions for plant MAP kinase phosphatases. Trends Plant Sci 15(6):322–329

Bhaskara GB, Wen TN, Nguyen TT, Verslues PE (2017) Protein phosphatase 2Cs and microtubule-associated stress protein 1 control microtubule stability, plant growth, and drought response. Plant Cell 29(1):169–191

Blume Y, Yemets A, Sulimenko V et al (2008b) Tyrosine phosphorylation of plant tubulin. Planta 229(1):143–150

Blume YB, Krasylenko YA, Yemets AI (2016) The role of the plant cytoskeleton in phytohormone signaling under abiotic and biotic stresses. In: Pandey GK (ed) Mechanism of plant hormone signaling under stress, vol 2. John Wiley and Sons, Inc., Hoboken, NJ, pp 127–185

Blume YB, Lloyd CW, Yemets AI (2008a) Plant tubulin phosphorylation and its role in cell cycle progression. In: Blume YB, Baird WV, Yemets AI, Breviario D (eds) The plant cytoskeleton: a key tool for agro-biotechnology. Springer, Dordrecht, pp 145–159

Bollen M, Gerlich DW, Lesage B (2009) Mitotic phosphatases: from entry guards to exit guides. Trends Cell Biol 19(10):531–541

Cho HP, Liu Y, Gomez M et al (2005) The dual-specificity phosphatase CDC14B bundles and stabilizes microtubules. Mol Cell Biol 25(11):4541–4551

Cho SH, Lee CH, Ahn Y et al (2004) Redox regulation of PTEN and protein tyrosine phosphatases in $H_{(2)}O_{(2)}$ mediated cell signaling. FEBS Lett 560(1–3):7–13

Chow EW, Clancey SA, Billmyre RB, Averette AF, Granek JA, Mieczkowski P, Cardenas ME, Heitman J (2017) Elucidation of the calcineurin-Crz1 stress response transcriptional network in the human fungal pathogen *Cryptococcus neoformans*. PLoS Genet 13(4):e1006667. https://doi.org/10.1371/journal.pgen.1006667

Cutler SR, Rodriguez PL, Finkelstein RR, Abrams SR (2010) Abscisic acid: emergence of a core signaling network. Annu Rev Plant Biol 61:651–679

De Wulf P, Montani F, Visintin R (2009) Protein phosphatases take the mitotic stage. Curr Opin Cell Biol 21(6):806–815

Ding Y, Li H, Zhang X et al (2015) OST1 kinase modulates freezing tolerance by enhancing ICE1 stability in Arabidopsis. Dev Cell 32(3):278–289

Ding Y, Lv J, Shi Y et al (2019) EGR2 phosphatase regulates OST1 kinase activity and freezing tolerance in Arabidopsis. EMBO J 38(1):e99819. https://doi.org/10.15252/embj.201899819

Duckett CM, Lloyd CW (1994) Gibberellic acid-induced microtubule reorientation indwarf peas is accompanied by rapid modification of anα-tubulin isotype. Plant J 5(3):363–372

Duerr B, Gawienowski M, Ropp T, Jacobs T (1993) MsERK1: a mitogen-activated protein kinase from a flowering plant. Plant Cell 5(1):87–96

Durian G, Rahikainen M, Alegre S et al (2016) Protein phosphatase 2A in the regulatory network underlying biotic stress resistance in plants. Front Plant Sci 7:812. https://doi.org/10.3389/fpls.2016.00812

Farkas I, Dombrádi V, Miskei M et al (2007) Arabidopsis PPP family of serine/threonine phosphatases. Trends Plant Sci 12(4):169–176

Feng J, Duan Y, Qin Y, Sun W, Zhuang Z, Zhu D, Jiang L (2017) The N-terminal pY33XL motif of CaPsy2 is critical for the function of protein phosphatase 4 in CaRad53 deactivation, DNA damage-induced filamentation and virulence in *Candida albicans*. Int J Med Microbiol 307(8):471–480

Fordham-Skelton AP, Skipsey M, Eveans IM et al (1999) Higher plant tyrosine-specific protein phosphatases (PTPs) contain novel amino-terminal domains: expression during embryogenesis. Plant Mol Biol 39(3):593–605

Foster R, Mattsson O, Mundy J (2003) Plants flex their skeletons. Trends Plant Sci 8(5):202–204

Fu C, Donadio N, Cardenas ME, Heitman J (2018) Dissecting the roles of the calcineurin pathway in unisexual reproduction, stress responses, and virulence in *Cryptococcus deneoformans*. Genetics 208(2):639–653

Fuchs S, Tischer SV, Wunsche C et al (2014) Abscisic acid sensor RCAR7/PYL13, specific regulator of protein phosphatase coreceptors. Proc Natl Acad Sci U S A 111(15):5741–5746

Fujita M, Fujita Y, Noutoshi Y et al (2006) Crosstalk between abiotic and biotic stress responses: a current view from the points of convergence in the stress signaling networks. Curr Opin Plant Biol 9(4):436–442

Fujita S, Pytela J, Hotta T et al (2013) An atypical tubulin kinase mediates stress-induced microtubule depolymerization in Arabidopsis. Curr Biol 23(20):1969–1978

Gadadhar S, Bodakuntla S, Natarajan K, Janke C (2017) The tubulin code at a glance. J Cell Sci 130(8):1347–1353

Ghelis T (2011) Signal processing by protein tyrosine phosphorylation in plants. Plant Sig Behav 6(7):942–951

Ghelis T, Bolbach G, Clodic G et al (2008) Protein tyrosine kinases and protein tyrosine phosphatases are involved in abscisic acid-dependent processes in Arabidopsis seeds and suspension cells. Plant Physiol 148(3):1668–1680

Ghorbel M, Zaidi I, Robe E et al (2015) The activity of the wheat MAP kinase phosphatase 1 is regulated by manganese and by calmodulin. Biochimie 108:13–19

Gimona M (2008) Protein linguistics and the modular code of the cytoskeleton. In: Barbieri M, Hoffmeyer J (eds) The codes of life. Biosemiotics, vol 1. Springer, Dordrecht, pp 189–206

Gonzalez-Quevedo R, Shoffer M, Horng L, Oro AE (2005) Receptor tyrosine phosphatase-dependent cytoskeletal remodeling by the hedgehog-responsive gene MIM/BEG4. J Cell Biol 168(3):453–463

González-Rubio G, Fernández-Acero T, Martín H, Molina M (2019) Mitogen-activated protein kinase phosphatases (MKPs) in fungal signaling: conservation, function, and regulation. Int J Mol Sci 20(7):E1709. https://doi.org/10.3390/ijms20071709

Goring DR, Glavin TL, Schafer U, Rothstein SJ (1993) An S receptor kinase gene in self-compatible *Brassica napus* has a 1-bp deletion. Plant Cell 5(5):531–539

Gou M, Shi Z, Zhu Y et al (2012) The F-box protein CPR1/CPR30 negatively regulates R protein SNC1 accumulation. Plant J 69(3):411–420

Gupta R, Huang Y, Kieber J, Luan S (1998) Identification of a dual-specificity protein phosphatase that inactivates a MAP kinase from Arabidopsis. Plant J 16(5):581–589

Hirayama T, Shinozaki K (2007) Perception and transduction of abscisic acid signals: keys to the function of the versatile plant hormone ABA. Trends Plant Sci 12(8):343–351

Ho HL (2015) Functional roles of plant protein kinases in signal transduction pathways during abiotic and biotic stress. J Biodivers Bioprospect Develop 2:147. https://doi.org/10.4172/2376-0214.1000147

Hua J (2013) Modulation of plant immunity by light, circadian rhythm, and temperature. Curr Opin Plant Biol 16(4):406–413

Huber SC, Huber JL (1991) Regulation of maize leaf sucrose-phosphate synthase by protein phosphorylation. Plant Cell Physiol 32(3):319–326

Hunter T (2009) Tyrosine phosphorylation: thirty years and counting. Curr Opin Cell Biol 21(2):140–146

Hunter T, Sefton BM (1980) Transforming gene product of Rous sarcoma virus phosphorylates tyrosine. Proc Natl Acad Sci U S A 77(3):1311–1315

Jaeger KE, Graf A, Wigge PA (2006) The control of flowering in time and space. J Exp Bot 57(13):3415–3418

Janke C (2014) The tubulin code: molecular components, readout mechanisms, and functions. J Cell Biol 206(4):461–472

Jenks MA, Wood AJ (2009) Genes for plant abiotic stress. John Wiley & Sons, New York, 344 p

Jiang L, Chen Y, Luo L, Peck SC (2018) Central roles and regulatory mechanisms of dual-specificity MAPK phosphatases in developmental and stress signaling. Front Plant Sci 9:1697. https://doi.org/10.3389/fpls.2018.01697

Jones C, Anderson S, Singha UK, Chaudhuri M (2008) Protein phosphatase 5 is required for Hsp90 function during proteotoxic stresses in *Trypanosoma brucei*. Parasitol Res 102(5):835–844

Karpov P, Raevsky A, Korablyov M, Blume Y (2014) Identification of plant homologues of dual specificity Yak1-related kinases. Computat Biol J 2014:909268. https://doi.org/10.1155/2014/909268

Karpov PA, Nadezhdina ES, Yemets AI, Blume YB (2010a) Results of the clusterization of human microtubule and cell cycle related serine/threonine protein kinases and their plant homologues. Mosc Univ Biol Sci Bull 65(4):213–216

Karpov PA, Nadezhdina ES, Yemets AI et al (2010b) Bioinformatic search of plant microtubule- and cell cycle related serine-threonine protein kinases. BMC Genomics 11(Suppl 1):S14. https://doi.org/10.1186/1471-2164-11-S1-S14

Katou S, Kuroda K, Seo S et al (2007) A calmodulin-binding mitogen-activated protein kinase phosphatase is induced by wounding and regulates the activities of stress-related mitogen-activated protein kinases in rice. Plant Cell Physiol 48(2):332–344

Kerk D, Templeton G, Moorhead GB (2008) Evolutionary radiation pattern of novel protein phosphatases revealed by analysis of protein data from the completely sequenced genomes of humans, green algae, and higher plants. Plant Physiol 146(2):351–367

Keyse SM, Emslie EA (1992) Oxidative stress and heat shock induce a human gene encoding a protein-tyrosine phosphatase. Nature 359(6396):644–646

Kirik A, Ehrhardt DW, Kirik V (2012) TONNEAU2/FASS regulates the geometry of microtubule nucleation and cortical array organization in interphase Arabidopsis cells. Plant Cell 24(3):1158–1170

Klyachko NL (2005) The cytoskeleton and intracellular motility in plants. Russ J Plant Physiol 52(5):700–708

Kobayashi I, Kobayashi Y (2007) Microtubules and pathogen defense. In: Nick P (ed) Plant microtubules. Springer, Heidelberg, pp 121–140

Kranner I, Minibayeva FV, Beckett RP, Seal CE (2010) What is stress? Concepts, definitions and applications in seed science. New Phytol 188(3):655–673

Kraus PR, Heitman J (2003) Coping with stress: calmodulin and calcineurin in model and pathogenic fungi. Biochem Biophys Res Commun 311(4):1151–1157

Kültz D, Burg M (1998) Evolution of osmotic stress signaling via MAP kinase cascades. J Exp Biol 201(Pt 22):3015–3021

Kumar R, Musiyenko A, Oldenburg A et al (2004) Post-translational generation of constitutively active cores from larger phosphatases in the malaria parasite, *Plasmodium falciparum*: implications for proteomics. BMC Mol Biol 5:6. https://doi.org/10.1186/1471-2199-5-6

Lahiri S, Banerjee S, Dutta T, Sengupta S, Dey S, Roy R, Sengupta D, Chattopadhyay K, Ghosh AK (2014) Enzymatic and regulatory attributes of trehalose-6-phosphate phosphatase from *Candida utilis* and its role during thermal stress. J Cell Physiol 229(9):1245–1255

Lee JS, Ellis BE (2007) Arabidopsis MAPK phosphatase 2 (MKP2) positively regulates oxidative stress tolerance and inactivates the MPK3 and MPK6 MAPKs. J Biol Chem 282(34):25020–25029. https://doi.org/10.1074/jbc.M701888200

Lee JS, Wang S, Sritubtim S et al (2009) Arabidopsis mitogen-activated protein kinase MPK12 interacts with the MAPK phosphatase IBR5 and regulates auxin signaling. Plant J 57(6):975–985

Lee K, Song EH, Kim HS et al (2008) Regulation of MAPK phosphatase 1 (AtMKP1) by calmodulin in Arabidopsis. J Biol Chem 283(35):23581–23588

Li H, Roux SJ (1992) Casein kinase II protein kinase is bound to lamina-matrix and phosphorylates lamin-like protein in isolated pea nuclei. Proc Natl Acad Sci U S A 89(18):8434–8438

Lichtenthaler HK (1996) Vegetation stress: an introduction to the stress concept in plants. J Plant Physiol 148(1–2):4–14
Lillo C, Kataya AR, Heidari B et al (2014) Protein phosphatases PP2A, PP4 and PP6: mediators and regulators in development and responses to environmental cues. Plant Cell Environ 37(12):2631–2648
Lindqvist A, Källström H, Lundgren A et al (2005) Cdc25B cooperates with Cdc25A to induce mitosis but has a unique role in activating cyclin B1-Cdk1 at the centrosome. J Cell Biol 171(1):35–45
Liu J, Yang H, Bao F et al (2015a) IBR5 modulates temperature-dependent, R protein CHS3-mediated defense responses in Arabidopsis. PLoS Genet 11(10):e1005584. https://doi.org/10.1371/journal.pgen.1005584
Liu R, Liu Y, Ye N et al (2015b) AtDsPTP1 acts as a negative regulator in osmotic stress signaling during Arabidopsis seed germination and seedling establishment. J Exp Bot 66(5):1339–1353
Liu X, Zheng H, Qu CK (2012) Protein tyrosine phosphatase Shp2 (Ptpn11) plays an important role in maintenance of chromosome stability. Cancer Res 72(20):5296–5306
Liu Y, Yan J, Qin Q, Zhang J, Chen Y, Zhao L, He K, Hou S (2020) Type one protein phosphatases (TOPPs) contribute to the plant defense response in Arabidopsis. J Integr Plant Biol 62(3):360–377. https://doi.org/10.1111/jipb.12845
Lu Y, Qin Y, Zhu D, Shan A, Feng J (2018) Identification and characterization of PP2C phosphatase SjPtc1 in *Schistosoma japonicum*. Parasitol Int 67(2):213–217
Luan S (2002) Tyrosine phosphorylation in plant cell signaling. Proc Natl Acad Sci U S A 99(18):11567–11569
Lumbreras V, Vilela B, Irar S et al (2010) MAPK phosphatase MKP2 mediates disease responses in Arabidopsis and functionally interacts with MPK3 and MPK6. Plant J 63(6):1017–1030
MacKintosh RW, Davies SP, Clarke PR et al (1992) Evidence for a protein kinase cascade in higher plants: 3-Hydroxy-3-methylglutaryl-CoA reductase kinase. Eur J Biochem 209(3):923–931
MacRobbie EA (2002) Evidence for a role for protein tyrosine phosphatase in the control of ion release from the guard cell vacuole in stomatal closure. Proc Natl Acad Sci U S A 99(18):11963–11968
Martínez-Esparza M, Martínez-Vicente E, González-Párraga P, Ros JM, García-Peñarrubia P, Argüelles JC (2009) Role of trehalose-6P phosphatase (TPS2) in stress tolerance and resistance to macrophage killing in *Candida albicans*. Int J Med Microbiol 299(6):453–464
Máthé C, Garda T, Freytag C, M-Hamvas M (2019) The role of serine-threonine protein phosphatase pp2a in plant oxidative stress signaling—facts and hypotheses. Int J Mol Sci 20(12):E3028. https://doi.org/10.3390/ijms20123028
McCurdy DW, Harmon AC (1992) Calcium-dependent protein kinase in the green alga Chara. Planta 188(1):54–61
Melcher K, Xu Y, Ng LM et al (2010) Identification and mechanism of ABA receptor antagonism. Nat Struct Mol Biol 17(9):1102–1108
Mizuno T, Nakamura M, Irie K (2018) Induction of Ptp2 and Cmp2 protein phosphatases is crucial for the adaptive response to ER stress in *Saccharomyces cerevisiae*. Sci Rep 8(1):13078. https://doi.org/10.1038/s41598-018-31413-6
Monroe-Augustus M, Zolman BK, Bartel B (2003) IBR5, a dual-specificity phosphatase-like protein modulating auxin and abscisic acid responsiveness in Arabidopsis. Plant Cell 15(12):2979–2991
Moorhead GB, De Wever V, Templeton G, Kerk D (2009) Evolution of protein phosphatases in plants and animals. Biochem J 417(2):401–409
Moorhead GB, Trinkle-Mulcahy L, Ulke-Lemée A (2007) Emerging roles of nuclear protein phosphatases. Nat Rev Mol Cell Biol 8(3):234–244
Mosa KA, Ismail A, Helmy M (2017) Plant stress tolerance. In: Springer briefs in systems biology. Springer, Cham. https://doi.org/10.1007/978-3-319-59379-1

Muszkieta L, Carrion Sde J, Robinet P, Beau R, Elbim C, Pearlman E, Latgé JP (2014) The protein phosphatase PhzA of A. fumigatus is involved in oxidative stress tolerance and fungal virulence. Fungal Genet Biol 66:79–85

Nick P (2008) Microtubules as sensors for abiotic stimuli. In: Nick P (ed) Plant microtubules. Springer, Heidelberg, pp 175–203

Nick P (2013) Microtubules, signaling and abiotic stress. Plant J 75(2):309–323

Norris-Mullins B, Krivda JS, Smith KL, Ferrell MJ, Morales MA (2018) Leishmania phosphatase PP5 is a regulator of HSP83 phosphorylation and essential for parasite pathogenicity. Parasitol Res 117(9):2971–2985

Olsen JV, Blagoev B, Gnad F, Macek B et al (2006) Global, *in vivo*, and site-specific phosphorylation dynamics in signaling networks. Cell 127(3):635–648

Opdenakker K, Remans T, Vangronsveld J, Cuypers A (2012) Mitogen-activated protein (MAP) kinases in plant metal stress: regulation and responses in comparison to other biotic and abiotic stresses. Int J Mol Sci 13(6):7828–7853

País SM, Téllez-Iñón MT, Capiati DA (2009) Serine/threonine protein phosphatases type 2A and their roles in stress signaling. Plant Signal Behav 4(11):1013–1015

Park HS, Chow EW, Fu C, Soderblom EJ, Moseley MA, Heitman J, Cardenas ME (2016) Calcineurin targets involved in stress survival and fungal virulence. PLoS Pathog 12(9):e1005873. https://doi.org/10.1371/journal.ppat.1005873

Peti W, Nairn AC, Page R (2013) Structural basis for protein phosphatase 1 regulation and specificity. FEBS J 280(2):596–611

Qin L, Yang Y, Yang Y, Zhu N, Gong Y, Zhang C, Li S, Liao D (2014) Ezetimibe suppresses cholesterol accumulation in lipid-loaded vascular smooth muscle cells *in vitro* via MAPK signaling. Acta Pharmacol Sin 35(9):1129–1136. https://doi.org/10.1038/aps.2014.10

Quettier AL, Bertrand C, Habricot Y et al (2006) The phs1-3 mutation in a putative dual-specificity protein tyrosine phosphatase gene provokes hypersensitive responses to abscisic acid in Arabidopsis *thaliana*. Plant J 47(5):711–719

Rahikainen M, Pascual J, Alegre S et al (2016) PP2A phosphatase as a regulator of ROS signaling in plants. Antioxidants 5(1):E8. https://doi.org/10.3390/antiox5010008

Reiter W, Klopf E, De Wever V, Anrather D, Petryshyn A, Roetzer A, Niederacher G, Roitinger E, Dohnal I, Görner W, Mechtler K, Brocard C, Schüller C, Ammerer G (2013) Yeast protein phosphatase 2A-Cdc55 regulates the transcriptional response to hyperosmolarity stress by regulating Msn2 and Msn4 chromatin recruitment. Mol Cell Biol 33(5):1057–1072

Rigoulot SB, Petzold HE, Williams SP et al (2019) *Populus trichocarpa* clade A PP2C protein phosphatases: their stress-induced expression patterns, interactions in core abscisic acid signaling, and potential for regulation of growth and development. Plant Mol Biol 100(3):303–317

Romá-Mateo C, Sacristán-Reviriego A, Beresford NJ, Caparrós-Martín JA, Culiáñez-Macià FA, Martín H, Molina M, Tabernero L, Pulido R (2011) Phylogenetic and genetic linkage between novel atypical dual-specificity phosphatases from non-metazoan organisms. Mol Gen Genomics 285(4):341–354

Rudrabhatla P, Rajasekharan R (2002) Developmentally regulated dual-specificity kinase from peanut that is induced by abiotic stresses. Plant Physiol 130(1):380–390

Rudrabhatla P, Reddy MM, Rajasekharan R (2006) Genome-wide analysis and experimentation of plant serine/threonine/tyrosine-specific protein kinases. Plant Mol Biol 60(2):293–319

Samofalova DA, Karpov PA, Blume YB (2015) Bioinformatic comparison of human and higher plant phosphatomes. Cytol Genet 49(4):207–219

Samofalova DO, Karpov PA, Nyporko AY, Blume YB (2011) Reconstruction of the spatial structure of plant phosphatases types 1 and 2a in complexes with okadaic acid. Cytol Genet 45(3):153–162

Samofalova DO, Karpov PA, Raevsky AV, Blume YB (2019) Protein phosphatases potentially associated with regulation of microtubules, their spatial structure reconstruction and analysis. Cell Biol Int 43(9):1081–1090

Santhanam A, Hartley A, Düvel K, Broach JR, Garrett S (2004) PP2A phosphatase activity is required for stress and Tor kinase regulation of yeast stress response factor Msn2p. Eukaryot Cell 3(5):1261–1271
Schmelzer E (2002) Cell polarization, a crucial process in fungal defense. Trends Plant Sci 7(5):411–415
Seo S, Katou S, Seto H et al (2007) The mitogen-activated protein kinases WIPK and SIPK regulate the levels of jasmonic and salicylic acids in wounded tobacco plants. Plant J 49(5):899–909
Sheremet YA, Yemets AI, Blume YB (2012) Inhibitors of tyrosine kinases and phosphatases as a tool for the investigation of microtubule role in plant cold response. Cytol Genet 46(1):1–8
Shi WL, Liu X, Jia WS, Zhang SQ (2005) Protein tyrosine phosphatases mediate the signaling pathway of stomatal closure in *Vicia faba* L. J Integr Plant Biol 47(3):319–326
Sines T, Granot-Attas S, Weisman-Welcher S, Elson A (2007) Association of tyrosine phosphatase epsilon with microtubules inhibits phosphatase activity and is regulated by the epidermal growth factor receptor. Mol Cell Biol 27(20):7102–7112
Soda N, Singla-Pareek SL, Pareek A (2016) Abiotic stress response in plants: role of cytoskeleton. In: Tuteja N, Gill SS (eds) Abiotic stress response in plants. John Wiley and Sons, Inc., Hoboken, NJ, pp 107–134
Sugimoto H, Kondo S, Tanaka T et al (2014) Overexpression of a novel Arabidopsis PP2C isoform, AtPP2CF1, enhances plant biomass production by increasing inflorescence stem growth. J Exp Bot 65(18):5385–5400
Sun H, Charles CH, Lau LF, Tonks NK (1993) MKP1(3CH134), an immediate early gene product, is a dual specificity phosphatase that dephosphorylates MAP kinase *in vivo*. Cell 75(3):487–493
Takemoto D, Hardham AR (2004) The cytoskeleton as a regulator and target of biotic interactions in plants. Plant Physiol 136(4):3864–3876
Tamnanloo F, Damen H, Jangra R, Lee JS (2018) MAP KINASE PHOSPHATASE1 controls cell fate transition during stomatal development. Plant Physiol 178(1):247–257
Tamura S, Hanada M, Ohnishi M et al (2002) Regulation of stress-activated protein kinase signaling pathways by protein phosphatases. Eur J Biochem 269(4):1060–1066
Tominga Y, Wayne R, Tung HYL, Tazawa M (1987) Phosphorylation-dephosphorylation is involved in Ca^{2+}-controlled cytoplasmic streaming in Characean cells. Protoplasma 136(2-3):161–169
Tournebize R, Andersen SSL, Verde F et al (1997) Distinct roles of PP1 and PP2A-like phosphatases in control of microtubule dynamics during mitosis. EMBO J 16(18):5537–5549
Tran HT, Nimick M, Uhrig RG et al (2012) Arabidopsis *thaliana* histone deacetylase 14 (HDA14) is an α-tubulin deacetylase that associates with PP2A and enriches in the microtubule fraction with the putative histone acetyltransferase ELP3. Plant J 71(2):263–272
Trush VV, Tanchuk VY, Cherenok SO et al (2014) Calix[4]arene α-hydroxymethylphosphonic acids as potential inhibitors of protein tyrosine phosphatases. J Org Pharmceut Chem 45(1):39–42
Uhrig RG, Labandera AM, Moorhead GB (2013) Arabidopsis PPP family of serine/threonine protein phosphatases: many targets but few engines. Trends Plant Sci 18(9):505–513
Ulm R, Ichimura K, Mizoguchi T et al (2002) Distinct regulation of salinity and genotoxic stress responses by Arabidopsis MAP kinase phosphatase 1. EMBO J 21(23):6483–6493
Ulm R, Revenkova E, di Sansebastiano GP et al (2001) Mitogen-activated protein kinase phosphatase is required for genotoxic stress relief in Arabidopsis. Genes Dev 15(6):699–709
Umezawa T, Nakashima K, Miyakawa T et al (2010) Molecular basis of the core regulatory network in ABA responses: sensing, signaling and transport. Plant Cell Physiol 51(11):1821–1839
Vedula P, Kashina A (2018) The makings of the 'actin code': regulation of actin's biological-function at the amino acid and nucleotide level. J Cell Sci 131(9):jcs215509. https://doi.org/10.1242/jcs.215509
Virshup DM, Shenolikar S (2009) From promiscuity to precision: protein phosphatases get a makeover. Mol Cell 33(5):537–545
Walia A, Lee JS, Wasteneys G, Ellis B (2009) Arabidopsis mitogen-activated protein kinase MPK18 mediates cortical microtubule functions in plant cells. Plant J 59(4):565–575

Wang R, Kaul A, Sperry AO (2010) TLRR (lrrc67) interacts with PP1 and is associated with a cytoskeletal complex in the testis. Biol Cell 102(3):173–189

Wang S, Liu X, Qian H, Zhang S, Lu L (2016) Calcineurin and calcium channel CchA coordinate the salt stress response by regulating cytoplasmic Ca^{2+} homeostasis in *Aspergillus nidulans*. Appl Environ Microbiol 82(11):3420–3430

Wang Z, Ji H, Yuan B et al (2015) ABA signaling is fine-tuned by antagonistic HAB1 variants. Nat Commun 6:8138. https://doi.org/10.1038/ncomms9138

Ward Y, Gupta S, Jensen P et al (1994) Control of MAP kinase activation by the mitogen induced threonine/tyrosine phosphatase PAC-1. Nature 367(6464):651–654

Wong MM, Bhaskara GB, Wen T-N et al (2019) Phosphoproteomics of Arabidopsis highly ABA-induced1 identifies AT-Hook–Like10 phosphorylation required for stress growth regulation. Proc Natl Acad Sci U S A 116(6):2354–2363

Xu Q, Fu HH, Gupta R, Luan S (1998) Molecular characterization of a tyrosine-specific protein phosphatase encoded by a stress-responsive gene in Arabidopsis. Plant Cell 10(5):849–857

Yaish MW (2017) Epigenetic modifications associated with abiotic and biotic stresses in plants: an implication for understanding plant evolution. Front Plant Sci 8:1983. https://doi.org/10.3389/fpls.2017.01983

Yang Q, Jiang J, Mayr C, Hahn M, Ma Z (2013a) Involvement of two type 2C protein phosphatases BcPtc1 and BcPtc3 in the regulation of multiple stress tolerance and virulence of *Botrytis cinerea*. Environ Microbiol 15(10):2696–2711

Yang Q, Tonks NK (1991) Isolation of a cDNA clone encoding a human protein-tyrosine phosphatase with homology to the cytoskeletal-associated proteins band 4,1, ezrin and talin. Proc Natl Acad Sci U S A 88(14):5949–5953

Yang Q, Yu F, Yin Y, Ma Z (2013b) Involvement of protein tyrosine phosphatases BcPtpA and BcPtpB in regulation of vegetative development, virulence and multi-stress tolerance in *Botrytis cinerea*. PLoS One 8(4):e61307

Yang S, Hua J (2004) A haplotype-specific resistance gene regulated by *BONZAI1* mediates temperature-dependent growth control in Arabidopsis. Plant Cell 16(4):1060–1071

Yemets A, Sheremet Y, Vissenberg K et al (2008) Effects of tyrosine kinase and phosphatase inhibitors on microtubules in Arabidopsis root cells. Cell Biol Int 32(6):630–637

Yoon JT, Ahn HK, Pai HS (2018) The subfamily II catalytic subunits of protein phosphatase 2A (PP2A) are involved in cortical microtubule organization. Planta 248(6):1551–1567

Yoshida R, Hobo T, Ichimura K et al (2002) ABA-activated SnRK2 protein kinase is required for dehydration stress signaling in Arabidopsis. Plant Cell Physiol 43(12):1473–1483

Zaïdi I, Ebel C, Touzri M et al (2010) TMKP1 is a novel wheat stress responsive MAP kinase phosphatase localized in the nucleus. Plant Mol Biol 73(3):325–338

Zeng K, Bastos RN, Barr FA, Gruneberg U (2010) Protein phosphatase 6 regulates mitotic spindle formation by controlling the T-loop phosphorylation state of Aurora A bound to its activator TPX2. J Cell Biol 191(7):1315–1332

Zhang M, Joyce BR, Sullivan WJ Jr, Nussenzweig V (2013) Translational control in plasmodium and toxoplasma parasites. Eukaryot Cell 12(2):161–167

Zhang S (2008) Mitogen-activated protein kinase cascades in plant intracellular signaling. Ann Plant Rev 33:100–136

Zhou S, Chen Q, Sun Y, Li Y (2017) Histone H2B monoubiquitination regulates salt stress-induced microtubule depolymerization in Arabidopsis. Plant Cell Environ 40(8):1512–1530

Zhu Y, Qian W, Hua J (2010) Temperature modulates plant defense responses through NB-LRR proteins. PLoS Pathogens 6(4):e1000844. https://doi.org/10.1371/journal.ppat.1000844

Zou B, Yang DL, Shi Z et al (2014) Monoubiquitination of histone 2B at the disease resistance gene locus regulates its expression and impacts immune responses in Arabidopsis. Plant Physiol 165(1):309–318

Chapter 15
Protein Phosphatase Mediated Responses in Plant Host-Pathogen Interactions

Anjan Barman and Suvendra Kumar Ray

15.1 Prologue

Autotrophic green plants are generally self-sufficient organisms, which lead a complete sessile life (Li and Cui 2014; Žádníková et al. 2015). Owing to their inherent ability of synthesizing and storing energy-rich biomolecules, they attract variety of heterotrophs (that include prokaryotic microbes, protozoans as well as metazoans). Inability to move and nutrient richness do pose some disadvantages as it makes plants readily vulnerable to both abiotic and biotic agents (Raghavendra et al. 2010; Macho and Zipfel 2014). To sustain and grow amidst clueless vulnerabilities posed by altered environmental perturbations, biotic interventions, etc. necessitate prolific tenacity and endurance. Propitiously, plants have been bestowed with abilities to sense and respond to these cues elegantly and thrive with much resilience. Presence of efficient but complicated cascades of evolutionarily conserved signaling networks involving specialized molecular arsenals that remain active both on surface and inside the cell is the primary basis of plant's competence (Jones and Dangl 2006; Dodds and Rathjen 2010; Schwessinger and Ronald 2012).

Apart from other opposing factors, negative biotic interventions (e.g., due to pathogens, parasites, herbivores) pose significant threat to plant's survivability and productivity (Maron and Crone 2006; Maron and Kauffman 2006; Brown and Hovmøller 2002; Mordecai 2011). For instance, microbial pathogens comprising of viruses, bacteria, fungi, and nematodes affect plant health significantly (Williamson and Gleason 2003; Berger et al. 2007; Scholthof et al. 2011; Dean et al. 2012; Mansfield et al. 2012; Boyd et al. 2013; Malcolm et al. 2013). The extent of distress caused by pathogens is, however, resultant of plant's inherent capacity (conferred

A. Barman (✉)
Department of Biotechnology, Pandu College, Guwahati, Assam, India

S. K. Ray
Department of Molecular Biology & Biotechnology, Tezpur University, Tezpur, Assam, India

G. K. Pandey (ed.), *Protein Phosphatases and Stress Management in Plants*,
https://doi.org/10.1007/978-3-030-48733-1_15

by plant's innate immunity) to circumvent the former's assault (Veronese et al. 2003; de Wit 2007; Dodds and Rathjen 2010). In a way, it's similar to "tug-of-war" episodes between the pathogen and the plant's defense ploys where victory may be on either side (Boller and He 2009). Plants explicitly lack a circulatory immune system comprising of specialized cells to fend themselves, which is unlike in higher eukaryotic organisms, e.g., mammals (Macho and Zipfel 2014; Couto and Zipfel 2016). Nevertheless, adroit innate immune responses initiating at their cell surfaces significantly guard plants from numerous invading pathogens (Dodds and Rathjen 2010). Interestingly, we have just begun to unravel fragments of such a dexterous mechanism awake in plant's defense.

Innate immune responses are clearly eloquent signaling networks that initiate at plant's cell surfaces (Faulkner and Robatzek 2012; Macho and Zipfel 2014; Mithoe and Menke 2018). These responses may proceed *via* two distinct branches, i.e., PAMP (pathogen-associated molecular patterns)-triggered immunity (PTI) and effector-triggered immunity (ETI), depending on the nature of pathogenic elicitor/factors involved (Muthamilarasan and Prasad 2013; Thomma et al. 2011; Macho and Zipfel 2014). At the initial phases, PAMP (or MAMP) and effector-mediated elicitations are said to have distinct differences (Jones and Dangl 2006; Tsuda and Katagiri 2010; Zhang et al. 2010; Peng et al. 2018); however, this dichotomy may not be so rigid due to considerable resemblances in both (Thomma et al. 2011; Peng et al. 2018; Adachi and Tsuda 2019; Kadota et al. 2019). Importantly, downstream events of PTI and ETI do converge at some points resulting in identical defense responses (Navarro et al. 2004; Peng et al. 2018; Kadota et al. 2019). The signals of pathogenic origin are perceived by host of receptor molecules, called pattern recognition receptors (PRRs), spanning the plant cell membranes, which have an extracellular domain and a transmembrane domain and may have an inner cytoplasmic domain (Macho and Zipfel 2014; Saijo et al. 2018). Extracellular domain of the receptors can recognize specific conserved molecular patterns present on microbial cell surfaces (called microbe-associated molecular patterns, MAMPs or PAMPs) (Felix et al. 1993, 1999; Newman et al. 1995; Meyer et al. 2001; Kunze et al. 2004; Gust et al. 2007; Miya et al. 2007) or fragmented components of pathogenic origin, pathogen-driven wounds of the plant itself (also called damage-associated molecular patterns) (Matzinger 2002; Boller and Felix 2009). Extracellular domains of the receptors can transmit recognition signals through cytoplasmic domains or through other transitional mediators into interior of the cell (Zipfel 2009; Monaghan and Zipfel 2012). Cytoplasmic domains of PRRs are generally associated with kinase functions as they phosphorylate target sites on signal transducer molecules downstream, before immunogenic responses emanate (Becraft 2002; Monaghan and Zipfel 2012). All PRRs may not possess a kinase domain; PRRs with associated cytoplasmic kinase domains are called Receptor-Like Kinases (RLKs), and those lacking a kinase domain are referred to as Receptor-Like Proteins (RLPs) (Couto and Zipfel 2016; Boutrot and Zipfel 2017). RLPs are said to rely on RLKs for intracellular signaling as they are devoid of own kinase domains (Tor et al. 2009).

Phosphorylation events mediated by protein kinases (PKs) are indispensable for triggering plant immune responses along with several other physiological processes in plants (Luan 1998; Becraft 2002; Tena et al. 2011; Xu and Zhang 2015). PKs generally phosphorylate hydroxyl groups on different amino acids in downstream signal proteins, although some PKs may need to undergo autophosphorylations for their own activation (Johnson et al. 1996). PKs associated with PRRs of plants and animals are largely members of non-RD kinase family which may not require autophosphorylation for their activation (Dardick and Ronald 2006). RD refers to amino acids arginine (R) and aspartate (D), respectively. Notably, depending on the presence or absence of arginine residue before conserved aspartate group in their catalytic groove, PKs are categorized into RD and non-RD types, respectively. Findings indicate arginine residue preceding aspartate, facilitate proper phosphotransfer (Dardick and Ronald 2006). In eukaryotes, serine, threonine and tyrosine residues are preferred targets of PK aided phosphorylations (Luan 2003; Jha et al. 2017). PKs are classified according to specific amino acid(s) residues they phosphorylate, i.e., serine/threonine kinases (add phosphates at serine and threonine groups), tyrosine kinases (add phosphates at tyrosine residues), and dual-specificity kinases (phosphorylate at both serine/threonine and tyrosine moieties) (Luan 2003; Jha et al. 2017). During PRR stimulation by PAMP/effectors, activated kinases phosphorylate at specific sites of signal relay proteins, thereby inducing downstream signaling events (Chinchilla et al. 2007; Zipfel 2009; Roux et al. 2011). This may follow recruitment of cellular responses releasing defense-related compounds, antimicrobials for arresting pathogenic progression, with concomitant inhibition of plant growth, sometimes leading to programmed cell death (Luan 1998; Taguchi et al. 2003; Chinchilla et al. 2007; Naito et al. 2008; Park et al. 2008). The arrest of plant growth and cell death is explained by the fact that there is significant overlap amidst the signaling machinery involved in defense as well as many physiological responses, and as such continued inappropriate trigger of defense responses can lead to deleterious consequences (Lang and Mansell 2007; Park et al. 2008; Segonzac et al. 2014). Occasionally, over-intensification of such defense responses may induce characteristic hypersensitive reactions (HR) in plants, manifested by irreversible regressive changes in cellular integrity (Morel and Dangl 1997; Coll et al. 2011; Schwessinger et al. 2011; Camagna and Takemoto 2018). In this context, it is extremely important that during defense responses, kinase-mediated phosphorylations are strictly regulated (Park et al. 2008; Virshup and Shenolikar 2009; Segonzac et al. 2014; Couto et al. 2016; Withers and Dong 2017).

Removal of phosphate moieties is an important means for regulation of kinase-activated immune triggers (Keyse 2000; Liu et al. 2007; Park et al. 2008). A class of enzymes, called protein phosphatases (PPs), performs dephosphorylation of signal relay proteins and hence regulates kinase activities, thus interrupting immune signaling in plants (Gómez-Gómez et al. 2001; Couto et al. 2016; Rahikainen et al. 2016; Shubchynskyy et al. 2017). Protein phosphatases are prevalent across all domains of life (Luan 1998; Keyse 2000; Kennelly 2002; Shi 2009) and possess intrinsic capacity to counteract kinase functions by dephosphorylating at specific target sites (Luan 2003; Bartels et al. 2010; Uhrig et al. 2013). Protein phosphatase functions are recognized as crucial for regulation of several signaling cascades involved in physiological as well as developmental pathways in living organisms

(Shi 2009; Virshup and Shenolikar 2009; Chen et al. 2017). Their pivotal roles in regulation of plant innate immune responses have been reinforced by several reports in recent times (Gómez-Gómez et al. 2001; Park et al. 2008; Segonzac et al. 2014; Shubchynskyy et al. 2017; Liu et al. 2018).

Plant protein phosphatases (plant PPs) being an inbuilt regulatory component play decisive role in culmination of plant immune responses stimulated by protein kinases (Schweighofer and Meskiene 2015). The precision with which kinase activities are impeded by plant PPs decides duration of immune reaction and hence degree of injuries caused to the plant tissues (Park et al. 2008; Segonzac et al. 2014). Inactivation of PRR signaling in the absence of relevant elicitors is another important task carried out by plant PPs (Couto et al. 2016; Liu et al. 2018). Plant PP functions in turn may be regulated by certain known or unknown plant metabolites during immune responses. For instance, signaling cascades of abscisic acid, jasmonic acid and salicylic acid were shown to influence specific phosphatase activities in different plants earlier (Kenton et al. 1999; Lorenzo et al. 2001; Wang et al. 2012) indicating intriguing crosstalks. Comprehensive studies have revealed high specificity of phosphatases for substrate recognition (Schweighofer and Meskiene 2015) which might explain for their precise regulatory role. Further, inhibitions or defects in PP activities have manifested conspicuous effects on plant's defense as well as survivability (Luan 1998; Park et al. 2008; Wang et al. 2012). Considering these, relevance of plant PPs in plant defense programming impels for a close glance.

Reported studies on phytopathogenic microbe interaction with plant hosts have highlighted decisive role of plant PPs in immune signaling. Few notable illustrations are representative Arabidopsis *thaliana* PPs involved in the former's defense responses against *Pseudomonas syringae* (Couto et al. 2016; Shubchynskyy et al. 2017), *Botrytis cinerea* (Schweighofer et al. 2007), *Golovinomyces* spp. (Wang et al. 2012), and *Heterodera schachtii* (Sidonskaya et al. 2016); *Oryza sativa* PP responding to *Xanthomonas oryzae* (Park et al. 2008); etc. In the same manner, protein phosphatase (PP) equivalents in pathogenic microbes may play crucial role in aiding their successful interaction with respective host (Underwood et al. 2007; Jiang et al. 2011; Macho et al. 2014; Liu et al. 2016; Yang et al. 2018). Instances of pathogenic PP equivalents targeting plant immune signaling network have been evident from different studies (Espinosa et al. 2003; Macho et al. 2014). The intriguing interplay between components of pathogen origin and plant immune complexes with necessary involvement of PPs highlights a sophisticated control mechanism of plant immune system. Further, surfacing innovations like PPs in both host and pathogen reinstate dynamic paradigms in these interaction processes, providing new grounds for scientific explorations.

In view of emerging significances of plant PPs, in this chapter, we focus our discussion on important features of known plant protein phosphatases with a primary emphasis on their roles in the plant-microbial pathogen interaction course. We are also stating about PP equivalents in few known pathogen species and citing our present understandings on involvement of these PP equivalents in plant host-pathogen communications.

15.2 Protein Phosphatases in Plant

Reversible phosphorylation events are central to numerous cellular signal transduction processes including immune signaling in plants. PKs and PPs are indispensable facets of these processes. Genome sequencing initiatives have afforded significant details of protein kinases (Kinome) as well as PPs in several plant species (Krupa et al. 2006; Rudrabhatla et al. 2006; Wang et al. 2007; Schweighofer and Meskiene 2015). In Arabidopsis *thaliana* itself, nearly 112 genes encode PPs, whereas for PKs, there are above 1000 genes (Kerk et al. 2002; Wang et al. 2003). There appears distinct variation amidst numbers of kinases as against phosphatases in plants as in case of other eukaryotes (Wang et al. 2007; Shi 2009; Smoly et al. 2017). Also, PPs have been found to be far more diverse in terms of structural resemblances, catalytic properties, substrate specificities, etc. than that of PKs (Luan 2003).

15.2.1 Types, Key Features

PPs are classified into three major families according to target amino acid moieties they dephosphorylate: serine/threonine-specific phosphoprotein phosphatase [removes phosphates from Ser/Thr residues], phosphotyrosine phosphatase (PTP) [removes phosphates from Tyr residues], and dual-specificity phosphatase (DSP) [removes phosphates from Ser/Thr and Tyr residues] (Luan 2003; Schweighofer and Meskiene 2015). Further categorization of PPs into subfamilies can be on the basis of occurrence of peculiar regulatory and functional domains or accessory regulatory subunits associated with their catalytic sites (Luan 2003). In fact, the numerous regulators that associate with these enzymes also serve as crucial determinants for their subcellular localization, substrate specificity, and alteration of phosphatase activity (Virshup and Shenolikar 2009).

The Protein Ser/Thr Phosphatase (PSP) group of enzymes include three broad categories, namely, phospho protein phosphatases (PPP), metal-dependent phosphatases (PPM), and aspartate-based phosphatases (FCP and SCP) (Uhrig et al. 2013; Schulman 2014). The members of PPP group harbor a common catalytic domain (about 280 residues) although their noncatalytic N- and C-termini may manifest marked diversity. Further, they can form diverse kinds of holoenzymes by associating with different regulatory subunits (Barford et al. 1998). PPP family embraces protein phosphatase 1 (PP1), PP2A, PP2B (also called calcineurin), PP4, PP5, PP6, and PP7 (Shi 2009). The metal-dependent phosphatase (PPM) group includes those members depending on divalent cations (e.g., Ca^{2+}, Mg^{2+}, Mn^{2+}, etc.) for their activities (Luan 2003). PP2C and pyruvate dehydrogenase phosphatase are representatives of PPM family as they require manganese/magnesium ions (Mn^{2+}/Mg^{2+}) for their functioning (Shi 2009). Historically, PSPs were classified as per their substrate specificities and pharmacological properties into two types, namely, Type I phosphatase (PP1) and Type 2 phosphatase (PP2) (Luan 2003). PP1 targets beta subunit of phosphorylase kinase, whereas PP2 prefers alpha subunit of their target enzymes. PP2 type further comprises members which may require divalent cations for their functions (like the PPM type) (Luan 2003). Aspartate-based phosphatases (FCP and SCP) are found to have only a single target site—the C-terminal domain of RNA polymerase II, which carries tandem repeats of a serine-rich heptapeptide (Shi 2009). FCP and SCP group of phosphatases are called nuclear localized PPs having crucial role in RNA polymerase II recycling during transcription processes, and their mutations have been often associated with survivability defects in cells (Archambault et al. 1998; Cho et al. 2001; Moorhead et al. 2007). They were characterized initially in animal cells, and their homologues are anticipated to occur in plant cells too (Cho et al. 2001). The mechanisms of dephosphorylation in PPP and PPM types have common features which markedly differ from aspartate-based phosphatases (Shi 2009).

Phosphotyrosine phosphatases (PTPs) were initially considered to be limited to animal kingdom, where they serve diverse functions (Luan 2003). The first plant PTP (AtPTP1) to be characterized was in Arabidopsis back in 1998 (Xu et al. 1998). Following that, vital functions of PTPs in plants have been critically realized (Luan et al. 2001; Shankar et al. 2015). Based on selectivity for amino acid residues, PTPs can be assorted into two distinct groups: tyrosine-specific PTPs and **dual-specificity PTPs (DsPTPs)**. Tyrosine-specific PTP dephosphorylates only tyrosine residues but not phosphoserine/threonine, whereas the **DsPTPs** remove phosphate from both (Stone and Dixon 1994; Tonks and Neel 1996). PTPs harbor a preserved catalytic domain in their active site consisting of [H/VCX_5RS/T] as signature motif (Shankar et al. 2015). Cysteine in this domain has active role in the reaction path (Zhang 1998). Tyrosine-specific PTPs embrace receptor-like PTPs and intracellular PTPs of which only the latter is reported from plants (Luan et al. 2001). PTPs are further grouped into four clusters, on the grounds of amino acid residues present in their catalytic domains, as type 1, 2, and 3 (cysteine-based) PTPs and Asp-based phosphatases (Neel and Tonks 1997; Alonso et al. 2004). In terms of protein sequence content, DsPTPs share modest homology with Tyr-specific PTP but possess the common CX_5R motif in their catalytic site [V/IHCXAGXGRS/T] (Onoda et al. 1989). Members of DsPTPs include MKP, PTEN cluster of phosphatases in plants (Luan et al. 2001). Genome-wide studies across different plant species however indicate lower number of tyrosine-specific PTPs than DsPTPs (Shankar et al. 2015). One of the indications might be DsPTPs are more specific than tyrosine-specific PTPs or tyrosine-specific PTPs may have very limited targets.

While all the three PPs are important for cellular dynamics, PSPs particularly seem to dominate in most dephosphorylation events in Arabidopsis (Wang et al. 2007), rice (Singh et al. 2010), and maize (Wei and Pan 2014), which might probably be the case for plants as a whole. Arabidopsis genome has 92 PSPs (off 112 PPs in total), rice contains 107 PSPs (out of 132 PPs in total), and maize genome represents 130 PSPs out of 159 total PPs altogether. Interestingly, in human (Liberti et al. 2013) and mouse (Forrest et al. 2003) genomes, PTPs predominate in contrast to PSPs.

15.2.2 Functional Attributes

PPs regulate key cellular processes principally *via* reversal of phosphorylations carried out by numerous PKs in animal as well as plant cells. Their crucial role in growth, development, differentiation, defense, and stress management across living organisms is palpable from a myriad of research descriptions. Since phosphorylation-dephosphorylation cycles form heart of most biochemical reactions within cell, significance of PPs can be precisely envisaged. The fundamental processes of cell such as chromosomal DNA replication, transcription, translation, DNA repair, etc. rely on functions of specific PPs. For instance, chromosomal DNA replication essentially requires PP2A activity during initiation stages (Lin et al. 1998). Likewise, involvement of PP1 (a PSP) during RNA polymerase II-driven transcription (Washington et al. 2002), regulation of protein translation and cell growth by PPM1G (another member of PSP) (Liu et al. 2013) and PP2A engagement in DNA repair dictated by replication stress (Feng et al. 2009) are few examples among many. For an extensive description on functions of animal PPs, a number of literatures are available (Millward et al. 1999; Barr et al. 2011; Chen et al. 2017). In plants, apart from fundamental processes, regulatory role of PPs in cell division and differentiation, photosynthesis, hormonal signaling, growth, metabolism, etc. has been well established (Schweighofer and Meskiene 2015). Besides these, PPs play critical role in the regulation of plant's responses to stress brought about by abiotic as well as biotic factors. Regarding PP-mediated abiotic stress management in plants, some excellent reviews may be referred (Bartels and Sunkar 2005; País et al. 2009). Corresponding to plant's responses towards negative biotic interventions, exquisite mechanisms for regulation of plant immune signaling have become gradually obvious from several notable findings (Gómez-Gómez et al. 2001; Park et al. 2008; Segonzac et al. 2014; Shubchynskyy et al. 2017; Liu et al. 2018). These estimations indicate PP's negative regulatory role in plant immune reactions towards phytopathogens which, on the contrary, remarkably alleviate host damages. Some members of PSP family were earlier shown to impede immune responses in several plant species, e.g., *Glycine max*, *Solanum tuberosum*, *Solanum lycopersicum*, *Nicotiana tabacum*, etc., when pathogenic stress was absent; it eventually prevented redundant immune signaling in these hosts (País et al. 2009). A range of PP2A members downregulates fungal component-induced immune responses in *S. tuberosum*, *S. lycopersicum*, and *Helianthus annuus* by desensitizing phosphorylation cascades as a protective measure for the host (País et al. 2009). Xb15, a PP2C-type phosphatase in rice, keeps a check on defense responses induced by a serious bacterial pathogen in rice (Park et al. 2008) about which we are elaborating in the following section. Importantly, PP2C group of enzymes are said to constitute the largest group of PSP cluster executing a myriad of signal regulations in plants (Kerk et al. 2008; Singh et al. 2010). The *A. thaliana* MKP2, a DsPTP-type phosphatase, exhibits differential mode of immune signal regulation in response to a biotrophic and a necrotrophic pathogen, respectively (Lumbreras et al. 2010), which further suggests inherent flexibility in functions of PPs.

New reports on plant PP functions have been quite frequent, thanks to the advancements in the genomics as well as phosphoproteomics research. The next section of this chapter illustrates plant PPs specifically involved in plant immune function regulations against a variety of phytopathogenic microbes.

15.3 Response of Plant Protein Phosphatases (PPs) Towards Varied Pathogens

Plant RLKs and RLPs located at the cell surfaces play prominent role in activating immune responses *via* phosphorylation(s) at target components as and when pathogen derived PAMPs or MAMPs are encountered. The activated signal cascades lead to a set of cellular responses for dampening pathogen progression. A host of inhibitory substances that are dispensed out for the purpose can however injure host cells/ tissues in the vicinity if immune triggers continue for long duration unchecked (Lang and Mansell 2007). Since PKs can't reverse phosphorylations, therefore PPs must act to diffuse immune triggers.

Dephosphorylation is also essential for regenerating active RLKs/RLPs to prepare for next set of immune signaling at cell surfaces. Although PPs' function opposes that of PKs, the former maintain a proper balance of the immune response threshold necessary for pathogen inhibition while also restoring safety of host cells/ tissues. This section takes into account of some known plant PPs (as appears in Table 15.1) and their roles in regulating host defense responses to different microbial pathogens.

15.3.1 Plant PPs Responding Towards Bacterial Pathogens

Several treacherous bacterial pathogens are known to negatively impact plant's health and survivability. Evolutionarily, selection and maintenance of resistance genes by plants to combat most pathogenic invasion has been in continuum. The sophisticated innovations in plant innate immune system are an outcome of such a gradual process. Here are few illustrations of bacterial phytopathogens whose elicitations of plant immune signals are known to be precisely regulated by PPs.

Rod-shaped, Gram-negative bacterium *Xanthomonas oryzae* pv. *oryzae* (*Xoo*) is the causal agent of lethal bacterial leaf blight (BLB) disease in rice. Rice resistance to *Xoo* is known to be conferred by the *xa21* gene in rice that encodes a cell surface localized RLK engaged in the downstream defense signaling processes (Song et al. 1995). The cytoplasmic kinase of Xa21 is a ser/thr kinase (Song et al. 1995) that is predicted to phosphorylate at serine/threonine residues of downstream relay proteins. At the membrane, Xa21 associates with XB24 (called Xa21 binding protein 24) protein (Chen et al. 2010). XB24 generally ensures Xa21 inactive state by

Table 15.1 List of some of the reported plant protein phosphatases regulating host defense responses against various phytopathogenic microorganisms

Serial number	Name of the plant protein phosphatase	Type	Plant origin	Target PRR (RLK/RLP)/molecules	Pathogenic elicitors/molecules	References
1	Xb15	PP2C (PSP)	*Oryza sativa*	Xa21	*Xanthomonas oryzae* pv. *oryzae* (*Xoo*)	Park et al. (2008)
2	AP2C1	PP2C (PSP)	*A. thaliana*	FLS2, EFR	*P. syringae* pv. *tomato* strain DC3000 (*Pto* DC3000)	Shubchynskyy et al. (2017)
3	PP2C38	PP2C (PSP)	*A. thaliana*	BIK1 (RLCK)	*Pto* DC3000	Couto et al. (2016)
4	MKP1 (MAP Kinase Phosphatase 1)	DsPTP phosphatase	*A. thaliana*	MPK6	*Pto* DC3000	Bartels et al. (2009)
5	PP2A	PSP	*A. thaliana*	BAK1	*P. syringae* PAMPs	Segonzac et al. (2014)
6	PIA1	PP2C	*A. thaliana*	RPM1 (NB-LRR)	AvrRpm1 [*P. syringae* (DC3000) T3E molecule]	Widjaja et al. (2008)
7	MKP2	DsPTP	*A. thaliana*	MPK3	*Ralstonia solanacearum* (GMI1000)	Lumbreras et al. (2010)
				MPK6	*B. cinerea*	
8	KAPP (Kinase Associated Protein Phosphatase)	PP2C	*A. thaliana*	FLS2	Bacterial flagellin MAMP flg-22	Stone et al. (1994); Gómez-Gómez et al. (2001)
9	AtPP2C62 and AtPP2C26	PP2C	*A. thaliana*	Not known (??)	*Xanthomonas campestris* pv. *campestris* (*Xcc*) T3E molecules	Akimoto-Tomiyama et al. (2018)
10	AP2C1	PP2C	*A. thaliana*	MPK6, MPK4	Oligogalacturonides released by *Botrytis cinerea* infection	Schweighofer et al. (2007)
11	PP2A (carrying catalytic subunit PP2Ac)	PSP	*Nicotiana benthamiana*	cf9 [the known resistance gene of tomato (R-gene)]	Avr9 effector (from *Cladosporium fulvum*)	He et al. (2004)

(continued)

Table 15.1 (continued)

Serial number	Name of the plant protein phosphatase	Type	Plant origin	Target PRR (RLK/RLP)/molecules	Pathogenic elicitors/molecules	References
12	TaPP2Ac-4B and TaPP2Ac-4D (PP2A orthologues)	PSP	*Triticum aestivum*	Not known (??)	*Rhizoctonia cerealis*	Zhu et al. (2018)
13	CIPP1	PSP	*A. thaliana*	CERK1	Fungal chitin MAMP	Liu et al. (2018)
14	PAPP2C (phytochrome-associated protein phosphatase type 2C)	PP2C	*A. thaliana*	RPW8.2	*Golovinomyces* spp.	Wang et al. (2012)
15	AP2C1	PP2C	*A. thaliana*	MPK3 and MPK6	*Heterodera schachtii* (cyst nematode)	Sidonskaya et al. (2016)

assisting in autophosphorylation of serine and threonine residue(s) on Xa21 (Chen et al. 2010). When *Xoo* is sensed, association of XA21 kinase and XB24 is relieved leading to Xa21 activation (Chen et al. 2010). Xa21 activation relays a series of downstream processes culminating in vigorous resistance response. Recently, a tyrosine-sulfated protein (RaxX) of *Xoo* origin was identified that might be one of the ligand activating Xa21-mediated defense responses in rice (Pruitt et al. 2015). The Xa21-mediated trigger is finally dampened by a PP called Xb15 (Park et al. 2008) [an illustration appears in Fig. 15.1]. Xb15 (the Xa21 binding protein 15) belongs to PP2C cluster of PSP, which requires metal ion (Mg^{2+}) for their activities (Park et al. 2008). It is plasma membrane localized as is Xa21 and both are said to interact there (Park et al. 2008). Interestingly, Xb15 requires a specific binding site at Xa21 juxtamembrane (JM) domain that carries a serine residue at 697th position (Park et al. 2008). Binding of Xb15 to Xa21 *via* JM domain didn't inhibit latter's kinase activity, and probably dephosphorylations at other sites might be required (Park et al. 2008). Park and associates (2008) eventually proved requirement of Xb15 in termination of Xa21-mediated defense responses against *Xoo*. Their study

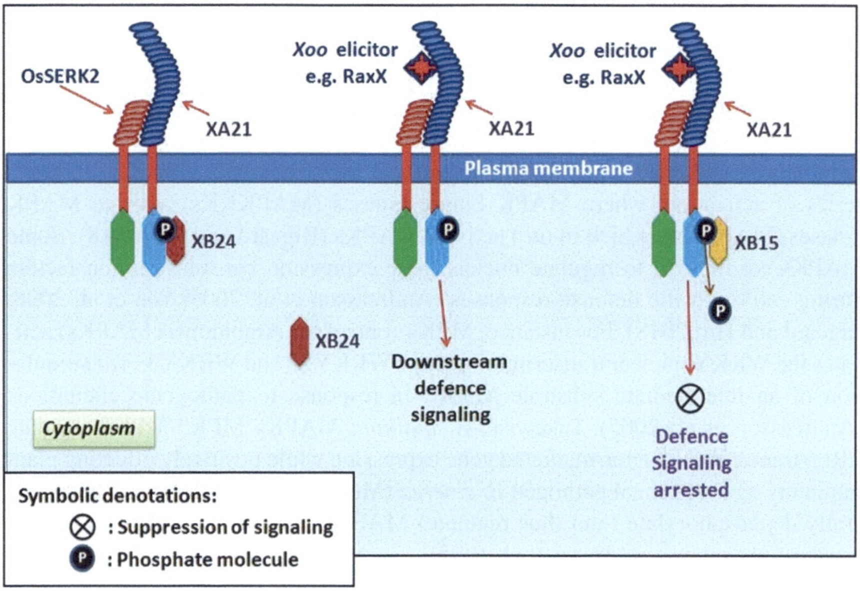

Fig. 15.1 Schematic representation of *O. sativa* PP2C-type Protein Phosphatase, XB15 regulating XA21 (a typical RLK)-mediated immune responses towards pathogenic *Xoo* strains; redrawn after Macho and Zipfel (2014), with some modifications. The well-known rice RLK, XA21 along with co-receptor OsSERK2 forms the PRR complex that senses *Xoo* PAMPs at the cell surface. Activated PRRs subsequently initiate intracellular defense signaling cascades, e.g., PTI in its downstream. XB24, an ATPase which generally remains bound to XA21, detaches from XA21 upon PRR activation (Chen et al. 2010). XB15 phosphatase next interacts with the activated XA21 and dephosphorylates it, thereby halting XA21-mediated defense signaling (Park et al. 2008). All these interactions are said to occur in the plasma membrane (Park et al. 2008)

also indicated significance of Xb15 function in negatively regulating cell death in *O. sativa* during Xa21-triggered immunity.

Xb15 has significant similarity with Arabidopsis PP2C members, namely, POL, PLL4, and PLL5 (Park et al. 2008). However, the role of these Arabidopsis PP2Cs (Song and Clark 2005; basically in plant organ development) seems distinctly different from that of Xb15 (Park et al. 2008) suggesting occurrence of peculiar functions even within highly similar PP2C representatives.

Pseudomonas syringae, the rod-shaped Gram-negative bacterium, infects almost all economically important crop plant species. It is said to be most usual phytopathogen and has more than 60 pathovars constituting the *P. syringae* species complex (Xin et al. 2018). In the model plant *A. thaliana*, a set of virulence experiments with *P. syringae* pv. *tomato* strain DC3000 (*Pto* DC3000) gave insight into the role of AP2C1 (a PP2C phosphatase) in downregulation of defense response to *Pto* DC3000 (Shubchynskyy et al. 2017). Upon treatment with elf18 and flg22 (MAMPs), induction of MAPK or MPK (mitogen-activated protein kinase) activity was much pronounced in *A. thaliana ap2c1* mutant line (Shubchynskyy et al. 2017). In fact, out of MPK3, MPK4, and MPK6 members, MPK6 was maximally activated in the absence of functional AP2C1 (Shubchynskyy et al. 2017) suggesting high affinity of AP2C1 for MPK6. Additionally, *ap2c1* mutation resulted in enhanced callose deposition in response to *Pto* DC3000 (Shubchynskyy et al. 2017). This may imply that AP2C1 phosphatase might be regulating *Pto*-induced callose formation in *A. thaliana* through a yet uncovered mechanism. MAPK cascades are vital in plant stress responses and development. They form crucial arsenals in plant's innate immunity (Lumbreras et al. 2010). In general, MAPK signal cascade follows a series of activation, where MAPK kinase kinases (MAPKKKs) activate MAPK kinases (MAPKKs), which in turn activate MAPKs (Bigeard and Hirt 2018). Some MAPKs are thought to regulate nuclear gene expression *via* transcription factors during cell's specific defense responses (Andreasson et al. 2005; Yoo et al. 2008; Bigeard and Hirt 2018). For instance, MPK4 (one of the Arabidopsis MAPKs) activates the WRKY nuclear transcription factors WRKY25 and WRKY33 *via* stimulation of an intermediate substrate MKS1, in response to pathogenic elicitations (Andreasson et al. 2005). Likewise, *A. thaliana* MAPKs MPK3/MPK6 regulate ERF6 transcription factor-mediated gene expression while positively effecting plant immunity against fungal pathogen *B. cinerea* (Meng et al. 2013). PSPs, which generally dephosphorylate (and thus regulate) MAPKs, incorporate PP2A and PP2C group of phosphatases (Alessi et al. 1995).

In an earlier study, MKP1 (MAP kinase phosphatase 1) in *A. thaliana* was shown to negatively regulate MPK6 activation in response to *Pto* DC3000 (Bartels et al. 2009). MKP1 is a member of DsPTP phosphatase cluster (Bartels et al. 2009). Apart from some PSPs, PTPs and DsPTPs too can negatively regulate MAPKs (Kondoh and Nishida 2007). Generally, MKPs regulate MAPKs *via* dephosphorylations at both phosphothreonine and phosphotyrosine residues concurrently within the MAPK TXY (Thr-Xaa-Tyr) activation motif, thereby thwarting downstream signaling cascades (Patterson et al. 2009). In their study, Bartels et al. (2009) observed that mutation of *mkp1* gene in the *A. thaliana* Columbia (Col) accession caused growth defects and uninterrupted immune signaling in response to *Pto* DC3000. Further, increased level of pathogenesis-related (PR) gene expression along with compounds like salicylic acid and camalexin followed by enhanced resistance to *Pto* DC3000 observed in the same experiments demonstrated conspicuous regulatory role of MKP1 (Bartels et al. 2009). Their work also highlighted cooperative antagonistic function of Protein Tyrosine Phosphatase 1 (PTP1; a member of PTP) along with MKP1 towards Arabidopsis MPK6, although PTP1 alone may not be sufficient to repress MPK6 signaling (Bartels et al. 2009). MKP1 was again investigated for its competence to diminish *Pto* DC3000 PAMP-induced defense responses in *A. thaliana* Wassilewskija (Ws) accession (Anderson et al. 2011). Genetic analyses revealed that MKP1 was more specifically antagonizing activation of MPK6 by *Pto* DC3000 PAMP than that of an earlier predicted MAP kinase, MPK3 (Anderson et al. 2011). The study also established predominant role of MPK6 in the host's resistance response to *P. syringae* when *mkp1* was mutated (Anderson et al. 2011). Independent studies have thus unveiled *A. thaliana* MPK6 to be the common substrate for PSP as well as DsPTP phosphatases although their mechanisms of regulation are distinct. In a seminal work (Anderson et al. 2014) involving metabolic profiling of *mkp1* mutant versus wild-type *A. thaliana,* intact MKP1 activity was correlated with the ability of *A. thaliana* to induce type III secretion system-mediated *Pto* DC3000 virulence in contrast to *mkp1* mutant line. This uncovered yet another intriguing role of PPs in regulating resistance response of plant towards pathogen.

Delving into molecular detail of immunogenic elicitations owing to *P. syringae* has provided some crucial information. *P. syringae* PAMPs induce PRR complex FLS2/BAK1 or EFR/BAK1 (as well as others) on cell surface during its interaction with *A. thaliana* (Chinchilla et al. 2007; Heese et al. 2007; Schulze et al. 2010; Schwessinger et al. 2011; Roux et al. 2011; Sun et al. 2013). In the absence of elicitors, PRRs also associate with a receptor-like cytoplasmic kinase (RLCK), BIK1, in the cytoplasm along with other signaling proteins (Lu et al. 2010; Zhang et al. 2010). BIK1 (Botrytis-Induced Kinase1), which is a member of serine/threonine kinase family, associates with PRRs and plays a critical role in defense signaling (like PTI) against a variety of pathogens in *A. thaliana* (Lu et al. 2010; Zhang et al. 2010; Laluk et al. 2011; Couto et al. 2016). Post PAMP perception, PRR-BAK1 complex phosphorylates BIK1 leading to its release (Lu et al. 2010; Zhang et al. 2010). Activated BIK1 next phosphorylates proteins involved in the cascade of immediate PTI responses such as reactive oxygen species (ROS) liberation to restrict pathogenic progression (Kadota et al. 2014; Li et al. 2014). Regulation of the immune triggers due to BIK1 has been found to be under strict control of a plasma membrane-localized PP2C-type phosphatase, PP2C38, in *A. thaliana* (Couto et al. 2016). PP2C38 dephosphorylates BIK1 and inhibits its downstream signal relay, thereby putting a halt on PAMP-induced ROS burst (Couto et al. 2016) [as depicted in Fig. 15.2]. As per findings of Couto and associates (2016), PP2C38 remains linked to BIK1 (in plasma membrane) until it (PP2C38) is phosphorylated at its 77th serine residue by BIK1 (upon stimulated *via* PAMP-induced PRR-BAK1 complex), finally releasing activated BIK1 from association. Their work also

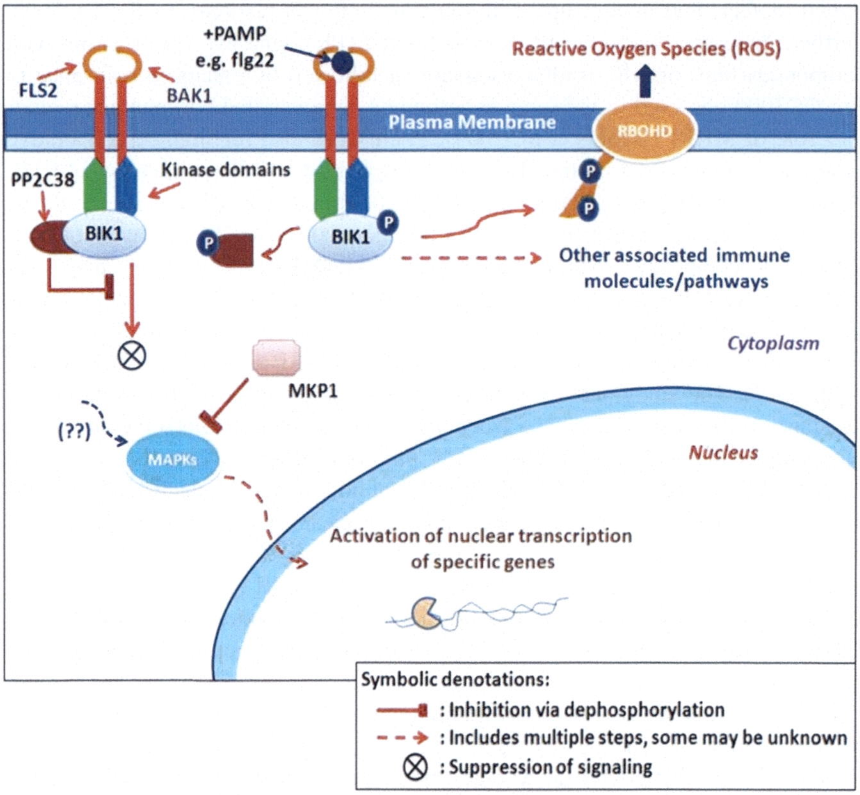

Fig. 15.2 Schematic representation of plant protein phosphatase-mediated regulation of immune signaling elicited by bacterial PAMP such as flagellin. In the event of flagellin perception (e.g., *P. syringae* flg22), stimulated PRR complex (FLS2/BAK1) in turn activates cytoplasmic RLCK, BIK1, *via* phosphorylation (Couto et al. 2016). Prior to PAMP perception, a PP2C phosphatase, PP2C38 is said to maintain dephosphorylation status of BIK1 and thus inhibit defense signaling downstream. Upon PAMP perception, PP2C38 dissociates from BIK1, allowing BIK1 to trigger subsequent defense responses like generation of ROS *via* activation of membrane-localized NADPH oxidase, RBOHD (Couto et al. 2016; Zipfel and Oldroyd 2017). Flagellin is also reported to activate MAPK kinase pathway *via* BAK1 (Heese et al. 2007), although the appropriate link between events of MAPK pathway and upstream PRR dynamics is still elusive (Yan et al. 2018). Activated MAPKs can induce subsequent defense responses *via* activation of nuclear transcription factors to promote expression of defense-related genes. In connection with findings of Bartels et al. (2009), MKP1, a PTP-type phosphatase, can be predicted to dampen MAPK signaling *via* its phosphatase-mediated functions. The figure has been redrawn from Zipfel and Oldroyd (2017), with added modifications. Findings of Bartels et al. (2009), Heese et al. (2007), and Yan et al. (2018) have significantly inspired the modifications incorporated

indicated that PP2C38 inhibits BIK1 trigger (by remaining bound to it) in the absence of threshold elicitations by PAMPs, thus maintaining basal resistance in *A. thaliana*. Akin to BIK1, at the upstream of PRR signaling, BAK1 (the RLK that forms PRR complex with variety of PRRs like FLS2, EFR, PEPRs, etc.) is also under the control of a different protein phosphatase, PP2A (Segonzac et al. 2014). PP2A associates with BAK1 and negatively regulates its phosphostatus in the absence of PAMP elicitations, thus regulating basal resistance response (Segonzac et al. 2014). Once PRR complex (along with BAK1) perceives *P. syringae* PAMPs and gets triggered, PP2A phosphatase functions undergo suppression allowing BAK1 to initiate signaling cascades downstream through phosphorylation of BIK1 and so on. Interestingly, unlike PP2C38, PP2A remains constitutively bound to BAK1 during PRR signaling downstream, without affecting BAK1 kinase function (Segonzac et al. 2014). PP2A enzymes are generally trimeric and comprise variants of all the constituent subunits, i.e., C (catalytic), A (scaffold), and B (regulatory subunit) (Trotta et al. 2011). PP2A in Arabidopsis consists subunit A1, the regulatory B subunits B'η/ζ and the catalytic subunit C4, respectively (Segonzac et al. 2014). Segonzac and associates (2014) generated Arabidopsis lines carrying mutants of *pp2a-c4* and *pp2a-a1* and found enhanced resistance of these mutant lines towards *Pto* DC3000 demonstrating critical role of specific subunits in the defense responses against the pathogen.

Unlike the plant PPs discussed above, WIN2, a PP2C member belonging to *A. thaliana*, was demonstrated to positively regulate host resistance responses against *Pto* DC3000/hopW1-1 strain (*P. syringae* strain expressing hopW1-1 effector) infection (Lee et al. 2007). Interestingly, *P. syringae* strains expressing functional hopW1-1 were found to have restricted host range with respect to certain accessions of *A. thaliana*. Yeast two-hybrid screening has manifested interaction of WIN2 and hopW1-1 effector, and it is predicted that hopW1-1 may be regulated *via* dephosphorylation mechanism carried out by WIN2. Another protein WIN3 (which is not a phosphatase) was also referred to assist in the resistance response against *Pto* DC3000/hopW1-1. The resistance due to WIN2 further correlated with elevation of salicylic acid (SA) level in the host pointing towards some role of WIN2 in SA synthesis process (Lee et al. 2007). Further, disabling WIN2 expression *via* RNA interference resulted in diminished resistance of Arabidopsis towards HopW1-1 effector-mediated induction (Lee et al. 2007). Although precise mechanism of WIN2 and HopW1 interplay remains to be investigated fully, WIN2 represents a suitable illustration of a plant PP regulating functions of a specific pathogenic effector.

An Arabidopsis PP2C-type phosphatase (PIA1) was reported to specifically recognize a *P. syringae* (DC3000)T3E molecule AvrRpm1 and induce NB-LRR (nucleotide-binding leucine-rich receptor) protein RPM1 for subsequent signaling downstream (Widjaja et al. 2008). However, *Pto* DC3000 strain secreting another T3E AvrB escaped PIA1 recognition. PIA1 was shown to be negatively regulating AvrRpm1-induced defense responses in the host, indicating that PIA1 might be negatively controlling RPM1-mediated immune responses which may unanimously benefit *Pto* DC3000 (AvrRmp1) strains for proliferation within host (Widjaja et al.

2008). *pia1* mutation was demonstrated to enhance host resistance towards *Pto* DC3000 (AvrRmp1), while AvrRmp1-stimulated expression of pathogenesis-related genes was reduced (Widjaja et al. 2008). PIA1 thus displays differential mode of immune regulation relying on the kind of *P. syringae* effectors, further illustrating its specialized functionality.

Immune responses in Arabidopsis towards bacterial wilt pathogen, *Ralstonia solanacearum* (GMI1000), were shown to involve regulatory function of the MKP2 phosphatase (a DsPTP member) (Lumbreras et al. 2010). *mkp2* mutant Arabidopsis displayed delayed wilt symptom development in response to GMI1000 infection, suggestive of negative regulatory role of MKP2 in the pathogen-induced defense signaling (Lumbreras et al. 2010). Further, out of MPK3 and MPK6 members in the MAPK cascade elicited by the pathogen, MKP2 interacts with MPK3 specifically during GMI1000 infection and regulates HR (Lumbreras et al. 2010). A differential mode of MKP2 regulation in response to another necrotrophic pathogen was reported, in the same work of these authors (Lumbreras et al. 2010).

In solanaceous crop tomato, a variant of FLS2 pattern recognition receptor, FLS3 was found to recognize another epitope of *P. syringae* flagellin called flgII-28 (Cai et al. 2011; Clarke et al. 2013; Hind et al. 2016). FLS3 associates with and depends on BAK1 (tomato orthologue) for downstream immune signal transduction in similar way as FLS2 does, although distinct players may engage in signaling events at some points (Hind et al. 2016). It can be predicted that tomato PP2A orthologue might be regulating BAK1 in tomato in an analogous way (Segonzac et al. 2014) in response to *P. syringae* MAMP elicitations.

Interestingly, BAK1 as an RLK serves as a co-receptor (in fact as the central kinase) of several PRRs across different plant species and assists in perception of varied PAMPs of phytopathogenic bacteria, finally triggering immune responses like PTI downstream (Yasuda et al. 2017). BAK1 regulation in these cases also can be thought of to be *via* phosphatases like PP2A (Segonzac et al. 2014) in similar manner.

The *A. thaliana* FLS2 receptor (a PRR with cytoplasmic Ser/Thr kinase domain) that senses bacterial flagellin MAMP flg-22 and initiates immune signaling downstream is reported to be negatively regulated by KAPP (Kinase Associated Protein Phosphatase), a member of PP2C phosphatase cluster (Stone et al. 1994; Gómez-Gómez et al. 2001). KAPP remains bound to kinase domain of FLS2, maintaining dephosphorylated status of the latter in the absence of flagellin MAMP elicitor (Gómez-Gómez et al. 2001). Overexpression of KAPP was correlated with reduced flagellin perception by FLS2, a condition resembling *fls2* null mutation in Arabidopsis (Gómez-Gómez et al. 2001).

The causative agent of destructive "black rot" disease, *Xanthomonas campestris* pv. *campestris* (*Xcc*), usually infects most cruciferous vegetables including *A. thaliana* (Meyer et al. 2005; Akimoto-Tomiyama et al. 2018). Recently, the role of two chloroplast-localized PP2C-type representatives (belonging to subgroup K) has been correlated with regulation of *Xcc*-mediated virulence in Arabidopsis (Akimoto-Tomiyama et al. 2018). The Arabidopsis PP2C subgroup K orthologues, AtPP2C62 and AtPP2C26, were shown to negatively regulate immune responses towards *Xcc* infection. In fact, AtPP2C62 and AtPP2C26 double mutant Arabidopsis lines significantly prevented pathogenic proliferation in the host revealing the role of these PP2Cs in Arabidopsis immunity (Akimoto-Tomiyama et al. 2018). Further investigation suggested that both these PP2Cs respond when *Xcc*'s type III secretion system (TTSS) was intact. Growth of *Xcc* TTSS mutants were not affected in the PP2C mutant Arabidopsis host fostering possibility of PP2Cs being target of some *Xcc* T3E (TTSS released effector) executing intriguing counter-defense tactic. Additionally, PR1 gene activation and expression in the PP2C-disabled host in response to *Xcc*s (carrying intact TTSS as well as lacking TTSS) indicated precise regulation of initial defense responses by these PP2Cs in Arabidopsis (Akimoto-Tomiyama et al. 2018).

15.3.2 Plant PPs Responding Towards Fungal Pathogens

Of all known phytopathogens, fungal pathogens are said to impact plants most profusely leading to considerable losses in plant productivity (Knogge 1996). In fact all the flowering plants of the planet can be infected by fungal pathogens (Knogge 1996). Pathogenic fungi have evolved refined arsenals to evade physical barriers while intruding plant tissues (Ferreira et al. 2006; Rodriguez-Moreno et al. 2018). Likewise, plant hosts have developed defense strategies to counteract pathogenic infringements by employing cascades of armories which can be triggered and regulated as and when necessary (Hammond-Kosack and Jones 1996; Ferreira et al. 2006). Plant PPs being an important part of defense contingent play decisive role in regulating plant immune triggers against fungal pathogen perception like we have mentioned above for bacterial pathogens.

Botrytis cinerea commonly referred to as gray mold is a necrotrophic ascomycete infesting up on more than 200 plant species (Dean et al. 2012). *B. cinerea* may preferentially induce programmed cell death in the host prior to infection (van Baarlen et al. 2007). In *A. thaliana*, MAPK signaling cascade is triggered, once upstream PRRs perceive DAMPs such as oligogalacturonides (OGs; polysaccharides released by fungal polygalacturonases acting on plant cell wall) (Hahn et al. 1981; Ridley et al. 2001; Ferrari et al. 2007). Importantly, release of OGs during *B. cinerea* infection in Arabidopsis and other hosts is also reported (Poinssot et al. 2003; Zhang et al. 2014). Activation of MAPK members down the lane *via* series of phosphorylations finally leads to stimulation of PTI-related events which may accompany ROS release, ion fluxes, induction of defense-related genes, accumulation of antimicrobials, etc. as described in various studies involving DAMP-mediated elicitations in different plants (Davis et al. 1986; Mathieu et al. 1991; Thain et al. 1995; Galletti et al. 2008, 2011; Ferrari et al. 2013). Among the participating MAPKs, MPK6 was found to contribute greatly towards immune signaling (Galletti et al. 2011). Arabidopsis AP2C1 phosphatase (a PP2C member) was found to negatively regulate MPK6 activity along with MPK4 (Schweighofer et al. 2007). A physical association between AP2C1 and MAPKs (in fact AP2C1/MPK4 and

AP2C1/MPK6) indicated possibilities of direct regulation by AP2C1 (Schweighofer et al. 2007). AP2C1 suppresses resistance to *B. cinerea*-induced immune response supported by the fact that in *ap2c1* mutant background, *B. cinerea* proliferation was highly inhibited as against in wild-type Arabidopsis (Schweighofer et al. 2007). MAPK activities were highly pronounced in the absence of functional AP2C1, whereas *ap2c1* expression level was elevated post *B. cinerea* infection (Schweighofer et al. 2007). Since MAPKs are multitasking signaling machineries involved in numerous cellular functions, a tight control of MAPK immune signaling *via* phosphatases like AP2C1 might be inevitable to make them available for other tasks.

In one study, *pp2a-b′γ* (gene encoding PP2A-B'γ) mutant *A. thaliana* was demonstrated to be more susceptible to *B. cinerea* infection as compared to wild-type plant (Trotta et al. 2011). PP2A being hetero-trimeric, its regulatory subunit B has been found to be highly variable. The A and C subunits associating with distinct B subunits are thought to confer substrate specificity to PP2A and also dictate its subcellular localization (Goldberg 1999). Probably, combinations of these variable subunits in different ways endow diversity as well as specificity in PP2A for regulating versatile signaling cascades (Trotta et al. 2011). Notably, B'*γ* is a subfamily of B-regulatory subunit, and along with PP2A holoenzyme, it has been implicated in negative regulation of untimely death and inappropriate immune responses in Arabidopsis (Trotta et al. 2011).

In response to *B. cinerea*, a DsPTP member of Arabidopsis, MKP2 was demonstrated to regulate defense responses in host (Lumbreras et al. 2010). Interestingly, *mkp2* mutant Arabidopsis host displayed enhanced susceptibility towards necrotrophic *B. cinerea* infection (Lumbreras et al. 2010). This is in contrast to the response manifested by the same MKP2 phosphatase towards a biotrophic bacterial pathogen communicated in the same report (Lumbreras et al. 2010). MKP2 selectively interacted with MPK6 member during HR reaction elicited by *B. cinerea* infection (Lumbreras et al. 2010). MKP2 is a classic example of plant PP possessing inherent potential to differentially regulate variable pathogen-induced defense responses in host.

Another notable fungal pathogen *Cladosporium fulvum* is responsible for tomato leaf mould disease and generally affects the foliage tissues. Sometimes, stems, flowers and petals may be also infested by this biotrophic fungus (Butler and Jones 1949; Jones et al. 1997). In several occasions, *Nicotiana benthamiana* plant has been recruited for *C. fulvum* and tomato "gene to gene" interaction studies by transient expression methodologies in the wild-type or virus-induced mutant host (Van der Hoorn et al. 1999; He et al. 2004; Gabriels et al. 2007; Chakrabarti et al. 2009). In one instance, immune function of PP2A catalytic subunit PP2Ac in *N. benthamiana* could be ascertained by studying host response towards fungal pathogen *C. fulvum* in the PP2Ac orthologue mutant background (He et al. 2004). Generally, two subfamilies of PP2Ac subunits have been recognized, namely, I and II (He et al. 2004). In the work of He et al. (2004), the subfamily I PP2Ac silencing was achieved using a Potato virus X (PVX) vector construct. Researchers co-expressed an effector Avr9 (from *C. fulvum*) along with cf9 [the known resistance gene of tomato (R-gene)] in the leaf tissues of *N. benthamiana* carrying mutation in PP2Ac orthologue and distinctly observed strong, rapid HR (hypersensitivity response) accompanied by full disintegration of infiltrated leaf areas (He et al. 2004). A mechanistic role of PP2Ac in the regulation of HR phenotype during effector-R gene interaction is thus highlighted by this study. Additionally, PVX-induced PP2Ac mutation in *N. benthamiana* was observed to cause localized cell death as well as elevated expression of PR genes, even in the absence of other pathogenic invasion, suggesting pivotal immunoregulatory role of PP2Ac in the normal plant (He et al. 2004).

The causal agent of sharp eyespot disease in wheat (*Triticum aestivum*) is a lethal necrotrophic fungus known as *Rhizoctonia cerealis*, which has worldwide distribution (Chen et al. 2008; Hamada et al. 2011; Lemañczyk and Kwaśna, 2013). The catalytic subunits of wheat PP2A orthologue, namely, TaPP2Ac-4B and TaPP2Ac-4D, were recently correlated with regulation of defense functions against *R. cerealis* pathogen (Zhu et al. 2018). Both TaPP2Ac-4B and TaPP2Ac-4D catalytic subunits were determined to be members of subfamily II type. Expression levels of both these subunits were found to be elevated in response to *R. cerealis* in a susceptible wheat cultivar 'Wenmai 6' as against the resistant cultivar line 'CI12633'. Barley stripe mosaic virus (BSMV)-based gene silencing of TaPP2Ac-4B and TaPP2Ac-4D subunits in the host manifested enhanced resistance of wheat plant towards *R. cerealis* infection. A marked increment in the expression levels of some PR genes as well as ROS-quenching enzymes in TaPP2Ac-mutant wheat plants indicated appreciable antagonistic role of TaPP2Ac-4B and TaPP2Ac-4D subunits during PTI-related responses against *R. cerealis* infection (Zhu et al. 2018).

Chitin being cell wall constituent of all pathogenic fungi, in general, serves as a primordial elicitor (PAMP or MAMP) of plant immune signaling by activating RLKs (Kaku et al. 2006; Sánchez-Vallet et al. 2015). In the event of fungal chitin perception, a study revealed that the phosphorylation cycle that regulates chitin elicitor receptor kinase 1 (CERK1; an RLK)-mediated immune signaling in Arabidopsis required crucial function of a predicted PSP called CIPP1 (CERK1-interacting protein phosphatase 1) (Liu et al. 2018). The seminal work of Liu et al. (2018) has highlighted a suitable mechanism for protein phosphatase-mediated regulation of a chitin-induced RLK. The mechanism forwarded by Liu et al. (2018) is as follows [*as can be seen through* Fig. 15.3]: CERK1 in unexcited state generally undergoes autophosphorylation at its 428th Tyrosine residue and forms complex with BIK1. On perceiving chitin, CERK1 is triggered and activates BIK1 next by phosphorylating it and releasing it from the complex. Along with BIK1, stimulated CERK1 is also said to induce MAPKs for subsequent immune responses downstream. On the other hand, triggered CERK1 is also thought to deploy CIPP1 for removing phosphate at 428th Tyrosine residue which ultimately dampens CERK1 stimulation. Dephosphorylated CERK1 detaches from CIPP1, and consequently CERK1-triggered immune responses are inhibited. CIPP1 phosphatase activity thus interrupts continued immune signaling due to CERK1, and subsequent damages to host are prevented. Meanwhile, CERK1 can again autophosphorylate itself at Tyr^{428} and participate readily in the next cycle of chitin-induced immune signaling.

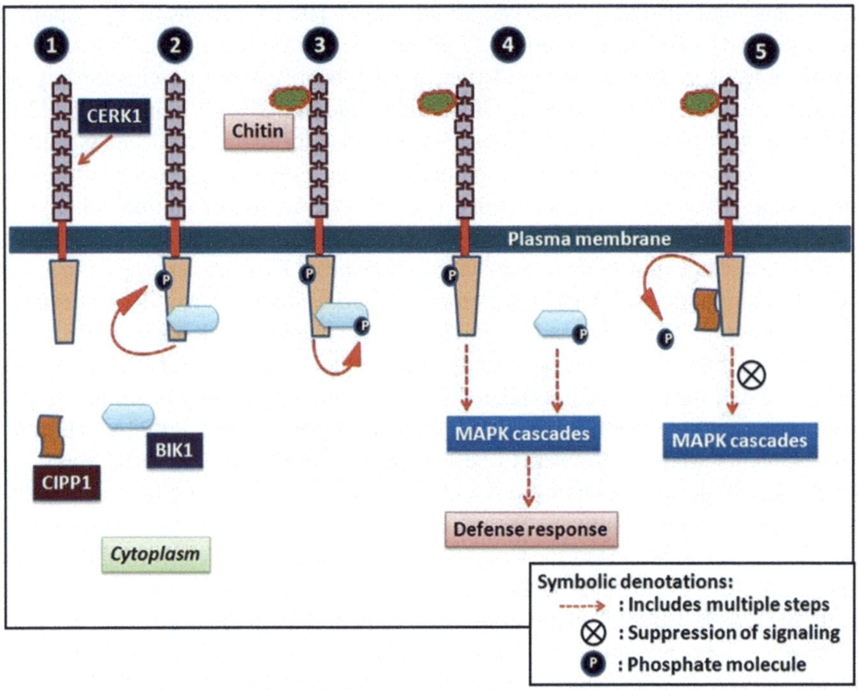

Fig. 15.3 Schematic showing how the PSP-type phosphatase, CIPP1 (belonging to *A. thaliana*), dampens chitin (fungal MAMP)-elicited immune trigger *via* a chitin specific RLK, CERK1 (Liu et al. 2018). The scheme of the CIPP1 function follows like this: (1) The unexcited state of CERK1, a RLK (in *A. thaliana* cell membrane), in the absence of chitin MAMP; CIPP1 (the CERK1-interacting protein phosphatase 1; a PSP phosphatase) and BIK1 (a RLCK) remains unbound to CERK1 at this state. (2) CERK1 undergoes autophosphorylation at its 428th Tyrosine residue and forms complex with BIK1. (3) On sensing chitin, triggered CERK1 phosphorylates BIK1 and activates it; activated BIK1 then detaches from CERK1. (4) Stimulated CERK1 as well as BIK1 may further induce MAPKs for subsequent defense responses downstream. (5)Triggered CERK1 is said to deploy CIPP1 for dephosphorylating its 428th Tyrosine residue which ultimately diminishes CERK1 stimulation. Dephosphorylated CERK1 then detaches from CIPP1 and consequently CERK1-triggered immune responses are halted (adapted from Liu et al. 2018)

In respect of defense response towards powdery mildew pathogen, *Golovinomyces* spp., one study (Wang et al. 2012) assessed interaction of an atypical Arabidopsis resistance factor RPW8.2 with the PP2C member PAPP2C (phytochrome-associated protein phosphatase type 2C) in yeast and *in planta*. RNA interference-mediated downregulation of *papp2c* in RPW8.2 active Arabidopsis background manifested prominent HR-induced cell death (Wang et al. 2012) indicative of PAPP2C immune regulatory function. Investigators (Wang et al. 2012) concluded PAPP2C's role as a negative regulator of SA (salicylic acid)-mediated defense responses against the powdery mildew disease agent through association with RPW8.2 in some way, yet to be understood fully.

Utilization of protein phosphatase inhibitors has been another imperative approach for studying regulatory role of phosphostatus during signal transductions

(Weiser and Shenolikar 2003). Several key information regarding roles of PPs, e.g., during defense responses, have been elucidated using PP inhibitors (Felix et al. 1994; MacKintosh et al. 1994; Agrawal et al. 2000; Rakwal et al. 2008; Bajsa et al. 2011). For instance, in one study, soybean (*Glycine max* L.) cotyledons were subjected to treatment with a glucoside derivative obtained from the soybean pathogen *Phytophthora megasperma* f. sp. *glycinea*, which induced synthesis of some defense-related compounds (belonging to isoflavonoid group) as an initial host defense response (MacKintosh et al. 1994). It needs to mention that *P. megasperma* f. sp. *glycinea* causes root and stem rot of soybean and preferably attacks soybean plants in the water-logging field conditions (Hahn et al. 1985).

MacKintosh et al. (1994) could monitor production of similar isoflavonoid compounds in soybean cotyledons, when the cotyledons were initially cut and then treated with specific protein phosphatase inhibitors (PPIs). Use of specific PPIs (e.g., okadaic acid, acanthifolicin, etc.) indicated inhibition of PP1 and PP2A phosphatases (MacKintosh et al. 1994). Importantly, these PPIs were active inducers of isoflavonoid production too. An abrupt rise in alkalinization (may be a result of ionic flux) due to PPI treatment as against lowered alkalinity when kinase inhibitor was used suggested some regulatory role of protein phosphatases (MacKintosh et al. 1994). Further, sustained PPI treatment led to expression of particular defense-related proteins [like phenylalanine ammonia-lyase (PAL)], whereas application of kinase inhibitor diminished their expression, pointing towards PP-mediated regulation of defense responses (MacKintosh et al. 1994). This further implicate predictable role of PP1 and PP2A phosphatases in the regulation of defense response against *P. megasperma* in soybean.

15.3.3 Plant PPs Responding Towards Other Known Pathogens

Apart from bacterial and fungal pathogens, other eukaryotic microbial pathogens do impact plant health immensely rendering considerable economic losses. For example, parasitic nematodes as agricultural pathogen affect growth, survivability as well as productivity of diverse plant species (Abad et al. 2008; Lin et al. 2016). As an illustration, obligate biotrophic nematode *Meloidogyne* spp. infects over 3000 plant species from varied plant families worldwide and causes characteristic root-knot disease (Caboni et al. 2012). Likewise, cyst-forming nematodes representing the genera *Heterodera* and *Globodera*, respectively, also parasitize and negatively affect health of numerous plants (Williamson and Gleason 2003). Plant immune mechanisms can moreover recognize parasitic nematode insults through latter's characteristic nematode-associated molecular patterns (NAMPs) like in case of various microbial pathogens (such as PAMPs/MAMPs) (Manosalva et al. 2015; Choi and Klessig 2016). In fact, parasitic nematodes from several taxonomic groups exudes certain small molecules known as "ascarosides," which serve as potent elicitors of plant defense responses involving MAMP-triggered immunity and various downstream signaling pathways (Manosalva et al. 2015). The ascaroside-induced immune triggers in variety of plants are said to render them resistant to varied microbial pathogens including nematode infections (Manosalva et al. 2015). Apart from that, parasitic nematode penetration into plants during infection may release DAMPS, which in turn can stimulate PTI-like defense responses in the latter (Holbein et al. 2016). Importantly, activation of both PTI and ETI branches of immunity can be stimulated by parasitic nematode infestation in plant host. In fact, modulation in the ETI branch of immunity during nematode infection has been studied much extensively, although reports on specific cell surface-based PRRs recognizing NAMPs are rare (Holbein et al. 2016; Sidonskaya et al. 2016). Induction of hypersensitive responses (HR) as well as PCDs (programmed cell death) by parasitic nematodes has also been indicated in different studies (Melillo et al. 2006; Leonetti 2018; Matuszkiewicz et al. 2018). However, it needs mention that reports on protein phosphatase-mediated regulation of nematode-induced resistance responses in the plant hosts are quite meager (Holbein et al. 2016).

A recent study made by Sidonskaya et al. (2016) has enlightened the role of a protein phosphatase in the Arabidopsis defense response against the parasitic cyst nematode *Heterodera schachtii*. Pathogenic sugar beet cyst nematode *H. schachtii* (HS) can generally parasitize upon plant species belonging to Brassicaceae as well as Chenopodiaceae, including *A. thaliana* (Sijmons et al. 1991). Juvenile stages of HS penetrates entire root region of hosts manipulating its peculiar stylet (Wyss 1992; Goellner et al. 2001) and thereby reaches inside of the plant causing significant damages to host tissues in its path (Sidonskaya et al. 2016). The characteristic initial syncytial cell (ISC) within host where HS attaches serves as the early feeding site of the nematode before its subsequent developmental stages resume (Wyss 1992; Wyss and Grundler 1992; Wyss and Zunke 1986). In the entire course of infection and migration of HS inside the plant host, the high degree of mechanical injuries afflicted to host tissues with liberation of host cell damaging secretory products (like cellulases or other cell wall degrading enzymes/factors) by HS is said to trigger plant immune signaling significantly which may involve specific RLPs, MAPK cascades, etc. (Sidonskaya et al. 2016). In their study, Sidonskaya and associates (2016) witnessed transient but distinct expression levels of MAPKs, i.e., MPK3/MPK6, in *A. thaliana* root epidermal cells when stimulated by artificial wounding; cellulase treatment in the same cells resulted in the distinct expression of MPK6 kinases. This is suggestive of probable events ensuing out of HS-mediated wounding as well as secreted products of HS in *A. thaliana*. Arabidopsis MAPK phosphatase AP2C1 (a PP2C member) has been correlated with negative regulation of HS triggered MAPKs- MPK3 and MPK6 in the same study made by these authors. The same group reported that mutant *ap2c1 A. thaliana* lines, while responding to HS infection, manifested elevated levels of MPK3 and MPK6 expressions. This accompanied enhanced resistance to HS infections followed by drop in the progression of both syncytia and nematode developments in the *ap2c1* mutant Arabidopsis. Interestingly, their study has further indicated decisive role of AP2C1 in the negative regulation of plant defense during initial phases of HS infection as development of HS-induced syncytia was markedly inhibited in the *ap2c1* mutant Arabidopsis which eventually affected nematode development. Importantly, the evidence relating to activation of specific MAPKs in the Arabidopsis defense response to HS infection process has also been revealed through their work.

Other than the MAPK-AP2C1 partners, no such distinct PK-plant PP collaborations have been reported for plant immune function regulations against nematode. However, in a couple of occasions, some coordination of either PK or plant PPs in response to nematode infection has been appreciated. For instance, in *A. thaliana*, peroxisome localized PP2A phosphatase comprising B′θ subunit manifested elevated expression level in response to infection with root-knot nematode pathogen *Meloidogyne incognita* (Kataya et al. 2015). Whether PP2A-B′θ subunit had direct or indirect role in regulation of host immune responses against *M. incognita* is not clear, nor it is known regarding specific RLPs/RLKs involved in the nematode recognition.

In one instance, the role of Arabidopsis orthologues for co-receptor BAK1 in tomato (i.e., SlSERK3A and SlSERK3B) defense response towards *M. incognita* infection was ascertained by VIGS (Virus-Induced Gene Silencing) using tobacco rattle virus (TRV)-based vectors (Peng and Kaloshian 2014). Silencing of both SlSERK3A and SlSERK3B in tomato resulted in enhanced susceptibility of the host towards *M. incognita* infection (Peng and Kaloshian 2014) suggesting that tomato BAK1 orthologue might be an intermediary kinase component in the defense signaling pathway. A corresponding PP for regulation of defense responses to *M. incognita* has not been found out yet.

A recent study has identified an Arabidopsis RLK designated as NILR1 to be specifically responsive to both parasitic nematodes HS and *M. incognita* (Mendy et al. 2017). Further, dependence of NILR1 on BAK1 co-receptor for PTI signaling downstream was revealed in the same study. Authors have found Arabidopsis lines carrying mutation in *nilr1* to be highly susceptible to both HS and *M. incognita* infections (Mendy et al. 2017). However, no PP counterparts have yet been established with respect to role in regulation of NILR1-mediated immune signaling.

In addition to agents cited above, few unicellular protozoan species are recognized to negatively impact health of several plant species (Camargo et al. 1990; Dollet 1994; Camargo 1999; Santos et al. 2007). As opposed to that, significant beneficial effects of rhizospheric protozoan species towards plants have also been ascertained (Bonkowski 2003). In respect of protozoa-aided pathogenicity in plants, a diverse bunch of trypanosomatids specialized in infecting plants is known and constitutes the group *Phytomonas* (Jaskowska et al. 2015). Three species of *Phytomonas* are designated plant agonizers: e.g., *Phytomonas staheli*, which causes fatal wilt of coconut palm (*Cocos nucifera*) and sudden and slow wilt of oil palm (*Elaeis guineensis*); *P. leptovasorum* which infects coffee plants; and *P. françai* which causes empty root syndrome in cassava (*Manihot esculenta*) (Parthasarathy et al. 1976; Di Lucca et al. 2013). *P. staheli* and *P. leptovasorum* are said to reside in the phloem tissues of the host (Stahel 1931;Parthasarathy et al. 1976). In recent documentations, *P. francai* has been listed as not so lethal parasite and thus hardly may pose threat to food security (Jaskowska et al. 2015). However, clear understandings on various aspects relating to these phytomonads in terms of their mode of infection, adaptation to host, induction of host defense signals, etc. are still inadequate (Jaskowska et al. 2015). In this context, direct or indirect involvement of plant PPs in the regulation of immune responses towards phytomonads is yet to be unveiled. Whether phytomonad-conserved molecular patterns can equally serve as PAMPs or MAPMs still remains to be confirmed.

15.4 Pathogenic Protein Phosphatases Involved in Virulence

A number of plant protein phosphatases, though not exhaustive, like Xb15, PP2C38, PP2A homologues, etc. have been associated with regulation of plant immune functions induced by various types of microbial pathogens. Nevertheless, these enzymes have a significant host-protective function. In the same manner, plant pathogenic microbes too possess protein phosphatase equivalents executing diverse functions in them. PPs regulate essential cellular processes including growth, proliferation, metabolism as well as cell survivability in microbes too. Readers may refer some of the literatures for host of PP functions in microbes (Dickman and Yarden 1999; Lammers and Lavi 2007; Cutler et al. 2010; Ariño et al. 2011; Du et al. 2013; Chen et al. 2016). A number of animal pathogenic microbes rely on specific PP functions for their pathogenicity in target hosts (Chen et al. 2016). Likewise, protein phosphatases of plant pathogenic microbes have also been implicated in their virulence functions within specific hosts (Du et al. 2013; Liu et al. 2016). However, it needs mention that pathogenic microbes recruit a variety of virulence factors including T3Es, type 2 secretion system released proteins as well as several others targeting host defense (Abramovitch et al. 2006), although only some of those may have been fully characterized and demonstrated to have PP-related function. Bypassing the detailed illustration of PP types in plant pathogenic microbes, here we emphasize on some of the distinct PPs of phytopathogenic origin that have been shown to have crucial role in virulence function of the former. Extensive outlines on plant pathogen origin PP types and their role in other functions can be found elsewhere (Kennelly 1998, 2001, 2002; Deutscher and Saier 2005; Standish and Morona 2014).

15.4.1 Examples of Bacterial PPs Required for Virulence

Xanthomonas gardneri (*Xg*), which causes bacterial spot disease of pepper, expresses the T3E protein AvrBs7. AvrBs7 is known to be recognized by corresponding host dominant resistance gene product, Bs7, that initiates resistance response outcomes like HR. Potnis and associates (2012) in their study found AvrBs7 of *Xg* to contain a putative protein tyrosine phosphatase (PTP) active site domain which was essentially required for its cognition by host [*Capsicum annuum* cv. Early Calwonder (ECW)]-encoded Bs7. Mutation of the cysteine residue (at

265th position) to serine in the PTP domain of AvrBs7 resulted in abrogation of HR responses, indicating PTP domain to be crucial for host recognition *via* Bs7 for downstream defense signaling (Potnis et al. 2012). Another species of *Xanthomonas*, *X. euvesicatoria* (*Xe*), responsible for bacterial spot disease in pepper as well as tomato (Potnis et al. 2012; Kyeon et al. 2016), which expresses AvrBs1.1 (a *Xe* secreted T3E) is also recognized by the common host resistance protein Bs7 (Potnis et al. 2012). Interestingly, AvrBs1.1 and AvrBs7 both possess a consensus PTP domain with important role in virulence mechanism of *Xe* and *Xg*, respectively. Importantly, alteration in the ten amino acid sequence of AvrBs1.1 catalytic domain eliminated induction of HR suggesting distinct recognition site on AVrBs1.1 and AvrBs7, respectively, for Bs7 (Potnis et al. 2012). Likewise, other xanthomonads comprising *Xcc* strains 33,913, 8004 and B100 are known to produce AvrBs1.1 effector proteins (also referred to as XopH in recent classification; White et al. 2009) that share PTP domains essential for virulence in respective hosts (Potnis et al. 2012).

Genome of *P. syringae* pv. *tomato* DC3000 (Pto DC3000) encodes a T3E protein, HopPtoD2, that was determined to possess a PTP domain within (Bretz et al. 2003). HopPtoD2 was shown to be translocated into *A. thaliana* cells by TTSS machinery of *Pto* DC3000 during infection process. Mutation in the PTP domain of HopPtoD2 was found to significantly diminish *Pto* DC3000-mediated virulence in Arabidopsis (Bretz et al. 2003) entailing worth of PTP domain in this T3E. HopPtoD2 later also renamed HopAO1 (Underwood et al. 2007) was shown to suppress HR in certain non-plant hosts when expressed ectopically (Bretz et al. 2003; Espinosa et al. 2003). Arabidopsis plant expressing HopAO1 (with intact PTP domain) allowed TTSS mutant *Pto* DC3000 to proliferate remarkably in the transgenic plant as compared to wild-type host (Underwood et al. 2007).

Constitutively expressed HopAO1 could further block flg22 (PAMP)-mediated immune responses in transgenic Arabidopsis, permitting multiplication of *Pto* DC3000 therein (Underwood et al. 2007). Further investigations on HopAO1 have revealed a precise mechanism of its function. The Arabidopsis PRR, EF-Tu receptor (EFR), is usually activated by elf18 (a PAMP) *via* addition of phosphoryl group at a Tyrosine residue [situated at 836th position] (Macho et al. 2014). Activated EFR subsequently initiates downstream defense signaling towards *Pto* DC3000. In the seminal work of Macho et al. (2014), HopAO1 was shown to suppress host immune responses by minimizing EFR phosphorylation (Macho et al. 2014) probably through its phosphatase activity [Fig. 15.4]. HopAO1 certainly demonstrates an intriguing armory which evolved in *P. syringae* to evade host defenses.

15.4.2 Examples of Fungal PPs Required for Virulence

The fungal pathogen for rice blast disease, *Magnaporthe oryzae* (*Mo*), expresses a putative dual-specificity phosphatase (DsPTP), MoYVH1, which has been implicated in the vegetative growth and conidial production of *Mo* as well as in its

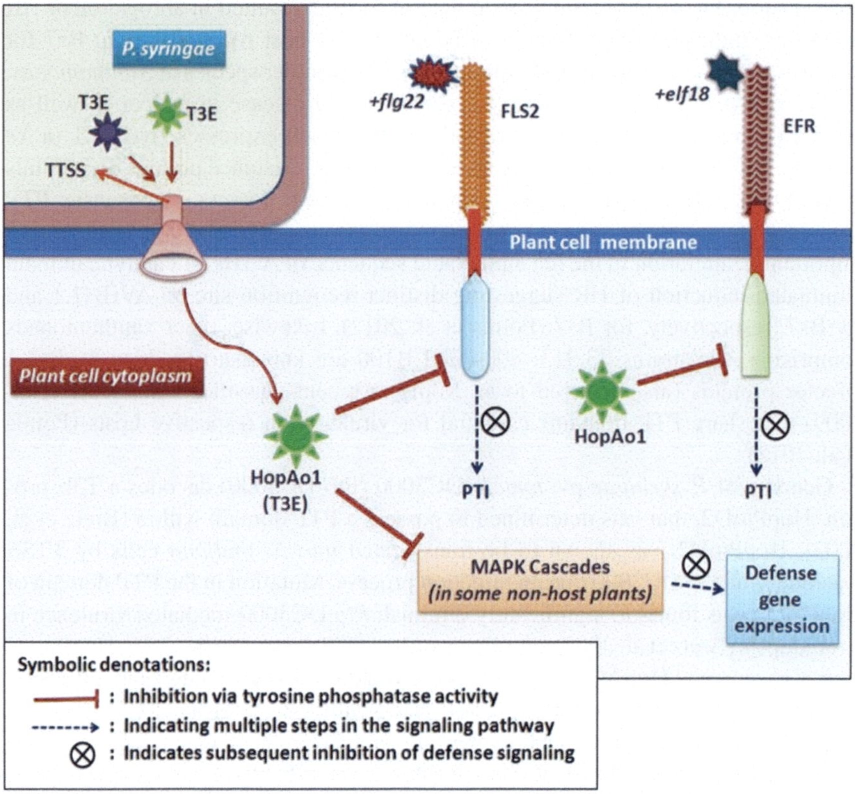

Fig. 15.4 Mechanism of *P. syringae* type 3 effector HopOA1 (possessing PTP catalytic domain) mediated suppression of plant cell immune responses. HopAO1 effector with tyrosine phosphatase activity is introduced by *P. syringae* into plant cell *via* TTSS (Bretz et al. 2003; Underwood et al. 2007). HopAO1 has been independently shown to inhibit RLKs FLS2 (that recognizes flagellin (MAMP), e.g., flg22 peptide) and EFR (that recognizes bacterial elongation factor (MAMP), e.g. elf18 peptide)-triggered signaling due to binding of respective MAMPs to them (Underwood et al. 2007; Macho et al. 2014). Inhibition of RLKs results in halting of subsequent PTI responses downstream. In some non-plant host species (like *Nicotiana benthamiana*), HopAO1 has been further shown to negatively affect MAPK signaling cascades (Bretz et al. 2003; Espinosa et al. 2003). Phosphatase function of HopAO1 is indeed a boon to *P. syringae* for restricting host defenses and thus advantageous for its own proliferation. *Abbreviations:* TTSS, type 3 secretion system; T3E, type 3 secreted effector; PTI, PAMP/MAMP triggered immunity

virulence mechanism (Liu et al. 2016). Mutation in MoYVH1 manifested defects in fungal hyphae growth and reduced pathogenicity in rice plant along with other cellular deficiencies in *Mo* (Liu et al. 2016). Liu and associates (2016) demonstrated ROS scavenging role of MoYVH1, the counter-mechanism thought to be crucial for successful establishment of *Mo* infection (Liu et al. 2016). A recent study has revealed MoYVH1 to be important for expression of several extracellular enzymes in *Mo*, needed for altering host defense (Liu et al. 2018). The same study indicated requisite of MoYVH1 translocation to nucleus and its interaction with other factors for modulating ROS pathway (Liu et al. 2018). Likewise in an earlier study, the role of MoPPG1 [which encodes the putative PP2A catalytic subunit (PP2Ac) in *Mo*] could be associated with *Mo* virulence in rice plant host (Du et al. 2013). *MoPPG1* mutation diminished penetrability of *Mo* into host and thus reduced its pathogenicity (Du et al. 2013). Deletion of *MoPPG1* further resulted in defective vegetative hyphal growth, conidiation as well as impaired expression levels of several *Mo* pathogenicity effectors (Du et al. 2013) indicating significance of PP2Ac in the blast pathogen.

Aspergillus flavus (*Afl*) is an opportunistic, saprophytic fungal pathogen of several crops that produces mycotoxin aflatoxin (Klich 2007; Yang et al. 2018). AflCDC14, a DSP-type phosphatase in *Afl*, has been connected with virulence function of this pathogen (Yang et al. 2018). *Afl* strains carrying *AflCDC14* deletion were compromised in growth and conidium morphology with associated reduction in amylase activity (Yang et al. 2018). *AflCDC14* deletion mutant manifested virulence deficiency, thus indicating requirement of AflCDC14 phosphatase function in *Afl* virulence (Yang et al. 2018).

The *Fusarium* head blight (FHB) disease agent of several cereal crops, *Fusarium graminearum* (*Fg*), was shown to depend on PP2C phosphatase functions for its virulence at least in two occasions (Jiang et al. 2010, 2011). The representative PP2C members, namely, FgPtc1p and FgPtc3, were independently demonstrated to be important for *Fg* pathogenicity functions (Jiang et al. 2010, 2011). *FgPtc1p* deletion resulted in delayed hyphal development as well as reduced virulence of the pathogen in wheat plants (Jiang et al. 2010). In the same manner, mutation in *FgPtc3* locus affected aerial hyphae development in the FHB pathogen and also obstructed the ability of *Fg* to infect wheat flowering head, thereby minimizing *Fg* virulence (Jiang et al. 2011).

15.5 Plant Protein Phosphatases as Targets of Pathogen Maneuvering

Plant protein phosphatases being at the heart of immune regulatory network seem to associate more often with respective PRRs, RLKs, RLCKs as well as intracellular receptors like NB-LRRs. Therefore, responses of plant PPs towards pathogens appear somewhat indirect. However, there are instances when pathogenic virulence factors are reported to target or manipulate plant PPs to their advantage. We highlight few of such known instances below.

TTSS-secreted T3Es belonging to AvrE family are virulence factors of some notable phytopathogenic bacteria such as *Pseudomonas*, *Pantoea*, *Ralstonia*, *Erwinia*, *Dickeya*, *Pectobacterium*, etc. (Jin et al. 2016). A yeast two-hybrid screen revealed interaction of an AvrE effector WtsE, from the causal agent of Stewart's wilt in maize, *Pantoea stewartii* subsp. *stewartii* (*Pss*), with PP2A phosphatase of maize plant host (Jin et al. 2016). More specifically, WtsE interacts with the B′ regulatory subunit of hetero-trimeric maize PP2A (Jin et al. 2016). Likewise, one more AvrE family member, AvrE1 secreted by *P. syringae* DC3000 strain (*Pto* DC3000), was shown to interact with B′ regulatory subunit of Arabidopsis PP2A homologue (Jin et al. 2016). Loss-of-function mutation in Arabidopsis PP2A homologue impaired *Pto* DC3000 virulence in the host indicating requirement of PP2A protein for pathogen's advantage (Jin et al. 2016). Chemical inhibition of PP2A activities further revealed halting of pathogenicity functions due to WtsE and AvrE1, respectively (Jin et al. 2016). Another *Pto* DC3000-secreted T3E member, HopM1, belonging to same AvrE family and sharing functional redundancy with AvrE1 was also found to have impaired virulence in the PP2A mutant host background (Jin et al. 2016). Interestingly, evolution of T3Es targeting plant PPs to get hold of its hosts may indicate that pathogen virulence functions are evolving rapidly with novel mechanisms to counteract plant host immune functions.

Dependence of *Pto* DC3000 on Arabidopsis MAPK phosphatase 1 (MKP1) for initial infection processes in *A. thaliana* was revealed from the work of Anderson et al. (2014). Their study demonstrated requirement of intact MKP1 in the host for induction of genes required for TTSS and release of effectors by *Pto* DC3000 leading to pathogenicity. Arabidopsis *mkp1* mutant lines were more resistant to *Pto* DC3000, and TTSS induction in the pathogen was also impaired in the mutant host (Anderson et al. 2014). Authors (Anderson et al. 2014) have connected some bioactive chemical signaling mechanism which was diminished in the *mkp1* mutant host to be responsible for inactivation of TTSS induction that is otherwise functional in the wild-type Arabidopsis expressing MKP1.

The late blight disease pathogen, *Phytophthora infestans*, was shown in a study to rely on host protein phosphatase (PP1c) isoforms for its virulence in potato and model plant *N. benthamiana* (Boevink et al. 2016). *P. infestans*-extruded effector Pi04314 was demonstrated to specifically target PP1c isoforms and cause repositioning of PP1cs for the pathogen's successful virulence establishment in the hosts. Loss-of-function mutation in the *pp1c* isoforms significantly diminished virulence function of *P. infestans* as well as pathogen colonization, suggesting critical dependence of the pathogen on this type of host PP. The investigators further hypothesize cooperative action of both Pi04314 and PP1c as a probable mechanism for progression of late blight disease (Boevink et al. 2016).

Plant pathogens have been known to manipulate host biochemical pathways such as hormone (e.g., abscisic acid, jasmonic acid, etc.)-mediated signaling networks to establish their pathogenicity. For instance, phytopathogenic *Pto* DC3000 can induce abscisic acid (ABA) signaling in its host (de Torres-Zabala et al. 2007). *Pto* DC3000 was also correlated with production of a phytotoxin coronatine which happens to be a jasmonate analogue enabling the pathogen to activate jasmonic acid (JA) pathway as well (Mine et al. 2017). Interestingly, ABA and JA pathways were shown to arrest signaling of MAPKs, MPK3 and MPK6, which form essential components of Arabidopsis defense response pathway (Mine et al. 2017). On the other hand, ABA was found to activate a group of PP2C phosphatases (HAI1, HAI2, HAI3, ABI1 clade) (de Torres-Zabala et al. 2007; Mine et al. 2017), some of which were instrumental in negatively regulating both MPK3 and MPK6 (Mine et al. 2017) [Fig. 15.5a]. In the absence of ABA/JA signaling in the host, *Pto* DC3000 could

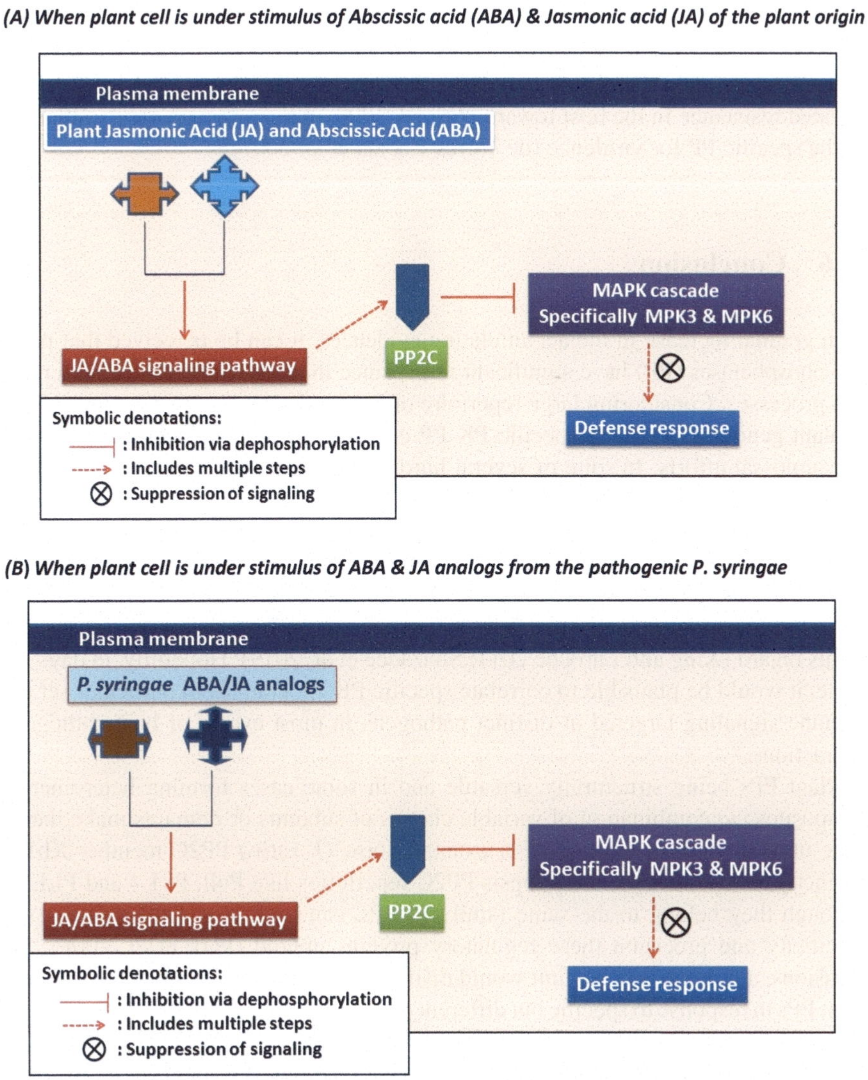

Fig. 15.5 Illustration of the mechanism through which *P. syringae* manipulates *A. thaliana's* ABA/JA signaling pathways to dampen host defense responses against the pathogen. (**a**) Generally, when *A. thaliana* cells are under stimulation of own ABA and JA molecules, subsequent induction of ABA/JA hormonal signaling pathways occurs within the cell. Interestingly, ABA and JA pathways are known to arrest signaling of MAPKs, MPK3 and MPK6 (which form important arsenals of Arabidopsis defense response pathway), *via* recruitment of some PP2C-type phosphatases (that may include HAI1, HAI2, HAI3, ABI1 clade) (de Torres-Zabala et al. 2007; Mine et al. 2017). (**b**) In the absence of *A. thaliana*'s own ABA/JA molecules, *P. syringae* deploys its own ABA/JA analogues and activates ABA/JA signaling within the host cell. Induction of ABA/JA signaling ultimately serves the pathogen in achieving suppression of host defense responses *via* recruitment of host PP2C phosphatase-mediated negative regulation of MAPKs, MPK3 and MPK6, as above (de Torres-Zabala et al. 2007; Mine et al. 2017)

induce both ABA and JA cascades to achieve PP2C-mediated suppression of MAPK defense signaling for its subsequent colonization and virulence in the host (Mine et al. 2017) [Fig. 15.5b]. Loss-of-function mutation in PP2C members resulted in elevated resistance in the host towards *Pto* DC3000, indicating pathogen's reliance on the specific PP for virulence (de Torres-Zabala et al. 2007).

15.6 Conclusion

With gradual increase in the accumulating evidences, it can be perceived that protein phosphatases (PP) have significant importance in plant host-pathogen interaction processes. Considering large repertoire of PKs and also diverse clusters of PPs in plant genomes, studying specific PK-PP couples in immune functions necessitates colossal efforts. In spite of several hurdles, genome-wide studies have, however, predicted homologues/orthologues of PKs as well as PPs in several plant species. Since the demonstration of the first purified plant protein phosphatase in the 1980s (Lin et al. 1980), after almost four decades, the amount of progress made in plant PP research has been encouraging, even though there is still a long way to go. The advances in the genomics and allied disciplines have been assisting immensely in this regard (Xing and Laroche 2011; Shazadee et al. 2019). Hopefully, in days to come, it would be plausible to correlate specific PK-PP couples to particular set of immune signaling targeted at distinct pathogens in most events of host-pathogen interactions.

Plant PPs being structurally versatile and in some cases forming heteromeric composites *via* combination of variable classes of subunits or domains make them quite unpredictable in function. For example, how *O. sativa* PP2C member Xb15 distinctly functions from Arabidopsis PP2C repertoires like Poll, PLL4 and PLL5, although they belong to the same family of PPs, can itself suggest the degree of specificity and precision these regulatory proteins uphold (Park et al. 2008). In occasions, alteration of a subunit would distinguish immune regulatory function of plant PPs in response to specific but different pathogens (Trotta et al. 2011; Segonzac et al. 2014).

Nonetheless, model plant Arabidopsis and its genome information as well as some of the pathogenic model organisms like *P. syringae*, *B. cinerea*, and *H. schachtii* have afforded paramount contributions towards understanding of plant immune functions and their role in immune regulations. Several intriguing aspects relating to plant PP functions have been realized progressively. Distinct role of PPs in both plant and pathogenic agents has further accentuated significances of PPs during host-pathogen communications. Interestingly, pathogenic effectors targeting plant PPs have also been revealed. Some pathogenic effectors with expanded PP functions have further intrigued researchers, for evolution of pathogenic counter-defense mechanisms akin to this seems quite extravagant, although much remains to be known in this regard.

Complexity of host genomes followed by difficulties in executing genetic studies such as gene silencing and gene knockout procedures due to apparent lack of suitable vectors/agents has considerably limited the extent of research to only model hosts. Hopefully, in days to come, intensive research in this theme would ameliorate these challenges to great extents.

Determination of pathogen PP equivalents and inventorying their roles in virulence mechanisms in hosts would be rewarding in the near future while designing suitable strategies to engineer virulence resistances in the hosts. Some of the plant PPs are serving as susceptible markers by being targets of pathogenic effectors. Engineering resistant isoforms/equivalents of such susceptible PPs from different cultivars/species may be one of the strategies that could benefit agronomy immensely in the near future. Albeit our discussion on PPs aligned primarily towards plant host-pathogen interaction events only, a substantial understanding on PPs as a whole will be indispensable to realize the intriguing signaling cascades regulated by them or yet unknown mechanisms that regulate them. In days to come, it may be speculated that PPs might serve as potential molecular tool for selecting superior cultivars of plants species and hence contribute towards agronomy and plant breeding significantly. We hope in the near future that continued explorations in regard to protein phosphatase repertoires (that may be from both plant and pathogen) would surely enrich our understandings on the complicated paradigms of plant-microbe interactions, more in detail.

Acknowledgement A. Barman humbly acknowledges the financial support received in the form of NECBH (NECBH/2019-20/127) from the Department of Biotechnology, Govt. of India (project no. BT/COE/34/SP28408/2018). S.K. Ray acknowledges the students and research scholars of MPMI laboratory (MBBT Department, Tezpur University, Assam, India), with whom he discusses plant-microbe interaction studies.

References

Abad P, Gouzy J, Aury JM et al (2008) Genome sequence of the metazoan plant-parasitic nematode *Meloidogyne incognita*. Nat Biotechnol 26(8):909–915

Abramovitch RB, Anderson JC, Martin GB (2006) Bacterial elicitation and evasion of plant innate immunity. Nat Rev Mol Cell Biol 7(8):601–611

Adachi H, Tsuda K (2019) Convergence of cell-surface and intracellular immune receptor signaling. New Phytol 221(4):1676–1678

Agrawal GK, Jwa N, Rakwal R (2000) A novel rice (*Oryza sativa* L.) acidic PR1 gene highly responsive to cut, phytohormones, and protein phosphatase inhibitors. Biochem Biophys Res Commun 274(1):157–165

Akimoto-Tomiyama C, Tanabe S, Kajiwara H, Minami E, Ochiai H (2018) Loss of chloroplast-localized protein phosphatase 2Cs in Arabidopsis *thaliana* leads to enhancement of plant immunity and resistance to *Xanthomonas campestris* pv. *campestris* infection. Mol Plant Pathol 19(5):1184–1195

Alessi DR, Gomez N, Moorhead G et al (1995) Inactivation of p42 MAP kinase by protein phosphatase 2A and a protein tyrosine phosphatase, but not CL100, in various cell lines. Curr Biol 5(3):283–295

Alonso A, Sasin J, Bottini N, Friedberg I, Osterman A, Godzik A, Hunter T, Dixon J, Mustelin T (2004) Protein tyrosine phosphatases in the human genome. Cell 117(6):699–711
Anderson JC, Bartels S, González Besteiro MA, Shahollari B, Ulm R, Peck SC (2011) Arabidopsis MAP Kinase Phosphatase 1 (AtMKP1) negatively regulates MPK6-mediated PAMP responses and resistance against bacteria. Plant J 67:258–268
Anderson JC, Wan Y, Kim YM, Pasa-Tolic L, Metz TO, Peck SC (2014) Decreased abundance of type III secretion system-inducing signals in Arabidopsis *mkp1* enhances resistance against *Pseudomonas syringae*. Proc Natl Acad Sci U S A 111:6846–6851
Andreasson E, Jenkins T, Brodersen P, Thorgrimsen S, Petersen NH, Zhu S et al (2005) The MAP kinase substrate MKS1 is a regulator of plant defense responses. EMBO J 24:2579–2589
Archambault J et al (1998) FCP1, the RAP74-interacting subunit of a human protein phosphatase that dephosphorylates the carboxyl-terminal domain of RNA polymerase IIO. J Biol Chem 273:27593–27601
Ariño J, Casamayor A, González A (2011) Type 2C protein phosphatases in fungi. Eukaryot Cell 10(1):21–33
Bajsa J, Pan Z, Duke SO (2011) Transcriptional responses to cantharidin, a protein phosphatase inhibitor, in Arabidopsis *thaliana* reveal the involvement of multiple signal transduction pathways. Physiol Plant 143(2):188–205
Barford D, Das AK, Egloff MP (1998) The structure and mechanism of protein phosphatases: insights into catalysis and regulation. Annu Rev Biophys Biomol Struct 27:133–164
Barr FA, Elliott PR, Gruneberg U (2011) Protein phosphatases and the regulation of mitosis. J Cell Sci 124(14):2323–2334
Bartels D, Sunkar R (2005) Drought and salt tolerance in plants. Crit Rev Plant Sci 24(1):23–58
Bartels S, Anderson JC, González Besteiro MA, Carreri A, Hirt H, Buchala A, Métraux JP, Peck SC, Ulm R (2009) MAP kinase phosphatase1 and protein tyrosine phosphatase1 are repressors of salicylic acid synthesis and SNC1-mediated responses in Arabidopsis. Plant Cell 21:2884–2897
Bartels S, Besteiro MAG, Lang D, Ulm R (2010) Emerging functions for plant MAP kinase phosphatases. Trends Plant Sci 15:322–329
Becraft PW (2002) Receptor kinase signaling in plant development. Annu Rev Cell Dev Biol 18:163–192
Berger S, Sinha AK, Roitsch T (2007) Plant physiology meets phytopathology: plant primary metabolism and plant–pathogen interactions. J Exp Bot 58(15/16):4019–4026
Bigeard J, Hirt H (2018) Nuclear signaling of plant MAPKs. Front Plant Sci 11(9):469
Boevink PC, Wang X, McLellan H et al (2016) A *Phytophthora infestans* RXLR effector targets plant PP1c isoforms that promote late blight disease. Nat Commun 7:10311
Boller T, Felix G (2009) A renaissance of elicitors: perception of microbe-associated molecular patterns and danger signals by pattern-recognition receptors. Annu Rev Plant Biol 60:379–406
Boller T, He SY (2009) Innate immunity in plants: an arms race between pattern recognition receptors in plants and effectors in microbial pathogens. Science 324:742–744
Bonkowski M (2003) Protozoa and plant growth: the microbial loop in soil revisited. New Phytol 162:617–631
Boutrot F, Zipfel C (2017) Function, discovery, and exploitation of plant pattern recognition receptors for broad-spectrum disease resistance. Annu Rev Phytopathol 55:257–286
Boyd LA, Ridout C, O'Sullivan DM, Leach JE, Leung H (2013) Plant–pathogen interactions: disease resistance in modern agriculture. Trends Genet 29(4):233–240
Bretz JR, Mock NM, Charity JC, Zeyad S, Baker CJ, Hutcheson SW (2003) A translocated protein tyrosine phosphatase of *Pseudomonas syringae* pv. *tomato* DC3000 modulates plant defense response to infection. Mol Microbiol 49:389–400
Brown JK, Hovmøller MS (2002) Aerial dispersal of pathogens on the global and continental scales and its impact on plant disease. Science 297(5581):537–541
Butler EJ, Jones SG (1949) Tomato leaf mould, *Cladosporium fulvum* Cooke. Macmillan, London

Caboni P, Ntalli NG, Aissani N, Cavoski I, Angioni A (2012) Nematicidal activity of (E, E)-2,4-decadienal and (E)-2-decenal from *Ailanthus altissima* against *Meloidogyne javanica*. J Agric Food Chem 60:1146–1151

Cai R, Lewis J, Yan S, Liu H, Clarke CR, Campanile F et al (2011) The plant pathogen *Pseudomonas syringae* pv. *tomato* is genetically monomorphic and under strong selection to evade tomato immunity. PLoS Pathog 7(8):e1002130

Camagna M, Takemoto D (2018) Hypersensitive response in plants. In: eLS. John Wiley & Sons Ltd, Chichester. https://doi.org/10.1002/9780470015902.a0020103.pub2

Camargo EP (1999) Phytomonas and other trypanosomatid parasites of plants and fruit. Adv Parasitol 42:29e112

Camargo EP, Kastelein P, Roitman I (1990) Trypanosomatid parasites of plants (Phytomonas). Parasitol Today 6:22e25

Chakrabarti A, Panter SN, Harrison K, Jones JD, Jones DA (2009) Regions of the Cf-9B disease resistance protein able to cause spontaneous necrosis in *Nicotiana benthamiana* lie within the region controlling pathogen recognition in tomato. Mol Plant-Microbe Interact 22(10):1214–1226

Chen E, Choy MS, Petrényi K, Kónya Z, Erdődi F, Dombrádi V, Peti W, Page R (2016) Molecular insights into the fungus-specific serine/threonine protein phosphatase Z1 in *Candida albicans*. MBio 7(4):e00872-16

Chen L, Zhang Z, Liang H, Liu H, Du L, Xu H et al (2008) Overexpression of TiERF1 enhances resistance to sharp eyespot in transgenic wheat. J Exp Bot 59:4195–4204

Chen MJ, Dixon JE, Manning G (2017) Genomics and evolution of protein phosphatases. Sci Signal 10:eaag1796

Chen X, Chern M, Canlas PE, Ruan D, Jiang C, Ronald PC (2010) An ATPase promotes autophosphorylation of the pattern recognition receptor XA21 and inhibits XA21-mediated immunity. Proc Natl Acad Sci U S A 107:8029–8034

Chinchilla D, Zipfel C, Robatzek S, Kemmerling B, Nürnberger T, Jones JD, Felix G, Boller T (2007) A flagellin-induced complex of the receptor FLS2 and BAK1 initiates plant defense. Nature 448(7152):497–500

Cho EJ, Kobor MS, Kim M, Greenblatt J, Buratowski S (2001) Opposing effects of Ctk1 kinase and Fcp1 phosphatase at Ser 2 of the RNA polymerase II C-terminal domain. Genes Dev 15:3319–3329

Choi HW, Klessig DF (2016) DAMPs, MAMPs, and NAMPs in plant innate immunity. BMC Plant Biol 16:232

Clarke CR, Chinchilla D, Hind SR et al (2013) Allelic variation in two distinct *Pseudomonas syringae* flagellin epitopes modulates the strength of plant immune responses but not bacterial motility. New Phytol 200(3):847–860

Coll NS, Epple P, Dangl JL (2011) Programmed cell death in the plant immune system. Cell Death Differ 18(8):1247–1256

Couto D, Niebergall R, Liang X, Bücherl CA, Sklenar J, Macho AP et al (2016) The Arabidopsis protein phosphatase PP2C38 negatively regulates the central immune kinase BIK1. PLoS Pathog 12(8):e1005811

Couto D, Zipfel C (2016) Regulation of pattern recognition receptor signaling in plants. Nat Rev Immunol 16(9):537–552

Cutler SR, Rodriguez PL, Finkelstein RR, Abrams SR (2010) Abscisic acid: emergence of a core signaling network. Annu Rev Plant Biol 61:651–679

Dardick C, Ronald P (2006) Plant and animal pathogen recognition receptors signal through non-RD kinases. PLoS Pathog 2(1):e2

Davis KR, Darvill AG, Albersheim P, Dell A (1986) Host-pathogen interactions. XXIX. Oligogalacturonides released from sodium polypectate by endopolygalacturonic acid lyase are elicitors of phytoalexins in soybean. Plant Physiol 80:568–577

de Torres-Zabala M, Truman W, Bennett MH et al (2007) *Pseudomonas syringae* pv. *tomato* hijacks the Arabidopsis abscisic acid signaling pathway to cause disease. EMBO J 26(5):1434–1443

de Wit PJGM (2007) How plants recognize pathogens and defend themselves. Cell Mol Life Sci 64(21):2726–2732
Dean R, Van Kan JA, Pretorius ZA et al (2012) The top 10 fungal pathogens in molecular plant pathology. Mol Plant Pathol 13(4):414–430
Deutscher J, Saier MH (2005) Ser/Thr/Tyr protein phosphorylation in bacteria—for long time neglected, now well established. J Mol Microbiol Biotechnol 9:125–131
Di Lucca AGT, Trinidad Chipana EF, Talledo Albújar MJ, Dávila Peralta W, Montoya Piedra YC et al (2013) Slow wilt: another form of Marchitez in oil palm associated with trypanosomatids in Peru. Trop Plant Pathol 38:522–533
Dickman MB, Yarden O (1999) Serine/threonine protein kinases and phosphatases in filamentious fungi. Fungal Genet Biol 26:99–117
Dodds PN, Rathjen JP (2010) Plant immunity: towards an integrated view of plant–pathogen interactions. Nat Rev Genet 11:539–548
Dollet M (1994) Identification and characterization of pest organisms; plant trypanosomes case study. In: HawksWorth DL (ed) The identification and characterization of Pest organisms. CAB Intal and the Systemic Association, Wallingford, pp 415–426
Du Y, Shi Y, Yang J, Chen X, Xue M, Zhou W, Peng YL (2013) A serine/threonine-protein phosphatase PP2A catalytic subunit is essential for asexual development and plant infection in *Magnaporthe oryzae*. Curr Genet 59(1–2):33–41
Espinosa A et al (2003) The *Pseudomonas syringae* type III-secreted protein HopPtoD2 possesses protein tyrosine phosphatase activity and suppresses programmed cell death in plants. Mol Microbiol 49(2):377–387
Faulkner C, Robatzek S (2012) Plants and pathogens: putting infection strategies and defense mechanisms on the map. Curr Opin Plant Biol 15:699–707
Felix G, Duran JD, Volko S, Boller T (1999) Plants have a sensitive perception system for the most conserved domain of bacterial flagellin. Plant J 18:265–276
Felix G, Regenass M, Boller T (1993) Specific perception of subnanomolar concentrations of chitin fragments by tomato cells: induction of extracellular alkalinization, changes in protein phosphorylation, and establishment of a refractory state. Plant J 4:307–316
Felix G, Regenass M, Spanu P, Boller T (1994) The protein phosphatase inhibitor calyculin A mimics elicitor action in plant cells and induces rapid hyperphosphorylation of specific proteins as revealed by pulse labeling with [^{33}P] phosphate. Proc Natl Acad Sci U S A 91(3):952–956
Feng J, Wakeman T, Yong S et al (2009) Protein phosphatase 2A-dependent dephosphorylation of Replication Protein A is required for the repair of DNA breaks induced by replication stress. Mol Cell Biol 29(21):5696–5709
Ferrari S, Galletti R, Denoux C, De Lorenzo G, Ausubel FM, Dewdney J (2007) Resistance to *Botrytis cinerea* induced in Arabidopsis by elicitors is independent of salicylic acid, ethylene, or jasmonate signaling but requires PHYTOALEXIN DEFICIENT3. Plant Physiol 144:367–379
Ferrari S, Savatin DV, Sicilia F et al (2013) Oligogalacturonides: plant damage-associated molecular patterns and regulators of growth and development. Front Plant Sci 4:49
Ferreira RB, Monteiro S, Freitas R et al (2006) Fungal pathogens: the battle for plant infection. Crit Rev Plant Sci 25(6):505–524
Forrest ARR, Ravasi T, Taylor D et al (2003) Phosphoregulators: protein kinases and protein phosphatases of mouse. Genome Res 13(6b):1443–1454
Gabriels SHEJ, Vossen JH, Ekengren SK et al (2007) An NB-LRR protein required for HR signaling mediated by both extra- and intracellular resistance proteins. Plant J 50:14–28
Galletti R, Denoux C, Gambetta S, Dewdney J, Ausubel FM, De Lorenzo G, Ferrari S (2008) The AtrbohD-mediated oxidative burst elicited by oligogalacturonides in Arabidopsis is dispensable for the activation of defense responses effective against *Botrytis cinerea*. Plant Physiol 148:1695–1706
Galletti R, Ferrari S, Lorenzo GD (2011) Arabidopsis MPK3 and MPK6 play different roles in basal and oligogalacturonide- or flagellin-induced resistance against *Botrytis cinerea*. Plant Physiol 157:804–814

Goellner M, Wang X, Davis EL (2001) Endo-beta-1,4-glucanase expression in compatible plant–nematode interactions. Plant Cell 13:2241–2255

Goldberg Y (1999) Protein phosphatase 2A: who shall regulate the regulator? Biochem Pharmacol 57:321–328

Gómez-Gómez L, Bauer Z, Boller T (2001) Both the extracellular leucine-rich repeat domain and the kinase activity of FLS2 are required for flagellin binding and signaling in Arabidopsis. Plant Cell 13:1155–1163

Gust AA, Biswas R, Lenz HD, Rauhut T, Ranf S, Kemmerling B, Gotz F, Glawischnig E, Lee J, Felix G, Nurnberger T (2007) Bacteria-derived peptidoglycans constitute pathogen-associated molecular patterns triggering innate immunity in Arabidopsis. J Biol Chem 282:32338–32348

Hahn MG, Bonhoff A, Grisebach H (1985) Quantitative localization of the phytoalexin glyceollin I in relation to fungal hyphae in soybean roots infected with *Phytophthora megasperma* f. sp. *glycinea*. Plant Physiol 77(3):591–601

Hahn MG, Darvill AG, Albersheim P (1981) Host-pathogen interactions. XIX. The endogenous elicitor, a fragment of a plant cell wall polysaccharide that elicits phytoalexin accumulation in soybeans. Plant Physiol 68:1161–1169

Hamada MS, Yin Y, Chen H, Ma Z (2011) The escalating threat of *Rhizoctonia cerealis*, the causal agent of sharp eyespot in wheat. Pest Manag Sci 67:1411–1419

Hammond-Kosack KE, Jones JDG (1996) Resistance gene-dependent plant defense responses. Plant Cell 8:1773–1791

He X, Anderson JC, del Pozo O, Gu YQ, Tang X, Martin GB (2004) Silencing of subfamily I of protein phosphatase 2A catalytic subunits results in activation of plant defense responses and localized cell death. Plant J 38(4):563–577

Heese A, Hann DR, Gimenez-Ibanez S et al (2007) The receptor-like kinase SERK3/BAK1 is a central regulator of innate immunity in plants. Proc Natl Acad Sci U S A 104(29):12217–12222

Hind SR, Strickler SR, Boyle PC et al (2016) Tomato receptor FLAGELLIN-SENSING 3 binds flgII-28 and activates the plant immune system. Nat Plants 2:16128

Holbein J, Grundler FMW, Siddique S (2016) Plant basal resistance to nematodes: an update. J Exp Bot 67(7):2049–2061

Jaskowska E, Butler C, Preston G, Kelly S (2015) Phytomonas: *Trypanosomatids* adapted to plant environments. PLoS Pathog 11(1):e1004484

Jha SK, Malik S, Sharma M, Pandey A, Pandey GK (2017) Recent advances in substrate identification of protein kinases in plants and their role in stress management. Curr Genomics 18:523–541

Jiang J, Yun Y, Yang Q, Shim W-B, Wang Z et al (2011) A type 2C protein phosphatase FgPtc3 is involved in cell wall integrity, lipid metabolism, and virulence in *Fusarium graminearum*. PLoS One 6(9):e25311

Jiang L, Yang J, Fan F, Zhang D, Wang X (2010) The type 2C protein phosphatase FgPtc1p of the plant fungal pathogen *Fusarium graminearum* is involved in lithium toxicity and virulence. Mol Plant Pathol 11(2):277–282

Jin L, Ham JH, Hage R, Zhao W, Soto-Hernández J, Lee SY et al (2016) Direct and indirect targeting of PP2A by conserved bacterial type-III effector proteins. PLoS Pathog 12(5):e1005609

Johnson LN, Noble MEM, Owen DJ (1996) Active and inactive protein kinases: structural basis for regulation. Cell 85:149–158

Jones JB, Jones JP, Stall RE, Zitter TA (1997) Compendium of tomato diseases. APS Press, St. Paul, MN

Jones JDG, Dangl JL (2006) The plant immune system. Nature 444:323–329

Kadota Y, Sklenar J, Derbyshire P et al (2014) Direct regulation of the NADPH oxidase RBOHD by the PRR-associated kinase BIK1 during plant immunity. Mol Cell 54(1):43–55

Kadota Y et al (2019) Quantitative phosphoproteomic analysis reveals common regulatory mechanisms between effector- and PAMP-triggered immunity in plants. New Phytol 221(4):2160–2175

Kaku H, Nishizawa Y, Ishii-Minami N, Akimoto-Tomiyama C et al (2006) Plant cells recognize chitin fragments for defense signaling through a plasma membrane receptor. Proc Natl Acad Sci U S A 103(29):11086–11091

Kataya ARA, Heidari B, Hagen L et al (2015) Protein phosphatase 2A holoenzyme is targeted to peroxisomes by piggybacking and positively affects peroxisomal b-oxidation. Plant Physiol 167:493–450

Kennelly PJ (1998) Prokaryotic protein-Serine/Threonine phosphatases. Methods Mol Biol 93:1–21

Kennelly PJ (2001) Protein phosphatases—a phylogenetic perspective. Chem Rev 101:2291–2312

Kennelly PJ (2002) Protein kinases and protein phosphatases in prokaryotes: a genomic perspective. FEMS Microbiol Lett 206(1):1–8

Kenton P, Mur LAJ, Draper J (1999) A requirement for calcium and protein phosphatase in the jasmonate-induced increase in tobacco leaf acid phosphatase specific activity. J Exp Bot 50:1331–1341

Kerk D, Bulgrien J, Smith DW, Barsam B, Veretnik S, Gribskov M (2002) The complement of protein phosphatase catalytic subunits encoded in the genome of Arabidopsis. Plant Physiol 129(2):908–925

Kerk D, Templeton G, Moorhead GBG (2008) Evolutionary radiation pattern of novel protein phosphatases revealed by analysis of protein data from the completely sequenced genomes of humans, green algae, and higher plants. Plant Physiol 146:351–367

Keyse SM (2000) Protein phosphatases and the regulation of mitogen-activated protein kinase signaling. Curr Opin Cell Biol 12:186–192

Klich MA (2007) *Aspergillus flavus*: the major producer of aflatoxin. Mol Plant Pathol 8(6):713–722

Knogge W (1996) Fungal infection of plants. Plant Cell 8:1711–1722

Kondoh K, Nishida E (2007) Regulation of MAP kinases by MAP kinase phosphatases. Biochim Biophys Acta 1773(8):1227–1237

Krupa A, Anamika K, Srinivasan N (2006) Genome-wide comparative analyses of domain organisation of repertoires of protein kinases of Arabidopsis *thaliana* and *Oryza sativa*. Gene 380:1–13

Kunze G, Zipfel C, Robatzek S, Niehaus K, Boller T, Felix G (2004) The N terminus of bacterial elongation factor Tu elicits innate immunity in Arabidopsis plants. Plant Cell 16:3496–3507

Kyeon M, Son S, Noh Y et al (2016) *Xanthomonas euvesicatoria* causes bacterial spot disease on pepper plant in Korea. Plant Pathol J 32(5):431–440

Laluk K, Luo H, Chai M, Dhawan R, Lai Z, Mengiste T (2011) Biochemical and genetic requirements for function of the immune response regulator BOTRYTIS-INDUCED KINASE1 in plant growth, ethylene signaling, and PAMP-triggered immunity in Arabidopsis. Plant Cell 23(8):2831–2849

Lammers T, Lavi S (2007) Role of type 2C protein phosphatases in growth regulation and in cellular stress signaling. Crit Rev Biochem Mol Biol 42:437–461

Lang T, Mansell A (2007) The negative regulation of Toll-like receptor and associated pathways. Immunol Cell Biol 85(6):425–434

Lee J, Felix G, Nurnberger T (2007) Bacteria-derived peptidoglycans constitute pathogen-associated molecular patterns triggering innate immunity in Arabidopsis. J Biol Chem 282:32338–32348

Lemañczyk G, Kwaśna H (2013) Effects of sharp eyespot (*Rhizoctonia cerealis*) on yield and grain quality of winter wheat. Eur J Plant Pathol 135:187–200

Leonetti P (2018) ROS and cell death in tomato roots infected by *Meloidogyne Incognita*. Methods Mol Biol 1743:87–95

Li L, Li M, Yu L et al (2014) The FLS2-associated kinase BIK1 directly phosphorylates the NADPH oxidase RbohD to control plant immunity. Cell Host Microbe 15(3):329–338

Li W, Cui X (2014) A special issue on plant stress biology: from model species to crops. Mol Plant 7:755–757

Liberti S, Sacco F, Calderone A et al (2013) HuPho: the human phosphatase portal. FEBS J 280:379–387
Lin B, Zhuo K, Chen S et al (2016) A novel nematode effector suppresses plant immunity by activating host reactive oxygen species-scavenging system. New Phytol 209:1159–1173
Lin PP, Mori T, Key JL (1980) Phosphoprotein phosphatase of soybean hypocotyls: purification, properties, and substrate specificities. Plant Physiol 66(3):368–374
Lin X, Walter J, Scheidtmann K et al (1998) Protein phosphatase 2A is required for the initiation of chromosomal DNA replication. Proc Natl Acad Sci U S A 95:14693–14698
Liu J, Liu B, Chen S, Gong BQ et al (2018) A tyrosine phosphorylation cycle regulates fungal activation of a plant receptor ser/thr kinase. Cell Host Microbe 23(2):241–253.e6
Liu J, Stevens PD, Eshleman NE, Gao T (2013) Protein phosphatase PPM1G regulates protein translation and cell growth by dephosphorylating 4E Binding Protein 1(4E-BP1). J Biol Chem 288(32):23225–23233
Liu X, Qian B, Gao C et al (2016) The putative protein phosphatase MoYvh1 functions upstream of MoPdeH to regulate the development and pathogenicity in *Magnaporthe oryzae*. Mol Plant-Microbe Interact 29(6):496–507
Liu Y, Shepherd EG, Nelin LD (2007) MAPK phosphatases—regulating the immune response. Nat Rev Immunol 7(3):202–212
Lorenzo O, Rodríguez D, Nicolás G, Rodríguez PL, Nicolás C (2001) A new protein phosphatase 2C (FsPP2C1) induced by abscisic acid is specifically expressed in dormant beechnut seeds. Plant Physiol 125:1949–1956
Lu D, Wu S, Gao X, Zhang Y, Shan L, He P (2010) A receptor-like cytoplasmic kinase, BIK1, associates with a flagellin receptor complex to initiate plant innate immunity. Proc Natl Acad Sci U S A 107(1):496–501
Luan S (1998) Protein phosphatases and signaling cascades in higher plants. Trends Plant Sci 3(7):271–275
Luan S (2003) Protein phosphatases in plants. Annu Rev Plant Biol 54:63–92
Luan S, Ting J, Gupta R (2001) Protein tyrosine phosphatases in higher plants. New Phytol 151:155–164
Lumbreras V, Vilela B, Irar S et al (2010) MAPK phosphatase MKP2 mediates disease responses in Arabidopsis and functionally interacts with MPK3 and MPK6. Plant J 63:1017–1030
Macho AP, Zipfel C (2014) Plant PRRs and the activation of innate immune signaling. Mol Cell 54:263–272
Macho AP et al (2014) A bacterial tyrosine phosphatase inhibits plant pattern recognition receptor activation. Science 343(6178):1509–1512
MacKintosh C, Lyon GD, MacKintosh RW (1994) Protein phosphatase inhibitors activate antifungal defense responses of soybean cotyledons and cell cultures. Plant J 5(1):137–147
Malcolm GM, Kuldau GA, Gugino BK, Jiménez-Gasco MM (2013) Hidden host plant associations of soilborne fungal pathogens: an ecological perspective. Phytopathology 103(6):538–544
Manosalva P, Manohar M, von Reuss SH et al (2015) Conserved nematode signaling molecules elicit plant defenses and pathogen resistance. Nat Commun 6:7795
Mansfield J, Genin S, Magori S et al (2012) Top 10 plant pathogenic bacteria in molecular plant pathology. Mol Plant Pathol 13(6):614–629
Maron JL, Crone E (2006) Herbivory: effects on plant abundance, distribution and population growth. Proc Biol Sci 273(1601):2575–2584
Maron JL, Kauffman MJ (2006) Habitat-specific impacts of multiple consumers on plant population dynamics. Ecology 87(1):113–124
Mathieu Y, Kurkdijan A, Xia H, Guern J, Koller A, Spiro M, O'Neill M, Albersheim P, Darvill A (1991) Membrane responses induced by oligogalacturonides in suspension-cultured tobacco cells. Plant J 1:333–343
Matuszkiewicz M, Sobczak M, Cabrera J et al (2018) The role of programmed cell death regulator LSD1 in nematode-induced syncytium formation. Front Plant Sci 9:314
Matzinger P (2002) The danger model: a renewed sense of self. Science 296:301–305

Melillo MT, Leonetti P, Bongiovanni M, Castagnone-Sereno P, Bleve-Zacheo T (2006) Modulation of reactive oxygen species activities and H_2O_2 accumulation during compatible and incompatible tomato-root-knot nematode interactions. New Phytol 170(3):501–512

Mendy B, Wang'ombe MW, Radakovic ZS et al (2017) Arabidopsis leucine-rich repeat receptor-like kinase NILR1 is required for induction of innate immunity to parasitic nematodes. PLoS Pathog 13(4):e1006284

Meng X, Xu J, He Y, Yang KY, Mordorski B, Liu Y et al (2013) Phosphorylation of an ERF transcription factor by Arabidopsis MPK3/MPK6 regulates plant defense gene induction and fungal resistance. Plant Cell 25:1126–1142

Meyer A, Puhler A, Niehaus K (2001) The lipopolysaccharides of the phytopathogen *Xanthomonas campestris* pv. *campestris* induce an oxidative burst reaction in cell cultures of *Nicotiana tabacum*. Planta 213:214–222

Meyer D, Lauber E, Roby D, Arlat M, Kroj T (2005) Optimization of pathogenicity assays to study the Arabidopsis *thaliana-Xanthomonas campestris* pv. *campestris* pathosystem. Mol Plant Pathol 6(3):327–333

Millward TA, Zolnierowicz S, Hemmings BA (1999) Regulation of protein kinase cascades by protein phosphatase 2A. Trends Biochem Sci 24(5):186–191

Mine A, Berens ML, Nobori T et al (2017) Pathogen exploitation of an abscisic acid- and jasmonate-inducible MAPK phosphatase and its interception by Arabidopsis immunity. Proc Natl Acad Sci U S A 114(28):7456–7461

Mithoe SC, Menke FLH (2018) Regulation of pattern recognition receptor signaling by phosphorylation and ubiquitination. Curr Opin Plant Biol 45:162–170

Miya A, Albert P, Shinya T, Desaki Y, Ichimura K, Shirasu K, Narusaka Y, Kawakami N, Kaku H, Shibuya N (2007) CERK1, a LysM receptor kinase, is essential for chitin elicitor signaling in Arabidopsis. Proc Natl Acad Sci U S A 104:19613–19618

Monaghan J, Zipfel C (2012) Plant pattern recognition receptor complexes at the plasma membrane. Curr Opin Plant Biol 15:349–357

Moorhead GB, Trinkle-Mulcahy L, Ulke-Lemee A (2007) Emerging roles of nuclear protein phosphatases. Nat Rev Mol Cell Biol 8:234–244

Mordecai EA (2011) Pathogen impacts on plant communities: unifying theory, concepts, and empirical work. Ecol Monogr 81:429–441

Morel JB, Dangl JL (1997) The hypersensitive response and the induction of cell death in plants. Cell Death Differ 4(8):671–683

Muthamilarasan M, Prasad M (2013) Plant innate immunity: an updated insight into defense mechanism. J Biosci 38:433–449

Naito K, Taguchi F, Suzuki T, Inagaki Y, Toyoda K, Shiraishi T, Ichinose Y (2008) Amino acid sequence of bacterial microbe associated molecular pattern flg22 is required for virulence. Mol Plant-Microbe Interact 21:1165–1174

Navarro L, Zipfel C, Rowland O, Keller I, Robatzek S, Boller T, Jones JD (2004) The transcriptional innate immune response to flg22. Interplay and overlap with Avr gene-dependent defense responses and bacterial pathogenesis. Plant Physiol 135:1113–1128

Neel BG, Tonks NK (1997) Protein tyrosine phosphatases in signal transduction. Curr Opin Cell Biol 9(2):193–204

Newman MA, Daniels MJ, Dow JM (1995) Lipopolysaccharide from *Xanthomonas campestris* induces defense-related gene expression in *Brassica campestris*. Mol Plant-Microbe Interact 8:778–780

Onoda T, Iinuma H, Sasaki Y et al (1989) Isolation of a novel tyrosine kinase inhibitor lavendustin A, from *Streptomyces griseolavendus*. J Nat Prod 52(6):1252–1257

País SM, Téllez-Iñón MT, Capiati DA (2009) Serine/threonine protein phosphatases type 2A and their roles in stress signaling. Plant Signal Behav 4(11):1013–1015

Park C, Peng Y, Chen X, Dardick C, Ruan D, Bart R, Canlas PE, Ronald PC (2008) Rice XB15, a protein phosphatase 2C, negatively regulates cell death and XA21-mediated innate immunity. PLoS Biol 6(9):e231

Parthasarathy MV, VAN Slobbe WG, Soudant C (1976) Trypanosomatid flagellate in the phloem of diseased coconut palms. Science 192:1346–1348

Patterson KI, Brummer T, O'brien PM, Daly RJ (2009) Dual-specificity phosphatases: critical regulators with diverse cellular targets. Biochem J 418:475–489

Peng HC, Kaloshian I (2014) The tomato leucine-rich repeat receptor-like kinases SlSERK3A and SlSERK3B have overlapping functions in bacterial and nematode innate immunity. PLoS One 9:e93302

Peng Y, Wersch RV, Zhang Y (2018) Convergent and divergent signaling in PAMP-triggered immunity and effector-triggered immunity. Mol Plant-Microbe Interact 31(4):403–409

Poinssot B et al (2003) The endopolygalacturonase 1 from *Botrytis cinerea* activates grapevine defense reactions unrelated to its enzymatic activity. Mol Plant-Microbe Interact 16:553–564

Potnis N, Minsavage G, Smith JK et al (2012) Avirulence proteins AvrBs7 from *Xanthomonas gardneri* and AvrBs1.1 from *Xanthomonas euvesicatoria* contribute to a novel gene-for-gene interaction in pepper. Mol Plant Microbe Interact 25(3):307–320

Pruitt RN et al (2015) The rice immune receptor XA21 recognizes a tyrosine-sulfated protein from a Gram-negative bacterium. Sci Adv 1(6):e1500245

Raghavendra AS, Gonugunta VK, Christmann A, Grill E (2010) ABA perception and signaling. Trends Plant Sci 15(7):395–401

Rahikainen M, Pascual J, Alegre S, Durian G, Kangasjärvi S (2016) PP2A phosphatase as a regulator of ROS signaling in plants. Antioxidants (Basel) 5(1):8

Rakwal R, Shii K, Agrawal GK, Yonekura M (2008) Protein phosphatase inhibitors activate defense responses in rice (*Oryza sativa*) leaves. Physiologia Plantaraum 111(2):151–157

Ridley BL, O'Neill MA, Mohnen D (2001) Pectins: structure, biosynthesis, and oligogalacturonide-related signaling. Phytochemistry 57:929–967

Rodriguez-Moreno L, Ebert MK, Bolton MD, Thomma BPHJ (2018) Tools of the crook- infection strategies of fungal plant pathogens. Plant J 93(4):664–674

Roux M, Schwessinger B, Albrecht C, Chinchilla D, Jones A, Holton N, Malinovsky FG, Tör M, de Vries S, Zipfel C (2011) The Arabidopsis leucine-rich repeat receptor-like kinases BAK1/SERK3 and BKK1/SERK4 are required for innate immunity to hemibiotrophic and biotrophic pathogens. Plant Cell 23(6):2440–2455

Rudrabhatla P, Reddy MM, Rajasekharan R (2006) Genome-wide analysis and experimentation of plant serine/threonine/tyrosine-specific protein kinases. Plant Mol Biol 60:293–319

Saijo Y, Loo EP, Yasuda S (2018) Pattern recognition receptors and signaling in plant-microbe interactions. Plant J 93(4):592–613

Sánchez-Vallet A, Mesters JR, Thomma BPHJ (2015) The battle for chitin recognition in plant-microbe interactions. FEMS Microbiol Rev 39(2):171–183

Santos ALS, d'Avila-Levy CM, Elias CGR et al (2007) *Phytomonas serpens*: immunological similarities with the human trypanosomatid pathogens. Microbes Infect 9:915e921

Scholthof KB, Adkins S, Czosnek H et al (2011) Top 10 plant viruses in molecular plant pathology. Mol Plant Pathol 12(9):938–954

Schulman H (2014) Intracellular signaling. In: Byrne JH, Heidelberger R, Waxham MN (eds) From molecules to networks, 3rd edn. Academic Press, Amsterdam, pp 119–148

Schulze B, Mentzel T, Jehle AK, Mueller K, Beeler S et al (2010) Rapid heteromerization and phosphorylation of ligand-activated plant transmembrane receptors and their associated kinase BAK1. J Biol Chem 285:9444–9451

Schweighofer A, Meskiene I (2015) Phosphatases in plants. In: Schulze WX (ed) Plant phosphoproteomics: methods and protocols, Methods in molecular biology series 1306. Springer, New York, pp 25–46

Schweighofer A et al (2007) The PP2C-type phosphatase AP2C1, which negatively regulates MPK4 and MPK6, modulates innate immunity, jasmonic acid, and ethylene levels in Arabidopsis. Plant Cell 19:2213–2224

Schwessinger B, Ronald PC (2012) Plant innate immunity: perception of conserved microbial signatures. Annu Rev Plant Biol 63:451–482

Schwessinger B, Roux M, Kadota Y et al (2011) Phosphorylation-dependent differential regulation of plant growth, cell death, and innate immunity by the regulatory receptor-like kinase BAK1. PLoS Genet 7(4):e1002046

Segonzac C, Macho AP, Sanmartín M, Ntoukakis V, Sánchez-Serrano JJ, Zipfel C (2014) Negative control of BAK1 by protein phosphatase 2A during plant innate immunity. EMBO J 33:2069–2079

Shankar A, Agrawal N, Sharma M, Pandey A, Pandey GKM (2015) Role of protein tyrosine phosphatases in plants. Curr Genomics 16(4):224–236

Shazadee H, Khan N, Wang J, Wang C, Zeng J, Huang Z, Wang X (2019) Identification and expression profiling of protein phosphatases (PP2C) gene family in *Gossypium hirsutum* L. Int J Mol Sci 20(6):E1395

Shi Y (2009) Serine/Threonine phosphatases: mechanism through structure. Cell 139(3):468–484

Shubchynskyy V, Boniecka J, Schweighofer A, Simulis J, Kvederaviciute K et al (2017) Protein phosphatase AP2C1 negatively regulates basal resistance and defense responses to *Pseudomonas syringae*. J Exp Bot 68(5):1169–1183

Sidonskaya E et al (2016) Plant resistance against the parasitic nematode *Heterodera schachtii* is mediated by MPK3 and MPK6 kinases, which are controlled by the MAPK phosphatase AP2C1 in Arabidopsis. J Exp Bot 67(1):107–118

Sijmons PC, Grundler FMW, von Mende N, Burrows PR, Wyss U (1991) Arabidopsis *thaliana* as a new model host for plant-parasitic nematodes. Plant J 1:245–254

Singh A, Giri J, Kapoor S et al (2010) Protein phosphatase complement in rice: genome-wide identification and transcriptional analysis under abiotic stress conditions and reproductive development. BMC Genomics 11:435

Smoly I, Shemesh N, Ziv-Ukelson M, Ben-Zvi A, Yeger-Lotem E (2017) An asymmetrically balanced organization of kinases versus phosphatases across eukaryotes determines their distinct impacts. PLoS Comput Biol 13(1):e1005221

Song SK, Clark SE (2005) POL and related phosphatases are dosage-sensitive regulators of meristem and organ development in Arabidopsis. Dev Biol 285(1):272–284

Song WY, Wang GL, Chen LL et al (1995) A receptor kinase-like protein encoded by the rice disease resistance gene, *Xa21*. Science 270(5243):1804–1806

Stahel G (1931) Zur Kenntis der Siebrohrenkrankheit (Phloemnekrose) des Kaffeebaumes in Surinam. I. Mikroskopische Untersuchungen und Infektionsversuche. Phytopathol Z 4:65–82

Standish AJ, Morona R (2014) The role of bacterial protein tyrosine phosphatases in the regulation of the biosynthesis of secreted polysaccharides. Antioxid Redox Signal 20(14):2274–2289

Stone JM, Collinge MA, Smith RD, Horn MA, Walker JC (1994) Interaction of a protein phosphatase with an Arabidopsis serine-threonine receptor kinase. Science 266(5186):793–795

Stone RL, Dixon JE (1994) Protein tyrosine phosphatases. J Biol Chem 269:31323–31326

Sun Y, Li L, Macho AP, Han Z, Hu Z et al (2013) Structural basis for flg22-induced activation of the Arabidopsis FLS2-BAK1 immune complex. Science 342:624–628

Taguchi F, Shimizu R, Nakajima R, Toyoda K, Shiraishi T, Ichinose Y (2003) Differential effects of flagellins from *Pseudomonas syringae* pv. *tabaci*, tomato and glycinea on plant defense response. Plant Physiol Biochem 41:165–174

Tena G, Boudsocq M, Sheena J (2011) Protein kinase signaling networks in plant innate immunity. Curr Opin Plant Biol 14(5):519–529

Thain JF, Gubb IR, Wildon DC (1995) Depolarization of tomato leaf cells by oligogalacturonide elicitors. Plant Cell Environ 18:211–214

Thomma BP, Nürnberger T, Joosten MH (2011) Of PAMPs and effectors: the blurred PTI-ETI dichotomy. Plant Cell 23(1):4–15

Tonks NK, Neel BG (1996) From form to function: signaling by protein tyrosine phosphatases. Cell 87:365–368

Tor M, Lotze MT, Holton N (2009) Receptor-mediated signaling in plants: molecular patterns and programmes. J Exp Bot 60(13):3645–3654

Trotta A, Wrzaczek M, Scharte J et al (2011) Regulatory subunit B'gamma of protein phosphatase 2A prevents unnecessary defense reactions under low light in Arabidopsis. Plant Physiol 156(3):1464–1480

Tsuda K, Katagiri F (2010) Comparing signaling mechanisms engaged in pattern-triggered and effector-triggered immunity. Curr Opin Plant Biol 13(4):459–465

Uhrig RG, Labandera AM, Moorhead GB (2013) Arabidopsis PPP family of serine/threonine protein phosphatases: many targets but few engines. Trends Plant Sci 18(9):505–513

Underwood W, Zhang S, He SY (2007) The *Pseudomonas syringae* type III effector tyrosine phosphatase HopAO1 suppresses innate immunity in *Arabidopsis thaliana*. Plant J 52(4):658–672

van Baarlen P, Woltering EJ, Staats M, van Kan JAL (2007) Histochemical and genetic analysis of host and non-host interactions of Arabidopsis with three *Botrytis* species: an important role for cell death control. Mol Plant Pathol 8:41–54

Van der Hoorn RAL, Laurent F, Roth R, de Wit PJGM (1999) Agroinfiltration is a versatile tool that facilitates comparative analyses of Avr9/Cf-9-induced and Avr4/Cf-4-induced necrosis. Mol Plant-Microbe Interact 13:439–446

Veronese P et al (2003) In defense against pathogens. Both plant sentinels and foot soldiers need to know the enemy. Plant Physiol 131:1580–1590

Virshup DM, Shenolikar S (2009) From promiscuity to precision: protein phosphatases get a makeover. Mol Cell 33:537–545

Wang D, Harper JF, Gribskov M (2003) Systematic trans-genomic comparison of protein kinases between Arabidopsis and *Saccharomyces cerevisiae*. Plant Physiol 132(4):2152–2165

Wang H, Chevalier D, Larue C et al (2007) The protein phosphatases and protein kinases of Arabidopsis *thaliana*. Arabidopsis Book 5:e0106

Wang WM et al (2012) PAPP2C interacts with the atypical disease resistance protein RPW8.2 and negatively regulates salicylic acid-dependent defense responses in Arabidopsis. Mol Plant 5(5):1125–1137

Washington K, Ammosova T, Beullens M et al (2002) Protein phosphatase-1 dephosphorylates the C-terminal domain of RNA polymerase-II. J Biol Chem 277(43):40442–40448

Wei K, Pan S (2014) Maize protein phosphatase gene family: identification and molecular characterization. BMC Genomics 15:773

Weiser DC, Shenolikar S (2003) Use of protein phosphatase inhibitors. Curr Protoc Mol Biol 62:18.10.1–18.10.13

White FF, Potnis N, Jones JB, Koebnik R (2009) The Type III effectors of *Xanthomonas*. Mol Plant Pathol 10:749–766

Widjaja I, Lassowskat I, Bethke G et al (2008) A protein phosphatase 2C, responsive to the bacterial effector AvrRpm1 but not to the AvrB effector, regulates defense responses in Arabidopsis. Plant J 61(2):249–258

Williamson VM, Gleason CA (2003) Plant–nematode interactions. Curr Opin Plant Biol 6(4):327–333

Withers J, Dong X (2017) Post-translational regulation of plant immunity. Curr Opin Plant Biol 38:124–132

Wyss U (1992) Observations on the feeding behavior of *Heterodera schachtii* throughout development, including events during moulting. Fundam Appl Nematol 15:75–89

Wyss U, Grundler FM (1992) Seminar: *Heterodera schachtii* and Arabidopsis *thaliana*, a model host–parasite interaction. Nematologica 38:488–493

Wyss U, Zunke U (1986) Observations on the behaviour of second stage juveniles of *Heterodera schachtii* inside host roots. Revue de Nematologie 9:153–166

Xin XF, Kvitko B, He SY (2018) *Pseudomonas syringae*: what it takes to be a pathogen. Nat Rev Microbiol 16(5):316–328

Xing T, Laroche A (2011) Revealing plant defense signaling. Plant Signal Behav 6(10):1469–1474

Xu J, Zhang S (2015) Mitogen-activated protein kinase cascades in signaling plant growth and development. Trends Plant Sci 20(1):56–64

Xu Q, Fu HH, Gupta R, Luan S (1998) Molecular characterization of a tyrosine specific protein phosphatase encoded by a stress-responsive gene in Arabidopsis. Plant Cell 10:849–857

Yan H, Zhao Y, Shi H, Li J, Wang Y, Tang D (2018) BRASSINOSTEROID-Signaling KINASE1 phosphorylates MAPKKK5 to regulate immunity in Arabidopsis. Plant Physiol 176(4):2991–3002

Yang G, Hu Y, Fasoyin OE et al (2018) The *Aspergillus flavus* phosphatase CDC14 regulates development, Aflatoxin biosynthesis and pathogenicity. Front Cell Infect Microbiol 8:141

Yasuda S, Okada K, Saijo Y (2017) A look at plant immunity through the window of the multitasking coreceptor BAK1. Curr Opin Plant Biol 38:10–18

Yoo SD, Cho YH, Tena G, Xiong Y, Sheen J (2008) Dual control of nuclear EIN3 by bifurcate MAPK cascades in C_2H_4 signaling. Nature 451:789–795

Žádníková P, Smet D, Zhu Q, Straeten DVD, Benková E (2015) Strategies of seedlings to overcome their sessile nature: auxin in mobility control. Front Plant Sci 6:218

Zhang J et al (2010) Effector-triggered and pathogen-associated molecular pattern–triggered immunity differentially contribute to basal resistance to *Pseudomonas syringae*. Mol Plant-Microbe Interact 23(7):940–948

Zhang L et al (2014) Fungal endopolygalacturonases are recognized as microbe-associated molecular patterns by the Arabidopsis receptor-like protein responsiveness to *Botrytis* Polygalacturonases1. Plant Physiol 164:352–364

Zhang ZY (1998) Protein-tyrosine phosphatases: biological function, structural characteristics, and mechanism of catalysis. Crit Rev Biochem Mol Biol 33(1):1–52

Zhu X, Wang Y, Su Z, Lv L, Zhang Z (2018) Silencing of the wheat protein phosphatase 2A catalytic subunit TaPP2Ac enhances host resistance to the necrotrophic pathogen *Rhizoctonia cerealis*. Front Plant Sci 9:1437

Zipfel C (2009) Early molecular events in PAMP-triggered immunity. Curr Opin Plant Biol 12:414–420

Zipfel C, Oldroyd GE (2017) Plant signaling in symbiosis and immunity. Nature 543(7645): 328–336

Chapter 16
Role of Dual Specificity Phosphatase in Stress and Starch Metabolism

Kanwaljeet Kaur, Manas Kumar Tripathy, and Girdhar K. Pandey

16.1 Introduction

All living organisms respond to adverse environment for their growth, reproduction, and survival and carry out numerous cellular processes and chemical reactions. Some of these are spontaneous, which release energy, and some are non-spontaneous, which require energy in order to proceed.

In eukaryotic cells, one-third of cellular proteins contain covalently bound phosphates (phosphorylated) (Mustelin 2007). Dephosphorylation (Removal of phosphate group) of phosphorylated cellular proteins are catalyzed by enzymes known as phosphatases. Protein phosphorylation is the most important and crucial post-translational event first reported by Edwin Krebs and Edmond Fisher almost 60 years ago (Olsen et al. 2006; Krebs et al. 1958). A protein phosphatase is an enzyme which removes a phosphate group from a phosphorylated amino acid residue of its substrate protein and produces a phosphate group and a molecule with a free hydroxyl group. During phosphorylation, kinases convert adenosine triphosphate (ATP) into adenosine diphosphate (ADP) or adenosine monophosphate (AMP) and use energy obtained from this to transfer a phosphate group to the proteins or other substrates. On the other hand, during dephosphorylation, phosphatases which function opposite to kinases release phosphate from phosphorylated protein/substrate into a solution as a free phosphate (Pi) and release energy (Kerk et al. 2002) (Fig. 16.1). Maintaining the balance between phosphorylation and dephosphorylation mediated by kinases and phosphatases is necessary for energy balance and also acts a biological switch in cellular processes (Smith and Walker

[SC]Kanwaljeet Kaur and Manas Kumar Tripathy contributed equally to this work.

K. Kaur · M. K. Tripathy · G. K. Pandey (✉)
Department of Plant Molecular Biology, University of Delhi South Campus, New Delhi, India
e-mail: gkpandey@south.du.ac.in

G. K. Pandey (ed.), *Protein Phosphatases and Stress Management in Plants*,
https://doi.org/10.1007/978-3-030-48733-1_16

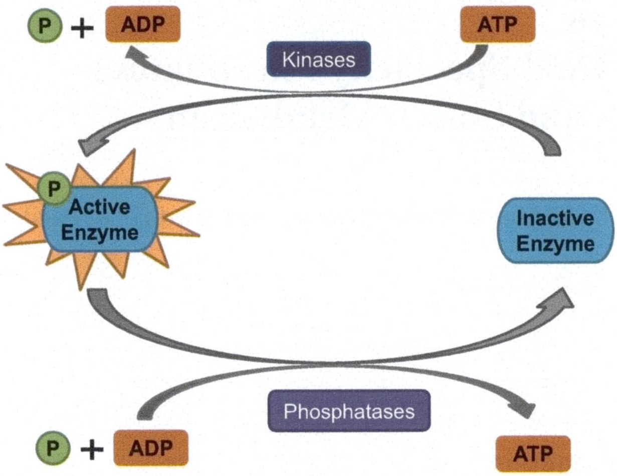

Fig. 16.1 Mechanism of action of protein phosphatases and protein kinases: protein phosphatases catalyze removal of a phosphate group from phosphorylated amino acid residue of its substrate protein and produces a phosphate group and a molecule with a free hydroxyl group. During dephosphorylation, phosphatases which function opposite to kinases release phosphate from a phosphorylated enzyme into a solution as free ions and release energy. During phosphorylation, kinases convert adenosine triphosphate (ATP) into adenosine diphosphate (ADP) or adenosine monophosphate (AMP) and use energy obtained from this to transfer a phosphate group to the enzyme

1996). The protein tyrosine kinases (PTKs) are well characterized as compared to protein tyrosine phosphatases (PTPs), whereas emerging evidences suggest PTPs regulate a variety of fundamental cellular processes such as cell growth, mitogenesis, metabolism, gene transcription, cell cycle control, and the immune response (Dunn et al. 1996). Disruption of balance between kinase and phosphatase activities may lead to an abnormal condition or diseased state (Lountos et al. 2011). During the process of phosphorylation and dephosphorylation mediated by kinases and phosphatases, many enzymes or regulators are activated and deactivated and act as molecular switches to modulate the structures and functions of many cellular proteins in prokaryotic and eukaryotic cells (Cheng et al. 2011; Singh and Pandey 2012; Singh et al. 2015; Ardito et al. 2017a, b; Bheri and Pandey 2019a, b).

The protein kinase structure, evolution, and function have been extensively investigated in a number of eukaryotes, and most of the kinases have a conserved functional domain. In contrast, protein phosphatases are less studied and displayed a great degree of diversity and harbor different catalytically important signature motifs and domains (Singh and Pandey 2012). During protein phosphorylation the three-dimensional structure, activity, cellular localization, and the stability of a protein changed, which in turn acts as an "on-or-off" switch in numerous pathways regulating growth, differentiation, and oncogenesis. In eukaryotes, more than 70% of all the known proteins are reversibly phosphorylated, and hence, the importance of reversible protein phosphorylation to cellular regulation cannot be overruled

(Shankar et al. 2015; Singh et al. 2015; Bheri and Pandey 2019a, b). Based on the amino acid residue dephosphorylated, plant protein phosphatases are broadly categorized into serine/threonine (Ser/Thr) and tyrosine (Tyr) phosphatases. Further the Ser/Thr phosphatases are categorized into two families: (i) phosphoprotein phosphatases (PPPs) comprising PP1, PP2A, PP2B, and other distantly related phosphatases (PP4, PP5, PP6, and PP7) and (ii) PPs requiring metal ion for catalysis (PPMs), including PP2C and other Mg^{2+}-dependent phosphatases (Lee et al. 2010; Singh et al. 2010; Singh et al. 2015; Bheri and Pandey 2019a, b). However, PP2B, which is a Ca^{2+}-dependent phosphatase, also known as calcineurin A (CNA), has not been identified so far in plants (Moorhead et al. 2007; Kerk et al. 2008; Singh et al. 2010). Protein tyrosine phosphatases (PTPs) characterized by a CX5R motif are composed of two groups, namely, protein Tyr-specific phosphatases (PTPs), which specifically act on phosphotyrosine, and dual specificity phosphatases (DSPs) which dephosphorylate phosphotyrosine as well as phosphoserine/phosphothreonine (Stone and Dixon 1994; Tonks and Neel 1996; Singh and Pandey 2012; Shankar et al. 2015) (Fig. 16.2).

Here, we provide an overview of the dual specificity phosphatase classification, structure, and functions and discuss their role in important processes such as in stress and starch metabolism.

16.2 Classification of Protein Phosphatases

Protein phosphorylation is the most common posttranslational event, which regulates many biological processes. It is reported that more than 30% of eukaryotic proteins are present in a phosphorylated form and this phosphorylation takes place most commonly on the hydroxyl group of Ser, Thr, and Tyr residues (Mailloux 2010). Nine amino acids (Tyr, Ser, Thr, Cys, Arg, Lys, Asp, Glu, and His) have been found to be phosphorylated in living organisms, while three among these are predominantly reported in eukaryotic cells, i.e., Ser, Thr, and Tyr (Moorhead et al. 2009). The distribution of protein phosphorylation is approximately 86.4%, 11.8%, and 1.8% on Ser, Thr, and Tyr, respectively, in eukaryotic cells (Olsen et al. 2006; Shankar et al. 2015; Ardito et al. 2017a, b).

On the basis of substrate dephosphorylated, phosphatases are broadly divided into two categories: Ser/Thr phosphatases and Tyr phosphatases. Phosphatases, which dephosphorylate both Ser and Thr amino acid residues, are classified into protein Ser/Thr phosphatases (PSTP); and phosphatases which dephosphorylate Tyr residues are categorized into protein Tyr phosphatases (PTPs) (Ardito et al. 2017a, b). Further PSTP is divided into three families:

1. Phosphoprotein phosphatases (PPPs): For several members of the PPP family, the catalytic subunit associates with a great variety of regulatory subunits. Representative members of the PPP family include protein phosphatases PP1, PP2A, PP2B (commonly known as calcineurin), PP4, PP5, PP6, and PP7.

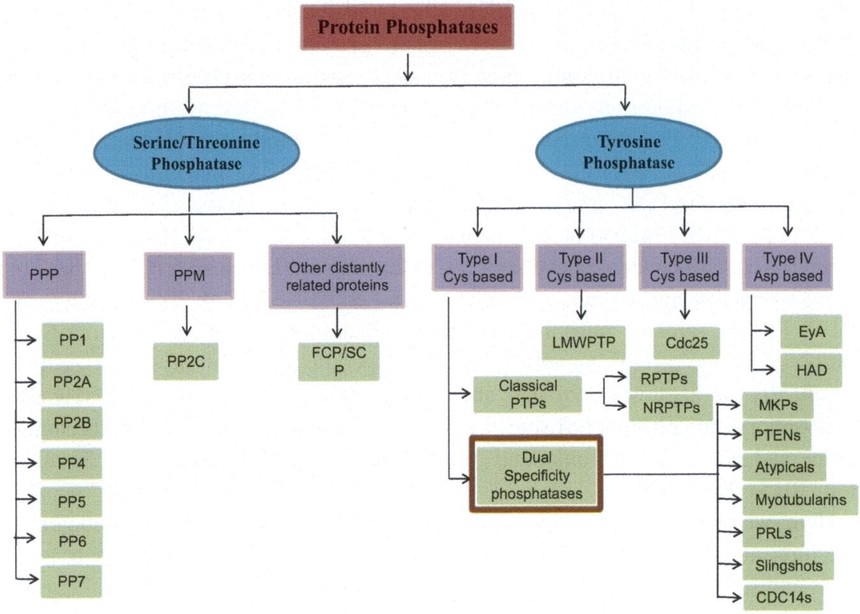

Fig. 16.2 Classification of protein phosphatases. Protein phosphatases can be classified into two groups on the basis of substrate specificity. (1) Ser/Thr phosphatases, which are further divided into three classes. (a) PPPs (phosphoprotein phosphatases): which include PP1, PP2A, PP2B, PP4, PP5, PP6, and PP7. (b) PPMs (metal-dependent protein phosphatases): which include PP2C. (c) Other distantly related proteins: which include FCP and SCP2. Tyr phosphatases are further divided into four classes on the bases of amino acid sequence of their catalytic domain. (1) *Class I phosphatases*, which include (A) Classical PTPs (these are further divided into two groups RPTPs (transmembrane, receptor-like enzymes) and NRPTPs (the intracellular, nonreceptor PTPs)) and (B) DSPs (dual specificity phosphatases) (these are further divided into (i) MAPKP (mitogen-activated protein kinase phosphatase), (ii) slingshot, (iii) PRLs (phosphatases of regenerating liver), (iv) atypical DSP, (v) CDC14, (vi) PTEN (phosphatase and tensin homologue deleted on chromosome 10), and (vii) MTMs (myotubularins)). (2) *Class II phosphatases*, which include LMWPTP low molecular weight PTPs. (3) *Class III phosphatases*, which include Cdc25 isoforms. (4) *Class IV phosphatases* are aspartic acid-based PTPs, which include EYA (Eyes Absents) phosphatase and HAD (Haloacid Dehalogenase) family of phosphatases in humans

2. Metal-dependent protein phosphatases (PPMs): Include protein phosphatases dependent on manganese/magnesium ions (Mn^{2+}/Mg^{2+}), such as PP2C and pyruvate dehydrogenase phosphatase. In contrast to PPP, members of the PPM family do not have regulatory subunits but contain additional domains and conserved sequence motifs that may determine substrate specificity.
3. Aspartate-based phosphatases: Such as FCP (TFIIF-associating component of RNA polymerase II CTD phosphatase) and SCP (small CTD phosphatase). FCP/SCP uses an Asp-based catalysis mechanism. The only known substrate for FCP/SCP is the C-terminal domain (CTD) of RNA polymerase II (Shi 2009; Singh and Pandey 2015).

The PTPs are characterized by their common signature motif CX_5R in the conserved catalytic domain. It is thought that this motif and catalytic mechanism evolved independently three times to yield three groups or classes of PTP (Kerk et al. 2008; Moorhead et al. 2009). Based on the amino acid sequences of their catalytic domain, PTPs are further divided into four families: class I, class II, class III, and class IV.

1. *Class I phosphatases* have a common PTP domain structural fold and are one of the largest groups of PTPs which are further divided into subfamilies. They consist of two groups based on the domain architecture and the degree of homology between catalytic domains. Classical PTPs (these are strictly specific for tyrosine) and dual specificity phosphatases (VH1-like and these are a diverse group in terms of specificity).

   (a) Classical PTPs (receptor and nonreceptor) are given this name as they defined the PTPs and they all dephosphorylate tyrosine residues. These classical PTPs are further divided into two groups transmembrane, receptor-like enzymes (RPTPs) and the intracellular, nonreceptor PTPs (NRPTPs).
   (b) Dual specificity phosphatases (DSPs) are a much more diverse family of intracellular proteins having a similar catalytic region but distinct noncatalytic regions and can be divided into several subgroups, which share less sequence identity with each other. The DSPs are further divided into:

      - *MAPKP* (mitogen-activated protein kinase phosphatase): These MAP kinase phosphatases (MKPs) are characterized by dual phosphothreonine and phosphotyrosine specificity and the presence of a CH_2 region and other MAP kinase targeting motifs (Bordo and Bork 2002; Alonso et al. 2003).
      - *Slingshot*: It has a minimal basic motif of classical PTPs with smaller size (~150 amino acids) and contains a central twisted five-stranded β-sheet surrounded by five or six helices on or beneath the β-sheet. Structural variations occur at the N-terminus and the C-terminus of the catalytic domain. In humans, three slingshots SSH1, SSH2, and SSH3 are known (Kim and Ryu 2012).
      - *PRLs* (phosphatases of regenerating liver): There are three PRLs (PRL-1, PRL-2, and PRL-3) present in human cells and very poorly known (Mustelin 2007).
      - *Atypical DSP*: Includes a number of poorly characterized enzymes that lack specific MAP kinase targeting motifs and tend to be much smaller enzymes. Examples of members are VHR, PIR, laforin, VHZ, and STYX (Mustelin 2007).
      - *CDC14*: It is involved in dephosphorylation of the Cdk activation loop phospho-Thr and inactivation of cyclin-dependent kinases. It plays role in cell division cycle. Some examples of members are CDC14A, KAP, and PTP9Q22 (Visintin et al. 1998; Mustelin 2007; Moorhead et al. 2009).

- *PTEN* (phosphatase and tensin homologue deleted on chromosome 10): It dephosphorylates 3′ position of the inositol sugar of phosphatidylinositol derivatives at the plasma membrane. In plants, PTEN is required for pollen maturation following cell division. AtPTEN1 encodes Arabidopsis PTEN 1, which shows similarity to a tumor suppressing phosphatase in human (Phosphatase and tensin homolog). Expresseing AtPTEN1 in pollen grains shows its requirement during pollen development (Zhang et al. 2011; Shankar et al. 2015). PTEN also dephosphorylates protein substrates including focal adhesion kinase (FAK) and regulates MAPK pathways (Luan et al. 2001). In *Drosophila* and *C. elegans*, PTENs play important role in development (Ogg and Ruvkun 1998; Huang et al. 1999).
- *MTMs (myotubularins)*: These primarily dephosphorylate phosphatidylinositol-3-phosphate on internal cell membranes (Wishart and Dixon 2002).

2. *Class II phosphatases* are low molecular weight PTPs and are tyrosine specific. Type-2 PTPs are frequently found in prokaryotes.
3. *Class III phosphatases* consist of lower molecular weight PTPs and Cdc25 isoforms and are thought to have evolved from a bacterial rhodanese-like enzyme and a bacterial arsenate reductase, respectively. The third subgroup of PTPs includes cell cycle regulators and shows specificity toward both Tyr and Thr residues. Type-3 PTPs are known to dephosphorylate cyclin-dependent kinases (CDKs) causing an inhibition of cell cycle progression.
4. *Class IV phosphatases* are aspartic acid-based PTPs, which comprise a heterogeneous group of phosphatases, and can be either Ser or Tyr specific. Examples of class IV phosphatases are EYA (Eyes Absents) phosphatase and HAD (Haloacid Dehalogenase) family of phosphatases in humans (Alonso et al. 2004a, b; Singh and Pandey 2012; Shankar et al. 2015).

16.3 Dual Specificity Phosphatases

As there is always an exception to the rule, certain phosphatases which recognize both phosphotyrosine and phosphothreonine/phosphoserine residues are known as dual specificity phosphatases (DSPs) and are implicated as major modulators of cellular signaling events via dephosphorylation (Meekins et al. 2016). Catalytic domain of DSPs shows similarity with Tyr phosphatases but does not share any similarity with catalytic domain of protein Ser/Thr phosphatases. In addition to active-site signature motif (H/V)C(X)5R(S/T) of catalytic domain, animal DSPs contain other domains like SH2 domains, PDZ domains, extracellular ligand-binding domains, and others (Zhang 2002). Plant DSPs mainly consist of three conserved domains, cTP (chloroplast transit peptide), DSP (dual specificity phosphatase) catalytic domain, and CBD (carbohydrate-binding domain) (Shankar et al. 2015;

Fig. 16.3 Schematic representation of the predicted domain architecture of dual specificity phosphatases. *cTP* chloroplast transit peptide, *DSP* dual specificity phosphatase catalytic domain, and *CBD* carbohydrate-binding domain

Huang et al. 2019) (Fig. 16.3). DSPs can act on both phosphothreonine/phosphoserine and phosphotyrosine and show limited sequence homology with PTPs, but still these are categorized into PTPs because they share catalytic site signature motif (H/V)C(X)5R(S/T) and they follow a similar mechanism of catalysis for the dephosphorylation of their substrate.

Substrate specificity in tyrosine phosphatases is determined by the depth of its catalytic site cleft. The active catalytic site cleft of DSPs is shallow having a depth of 5.5 Å, which enables its active site to accommodate both phosphotyrosine and phosphothreonine/phosphoserine. Therefore, these are known as dual specificity as they can dephosphorylate both phosphotyrosine and phosphoserine/threonine. In case of tyrosine phosphatases, the active-site cleft is much deeper as compared to DSPs, having a depth of around 10 Å. Therefore, it selects exclusively phosphotyrosine to access the active catalytic site and dephosphorylate only phosphotyrosine residues (Stewart et al. 1999; Luan et al. 2001).

DSPs are the most important regulators of a wide variety of protein kinases and known to regulate the signaling cascade in plants, animals, and fungi (Kerk et al. 2006). In animals DSPs regulate MAPK pathways and also play a role in tumor suppression, immune response, and many more similar processes. Role of DSPs has been extensively studied in animal system, but in plant system, their role still need to be explored.

16.3.1 Discovery and Structure of Dual Specificity Phosphatase

The first dual specificity phosphatase discovered from yeast, an eukaryotic system, is atypical phosphatase YVH1, isolated from the Hl open reading frame of Vaccinia virus. The YVH1 phosphatase displays a high evolutionary conservation with orthologs present in species ranging from yeasts to humans.

The crystal structure of a human DSP, vaccinia HI-related phosphatase (VHR), has a shallow active-site pocket which allows the hydrolysis of phosphorylated Ser, Thr, or Tyr protein residues, whereas the deeper active site of PTPs restricts substrate specificity to only phosphotyrosine (Yuvaniyama et al. 1996; Stewart et al. 1999; Luan et al. 2001). The structure of VHR consists of a single $\alpha + \beta$-type domain of dimensions having a length, width, and height of 50 Å, 40 Å, and 32 Å, respectively. The loop between the β8-strand and α5-helix contains the consensus active-site sequence His-Cys-X-X-Gly-X-X-Arg-(Ser or Thr) with the catalytic Cys^{124}

thiol at its center (X is any amino acid). In the crystal structure, the active site was bound by the competitive inhibitor. The oxygen atoms of sulfate form hydrogen bonds to the main chain amides of the active-site loop and to the Arg130 side chain, mimicking the corresponding interactions between the phosphate oxygens on a phosphoprotein substrate (Yuvaniyama et al. 1996). A water molecule is not present in the active site of the VHR structure and may be absent in other DSPs. Residues Ser^{129} and Asp^{92} occupy the position of the water molecule in the VHR structure and are conserved in other DSP sequences. Also, DSPs have a hydrophobic residue (Phe^{166} in VHR) in a position corresponding to that of the *Yersinia* PTP Gln^{450} (Yuvaniyama et al. 1996).

Although there is no sequence similarity between PTPs and DSPs, catalytic domain of both shares the same core structural features. Both contain catalytically essential Cys and Arg residues. The central part consists of four-stranded parallel β-sheet flanked by antiparallel β-sheets, surrounded by α-helices with four on one side and two on the other side. A signature motif is present in the single loop at the base of a catalytic site cleft of protein surrounded by four loops; three of them provide a residue necessary for catalysis and substrate specificity. The essential cysteine is in the position for nucleophilic attack on an incoming phosphotyrosyl residue. The remaining residues of the core motif function to increase the nucleophilicity of the catalytic cysteine and to bind to and position the phosphate group. The arginyl residue in the signature motif is particularly important for this process (Barford et al. 1998; Luan et al. 2001).

16.3.2 Catalytic Mechanism

DSPs employ a similar catalytic mechanism as PTPs. Essential cysteine is present in both PTPs and DSPs, which forms a covalent thiol-phosphate intermediate in the active site. The previous report suggest that when replacing Cys with Ser, the VHR is unable to dephosphorylate both phosphotyrosine and phosphoserine/threonine residues (Denu and Dixon 1995).

Dephosphorylation mechanism of PTPs takes place in a two-step process. The nucleophilic Cys, which is present in the active site, is specifically designed to bind a negatively charged substrate and maintains the negative charge on thiolate group. The first thiol group of active-site Cys residue attacks as a nucleophile and forms a thiophosphoryl enzyme intermediate by covalent catalysis. During this process conserved aspartic acid donates a proton and functions as a general acid to neutralize the negative charge on leaving group. Release of Tyr/Ser/Thr follows the donation of a proton by the aspartic acid acting as a general acid (Denu et al. 1996; Zhang 2002). Asp181, which behaves as a general acid during the formation of thiophosphoryl enzyme intermediate, acts as a general base during hydrolysis of formed intermediate and releases a proton from attacking water molecule. This released

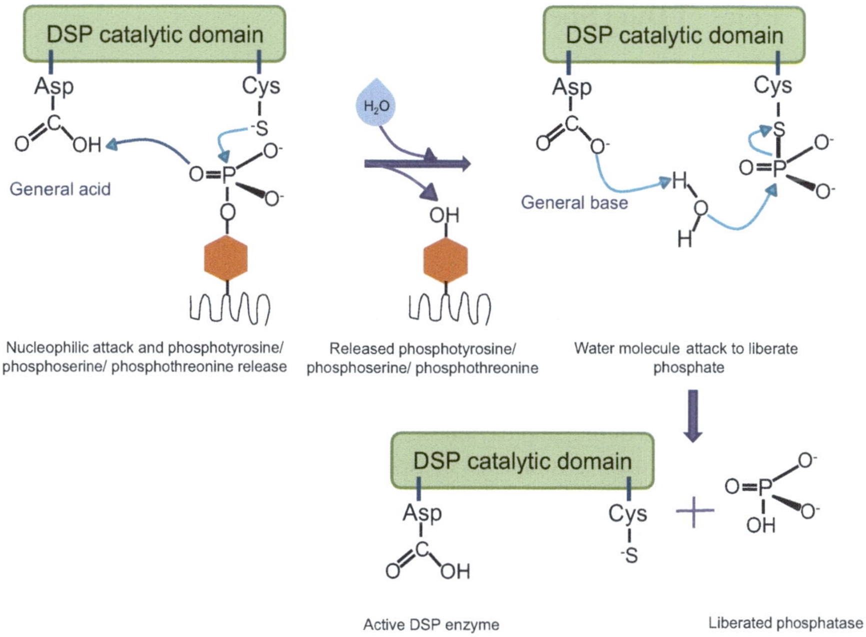

Fig. 16.4 Catalytic mechanism of dual specificity phosphatases: dephosphorylation mechanism takes place in two steps: (a) the first thiol group of active-site cysteine residue attacks as a nucleophile and forms a thiophosphoryl enzyme intermediate by covalent catalysis. Conserved aspartic acid in catalytic domain donates a proton and functions as a general acid to neutralize the negative charge on leaving group, which results in the release of Ser/Thr and Tyr. (b) Asp, which behaves as a general acid during the formation of thiophosphoryl enzyme intermediate, acts as a general base during hydrolysis of formed intermediate and releases a proton from attacking water molecule. This released proton in turn attacks the phospho-enzyme intermediate to eliminate phosphate and regenerate active DSP enzyme

proton in turn attacks the phospho-enzyme intermediate to eliminate phosphate and regenerate an active DSP enzyme (Zhang 2002) (Fig. 16.4). Laforin DSP is a common DSP in animals. Lafora disease is a rare, fatal autosomal recessive neurodegenerative disorder, which is characterized by the development of Lafora bodies in the cytoplasm of the brain, liver, skin, kidney, and skeletal and cardiac muscle cells. Defect in the *EPM2A* gene is a reason for the generation of this disease. The *EPM2A* gene product contains an amino-terminal carbohydrate-binding domain (CBD), and this CBD is critical for association with glycogen. The CBD domain localizes the phosphatase to specific subcellular compartments that correspond to the expression pattern of glycogen-processing enzyme, glycogen synthase. Mutations in the CBD result in mislocalization of the phosphatase, and thereby the CBD targets laforin to intracellular glycogen particles where it is likely to function. Thus, naturally occurring mutations in the CBD of laforin result in progressive myoclonus epilepsy due to mislocalization of phosphatase expression (Wang et al. 2002; Gentry et al. 2013).

16.4 Roles of Dual Specificity Phosphatase

DSPs have been extensively studied in animals, but not much work has been carried out in plants. The available literature suggests DSPs were involved in starch metabolism ROS homeostasis and abiotic and biotic stresses in plants (Fig. 16.5). In animals they are involved in immune responses, and growth and development.

16.4.1 Roles of Dual Specificity Phosphatases in Plants

16.4.1.1 Role in Starch Metabolism

In plants, carbon fixed during photosynthesis is majorly stored in the form of starch. This stored starch is mainly composed of two glucan polymers—amylopectin and amylose (Meekins et al. 2016; Sokolov et al. 2006). Amylopectin, by far the major component (70–90%), is a large molecule with a branched structure and is responsible for the granular nature of starch. Amylose is minor component (10–30%), essentially linear and synthesized within the matrix formed by amylopectin.

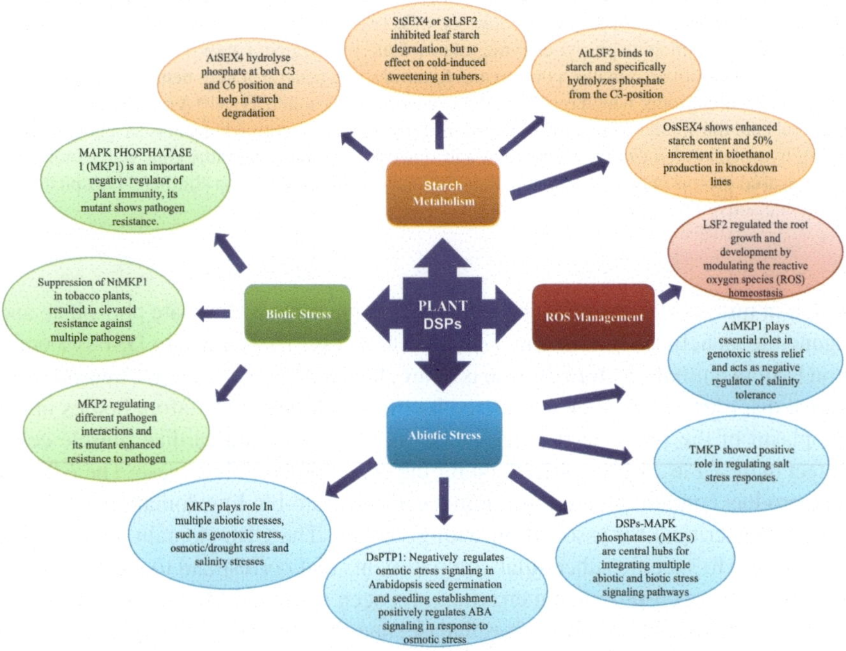

Fig. 16.5 Roles of DSPs (dual specificity phosphatases) in plants. Roles of phosphatases are mainly divided into four parts in plants: (1) starch metabolism, (2) ROS management, (3) abiotic stress, and (4) biotic stress

Depending on the source of starch, trace amounts of proteins, lipids, and ions are also present in starch granule (Meekins et al. 2016; Ritte et al. 2002, 2006). Energy produced during the day via photosynthesis is stored in the chloroplast in the form of granular starch, which is further hydrolyzed by a combination of different enzymes during the night or unavailability of photosynthetic carbon and exported to different plant organs and tissues for their growth and development (Sokolov et al. 2006). In general, starch molecule is phosphorylated at C3 and C6 positions (Ritte et al. 2002, 2006). For its degradation, removal of phosphate group from C3 and C6 positions is required. Arabidopsis starch excess protein 4 (SEX4) hydrolyzes the phosphate from both the positions and helps in normal starch degradation (Niittylä et al. 2006; Sokolov et al. 2006). Interestingly, SEX4 structurally and functionally resembles Laforin enzyme of animals, which is required in normal glycogen metabolism and dysregulation which leads to a neurodegenerative disease called Lafora disease (Gentry et al. 2013).

Bioinformatics analyses revealed there is another homologue of SEX4, LSF2 (like sex four 2) present in Arabidopsis. Carbohydrate-binding module present in SEX4 is absent in LSF2 while it binds to starch and specifically hydrolyzes the phosphate present at C3 position of starch molecule (Santelia et al. 2011). Later, it has been established that SEX4 preferentially removes phosphate at the C6 position while LSF2 removes phosphate mainly at the C3 position of glucose moieties (Kötting et al. 2009; Hejazi et al. 2010; Santelia et al. 2011; Meekins et al. 2014). In comparison to *sex4* single mutant, *sex4lsf2* double mutant has more severe starch excess phenotype, impaired growth, and changes in the ratio of C3- and C6-bound phosphate (Santelia et al. 2011). Ortholog of AtSEX4 in rice, i.e., OsSEX4, shows an enhanced starch content in suspension culture cells, leaves, and rice straw of knockdown lines as compared to wild type, whereas physical parameters of knockdown lines are similar to wild type and no yield penalty was observed. 50% increment in bioethanol production was observed when straw from knockdown lines was used for its production as compared to wild type (Huang et al. 2019).

Gentry et al. (2007) through bioinformatics screening found the kingdom Plantae, which lacks laforin, possesses a protein with laforin-like properties called starch excess protein 4 (SEX4). Mutations in the Arabidopsis *thaliana* SEX4 gene result in a starch excess phenotype reminiscent of Lafora disease and *Homo sapiens* laforin complements the *sex4* phenotype, suggesting that laforin and SEX4 are functional equivalents. The roles of starch phosphatases in potatoes were examined by developing and characterizing transgenic lines repressed in transcription of potato orthologs of SEX4 and LIKE SEX FOUR2 (LSF2). The repression of either SEX4 or LSF2 inhibited leaf starch degradation, whereas it had no effect on cold-induced sweetening in tubers. Starch amounts were unchanged in the tubers, but the amount of phosphate bound to the starch was significantly increased in transgenic lines, with phosphate bound at the C6 position of the glucosyl units increased in lines repressed in StSEX4 and C3 position in lines repressed in StLSF2 expression. Transgenic lines also showed reduction in starch granule size and an alteration in the constituent glucan chain lengths within the starch molecule, with no alteration in granule morphology, whereas the physical properties of the starch were changed with increased swelling power due to an enhanced capacity for hydration (Samodien et al. 2018).

16.4.1.2 Role in Reactive Oxygen Species Management and Abiotic and Biotic Stresses

Role in Reactive Oxygen Species Management

Plants take up water and nutrients from the soil through their root hairs and are also exposed to abiotic and biotic stresses such as limitation of water in soil (drought), soil salinity, heavy metals, and numerous pathogens in the soil. Reactive oxygen species (ROS) play critical roles in root growth and root hair development in plants (Causin et al. 2012; Zhao et al. 2016). LSF2 regulated the root growth and development by modulating the ROS homeostasis (Zhao et al. 2016). It has been established that *lsf2* mutants showed higher endogenous hydrogen peroxide, which leads to the inhibition of root growth and root hair development. LSF2 is a dual specificity protein phosphatase that displays high activity toward para-nitrophenyl phosphate (pNPP), and it dephosphorylates glucan to provide access for amylases that release maltose and glucose from starch in Arabidopsis (Meekins et al. 2015). Morphological study showed LSF2 null-mutation (*lsf2-1*) leaves showed severity of curly dwarf symptom under SA treatment. This suggests that LSF2 could improve the tolerance to oxidative stresses in plants, which was correlated that in wild-type leaves under oxidative stress conditions, LSF2 expression was upregulated. This report suggests that LSF2 not only plays a role in the root development but also functions in maintaining cellular ROS homeostasis in plants. Changes in redox are involved in controlling plant growth and development, and ROS production is associated with root development. NADPH oxidases and peroxidase have often been shown to be responsible for the production of endogenous O_2•— and H_2O_2 in plants. When H_2O_2 levels are reduced in plants, increased production of O_2•— is an important mechanism that maintains ROS homeostasis (Tsukagoshi et al. 2010). In *lsf2-1* mutant, under oxidative stress conditions, the generation rate of $O_2^{\bullet -}$ was lower, and the level of endogenous H_2O_2 was higher. These changes are consistent with the changes in Cu/Zn-SODs, CAT, and APX activity in the *lsf2-1* mutant, which led to more inhibition of root growth and less inhibition of root hair development under oxidative stress conditions compared to wild-type plants. No significant difference in the expression of the marker genes is involved in the regulation of root elongation, such as *SCN1*, *CPC*, and *PIN3*, between wild type and *lsf2-1* mutant. Furthermore, there was less reduced expression of marker genes *RHD6*, which is involved in the regulation of root hairs under less functional LSF2 under oxidative stress in Arabidopsis. This is consistent with the reduced inhibition of root hair development in the *lsf2-1* mutant compared with wild-type plants. In summary, LSF2 affects root development through modulating ROS homeostasis in Arabidopsis. Also ROS when generated in low quantity functions as signal molecules in cells, but when generated in large quantity, it became toxic to plants (Zhao et al. 2016).

Abiotic Stress

Plants are constantly exposed to diverse environmental stimuli and need to respond rapidly and effectively to these changes. DSP-MAPK phosphatases (MKPs) function as central hubs for integrating multiple abiotic and biotic stress signaling pathways (Jiang et al. 2018).

Earlier reports suggest MKPs play a role as an important component during multiple abiotic stresses, such as genotoxic stress, osmotic/drought stress, and salinity stress (Jiang et al. 2018). Arabidopsis *mkp1* mutant shows hypersensitivity to various genotoxic stresses (e.g., UV-C and ethyl methanesulfonate treatments), suggesting that MKP1 plays essential roles in genotoxic stress relief and this regulation appears to be inactivation of its interacting partner, MPK6 (Ulm et al. 2001, 2002). Recently, it was found that MKP1 negatively regulates the UV-B-induced stomatal closure whereas MPK6 positively regulates this process. Both aspects of regulation involve modulating hydrogen peroxide (H_2O_2)-induced nitric oxide (NO) production in guard cells (Li et al. 2017).

Loss-of-function of MKP1 increased plant tolerance to salt stress which suggests MKP1 also acts as a negative regulator of salinity tolerance (Ulm et al. 2002), whereas ectopic overexpression of the wheat ortholog TMKP1 in Arabidopsis *mkp1* results in enhanced salt stress tolerance, indicating a positive role of TMKP1 in regulating salt stress responses. The increase in salinity tolerance in TMKP1 transgenic seedlings was due to increased antioxidant enzyme activities and lower malondialdehyde (MDA), superoxide anion O_2^-, and H_2O_2 levels (Zaidi et al. 2016). This study suggests that, despite their significant homology, MKP1 and TMKP1 seem to act in an antagonistic manner to regulate salt stress responses. This might be due to distinct subcellular localization and differential catalytic regulation by Ca^{2+} (Jiang et al. 2018).

Arabidopsis dual specificity phosphatase (At DsPTP1) acts as a negative regulator in osmotic stress signaling in Arabidopsis seed germination and seedling establishment (Liu et al. 2015), whereas DsPTP1 positively regulates ABA accumulation and ABA signaling in response to osmotic stress (Liu et al. 2015). Studies show that null mutant *dsptp1* displayed less sensitivity to osmotic stress as shown by a higher seed germination rate and longer root length in response to osmotic stress, along with increased proline accumulation, reduced MDA content and ion leakage, and enhanced antioxidant enzyme activity (Liu et al. 2015). The ABA accumulation in *dsptp1* mutants decreased as compared to wild-type plants possibly by reducing the expression of ABA-biosynthesis gene *NCED3* and increasing the expression of ABA-catabolism gene *CYP707A4* under an osmotic stress condition. Consistently, downregulation of DsPTP1 also suppressed the expression of positive regulators of ABA signaling such as ABI3 and ABI5 while enhancing the expression of negative regulator ABI1 (Liu et al. 2015).

Biotic Stress

In Arabidopsis, several DSP-type phosphatases have been implicated in regulating pathogen-associated responses and resistance. MAPK PHOSPHATASE 1 (MKP1) is an important negative regulator of plant immunity. Diverse defense responses are hyperinduced in the Arabidopsis *mkp1* null mutant following pathogen-associated molecular pattern (PAMP) treatment, including activation of MPK6 and MPK3, production of reactive oxygen species (ROS), accumulation of a subset of PAMP-regulated transcripts, and inhibition of seedling growth (Anderson et al. 2011; Jiang et al. 2017). Consistent with enhanced PAMP responses, the *mkp1* mutant also displays enhanced resistance to the virulent pathogen *Pseudomonas syringae* pv. tomato (Pto) DC3000 (Anderson et al. 2011). Similar to the results from Arabidopsis, suppression of NtMKP1 in tobacco plants also resulted in elevated resistance against multiple pathogens including a necrotrophic pathogen, *Botrytis cinerea*, and *lepidopteran herbivores*, *Mamestra brassicae*, and *Spodoptera litura* (Oka et al. 2013). Interestingly, enhanced resistance against DC3000 in Arabidopsis *mkp1* can be explained by decreased abundance of specific extracellular plant metabolites that DC3000 uses as signals to activate its virulence program (Anderson et al. 2014). Thus, MKP1 seems to regulate a novel layer of immunity against pathogen infection. However, the molecular mechanisms by which MKP1 regulates the secretion of extracellular plant metabolites need to be further explored; and the knowledge of regulatory roles of metabolites on pathogen resistance can be applied to other pathogen species and crop species.

MAPK PHOSPHATASE 2 (MKP2) dephosphorylates phospho-MPK3 and phospho-MPK6 *in vitro* and has distinct functions in regulating different pathogen interactions (Lee and Ellis 2007; Lumbreras et al. 2010). Plants lacking MKP2 have enhanced resistance against *Ralstonia solanacearum*, a biotrophic pathogen, whereas increased susceptibility to *Botrytis cinerea*, a necrotrophic pathogen (Lumbreras et al. 2010). In addition, bimolecular fluorescence complementation (BiFC) experiments have shown that MKP2 interacts with MPK3 and MPK6 *in vivo* and fungal elicitors decreased the MKP2-MPK3 association but increased the MKP2-MPK6 interaction (Lumbreras et al. 2010). In agreement with enhanced MKP2-MPK6 interactions, co-infiltration of MKP2 and MPK6 into *N. benthamiana* leaves significantly reduced fungal elicitor-induced HR responses compared to infiltration with MPK6 alone. Interestingly, infiltration of MPK3 did not cause significant effects in these assays (Lumbreras et al. 2010). These results suggest that MKP2 exerts differential regulation on MPK3 and MPK6 during pathogen infection.

MKPs also contribute to the regulation of several resistance (R) proteins. IBR5 plays a positive role in regulating R protein CHS3, as evidenced by that mutation of *ibr5-7* suppresses the chilling-induced defense responses of *chs3-1* (Liu et al. 2015). Biochemical studies have shown that IBR5 interacts with CHS3 through the TIR domain of CHS3 *in vivo* and IBR5 forms a complex with chaperone proteins HSP90 and SGT1b (suppressor of the G2 allele of *skp1*) to stabilize CHS3 protein, thus increasing the accumulation of CHS3 (Liu et al. 2015a). Similarly, an *ibr5* mutant partially suppresses temperature-sensitive growth and autoimmune phenotypes resulting from constitutive activation of R protein SNC1 (suppressor of *npr1-1*, constitutive 1). IBR5 interacts with and promotes the accumulation of SNC1 (Liu et al. 2015). Additionally, IBR5 is also involved in controlling disease resistance mediated by R proteins RPM1 and RPS4. The *ibr5* mutants are more susceptible to avirulent bacterial pathogens DC3000 (avrRpm1) and DC3000 (avrRps4) (Liu et al. 2015). MKP1 has also been shown to play important roles in regulating the plant growth homeostasis by repressing inappropriate stress signaling mediated by SNC1. When the Arabidopsis *mkp1* mutation was introgressed into the Columbia ecotype from Wassilewskija, it showed weak dwarfism compared to wild-type plants under standard growth conditions, and such dwarfing was caused by constitutive activation of SNC1-mediated responses (Bartels et al. 2009). These studies demonstrate the roles of MKPs in regulating plant immunity against pathogen infection through modulating multiple signaling layers in PTI and ETI.

16.4.2 Roles of Dual Specificity Phosphatases in Animals

Dual specificity phosphatases (DSPs) regulate the activity of mitogen-activated protein kinases (MAPKs). The three main MAPKs, extracellular signal-regulated kinase (ERK), p38, and c-Jun N-terminal kinase (JNK), are found to be associated with immune responses. They are activated in response to certain stimuli. Their activation, intensity, and duration of activity determine the type of immune cell response that is specifically regulated by DSPs. DSPs have been demonstrated to have positive and negative control of immune responses such as sepsis, inflammatory arthritis, and experimental autoimmune encephalomyelitis; this suggests that targeting DUSPs can be a viable approach for an anti-inflammatory therapy (Jeffrey et al. 2007). In addition to its role in innate immunity, these also affect adaptive responses, as demonstrated by enhanced severity in a collagen-induced disease model for arthritis in knockout animals (Salojin et al. 2006). Knockout studies also suggest that DUSP1 protects mice from and endotoxin shock (Hammer et al. 2006). Knock out lines of DUSP1/MKP-1, controls innate immune responses and suppresses endotoxic shock and produces more TNF-α, IL-6, and IL-10 than wild-type (Zhao et al. 2006; Chi et al. 2006).

It has been shown that Dusp1 and Dusp10 null mice's macrophages produce more pro-inflammatory cytokines when exposed to LPS (Zhao et al. 2006; Zhang et al. 2004).

In zebrafish, DUSP4 was found to be essential for early development and endoderm specifications as loss-of-function of DUS4 results in the loss of foregut and pancreatic endoderm and necrosis of head tissues (Brown et al. 2008). DSP5, an inducible ERK-specific MAP kinase phosphatase of mammals, specifically interacts with extracellular signal-receptor kinase 2 (ERK-2) via a kinase interaction motif (KIM) and inactivates it. It also functions as a nuclear anchor of ERK-2 in mammalian cells. Earlier study demonstrated that the expression of DUSP5 in mammalian cell causes nuclear translocation and sequestration of inactive ERK-2 (Mandl et al. 2005). It has been also reported that they control angioblast populations in lateral plate mesoderm and are essential for vascular development (Pramanik et al. 2009). In *Drosophila*, loss-of-function mutant study of DUSP6 established that DUSP6 is important for photoreceptor cell differentiation, wing vein formation, and oogenesis (Gómez et al. 2005). Overexpression of DUSP6 in zebrafish results in ventralization of the embryo suggesting that it has a role in pattern formation in zebrafish (Tsang et al. 2004). Knockout studies established that they are involved in developmental processes as knockout mice suffer from dwarfism, premature fusion of cranial sutures (craniosynostoses), and defects in ear bones and otic capsules/osseous (bony) labyrinth, leading to hearing loss (Li et al. 2007). DUSP6 was also found to be a negative feedback modulator of FGF8 signaling in mammalian brain isthmic organizer (Echevarria et al. 2005). Hepatocyte dual specificity phosphatase 9 (DUSP9) is known to protect against non-alcoholic fatty liver disease (NAFLD) in mice by blocking apoptosis signal-regulating kinase 1 (ASK1) phosphorylation and the subsequent activation of p38 and c-Jun NH2-terminal kinase signaling. It is known to prevent NAFLD progression in mice, including lipid accumulation, glucose metabolism disorder, inflammation, and liver fibrosis in an ASK1-dependent manner (Ping et al., 2018). DUSP9/MKP-4 is also found to be essential for the placental organogenesis in mice as Dusp9 null mice die in utero due to placental insufficiency and failure of normal labyrinth development (Christie et al. 2005).

MPK3 inactivity leads to pancreatic and other tumor formation suggest that MPK3 is involved in tumor suppression (Warmka et al. 2004; Furukawa et al. 2003) and this mechanism was mediated by promoter hypermethylation (Xu et al. 2005) or chromosomal loss (Furukawa et al. 1998, 2003). MKP-3 (named DMKP-3) is also involved in *Drosophila* development (Kim et al. 2004). In zebrafish, MKP-3 controls Ras-MAPK signaling; Ras-MAPK downstream of the FGFR is important for axial polarity during development (Tsang et al. 2004). Calcitriol is hormonally active form of vitamin D, involved in inhibition of cancer. Calcitriol enhances expression of mitogen-activated protein kinase phosphatase 5 (MKP5) works as tumor suppressor by dephosphorylating and inactivating tumor promoter stress-activated protein kinase p38. Authors reported calcitriol as potential contributor as chemopreventive and therapeutic agent in prostate cancer (Krishnan et al. 2007). p.38 activation and interleukin 6 (IL-6) downstream production are proinflammatory, results in initiation and progression of prostate cancer. Prostate cell pretreatment with 1,25D (1,25dihydroxyvitamin-D3), inhibits both UV- and tumor necrosis factor A and stimulates IL-6 production in normal cells via p38 inhibition. MKP5 inactivates p38 and decreases IL-6 expression by 1,25D treatment in primary prostatic cultures of normal and adenocarcinoma cell. These results conclude 1,25D pretreatment decreases prostatic inflammation and helps in prostate cancer prevention (Nonn et al. 2006).

16.5 Conclusions and Future Prospects

This chapter summarizes the current knowledge of the role of DSP function through multiple signaling pathways. DSP is involved in many aspects of plant physiology such as starch metabolism, ROS homeostasis, as well as the adaptation to various biotic and abiotic stresses for survival in adverse situations. Also, in animal system, DSPs play crucial roles in innate immunity, growth, and development. As protein phosphorylation is the most important and crucial posttranslational modification, DSPs are important and indispensable components of different signaling pathways in plant and animal cells. A detailed and thorough investigation of regulatory mechanisms of DSP will provide insights into possible mechanisms that might explain how the phosphatases assist in transducing specific signals to generate diverse outputs or responses.

A better understanding of how plants coordinate and balance different signaling pathways in response to diverse environmental stimuli could lead to more rationally designed strategies for improving crop yield under changing environmental conditions. Promising results such as gain-of-function and loss-of-function of DSPs result in enhanced resistance to various biotic and abiotic stresses without compromising plant growth. This suggests that it may be possible to produce crops with elevated resistance against adverse environmental stresses. As an apparent phenomenon to initiate diverse signaling pathways, DSPs are important targets for modulating cross-talk to help overcome barriers for the improvement of plant resistance against a variety of stresses and hence improving crop yield and quality.

References

Alonso A, Rahmouni S, Williams S, van Stipdonk M, Jaroszewski L, Godzik A & Mustelin T (2003) Tyrosine phosphorylation of VHR phosphatase by ZAP-70. Nat Immun 4(1): 44–48

Alonso A, Burkhalter S, Sasin J, Tautz L, Bogetz J, Huynh H, Bremer MC, Holsinger LJ, Godzik A, Mustelin T (2004b) The minimal essential core of a cysteine-based protein-tyrosine phosphatase revealed by a novel 16-kDa VH1-like phosphatase, VHZ. J Biol Chem 279:35768–35774

Alonso A, Rojas A, Godzik A, Mustelin T (2004a) The dual-specific protein tyrosine phosphatase family. In: Protein phosphatases. Springer, Berlin, Heidelberg, pp 333–358

Anderson JC, Bartels S, González Besteiro MA, Shahollari B, Ulm R, Peck SC (2011) Arabidopsis MAP Kinase Phosphatase 1 (AtMKP1) negatively regulates MPK6-mediated PAMP responses and resistance against bacteria. Plant J 67:258–268

Anderson JC, Wan Y, Kim YM, Pasa-Tolic L, Metz TO, Peck SC (2014) Decreased abundance of type III secretion system-inducing signals in Arabidopsis mkp1 enhances resistance against Pseudomonas syringae. Proc Natl Acad Sci U S A 111:6846–6851

Ardito F, Giuliani M, Perrone D, Troiano G, Lo Muzio L (2017b) The crucial role of protein phosphorylation in cell signaling and its use as targeted therapy (Review). Int J Mol Med 40:271–280

Ardito F, Giuliani M, Perrone D, Troiano G, Muzio LL (2017a) The crucial role of protein phosphorylation in cell signaling and its use as targeted therapy. Int J Mol Med 40:271–280

Barford D, Das AK, Egloff MP (1998) The structure and mechanism of protein phosphatases: insights into catalysis and regulation. Annu Rev Biophys Biomol Struct 27:133–164
Bartels S, Anderson JC, González Besteiro MA, Carreri A, Hirt H, Buchala A et al (2009) MAP KINASE PHOSPHATASE1 and PROTEIN TYROSINE PHOSPHATASE1 are repressors of salicylic acid synthesis and SNC1-mediated responses in Arabidopsis. Plant Cell 21:2884–2897
Bheri M, Pandey GK (2019a) PP2A Phosphatases Take a Giant Leap in the Post-Genomics Era. Curr Genomics 20(3), 154–171
Bheri M, Pandey GK (2019b) Protein phosphatases meet reactive oxygen species in plant signaling networks. Environ Exp Bot 161, 26–40
Bordo D, Bork P (2002) The rhodanese/Cdc25 phosphatase superfamily. Sequence–structure–function relations. EMBO Rep 3:741–746
Brown JL, Snir M, Noushmehr H, Kirby M, Hong SK, Elkahloun AG, Feldman B (2008) Transcriptional profiling of endogenous germ layer precursor cells identifies dusp4 as an essential gene in zebrafish endoderm specification. Proc Natl Acad Sci U S A 105(34):12337–12342
Causin HF, Roqueiro G, Petrillo E, Láinez V, Pena LB, Marchetti CF, Gallego SM, Maldonado SI (2012) The control of root growth by reactive oxygen species in Salix nigra marsh seedlings. Plant Sci 183:197–205
Cheng HC, Qi RZ, Paudel H, Zhu HJ (2011) Regulation and function of protein kinases and phosphatases. Enzyme Res 2011:794089
Chi H, Barry SP, Roth RJ, Wu JJ, Jones EA, Bennett AM, Flavell RA (2006) Dynamic regulation of pro-and anti-inflammatory cytokines by MAPK phosphatase 1 (MKP-1) in innate immune responses. Proc Natl Acad Sci U S A 103(7):2274–2279
Christie GR, Williams DJ, MacIsaac F, Dickinson RJ, Rosewell I, Keyse SM (2005) The dual-specificity protein phosphatase DUSP9/MKP-4 is essential for placental function but is not required for normal embryonic development. Mol Cell Biol 25(18):8323–8333
Denu JM, Dixon JE (1995) A catalytic mechanism for the dual-specific phosphatases. Proc Natl Acad Sci U S A 92:5910–5914
Denu JM, Lohse DL, Vijayalakshmi J, Saper MA, Dixon JE (1996) Visualization of intermediate and transition-state structures in protein-tyrosine phosphatase catalysis. Proc Natl Acad Sci U S A 93:2493–2498
Dunn D, Chen L, Lawrence DS, Zhang ZY (1996) The active site specificity of the Yersinia protein-tyrosine phosphatase. J Biol Chem 271:168–173
Echevarria D, Martinez S, Marques S, Lucas-Teixeira V, Belo JA (2005) Mkp3 is a negative feedback modulator of Fgf8 signaling in the mammalian isthmic organizer. Develop Biol 277(1):114–128
Furukawa M, Iwahashi Y, Horita Z, Nemoto M, Langdon TG (1998) The shearing characteristics associated with equal-channel angular pressing. Mater Sci Eng A 257(2):328–332
Furukawa T, Sunamura M, Motoi F, Matsuno S, Horii A (2003) Potential tumor suppressive pathway involving DUSP6/MKP-3 in pancreatic cancer. Am J Pathol 162(6):1807–1815
Gentry MS, Dowen RH, Worby CA, Mattoo S, Ecker JR, Dixon JE (2007) The phosphatase laforin crosses evolutionary boundaries and links carbohydrate metabolism to neuronal disease. J Cell Biol 178:477–488
Gentry MS, Romá-Mateo C, Sanz P (2013) Laforin, a protein with many faces: glucan phosphatase, adapter protein, et alii. FEBS J 280(2):525–537
Gómez AR, López-Varea A, Molnar C, de la Calle-Mustienes E, Ruiz-Gómez M, Gómez-Skarmeta JL, de Celis JF (2005) Conserved cross-interactions in Drosophila and Xenopus between Ras/MAPK signaling and the dual-specificity phosphatase MKP3. Develop Dyn 232(3):695–708
Hammer M, Mages J, Dietrich H, Servatius A, Howells N, Cato AC, Lang R (2006) Dual specificity phosphatase 1 (DUSP1) regulates a subset of LPS-induced genes and protects mice from lethal endotoxin shock. J Exp Med 203(1):15–20
Hejazi M, Fettke J, Kötting O, Zeeman SC, Steup M (2010) The Laforin-like dual-specificity phosphatase SEX4 from Arabidopsis hydrolyzes both C6- and C3-phosphate esters introduced by starch-related dikinases and thereby affects phase transition of alpha-glucans. Plant Physiol 152:711–722

Huang H, Potter CJ, Tao W, Li DM, Brogiolo W, Hafen E, Sun H, Xu T (1999) PTEN affects cell size, cell proliferation and apoptosis during Drosophila eye development. Development 126(23):5365–5372

Huang LF, Liu YK, Su SC, Lai CC, Wu CR, Chao TJ, Yang YH (2019) Genetic engineering of transitory starch accumulation by knockdown of OsSEX4 in rice plants for enhanced bioethanol production. Biotechnol Bioeng 117:933–944

Jeffrey KL, Camps M, Rommel C, Mackay CR (2007) Targeting dual-specificity phosphatases: manipulating MAP kinase signalling and immune responses. Nat Rev Drug Discov 6:391–403

Jiang L, Chen Y, Luo L, Peck SC (2018) Central roles and regulatory mechanisms of dual-specificity MAPK phosphatases in developmental and stress signaling. Front Plant Sci 9:1697

Jiang L, Wan Y, Anderson JC, Hou J, Islam SM, Cheng J et al (2017) Genetic dissection of Arabidopsis MAP kinase phosphatase 1-dependent PAMP-induced transcriptional responses. J Exp Bot 68:5207–5220

Kerk D, Bulgrien J, Smith DW, Barsam B, Veretnik S, Gribskov M (2002) The complement of protein phosphatase catalytic subunits encoded in the genome of Arabidopsis. Plant Physiol 129:908–925

Kerk D, Conley TR, Rodriguez FA, Tran HT, Nimick M, Muench DG, Moorhead GB (2006) A chloroplast-localized dual specificity protein phosphatase in Arabidopsis contains a phylogenetically dispersed and ancient carbohydrate-binding domain, which binds the polysaccharide starch. Plant J 46:400–413

Kerk D, Templeton G, Moorhead GB (2008) Evolutionary radiation pattern of novel protein phosphatases revealed by analysis of protein data from the completely sequenced genomes of humans, green algae, and higher plants. Plant Physiol 146:351–367

Kim SJ, Ryu SE (2012) Structure and catalytic mechanism of human protein tyrosine phosphatome. BMB Rep 45:693–699

Kötting O, Santelia D, Edner C, Eicke S, Marthaler T, Gentry MS, Comparot-Moss S, Chen J, Smith AM, Steup M, Ritte G, Zeeman SC (2009) STARCH-EXCESS4 is a laforin-like phosphoglucan phosphatase required for starch degradation in Arabidopsis thaliana. Plant Cell 21:334–346

Krebs EG, Kent AB, Fischer EH (1958) The muscle phosphorylase b kinase reaction. J Biol Chem 231(1):73–83

Krishnan AV, Moreno J, Nonn L, Swami S, Peehl DM, Feldman D (2007) Calcitriol as a chemopreventive and therapeutic agent in prostate cancer: role of anti-inflammatory activity. J Bone Miner Res 22(S2):V74–V80

Lee C, Cheong YH, Kim KN et al (2010) Protein kinases and phosphatases for stress signal transduction in plants. In: Abiotic stress adaptation in plants. Springer, Dordrecht, pp 123–163

Lee JS, Ellis BE (2007) Arabidopsis MAPK phosphatase 2 (MKP2) positively regulates oxidative stress tolerance and inactivates the MPK3 and MPK6 MAPKs. J Biol Chem 282:25020–25029

Li C, Scott DA, Hatch E, Tian X, Mansour SL (2007) Dusp6 (Mkp3) is a negative feedback regulator of FGF-stimulated ERK signaling during mouse development. Development 134(1):167–176

Li FC, Wang J, Wu MM, Fan CM, Li X, He JM (2017) Mitogen-activated protein kinase phosphatases affect UV-B-induced stomatal closure via controlling NO in Guard Cells. Plant Physiol 173:760–770

Liu R, Liu Y, Ye N, Zhu G, Chen M, Jia L et al (2015) AtDsPTP1 acts as a negative regulator in osmotic stress signalling during Arabidopsis seed germination and seedling establishment. J Exp Bot 66:1339–1353

Lountos GT, Tropea JE, Waugh DS (2011) Structure of human dual-specificity phosphatase 27 at 2.38 Å resolution. Acta Crystallogr D Biol Crystallogr 67(5):471–479

Luan S, Ting J, Gupta R (2001) Protein tyrosine phosphatases in higher plants. New Phytol 151:155–164

Lumbreras V, Vilela B, Irar S, Solé M, Capellades M, Valls M et al (2010) MAPK phosphatase MKP2 mediates disease responses in Arabidopsis and functionally interacts with MPK3 and MPK6. Plant J 63:1017–1030

Mailloux C (2010) Structure-function characterization of the human dual specificity phosphatase hYVH1. Electronic Theses and Dissertations, p 307. https://scholar.uwindsor.ca/etd/307
Mandl M, Slack DN, Keyse SM (2005) Specific inactivation and nuclear anchoring of extracellular signal-regulated kinase 2 by the inducible dual-specificity protein phosphatase DUSP5. Mol Cell Biol 25(5):1830–1845
Meekins DA, Raththagala M, Auger KD, Turner BD, Santelia D, Kötting O, Gentry MS, Vander Kooi CW (2015) Mechanistic insights into glucan phosphatase activity against polyglucan substrates. J Biol Chem 290:23361–23370
Meekins DA, Raththagala M, Husodo S, White CJ, Guo HF, Kötting O, Vander Kooi CW, Gentry MS (2014) Phosphoglucan-bound structure of starch phosphatase Starch Excess4 reveals the mechanism for C6 specificity. Proc Natl Acad Sci U S A 111:7272–7277
Meekins DA, Vander Kooi CW, Gentry MS (2016) Structural mechanisms of plant glucan phosphatases in starch metabolism. FEBS J 283:2427–2447
Moorhead GB, De Wever V, Templeton G, Kerk D (2009) Evolution of protein phosphatases in plants and animals. Biochem J 417:401–409
Moorhead GB, Trinkle-Mulcahy L, Ulke-Lemee A (2007) Emerging roles of nuclear protein phosphatases. Nat Rev Mol Cell Biol 8:234–244
Mustelin T (2007) A brief introduction to the protein phosphatase families. Methods Mol Biol 365:9–22
Niittylä T, Comparot-Moss S, Lue WL, Messerli G, Trevisan M, Seymour MD, Gatehouse JA, Villadsen D, Smith SM, Chen J, Zeeman SC, Smith AM (2006) Similar protein phosphatases control starch metabolism in plants and glycogen metabolism in mammals. J Biol Chem 281:11815–11818
Nonn L, Peng L, Feldman D, Peehl DM (2006) Inhibition of p38 by vitamin D reduces interleukin-6 production in normal prostate cells via mitogen-activated protein kinase phosphatase 5: implications for prostate cancer prevention by vitamin D. Cancer Res 66(8):4516–4524
Ogg S, Ruvkun G (1998) The *C. elegans* PTEN homolog, DAF-18, acts in the insulin receptor-like metabolic signaling pathway. Mol Cell 2:887–893
Oka K, Amano Y, Katou S, Seo S, Kawazu K, Mochizuki A et al (2013) Tobacco MAP kinase phosphatase (NtMKP1) negatively regulates wound response and induced resistance against necrotrophic pathogens and lepidopteran herbivores. Mol Plant-Microbe Interact 26:668–675
Olsen JV, Blagoev B, Gnad F, Macek B, Kumar C, Mortensen P, Mann M (2006) Global, *in vivo*, and site-specific phosphorylation dynamics in signaling networks. Cell 127:635–648
Pramanik K, Chun CZ, Garnaas MK, Samant GV, Li K, Horswill MA et al (2009) Dusp-5 and Snrk-1 coordinately function during vascular development and disease. Blood 113(5):1184–1191
Ritte G, Heydenreich M, Mahlow S, Haebel S, Kötting O, Steup M (2006) Phosphorylation of C6- and C3-positions of glucosyl residues in starch is catalysed by distinct dikinases. FEBS Lett 580:4872–4876
Ritte G, Lloyd JR, Eckermann N, Rottmann A, Kossmann J, Steup M (2002) The starch-related R1 protein is an alpha glucan, water dikinase. Proc Natl Acad Sci U S A 99:7166–7171
Salojin KV et al (2006) Essential role of MAPK phosphatase-1 in the negative control of innate immune responses. J Immunol 176:1899–1907
Samodien E, Jewell JF, Loedolff B, Oberlander K, George GM, Zeeman SC, Damberger FF, van der Vyver C, Kossmann J, Lloyd JR (2018) Repression of *Sex4* and *Like Sex Four2* orthologs in potato increases tuber starch bound phosphate with concomitant alterations in starch physical properties. Front Plant Sci 9:1044
Santelia D, Kötting O, Seung D, Schubert M, Thalmann M, Bischof S, Meekins DA, Lutz A, Patron N, Gentry MS, Allain FH, Zeeman SC (2011) The phosphoglucan phosphatase like sex Four2 dephosphorylates starch at the C3-position in Arabidopsis. Plant Cell 23:4096–4111
Shankar A, Agrawal N, Sharma M, Pandey A, Pandey GK (2015) Role of protein tyrosine phosphatases in plants. Curr Genomics 16:224–236
Shi Y (2009) Serine/threonine phosphatases: mechanism through structure. Cell 139:468–484
Singh A, Giri J, Kapoor S, Tyagi AK, Pandey GK (2010) Protein phosphatase complement in rice: genome wide identification and transcriptional analysis under abiotic stress conditions and reproductive development. BMC Genomics 11:435

Singh A, Pandey GK (2012) Protein phosphatases: a genomic outlook to understand their function in plants. J Plant Biochem Biotechnol 21:S100–S107

Singh A, Jha SK, Bagri J, Pandey GK (2015) ABA inducible rice protein phosphatase 2C confers ABA insensitivity and abiotic stress tolerance in Arabidopsis. PloS one, 10(4), e0125168

Smith RD, Walker JC (1996) Plant protein phosphatases. Annu Rev Plant Physiol Plant Mol Biol 47:101–125

Sokolov LN, Dominguez-Solis JR, Allary AL, Buchanan BB, Luan S (2006) A redox-regulated chloroplast protein phosphatase binds to starch diurnally and functions in its accumulation. Proc Natl Acad Sci U S A 103:9732–9737

Stewart AE, Dowd S, Keyse SM, McDonald NQ (1999) Crystal structure of the MAPK phosphatase Pyst1 catalytic domain and implications for regulated activation. Nat Struct Biol 6:174–181

Stone RL, Dixon JE (1994) Protein-tyrosine phosphatases. J Biol Chem 269:31323–31326

Tonks NK, Neel BG (1996) From form to function: signaling by protein tyrosine phosphatases. Cell 87:365–368

Tsang M, Maegawa S, Kiang A, Habas R, Weinberg E, Dawid IB (2004) A role for MKP3 in axial patterning of the zebrafish embryo. Development 131(12):2769–2779

Tsukagoshi H, Busch W, Benfey PN (2010) Transcriptional regulation of ROS controls transition from proliferation to differentiation in the root. Cell 143:606–616

Ulm R, Ichimura K, Mizoguchi T, Peck SC, Zhu T, Wang X, Shinozaki K, Paszkowski J (2002) Distinct regulation of salinity and genotoxic stress responses by Arabidopsis MAP kinase phosphatase 1. EMBO J 21:6483–6493

Ulm R, Revenkova E, Sansebastiano G, Bechtold N, Paszkowski J (2001) Mitogen-activated protein kinase phosphatase is required for genotoxic stress relief in Arabidopsis. Genes Dev 15:699–709

Visintin R, Craig K, Hwang ES, Prinz S, Tyers M, Amon A (1998) The phosphatase Cdc14 triggers mitotic exit by reversal of Cdk-dependent phosphorylation. Mol Cell 2:709–718

Wang J, Stuckey JA, Wishart MJ, Dixon JE (2002) A unique carbohydrate binding domain targets the lafora disease phosphatase to glycogen. J Biol Chem 277:2377–2380

Warmka JK, Mauro LJ, Wattenberg EV (2004) Mitogen-activated protein kinase phosphatase-3 is a tumor promoter target in initiated cells that express oncogenic Ras. J Biol Chem 279(32):33085–33092

Wishart MJ, Dixon JE (2002) PTEN and myotubularin phosphatases: from 3-phosphoinositide dephosphorylation to disease. Trends Cell Biol 12:579–585

Xu S, Furukawa T, Kanai N, Sunamura M, Horii A (2005) Abrogation of DUSP6 by hypermethylation in human pancreatic cancer. J Hum Genet 50(4):159–167

Yuvaniyama J, Denu JM, Dixon JE, Saper MA (1996) Crystal structure of the dual specificity protein phosphatase VHR. Science 272:1328–1331

Zaidi I, Ebel C, Belgaroui N, Ghorbel M, Amara I, Hanin M (2016) The wheat MAP kinase phosphatase 1 alleviates salt stress and increases antioxidant activities in Arabidopsis. J Plant Physiol 193:12–21

Zhang Y, Blattman JN, Kennedy NJ, Duong J, Nguyen T, Wang Y et al (2004) Regulation of innate and adaptive immune responses by MAP kinase phosphatase 5. Nature 430(7001):793–797

Zhang Y, Li S, Zhou LZ, Fox E, Pao J, Sun W, Zhou C, McCormick S (2011) Overexpression of Arabidopsis *thaliana* PTEN caused accumulation of autophagic bodies in pollen tubes by disrupting phosphatidylinositol 3-phosphate dynamics. Plant J 68:1081–1092

Zhang ZY (2002) Protein tyrosine phosphatases: structure and function, substrate specificity, and inhibitor development. Annu Rev Pharmacol Toxicol 42:209–234

Zhao P, Sokolov LN, Ye J, Tang CY, Shi J, Zhen Y, Lan W, Hong Z, Qi J, Lu GH, Pandey GK, Yang YH (2016) The LIKE SEX FOUR2 regulates root development by modulating reactive oxygen species homeostasis in Arabidopsis. Sci Rep 6:28683

Zhao Q, Wang X, Nelin LD, Yao Y, Matta R, Manson ME, Chang CH (2006) MAP kinase phosphatase 1 controls innate immune responses and suppresses endotoxic shock. J Exp Med 203(1):131–140

Chapter 17
Protein Tyrosine Phosphatases: Implications in the Regulation of Stress Responses in Plants

Malathi Bheri and Girdhar K. Pandey

17.1 Introduction: Tyrosine Phosphorylation Machinery in Plants

Protein phosphorylation is a sine qua non of signaling pathways in prokaryotic and eukaryotic systems (Miller 2012). Prokaryotes utilize phospho-histidine (His) signaling, while eukaryotes utilize serine (Ser), threonine (Thr) and tyrosine (Tyr) phosphorylation majorly in signaling pathways (Hunter 2014; Manning et al. 2002). The phospho-Tyr signaling machinery of animal systems comprises of (a) protein tyrosine kinases (PTKs), (b) protein tyrosine phosphatases (PTPs), and (c) modular phospho-Tyr-binding domains, comparable to "writers," "erasers," and "readers," respectively (Lim and Pawson 2010; Pincus et al. 2008). The Tyr phosphorylation levels are maintained by a fine balance between the functioning of the PTKs and PTPs. PTKs phosphorylate the substrate proteins by transferring the γ-phosphate from adenosine triphosphate to Tyr residues, and PTPs dephosphorylate the phosphorylated proteins by removing the phosphate group(s) from phospho-Tyr residues. In contrast to phospho-Ser/Thr, phospho-Tyr is involved majorly in a regulatory role rather than a structural one. The high turnover rate of PTPs results in a short half-life of phospho-Tyr residues, which is overcome through an intramolecular interaction or by binding to SH2 or PTP domains (Hunter 2014). Phospho-Tyr residues make up for less than 0.1% of all phosphoresidues in mammalian cells. Their detection is hindered at a high signal-to-noise ratio in the background of crosstalk between different signaling molecules (Yoshimoto and Kuroda 2017).

The PTP superfamily comprises of the classical PTPs, the dual-specificity phosphatases (DsPTPs), the CDC25 and the low-molecular-weight PTPs (LWMPTPs) families, based on their sequence, structure, and function (Fauman and Saper 1996).

M. Bheri · G. K. Pandey (✉)
Department of Plant Molecular Biology, University of Delhi South Campus, New Delhi, India
e-mail: gkpandey@south.du.ac.in

G. K. Pandey (ed.), *Protein Phosphatases and Stress Management in Plants*,
https://doi.org/10.1007/978-3-030-48733-1_17

PTPs are Tyr-specific and DsPTPs dephosphorylate Ser and Thr residues, in addition to Tyr residues (Farkas et al. 2007). Cys-based PTPs are classified as type I, II, and III subfamilies. The type I subfamily includes classical Tyr-specific PTPs and DsPTPs (Tonks 2013). Class I DsPTPs include MAP kinase phosphatases (MKPs), phosphatase and tensin homologue deleted on chromose 10 (PTEN), phosphatases of regenerating liver (PRL), CDC14 phosphatases, slingshots, myotubularins (MTMs), and atypical DsPTPs in animals (Moorhead et al. 2009). Type II PTPs show Tyr- and Thr specificity and function in cell cycle regulation while type-III PTPs include prokaryotic Tyr-specific LMWPTPs (Neel and Tonks 1997; Alonso 2004; Mustelin 2007). PTPs are also categorized as receptor-like as well as intracellular PTPs, with the former possessing not more than two cytoplasmic catalytic domains and the latter possessing a single catalytic domain (Stone and Dixon 1994).

Tyr phosphorylation is employed as a regulatory mechanism involved in intra- and intercellular communication and coordination of cellular processes in animals (Hunter 2014). Abnormalities in this process are known to be involved in the development of inherited or acquired pathological conditions in humans (Alonso et al. 2004). Although PTKs are not reported in yeast, Tyr phosphorylation occurs in yeast, indicating that DSKs like MAPK kinases phosphorylate Tyr-containing proteins (Barizza et al. 1999; Lindberg et al. 1992; Shiozaki and Russell 1995). It has been reported in bacterial and archaeal species also, the genomes of which contain genes encoding PTPs (Mustelin 2007). The process is involved in regulation of processes like transcription, protein localization, nutrient perception, virulence, and stress responses, in addition to capsule and lipopolysaccharide (LPS) biogenesis (Getz et al. 2019). Tyr phosphorylation in bacteria and archaea as well as plants has received less attention, and that may be a reason behind the limited information available as compared to the mammalian systems (Ghelis et al. 2011; Getz et al. 2019). The identification of novel enzymes showing Tyr kinase function indicates the involvement of Tyr phosphorylation in the cellular regulation (Getz et al. 2019). Arabidopsis contains more proteins specifically phosphorylated on Tyr residues than in case of yeast (Carpi et al. 2002). However, only a few PTPs are reported and PTKs are still to be identified in plants (Kerk et al. 2008, 2002; Luan 2003; Shankar et al. 2015).

The initial perception about Tyr phosphorylation being less common in plants as compared to mammalian systems has been thwarted by the recent developments (de la Fuente van Bentem and Hirt 2009; Luan 2002). Tyr kinase(s) and phospho-Tyr proteins have been reported in higher plants like Arabidopsis (Barizza et al. 1999; Carpi et al. 2002; Fordham-Skelton et al. 1999; Ndimba et al. 2003), rice (Singh et al. 2010), pea (Fordham-Skelton et al. 1999; Torruella et al. 1986), soya bean (Fordham-Skelton et al. 1999), tobacco (Suzuki and Shinshi 1995), coconut (Islas-Flores et al. 1998), *Mimosa pudica* (Kameyama et al. 2000), *Scots pine* (*Pinus sylvestris L.*; Kovaleva et al. 2013), and peanut (*Arachis hypogaea*; Rudrabhatla and Rajasekharan 2002).

The process of Tyr phosphorylation is carried out through dual-specificity kinase (DSKs) in plants (Rudrabhatla et al. 2006). The existence of Tyr phosphorylation machinery in plants is supported by the presence of Tyr-like kinases (TK-like kinases; Martin et al. 2009), protein kinases (PKs) containing TK catalytic domains

(Rudrabhatla et al. 2006), calcium-dependent protein kinase (CDPK/CPK)-related PKs (CRKs) showing TK activity (Nemoto et al. 2015) as well as proteins containing C2 phospho-Tyr-binding domain (Miranda-Saavedra and Barton 2007) and SH2 (src homology 2) domain (de la Fuente van Bentem and Hirt 2009; Williams and Zvelebil 2004). PTPs regulate MAPK and hormonal signaling pathways, development as well as stress responses in plants (Luan 2003; Shankar et al. 2015).

A total of 107 PTPs are reported in humans that include 38 classical PTPs (17 non-receptor and 21 receptor-like PTPs), 61 DsPTPs, 3 CDC25 DsPTPs, 4 Asp-based PTPs, and a single LMW PTP (Alonso et al. 2004). Of these, 11 are inactive and 16 utilize glycogen, mRNA, or phosphoinositide substrates (Ghelis 2011). In contrast, fewer PTPs are identified in plants. Plants lack Tyr kinases and 67% of plant protein phosphatases (PPs) exhibit Ser/Thr specificity, indicating their preference for Ser/Thr phosphosites in comparison with human system (Wei and Pan 2014). Arabidopsis PTP: AtPTP1(Xu et al. 1998), and DsPTP: AtDsPTP1 (Gupta et al. 1998) were the first plant PTPs to be identified. The AtDsPTP1 protein sequence deduced from cDNA is 25–35% identical to eukaryotic DsPTPs with highly conserved active site signature motif. It is among the smallest compared to its counterparts from different organisms and possesses the phosphatase domains only (Gupta et al. 1998). The LIKE SEX FOUR4 1 (LSF1), LIKE SEX FOUR2 (LSF2), and STARCH EXCESS4 (SEX4) contain a DsPTP domain and are involved in starch degradation (Comparot-Moss et al. 2010; Gentry and Pace 2009; Santelia et al. 2011; Silver et al. 2014). SEX4 and LSF2 glucan phosphatases are closely related to Laforin, a phosphatase regulating glycogen metabolism in vertebrates and few protozoa (Gentry and Pace 2009; Tagliabracci et al. 2007; Worby et al. 2006). The glucan phosphatase characteristic motif, CζAGΨGR (ζ = hydrophilic residue; Ψ = long chain aliphatic), is based on the similarity of PTP-loop of SEX4, LSF2, and Laforin (Meekins et al. 2015).

Kerk and co-workers (2008) predicted 150 genes encoding PPs in Arabidopsis genome using human PP sequences as queries. These include 22 DsPTPs as well as 1 each of class I PTP and LMW-PTP. Similarly rice genome encodes 23 DsPTPs as well as 1 each of PTP and LMW-PTP (Singh et al. 2010). Maize genome encodes 159 PPs, of which 29 are PTPs (Wei and Pan 2014). The PTP family does not show an equal distribution of introns and includes ZmPP14 (LMWPTP), ZmPP47 (PTPL), and ZmPP145 (CDC25; Wei and Pan 2014). A single class I PTP was predicted in *Oryza sativa*, *Populus trichocarpa* as well as the green algae *Chlamydomonas reinhardtii* and *Ostreococcus tauri*. Single homologues of CDC14 and CDC25 were identified in *C. reinhardtii* and *O. tauri*, respectively. *C. reinhardtii* and *P. trichocarpa* encode one and two homologues of LMWPTP, respectively. No homologues of slingshots were identified in any of these species (Kerk et al. 2008). Plant and algal CDC25s may be arsenate reductases as they form a separate clade with arsenate reductases, unlike human and yeast CDC25s. The N-terminal regulatory domain characteristic of CDC25s is also absent. This along with the absence of CDC14 in plants indicates that their involvement as cell cycle regulators may have occurred after the divergence of plants from animals and yeast or may be lost during early evolution in plants (Moorhead et al. 2009). The classification of PTPs found in plants is shown in Fig. 17.1.

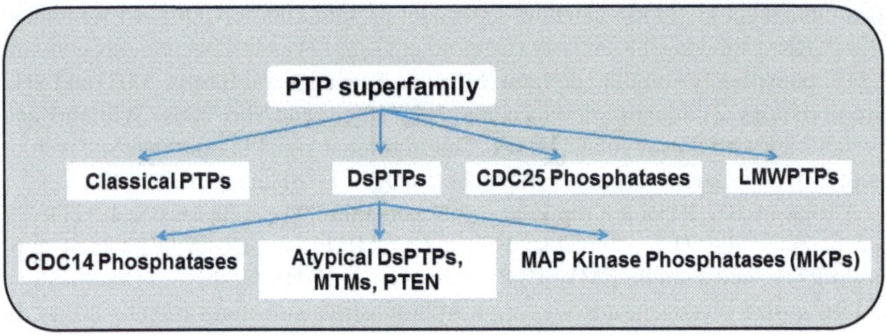

Fig. 17.1 Classification of PTP superfamily in plants. PTPs include classical PTPs (PTPs), dual-specificity phosphatases (DsPTPs), low-molecular-weight PTPs (LMWPTPs), and CDC25 phosphatases. The DsPTPs include CDC14 phosphatases, atypical DsPTPs, myotubularins (MTMs), phosphatase and tensin homologue deleted in chromosome 10 (PTEN), and MAPK phosphatases (MKPs)

17.2 Protein Tyrosine Phosphatase: Structure and Catalysis

The PTP superfamily has a highly conserved catalytic domain indicating a common hydrolytic mechanism (Fauman and Saper 1996). The lack of sequence homology with STPs and the unique three-dimensional structure of their catalytic domains indicate an independent evolution of PTPs. The eukaryotic PTPs have a signature -CX5R- motif in the catalytic domain: (V/I)HCXAGXGR(S/T) (Cohen 2010; Kerk et al. 2008). PTPs have the ability to hydrolyse *p*-nitrophenyl phosphate (pNPP) and do not require metal ions for functioning. They are sensitive to vanadate, a PTP inhibitor, and insensitive to okadaic acid (OA), a potent inhibitor of Ser/Thr phosphatases (STPs) (Fauman and Saper 1996). In accordance with this, the Arabidopsis PTP, AtPTP1, is inhibited by vanadate, while inhibitors of STP like OA do not inhibit AtPTP1 activity. The presence of divalent cations or EDTA does not influence AtPTP1 activity significantly (Xu et al. 1998). However, PTP activity may depend on cofactors, as in case of PTP1B, the hydrogen peroxide (H_2O_2)-induced inactivation requires calcium in animals (Skorey et al. 1997).

PTPs may have evolved from a common ancestor of related DsPTPs, also reported in most single-celled eukaryotes (Lim and Pawson 2010). The PTPs and DsPTPs are related but show distinct catalytic domains. They have a common fold and show homology in the catalytic motif HC(X)5R. The PTPs show highly similar secondary and tertiary structure in the catalytic site, with the substrate specificity being determined by sequences other than the catalytic domain (Luan 2000; Tonks and Neel 1996). The PTP catalytic domain is composed of ~250 amino acid residues. It has a central parallel β-sheet which is flanked by α-helices, with the -CX5R motif encompassed by a β-loop-α-loop/phosphate-binding loop (P-loop; Ruddraraju and Zhang 2017; Zhang 2002). The presence of conserved residues (Cys, Arg, Ser, Thr, and Asp) in the P-loop of the PTPs, DSPs as well as LMWPTPs indicates a common catalytic mechanism (Shankar et al. 2015). The PTPs also contain the WPD-, Q-, Lys-, and tyr(P) recognition loops (Brandão et al. 2010). The active site of DsPTPs is shallower than classical PTPs, facilitating the dephosphorylation of Tyr, Thr, or Ser residues (Dunn et al. 1996; Jia et al. 1995; Stewart et al. 1999; Yuvaniyama et al. 1996). DsPTPs have undergone functional divergence enabling them to target lipids and MTMs (Alonso et al. 2004).

The catalytic reaction involves two steps: (a) The catalytic Cys residue acts as a nucleophile that accepts the PO_3 moiety from the phosphoresidues and forms a phospho-Cys intermediate (Guan and Dixon 1991). (b) This is followed by a nucleophilic attack by a water molecule, transferring the PO_3 moiety to a water molecule (Hengge et al. 1995). The Arg residue stabilizes the transition state through formation of hydrogen bonds. Asp acts as a general acid which donates a proton to the oxygen of the leaving group (Hengge et al. 1995). It may also act as a general base, by accepting a proton from the attacking water molecule. The His and the Ser/Thr residues lower the pKa of the Cys residue, which stabilizes the negative charge on the Cys residue, thus facilitating its removal and promoting the cleavage of the phosphoenzyme intermediate (Denu and Dixon 1995). The nitrogen of the amide groups in the phosphate-binding loop also stabilize the Cys residue (Fauman and Saper 1996). A conserved Glutamine (Gln) is instrumental in the proper orientation of the nucleophilic water molecule. Asp181, Gln262, and Arg221 are involved in PTP1B-mediated catalysis. The phosphoenzyme intermediate formation and hydrolysis also involve binding of the guanidinium side chain of the Arg221 with the penta-coordinated transition states (Liang et al. 2007). The Cys265 and Asp234 conserved residues are involved in the catalytic activity of AtPTP1 phosphatase (Xu et al. 1998). The AtPTP1 and Tyr-specific PTPs from other organisms show high homology in structural domains including the 263–271 residues in the active site, charged residues between 90–104 position and the WPD motif (Xu et al. 1998).

The intracellular redox state and the extracellular ligand-induced reactive oxygen species (ROS) generation can result in oxidation of PTPs (Fig. 17.2). The catalytic Cys residue has an SH group, existing in the thiolate state (S^-; Peters et al. 1998). The PTP oxidation converts active site thiolates to sulfenic acid (–SOH), with further oxidation resulting in the formation of sulfinic (SO_2H) or sulfonic acid (SO_3H) derivatives. Oxidation to sulfenic acid inhibits the PTP activity due to the inability of Cys residue to act as a phosphate acceptor and is reversible, with further oxidation resulting in irreversible modification (Meng et al. 2002). The interaction of sulfenic acid intermediate with proximal amino acid residues results in formation of secondary products like sulfenylamides (SN) and intramolecular disulfides that prevent further oxidation (Östman et al. 2011).

The identification of PTPs is vital for further expansion of our understanding of Tyr phosphorylation in higher plants, particularly from a functional perspective. The post-genome era has allowed the application of whole-genome sequencing and global transcriptome profiling towards the identification and functional characterization of genes on a large scale. This has brought perspective to our understanding of the regulatory checks in developmental pathways and stress responses in plants. In the following sections, we discuss the role of PTPs in different signaling pathways.

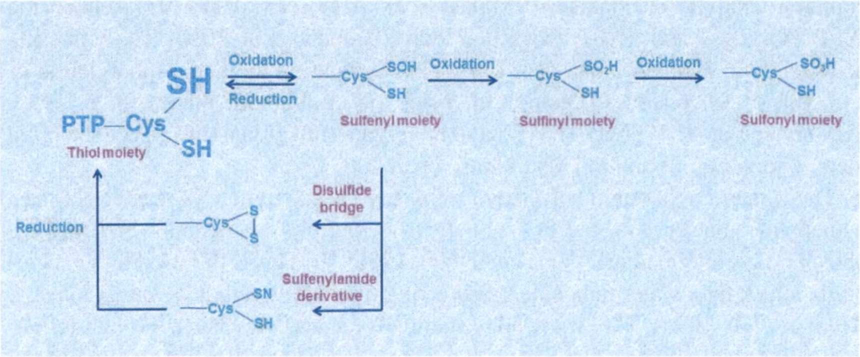

Fig. 17.2 Redox regulation of protein Tyr phosphatases (PTPs). The active thiol moiety (–SH) of the catalytic Cys residues undergoes reversible oxidation resulting in the formation of the inactive sulfenyl moiety (–SOH). The –SOH moiety forms either a disulfide bond or a sulfenylamide bond, both of which can undergo reduction. The –SOH moiety can undergo further oxidization to form the sulfinyl (–SO_2H) and sulfonyl (–SO_3H) moieties. These reactions are irreversible and render these derivatives inactive

17.3 Redox Signaling

The mechanisms regulating redox signaling in plants are not well understood (Apel and Hirt 2004; Forman et al. 2002; Oelze et al. 2008). The process of redox signaling alters the oxidation state of a protein and its behavior reversibly at the post-translational level. Proteins containing Cys residues are critical targets as thiol group of the Cys residue is prone to oxidation. The redox state of the cell affects PTP function. The molecular structure and function of PTPs is influenced by alterations in as much as a single Cys residue (Li et al. 2012a; Tonks 2005; Winterbourn and Hampton 2008). Reductants and oxidants exhibit contrasting effects on PTP activity. Oxidants like H_2O_2 oxidize the catalytic Cys residue of the PTPs, thus inactivating them. The thiol group of the catalytic Cys residue is modified by PAO, thus inhibiting PTPs. However, reductants like dithiothreitol (DTT) reduce the catalytic Cys residue, thus maintaining the PTPs in an active state (Luan et al. 2001; Smith and Walker 1996; Tonks 2005). Exposure to reductants and PAO is reported to inhibit abscisic acid (ABA) accumulation, as well as the gene expression of the stress-related genes indicating that PTPs are important components of stress-linked redox signaling (Jia and Zhang 2000; Li et al. 2012a).

PTPs that are inhibited by a reductant have not been identified in animal systems. However, maize cells have been identified to contain several PTPs that are inactivated by reductants, such as reductant-inhibited PTPase1 (ZmRIP1), identified in maize coleoptiles, thus exhibiting a unique mechanism of regulating redox status and signaling. ZmRIP1/ZmPP1 acts as a chloroplast-to-nucleus signaling messenger (Li et al. 2012a). It is 59% identical to SEX4 of Arabidopsis and 43% identical to the DsPTP8 of *C. reinhardtii*. The mutation of Cys181 to Arg181 led to activation by a reductant, while its translocation from chloroplast to the nucleus occurs on exposure to an oxidative stimulus. ZmRIP1 is unaffected by H_2O_2 *in vitro*. *ZmRIP1* expression in maize protoplasts and Arabidopsis plants resulted in a change in the expression of gene encoding glutathione transferase enzyme pertaining to antioxidant catabolism (Li et al. 2012a).

SEX4 phosphatase and LSF2 activity is regulated by redox status in plants (Sokolov et al. 2006; Zhao et al. 2016). SEX4 function is regulated through a disulfide bridge in the phosphatase domain that acts as a redox switch (Silver et al. 2013). LSF2 shows dual localization in the chloroplast and cytoplasm and is involved in transient starch metabolism (Santelia et al. 2011). LSF2 is also involved in the maintenance of ROS homeostasis in the regulation of root growth and root hair development in Arabidopsis. The generation of ROS in *lsf2-1* mutant occurs independently of the ROS regulator, respiratory burst oxidase homolog D (RbohD). The oxidant-induced expression of *MPK8* and the interaction of LSF2 with MPK8 in the cytoplasm implicate LSF2 in the regulation of ROS homeostasis under oxidative stress in Arabidopsis (Zhao et al. 2016).

17.4 MAPK Signaling

MAPK cascade involves sequential phosphorylation by upstream MAPKK kinases (MAPKKKs), MAPK kinases (MAPKKs), and MAPKs. MAPK activation occurs through dual phosphorylation of the T-*X*-Y motif located in their activation loop. Activated MAPKs phosphorylate downstream targets. The MAPK signaling pathways are regulated by specific MKPs (Rodriguez et al. 2010). They are potent antagonists of MAP kinases, acting as regulators of MAP kinase signaling pathways (Ulm et al. 2002). The Arabidopsis genome encodes five MKPs: AtMKP1 (Ulm et al. 2002, 2001), AtMKP2 (Lee and Ellis 2007), AtDsPTP1 (Gupta et al. 1998), PROPYZAMIDE HYPERSENSITIVE 1 (PHS1) (Quettier et al. 2006), and INDOLE-3-BUTYRIC ACID RESPONSE 5 (IBR5) (Monroe-Augustus et al. 2003). The involvement of these MKPs in the regulation of plant responses under different stresses is summarized in Fig. 17.3.

The Arabidopsis MKP1 has a long C-terminal extension, containing a domain observed in the actin-binding gelsolin family (323–391 residues). Although it may be classified as a member of the DsPTPs, it contains the conserved residues of the extended catalytic site required for the MKP activity as well as plant-specific characteristics, like a gelsolin motif. Such characteristics were observed in MKPs of *M. truncatula*, tomato and maize (Ulm et al. 2001). AtMKP1 interacts strongly with MPK6, in addition to MPK3 and MPK4 (degree of interaction: MPK6>> MPK3 = MPK4), similar to differential interaction between some mammalian MKPs and MAPKs (Ulm et al. 2002). The AtMKP1 activity is regulated by calmodulin (CaM) and AtMPK6 (Park et al. 2011). MKP1 is reported to exist in two phosphorylation states *in vivo*. Though the constitutively phosphorylated form of MKP1 undergoes constant degradation in the absence of stress, the phospho-MKP1 undergoes rapid stabilization under UV-B stress, indicating its post-translational regulation *in vivo* (González Besteiro and Ulm 2013). AtMKP1 is involved in the

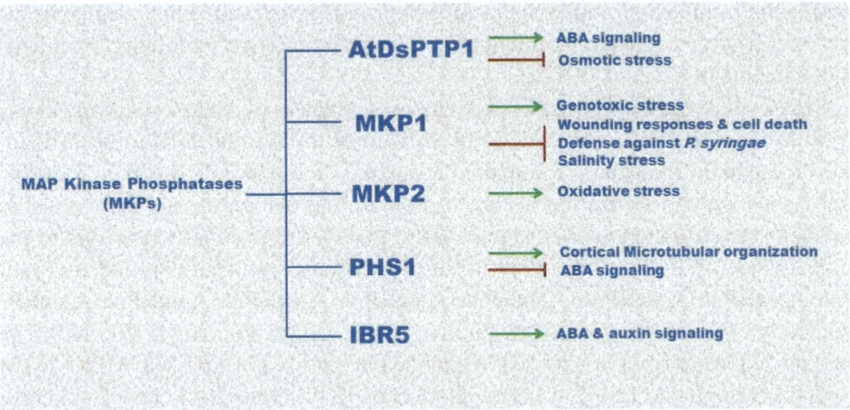

Fig. 17.3 Different signaling responses regulated by the five MAP Kinase Phosphatases (MKPs): AtDsPTP1, AtMKP1, AtMKP2, PROPYZAMIDE HYPERSENSITIVE 1 (PHS1), and INDOLE-3-BUTYRIC ACID RESPONSE 5 (IBR5). The green arrows represent positive regulation while the blocked arrows represent the negative regulation

regulation of stress responses to genotoxic stress as evident from the hypersensitivity to UV-C and methyl methanesulfonate of the Arabidopsis *mkp1* mutant (Ulm et al. 2001). The regulation occurs through inactivation of MPK6 *in planta*. The transcriptional induction of *MPK3* and *MPK4* in response to genotoxic stress in Arabidopsis occurs independent of MKP1 (Ulm et al. 2002). The *mkp1* plants also exhibit enhanced tolerance to salinity stress, thus indicating that MKP1 is instrumental in the integration and refining of different stress responses (Ulm et al. 2002). AtDsPTP1 dephosphorylates MPK4 *in vitro* (Gupta et al. 1998).

AtMKP2 interacts with oxidant-activated MAPKs, AtMPK3 and -6 *in vitro*. *AtMKP2*-silenced plants showed hypersensitivity and delayed activation of MPK3 and MPK6 on exposure to ozone, thus indicating that AtMKP2 may be involved in the positive regulation of cellular responses to oxidative stress (Lee and Ellis 2007).

The Arabidopsis *PHS1* (PROPYZAMIDE-HYPERSENSITIVE 1) gene consists of 11 exons (Quettier et al. 2006) and encodes an MKP (Kerk et al. 2002; Naoi and Hashimoto 2004; Quettier et al. 2006). The PHS1 phosphatase is involved in the regulation of MAPKs, some of which regulate organization of cortical microtubules. The semi-dominant *phs1-1* mutation compromises the functioning of cortical microtubules. The *phs1-1* mutation acts in a dominant negative manner, while the null *phs1-2* allele shows embryonic lethality in a recessive manner (Naoi and Hashimoto 2004). The *phs1-1* mutant shows a point mutation, which results in hypersensitivity to propyzamide (microtubule destabilizer) in seedlings. The *phs1-1* mutant shows disorganization of the cortical microtubule arrays and the formation of left-hand helices by root epidermal cell files. The *phs1-2* allele shows a T-DNA insertion in the third exon. The heterozygosity in *phs1-2* plants does not result in hypersensitivity to propyzamide and microtubular disorganization (Naoi and Hashimoto 2004). The recessive *phs1-3* mutation results from a T-DNA insertion in the promoter, 391-bp upstream of the start codon. The *phs1-3* seedlings did not show hypersensitivity to propyzamide and did not show phenotypes pertaining to microtubule disorganization (Quettier et al. 2006).

IBR5 contains the highly conserved DsPTP active site motif. The Arabidopsis *IBR5* gene shows ubiquitous expression in plants, with its function conserved among angiosperms. IBR5 protein comprises of 257 residues and is ~35% identical to catalytic domain of AtDsPTP1, with dissimilar N- and C-termini. The *ibr5-1* mutation results in a premature stop codon, which leads to formation of a truncated protein lacking the conserved phosphatase domain (Monroe-Augustus et al. 2003). *IBR5* generates two isoforms, IBR5.1 and IBR5.3. The *ibr5-4* mutant contains a substitution mutation in the catalytic site, and *ibr5-5* mutant contains an unspliced fourth intron, resulting in 27 extra amino acid residues in the predicted IBR5.1 polypeptide (Jayaweera et al. 2014). IBR5 interacts with and dephosphorylates MPK12 in Arabidopsis. Arabidopsis MPK12 is activated by auxin *in vivo* and acts as a negative regulator of auxin signaling pathway (Lee et al. 2009).

17.5 Hormone Signaling Pathways

17.5.1 Abscisic Acid (ABA) Signaling

Tyr phosphorylation is involved in abscisic acid (ABA) signaling. PTP inhibitors such as PAO and 3,4 dephosphatin (3,4 DP) inhibit stomatal closure induced by ABA, H_2O_2, dark conditions, and high external calcium (Ca^{2+}) levels, thus implicating PTPs in stomatal movements in *Commelina communis*. PTPs work at or downstream of the Ca^{2+} signal involved in the initiation of vacuolar ion efflux (MacRobbie 2002). PAO inhibits the bending of the petiole in *Mimosa pudica*. The high levels of Tyr phosphorylation of actin and the alterations in its phosphorylation levels correlate with the bending of petioles (Kameyama et al. 2000).

The Tyr phosphorylation levels showed alterations in Arabidopsis seeds on ABA exposure (Ghelis et al. 2008). The Tyr-phosphorylated proteins include those implicated in germination, particularly in the mobilization of storage proteins and lipid reserves. PAO inhibits the ABA-regulated accumulation of *RAB18* (*responsive to ABA 18*) transcripts. PTK inhibitors (tyrphostin A23, erbstatin, and genistein) inhibited the ABA-induced *RAB18* expression in Arabidopsis. PAO and genistein inhibit ABA-induced stomatal closure. Genistein inhibits ABA-regulated Tyr phosphorylation as well (Ghelis et al. 2008). PAO inhibits the ABA-induced MAPK activation as well as *RAB16* expression in barley aleurone protoplasts, thus indicating that ABA activates MAP kinase signaling in a rapid and transient manner through Tyr dephosphorylation (Knetsch 1996). Tyr dephosphorylation is involved in the ABA-mediated post-germination arrest of seed development in Arabidopsis. PAO increases the inhibition of ABA-induced seed germination as well as *RAB18* expression in seeds (Reyes et al. 2006), which may be due to crosstalk between signaling pathways (Ghelis et al. 2008).

AtDsPTP1 is involved in the positive regulation of ABA accumulation and signaling under osmotic stress. The ABA accumulation under osmotic stress is regulated by the downregulation of the expression *AtDsPTP1* gene, which lowers the expression of genes involved in ABA biosynthesis and enhances the expression of genes involved in ABA catabolism. The downregulation of the rate-limiting gene of the *de novo* ABA biosynthesis pathway, *9-cis-epoxycarotenoid dioxygenase 3* (*NCED3*), and the upregulation of the expression of the gene encoding the ABA catabolic enzyme, *CYP707A4*, result in regulation of ABA accumulation. The downregulation of *AtDsPTP1* gene leads to downregulation of *ABI3* and *ABI5* expression and upregulation of the expression of *ABI1*, under osmotic stress. DsPTP1 regulates the expression of a series of dehydration-responsive genes under osmotic stress (Liu et al. 2015).

ZmRIP1 is not involved in the regulation of stress-induced ABA accumulation (Wei and Pan 2014). Exposure to estradiol led to enhanced *ZmRIP1* expression, while the expression of genes involved in ABA biosynthesis remained unaffected in estradiol-inducible *ZmRIP1*-expressing Arabidopsis lines, indicating that ZmRIP1 is not involved in stress-induced ABA accumulation (Li et al. 2012a).

PHS1 also acts as a negative regulator of ABA signaling. The *phs1-3* mutant showed ABA hypersensitivity in gene expression during early phases of development as well as in germination and stomatal responses. The *phs1-3* mutant showed enhanced transcript levels of the ABA-induced genes, *At5g06760* and *RAB18*, and reduced transcript levels of the ABA-repressed genes, *AtCLC-A* and *ACL*, as compared to wild-type plants. The *phs1-3* mutation affects ABA signaling and not ABA metabolism as seeds and seedlings of the *phs1-3* mutant showed ABA levels similar to those in the wild-type plants. The *phs1-3* mutant showed reduction in the stomatal aperture *in vivo* as well (Quettier et al. 2006).

FsPTP1, identified in *Fagus sylvatica* seeds, acts as a negative regulator of ABA signaling pathway. The Arabidopsis lines overexpressing *FsPTP1* gene resulted in ABA insensitivity, decrease in seed dormancy as well as decreased sensitivity to osmotic stress. The expression of ABA marker genes, *RAB18* and *RD29* (*Responsive to Desiccation29*), was downregulated in the OE lines (Alonso-Ramírez et al. 2011). The phenotype of *FsPTP1*-OE plants is similar to that of ethylene constitutive mutants. The higher expression of the *EIN2* gene in the *FsPTP1*-OE lines indicate a crosstalk between PTP1 and the ethylene signaling pathways (Alonso-Ramírez et al. 2011). The MKK9-MPK3/MPK6 modules act downstream of the ethylene receptors and regulate ethylene signaling pathway (Yoo and Sheen 2008).

17.5.2 Auxin Signaling

The IBR5.1 isoform regulates auxin-induced gene expression as well as degradation of Aux/IAA proteins independently (Jayaweera et al. 2014). The Arabidopsis *ibr5-1* mutant shows aberrations in vascular patterning, increased leaf serration, and reduction in number of lateral roots as well as height. It also shows reduced sensitivity to auxin and ABA, thus indicating a connect between auxin and ABA signaling (Monroe-Augustus et al. 2003). Double-mutant analyses using an auxin receptor mutant, *tir1* and *ibr5* mutant, promoted an increase in auxin resistance. The *ibr5* mutant showed a decreased accumulation of auxin-responsive reporter, destabilization of the Aux/IAA repressor reporter protein, AXR3NT-GUS, as well as decrease in levels of Aux/IAA repressor, IAA28. The *ibr5* defects were partly rescued by overexpression of $IBR5^{C129S}$ mutant form. Thus, IBR5 enhances auxin responses like auxin-inducible transcription, without causing destabilization of the Aux/IAA repressor proteins, independent of the TIR1 receptor (Strader et al. 2008). The three *ibr5* alleles show many similar phenotypic characteristics as well as distinct aberrations indicating isoform-specific roles. Some functions of IBR5 are independent of IBR5.1 catalytic function. *IBR5* may connect the ABP1 (Auxin Binding Protein1) and $SCF^{TIR1/AFBs}$ dependent pathways (Jayaweera et al. 2014).

17.5.3 Brassinosteroid Signaling

Tyr phosphorylation may be involved in the regulation of membrane-localized receptor signaling in plants as well as in metazoans. Botrytis-induced kinase 1 (BIK1) is a receptor-like cytoplasmic kinase (RLCK) that is phosphorylated by Brassinosteroid insensitive 1-associated kinase 1 (BAK1) at both Tyr and Ser/Thr residues. The BIK1 Tyr phosphorylation is important for regulation of BIK1-mediated innate immunity in plants. BIK1 acts as a non-receptor dual-specificity kinase (DSK; Lin et al. 2014). BSU1 (bri1 Suppressor 1) phosphatase acts as a positive regulator of BR signaling pathway (Mora-Garcia et al. 2004). BSU1-regulated pTyr200 dephosphorylation regulates GSK3-like kinase, BIN2. It dephosphorylates phospho-Ser/Thr as well as phospho-Tyr residues (Kim et al. 2009).

17.6 Developmental Pathways

Tyr phosphorylation patterns exhibit variation in different adult tissues and somatic embryogenesis (Barizza et al. 1999; Fordham-Skelton et al. 1999) as well as during seed germination (Kovaleva et al. 2013). The loss-of-*AtMKP2* function results in the progression of senescence. The RNAi lines of *AtMKP2* (*AtMKP2i* lines) showed an early onset of senescence in comparison with wild-type and *AtMKP2*-OE lines. However, the latter did not show an increase in lifespan (Li et al. 2012b). MAPK signaling pathway involves MKK9-MPK6 in the regulation of leaf senescence in Arabidopsis (Zhou et al. 2009).

ZmPP133 may be involved in the regulation of carbon fixation pathways at the transcriptional level as its expression levels were high in the maturing and mature regions. However, these regions witnessed a downregulation of *ZmPP48* and *ZmPP120* genes, and their role in leaf development is not yet reported. *ZmPP67* gene may be involved in floral transition as evident from its high expression in both root and shoot apical meristem. The ZmPP67 is a homologue of FLOWERING ASSOCIATED PTPase1 (FPTP1) in Arabidopsis. Mutation in *FPTP1* result in enhanced expression of *FLOWERING LOCUS T* (*FT*) and no visible effects on the expression of circadian-regulated gene, *Constance* (*CO*). The *FPTP1* overexpression resulted in delayed flowering, indicating that ZmPP67 may be acting upstream of FT (Wei and Pan 2014).

LSF2 and SEX4 are conserved across the plant genomes ranging from single-cell green algae to land plants (Gentry and Pace 2009; Santelia et al. 2011). SEX4 is a trimodular protein comprising of a carbohydrate-binding module (CBM), a chloroplast-targeting peptide (cTP), and a DsPTP domain (Meekins et al. 2016). It preferentially dephosphorylates the C6 position, although it can dephosphorylate the C3 position of starch as well (Hejazi et al. 2010; Meekins et al. 2014). LSF2 lacks a CBM and it specifically dephosphorylates the C3 position of starch glucose moieties (Santelia et al. 2011). It also has two non-catalytic or secondary binding sites (SBSs; Meekins et al. 2013). LSF1 is not an active glucan phosphatase (Schreier et al. 2019) and may function as a putative inactive scaffold protein that regulates starch-degrading enzymes (Silver et al. 2014). It contains a CBM and a DsPTP domain. Though it is involved in starch degradation, it lacks phosphatase activity (Comparot-Moss et al. 2010).

The Arabidopsis *AtPTEN1* gene encodes a protein and lipid dual phosphatase. The AtPTEN1 is pollen-specific and closely related to tumor suppressor, PTEN. The AtPTEN1 protein, encoded from cDNA, comprises of 412 residues (calculated molecular mass = 47.3 kD), with a pI of 7.5. The recombinant AtPTEN1, similar to its counterparts in animals, acts as an active phosphatase that targets phospho-Tyr and phosphatidylinositol substrates. The *AtPEN* gene shows several homologues in other plants as well. The *AtPTEN1* gene expression is limited to pollen grains, with RNA interference resulting in pollen cell death post-mitosis (Gupta et al. 2002). It is involved in the regulation of autophagy in pollen tubes. The *AtPTEN*-OE resulted in accumulation of autophagic bodies in tobacco and gametophytic male sterility in Arabidopsis. Plant PTENs differ from their animal counterparts in the presence of regulatory sequences which leads to difference in substrate specificities and membrane localizations (Zhang et al. 2011).

17.7 Abiotic Stress Tolerance

AtPTP1 dephosphorylates AtMAPK4 and AtMAPK6 *in vitro*, that are involved in regulation of osmotic stress signaling (Huang et al. 2000; Gupta and Luan 2003). AtPTP1 may be involved in regulation of stress responses in higher plants (Xu et al. 1998). The expression levels of *AtPTP1* showed upregulation under salinity stress and downregulation on exposure to cold temperatures, indicating variance in plant responses. Stress factors like drought, wounding, and heat shock did not affect *AtPTP1* expression significantly (Xu et al. 1998). *AtDsPTP1* acts as a negative regulator in growth and developmental processes under stress. The hyposensitive Arabidopsis *dsptp1* mutant exhibits enhanced tolerance during seed germination and root elongation of seedlings under osmotic stress. *AtDsPTP1*-OE line showed attenuation of tolerance in comparison with the wild-type plants, while the *AtDsPTP1*-complemented lines showed restoration of tolerance in the *dsptp1* mutant plants similar to the wild-type plants (Liu et al. 2015). OsPFA-DSP1, identified in rice, is a functional PTP, belonging to plant and fungi atypical DsPTP (PFA-DSP) subfamily, which exhibits high homology with Arabidopsis AtPFA-DSP1. Its overexpression in rice and tobacco showed enhanced sensitivity to drought stress, indicating that it acts as a negative regulator under drought stress conditions (Liu et al. 2012).

DSP4 is involved in diurnal starch degradation in leaves of Arabidopsis (Hejazi et al. 2010; Kötting et al. 2009; Niittylä et al. 2006). The *CsDSP4* gene encoding the DSP4 homologue in chestnut (*Castanea sativa* Mill.) is expressed in both leaves and stems. Its expression is induced under low temperatures and winter dormancy indicating its potential involvement in starch degradation as well as cold acclimation, particularly during the transitory stages between active and dormant states (Berrocal-Lobo et al. 2011).

17.8 Biotic Responses

MPK6 and MPK3 are involved in the positive regulation of defense responses through rapid activation on exposure to pathogen-associated molecular patterns (PAMPs). The Arabidopsis MKP1 is involved in the negative regulation of different PAMP responses, like activation of MPK3 and MPK6, transient generation of extracellular ROS, higher levels of PAMP-regulated transcripts, and inhibition of growth in seedlings. *MKP1* is involved in the negative regulation of *MPK6*-regulated PAMP responses. The *mkp1* mutant plants show PAMP response phenotypes and higher resistance to the *Pseudomonas syringae* pv. *tomato* DC3000 (*Pst* DC3000). *MPK6* is vital for the increase in *mkp1*-dependent resistance as well as PAMP-induced growth inhibition in the *mkp1* mutant seedlings (Anderson et al. 2011).

*At*MKP1 and PTP1 show redundancy in the suppression of salicylic acid (SA) and camalexin biosynthesis as well as defense responses. The *mkp1* (Col-0) mutant plants show constitutive defense responses and resistance to *P. syringae*, in addition to growth defects. The *ptp1* mutant plants do not show growth defects. The *mkp1ptp1* double-mutant plants also show constitutive defense responses indicating that MKP1 and PTP1 are involved in repression of defense responses. The suppression of *mkp1* and *mkp1ptp1* phenotypes is due to mutations in MPK3 and MPK6, indicating that MKP1 and PTP1 are involved in the repression of inappropriate MPK3/MPK6-regulated stress signaling pathways. MKP1 and PTP1 may be involved in the repression of SA biosynthesis in the SNC1 (SUPPRESSOR OF npr1-1, CONSTITUTIVE 1)-induced autoimmune-like responses (Bartels et al. 2009).

The NtMKP1 interacts with its target MAPKs and CaM, although CaM did not activate NtMKP1. NtMKP1 inactivates SA-induced protein kinase (SIPK) through dephosphorylation of its TEY motif. NtMKP1 shows four characteristic domains: a DsPTP catalytic domain, a gelsolin homology domain, a CaM-binding domain, and a C-terminal domain. Its N-terminal non-catalytic region binds SIPK and is required for the inactivation of SIPK. The NtMKP1 activity is higher in case of association with SIPK than with wound-induced protein kinase (WIPK; Katou et al. 2005). The *NtMKP1*-OE lines showed decreased JA generation in response to wounding. The *WIPK/SIPK*-silenced and *NtMKP1*-OE plants showed abnormal levels of SA as well as transcripts of SA-responsive genes, thus indicating the importance of NtMKP, WIPK and SIPK interactions in JA generation and SA biosynthesis (Seo et al. 2007).

AtMKP2 is involved in the regulation of plant responses to pathogen attacks, in addition to oxidative stress. The Arabidopsis *mkp2* plants show delayed wilting in response to *Ralstonia solanacearum* and enhanced disease progression in case of *Botrytis cinerea* infection, indicating a differential response to biotrophic and necrotrophic pathogens. MKP2 interacts with MPK3 and MPK6 *in vivo* and shows differential interactions in response to fungal elicitors. MKP2 and MPK6 interact in HR-like responses in case of fungal infection. Thus, MKP2 acts as a key regulator of MPK3 and MPK6 signaling underlying pathogen-specific responses, in addition to abiotic stress in plants (Lumbreras et al. 2010).

OsPFA-DSP2 and its homologue, AtPFA-DSP4, are involved in the negative regulation of the plant responses to pathogen attacks. *OsPFA-DSP2* expression occurs in calli, seedlings, young panicles and roots. It localizes in nucleus and cytoplasm. The *OsPFA-DSP2*-OE lines in rice show enhanced sensitivity to *Magnaporthe grisea* (Z1 strain), inhibition of the accumulation of H_2O_2 and suppression of the gene expression of pathogenesis-related (*PR*) genes post-infection. Arabidopsis *AtPFA-DSP4*-OE plants also showed sensitivity to *Pst* DC3000, in addition to reduction in accumulation of H_2O_2 and photosynthetic capacity post-infection (He et al. 2012).

17.9 Abiotic Stress

The identification of signaling pathways activated under various stress conditions like drought and salinity (Wang et al. 2011; Zhou et al. 2007), potassium (K^+) deficiency (Ma et al. 2012; Shankar et al. 2013; Singh et al. 2010), low Ca^{2+} (Shankar et al. 2014), low nitrogen (Lian et al. 2006), low phosphorus (Wasaki et al. 2003, 2006; Li et al. 2010), and low iron (Zheng et al. 2009) have been reported. Rice DsPTPs show differential expression under different stress conditions, as inferred from Genevestigator database. The expression of *OsPP3* gene, which encodes PHS1, showed upregulation in leaves and roots under drought and cold stress conditions. However, it was not expressed under heat stress. The *OsPP6* and *OsPP126* expression showed downregulation under all the three stresses. *OsPP14* expression showed upregulation under drought stress and downregulation under cold and heat stress conditions. *OsPP39*, *OsPP77*, and *OsPP85* expression showed upregulation under cold and drought stresses (Shankar et al. 2015).

Several DsPTPs are unaffected under K^+- and Ca^{2+}-deficient conditions in rice (Shankar et al. 2014, 2013). The *OsPP3* expression showed upregulation under low phosphorus and ammonium conditions while showing downregulated expression in low K^+, phosphate, and iron conditions (Shankar et al. 2015). *OsPP14* expression was downregulated under K^+-deficient conditions while showing upregulated expression under Fe^{2+}-deficient conditions (Shankar et al. 2015). *OsPP39* expression showed downregulation under low phosphorus conditions, and *OsPP105* expression was upregulated (Shankar et al. 2015). *OsPP42*, *OsPP82*, and *OsPP100* gene expression showed downregulation under low Fe^{2+} and phosphate conditions. *OsPP6* gene expression was downregulated in case of hormonal treatments (naphthalene acetic acid, kinetin, trans-zeatin, and gibberellic acid GA_3). The *OsPP14* and *OsPP39* gene expression was also downregulated in case of treatment with trans-zeatin. However, the *OsPP3* and *OsPP82* expression was upregulated in case of treatment with NAA and GA_3 (Shankar et al. 2015).

The genes encoding PPs in maize (except for *ZmPP13*, *ZmPP71*, *ZmPP125*) including those encoding for PTPs show promoters like A-box, ABRE, CCAAT-box, G-box, MBS, C-repeat/DRE, and W-box, implicating them in regulation of stress responses (Wei and Pan 2014). The microarray data of gene expression under water deficit in the maize inbred lines, Han21 (drought-tolerant) and Ye478 (drought-sensitive), as well as RNA-seq data from cob and leaf were analyzed to identify the role of ZmPTPs under drought stress. *ZmPP67*, encoding a PTP, showed upregulation in Han21 (twofold change, P value <0.05) as well as in both cob and leaf, thus indicating its involvement in plant responses to drought stress (Wei and Pan 2014). The genes encoding PTPs, *ZmPP54* and *ZmPP59*, showed upregulation in expression levels, while the genes encoding *ZmPP125*, *ZmPP133*, *ZmPP101*, and *ZmPP68* were downregulated under drought stress, although their role is still to be reported (Wei and Pan 2014).

17.10 Conclusion and Future Perspective

Protein Tyr phosphorylation plays a significant role in plant signaling pathways regulating metabolism, cytoskeletal reorganization, defense responses, growth, and development. PTPs are known to act as positive and negative regulators in mammalian signaling pathways. Abnormalities in PTP functioning result in the development of several human disorders, as in the case of kinases. Similar aberrations in their plant counterparts hold the potential to disturb homeostasis in innumerable ways. The research on PTPs in plants is still in the nascent stages and needs more efforts, particularly in terms of identification of novel substrates, cofactors, and regulators. On the lines of PTPs being viewed as novel targets for drug development in the human system, it will be interesting to see how they fit into the design of stress tolerance in plants in the coming years.

The post-genomic era has enabled the elucidation of the physiological functions of many genes. The use of mass spectrometric approaches aids quantitative profiling of phosphoproteome complement. The recent progress in the field of protein phosphatases in plants has opened up avenues to identify key regulatory molecules. However, the benefits have been limited in case of PTP superfamily in plants. Large-scale datasets can help understand the cellular regulation precisely and the subsequent alterations under different conditions (Hunter 2014). The presence of splice variants, mutations, effects of other post-translational modifications, protein-protein interactions, and redundancy limit our understanding (Alonso et al. 2004; Hunter 2014). The variable inactivation of PTPs (Ross et al. 2007) as well as their structures and catalytic activities also needs to be examined in plants and their impact in signaling pathways. The structure of oxidized enzymes may determine the degree of susceptibility to further oxidation which leads to irreversible inactivation of enzyme (Tanner et al. 2011). The levels of PTP inactivation and its impact on signaling pathways also need to be determined (Stoker 2005). It remains for us to see if indeed the plant PTPs are limited in numbers or still await identification. The cases of glucan phosphatases and inactive PTPs have brought perspective through difference in substrates and inability to exhibit catalytic activity. More efforts towards the functional characterization of identified PTPs and DsPTPs *in planta* will help in unraveling their involvement in the signaling pathways that have been identified to show the presence of phospho-Tyr-binding domains or peptides as well as tyrosine kinase activity. Their interactions in the context of substrates and regulators, particularly their 3D structures, also need to be evaluated. Their involvement in responses to various stresses and their spatial and temporal implications hold the key to their vitality in the design of stress-tolerant crops.

Acknowledgement We are thankful to the Department of Biotechnology (DBT), Science and Engineering Research Board (SERB), Council for Scientific and Industrial Research (CSIR), Delhi University, India, for research funding in GKP's lab.

References

Alonso A, Sasin J, Bottini N, Friedberg I, Friedberg I, Osterman A, Godzik A, Hunter T, Dixon J, Mustelin T (2004) Protein tyrosine phosphatases in the human genome. Cell 117:699–711. https://doi.org/10.1016/j.cell.2004.05.018

Alonso-Ramírez A, Rodríguez D, Reyes D, Jiménez JA, Nicolás G, Nicolás C (2011) Functional analysis in Arabidopsis of FsPTP1, a tyrosine phosphatase from beechnuts, reveals its role as a negative regulator of ABA signaling and seed dormancy and suggests its involvement in ethylene signaling modulation. Planta 234:589–597. https://doi.org/10.1007/s00425-011-1426-8

Anderson JC, Bartels S, Besteiro MAG, Shahollari B, Ulm R, Peck SC (2011) Arabidopsis MAP Kinase Phosphatase 1 (AtMKP1) negatively regulates MPK6-mediated PAMP responses and resistance against bacteria. Plant J 67:258–268. https://doi.org/10.1111/j.1365-313X.2011.04588.x

Apel K, Hirt H (2004) REACTIVE OXYGEN SPECIES: metabolism, oxidative stress, and signal transduction. Annu Rev Plant Biol 55:373–399. https://doi.org/10.1146/annurev.arplant.55.031903.141701

Barizza E, Lo Schiavo F, Terzi M, Filippini F (1999) Evidence suggesting protein tyrosine phosphorylation in plants depends on the developmental conditions. FEBS Lett 447:191–194. https://doi.org/10.1016/S0014-5793(99)00272-0

Bartels S, Anderson JC, Gonzalez Besteiro MA, Carreri A, Hirt H, Buchala A, Metraux J-P, Peck SC, Ulm R (2009) MAP KINASE PHOSPHATASE1 and PROTEIN TYROSINE PHOSPHATASE1 are repressors of salicylic acid synthesis and SNC1-mediated responses in Arabidopsis. Plant Cell 21:2884–2897. https://doi.org/10.1105/tpc.109.067678

Berrocal-Lobo M, Ibañez C, Acebo P, Ramos A, Perez-Solis E, Collada C, Casado R, Aragoncillo C, Allona I (2011) Identification of a homolog of Arabidopsis DSP4 (SEX4) in chestnut: its induction and accumulation in stem amyloplasts during winter or in response to the cold. Plant Cell Environ 34:1693–1704. https://doi.org/10.1111/j.1365-3040.2011.02365.x

Brandão TAS, Hengge AC, Johnson SJ (2010) Insights into the reaction of protein-tyrosine phosphatase 1B: crystal structures for transition state analogs of both catalytic steps. J Biol Chem 285:15874–15883. https://doi.org/10.1074/jbc.M109.066951

Carpi A, Di Maira G, Vedovato M, Rossi V, Naccari T, Floriduz M, Terzi M, Filippini F (2002) Comparative proteome bioinformatics: identification of a whole complement of putative protein tyrosine kinases in the model flowering plant Arabidopsis *thaliana*. Proteomics 2:1494–1503. https://doi.org/10.1002/1615-9861(200211)2:11<1494::AID-PROT1494>3.0.CO;2-#

Cohen PTW (2010) Phosphatase families dephosphorylating serine and threonine residues in proteins. In: Bradshaw RA, Dennis EA (eds) Handbook of cell signaling, 2nd edn. Academic Press, London, pp 659–675. https://doi.org/10.1016/B978-0-12-374145-5.00085-1

Comparot-Moss S, Kötting O, Stettler M, Edner C, Graf A, Weise SE, Streb S, Lue W-L, MacLean D, Mahlow S, Ritte G, Steup M, Chen J, Zeeman SC, Smith AM (2010) A putative phosphatase, LSF1, is required for normal starch turnover in Arabidopsis leaves. Plant Physiol 152:685–697. https://doi.org/10.1104/pp.109.148981

de la Fuente van Bentem S, Hirt H (2009) Protein tyrosine phosphorylation in plants: more abundant than expected? Trends Plant Sci 14:71–76. https://doi.org/10.1016/j.tplants.2008.11.003

Denu JM, Dixon JE (1995) A catalytic mechanism for the dual-specific phosphatases. PNAS 92:5910–5914. https://doi.org/10.1073/pnas.92.13.5910

Dunn D, Chen L, Lawrence DS, Zhang ZY (1996) The active site specificity of the Yersinia protein-tyrosine phosphatase. J Biol Chem 271:168–173. https://doi.org/10.1074/jbc.271.1.168

Farkas I, Dombrádi V, Miskei M, Szabados L, Koncz C (2007) Arabidopsis PPP family of serine/threonine phosphatases. Trends Plant Sci 12:169–176. https://doi.org/10.1016/j.tplants.2007.03.003

Fauman EB, Saper MA (1996) Structure and function of the protein tyrosine phosphatases. Trends Biochem Sci 21:413–417. https://doi.org/10.1016/S0968-0004(96)10059-1

Fordham-Skelton AP, Skipsey M, Eveans IM, Edwards R, Gatehouse JA (1999) Higher plant tyrosine-specific protein phosphatases (PTPs) contain novel amino-terminal domains: expression during embryogenesis. Plant Mol Biol 39:593–605. https://doi.org/10.1023/A:1006170902271

Forman H, Torres M, Fukuto J (2002) Redox signaling. Mol Cell Biochem 234–235:49–62

Gentry MS, Pace RM (2009) Conservation of the glucan phosphatase laforin is linked to rates of molecular evolution and the glucan metabolism of the organism. BMC Evol Biol 9:138. https://doi.org/10.1186/1471-2148-9-138

Getz LJ, Runte CS, Rainey JK, Thomasa NA (2019) Tyrosine phosphorylation as a widespread regulatory mechanism in prokaryotes. J Bacteriol 201:e00205–e00219. https://doi.org/10.1128/JB.00205-19

Ghelis T (2011) Signal processing by protein tyrosine phosphorylation in plants. Plant Signal Behav 6:942–951. https://doi.org/10.4161/psb.6.7.15261

Ghelis T, Bolbach G, Clodic G, Habricot Y, Miginiac E, Sotta B, Jeannette E (2008) Protein tyrosine kinases and protein tyrosine phosphatases are involved in abscisic acid-dependent processes in Arabidopsis seeds and suspension cells. Plant Physiol 148:1668–1680. https://doi.org/10.1104/pp.108.124594

González Besteiro MA, Ulm R (2013) Phosphorylation and stabilization of Arabidopsis MAP kinase phosphatase 1 in response to UV-B stress. J Biol Chem 288:480–486. https://doi.org/10.1074/jbc.M112.434654

Guan K, Dixon J (1991) Evidence for protein-tyrosine-phosphatase catalysis proceeding via a cysteine-phosphate intermediate. J Biol Chem 266:17026–17030

Gupta R, Luan S (2003) Redox control of protein tyrosine phosphatases and mitogen-activated protein kinases in higher plants. Plant Physiol 132:1149–1152. https://doi.org/10.1104/pp.103.020792

Gupta R, Huang Y, Kieber J, Luan S (1998) Identification of a dual-specificity protein phosphatase that inactivates a MAP kinase from Arabidopsis. Plant J 16:581–589. https://doi.org/10.1046/j.1365-313X.1998.00327.x

Gupta R, Ting JTL, Sokolov LN, Johnson SA, Luan S (2002) A tumor suppressor homolog, AtPTEN1, is essential for pollen development in Arabidopsis. Plant Cell 14:2495–2507. https://doi.org/10.1105/tpc.005702

He H, Su J, Shu S, Zhang Y, Ao Y, Liu B, Feng D, Wang J, Wang H (2012) Two homologous putative protein tyrosine phosphatases, OsPFA-DSP2 and AtPFA-DSP4, negatively regulate the pathogen response in transgenic plants. PLoS One 7:e34995. https://doi.org/10.1371/journal.pone.0034995

Hejazi M, Fettke J, Kötting O, Zeeman SC, Steup M (2010) The Laforin-like dual-specificity phosphatase SEX4 from Arabidopsis hydrolyzes both C6- and C3-phosphate esters introduced by starch-related dikinases and thereby affects phase transition of alpha-glucans. Plant Physiol 152:711–722. https://doi.org/10.1104/pp.109.149914

Hengge AC, Sowa GA, Wu L, Zhang Z-Y (1995) Nature of the transition state of the protein-tyrosine phosphatase-catalyzed reaction. Biochemistry 34:13982–13987. https://doi.org/10.1021/bi00043a003

Huang Y, Li H, Gupta R, Morris PC, Luan S, Kieber JJ (2000) ATMPK4, an Arabidopsis homolog of Mitogen-Activated Protein Kinase, is activated *in vitro* by AtMEK1 through threonine phosphorylation. Plant Physiol 122:1301–1310. https://doi.org/10.1104/pp.122.4.1301

Hunter T (2014) The genesis of tyrosine phosphorylation. Cold Spring Harb Perspect Biol 6:a020644–a020644. https://doi.org/10.1101/cshperspect.a020644

Islas-Flores I, Oropeza C, Hernández-Sotomayor SMT (1998) Protein phosphorylation during coconut zygotic embryo development. Plant Physiol 118:257–263. https://doi.org/10.1104/pp.118.1.257

Jayaweera T, Siriwardana C, Dharmasiri S, Quint M, Gray WM, Dharmasiri N (2014) Alternative splicing of Arabidopsis IBR5 pre-mRNA generates two ibr5 isoforms with distinct and overlapping functions. PLoS One 9:e102301. https://doi.org/10.1371/journal.pone.0102301

Jia W, Zhang J (2000) Water stress-induced abscisic acid accumulation in relation to reducing agents and sulfhydryl modifiers in maize plant. Plant Cell Environ 23:1389–1395. https://doi.org/10.1046/j.1365-3040.2000.00646.x

Jia Z, Barford D, Flint AJ, Tonks NK (1995) Structural basis for phosphotyrosine peptide recognition by protein tyrosine phosphatase 1B. Science 268:1754–1758. https://doi.org/10.1126/science.7540771

Kameyama K, Kishi Y, Yoshimura M, Kanzawa N, Sameshima M, Tsuchiya T (2000) Tyrosine phosphorylation in plant bending. Nature 407:37. https://doi.org/10.1038/35024149

Katou S, Karita E, Yamakawa H, Seo S, Mitsuhara I, Kuchitsu K, Ohashi Y (2005) Catalytic activation of the plant MAPK phosphatase NtMKP1 by its physiological substrate salicylic acid-induced protein kinase but not by calmodulins. J Biol Chem 280:39569–39581. https://doi.org/10.1074/jbc.M508115200

Kerk D, Bulgrien J, Smith DW, Barsam B, Veretnik S, Gribskov M (2002) The complement of protein phosphatase catalytic subunits encoded in the genome of Arabidopsis. Plant Physiol 129:908–925. https://doi.org/10.1104/pp.004002

Kerk D, Templeton G, Moorhead GBG (2008) Evolutionary radiation pattern of novel protein phosphatases revealed by analysis of protein data from the completely sequenced genomes of humans, green algae, and higher plants. Plant Physiol 146:351–367. https://doi.org/10.1104/pp.107.111393

Kim TW, Guan S, Sun Y, Deng Z, Tang W, Shang JX, Sun Y, Burlingame AL, Wang ZY (2009) Brassinosteroid signal transduction from cell-surface receptor kinases to nuclear transcription factors. Nat Cell Biol 11:1254–1260. https://doi.org/10.1038/ncb1970

Knetsch M (1996) Abscisic acid induces mitogen-activated protein kinase activation in barley Aleurone protoplasts. Plant Cell 8:1061–1067. https://doi.org/10.1105/tpc.8.6.1061

Kötting O, Santelia D, Edner C, Eicke S, Marthaler T, Gentry MS, Comparot-Moss S, Chen J, Smith AM, Steup M, Ritte G, Zeeman SC (2009) STARCH-EXCESS4 is a laforin-like Phosphoglucan phosphatase required for starch degradation in Arabidopsis *thaliana*. Plant Cell 21:334–346. https://doi.org/10.1105/tpc.108.064360

Kovaleva V, Cramer R, Krynytskyy H, Gout I, Gout R (2013) Analysis of tyrosine phosphorylation and phosphotyrosine-binding proteins in germinating seeds from Scots pine. Plant Physiol Biochem 67:33–40. https://doi.org/10.1016/j.plaphy.2013.02.008

Lee JS, Ellis BE (2007) Arabidopsis MAPK phosphatase 2 (MKP2) positively regulates oxidative stress tolerance and inactivates the MPK3 and MPK6 MAPKs. J Biol Chem 282:25020–25029. https://doi.org/10.1074/jbc.M701888200

Lee JS, Wang S, Sritubtim S, Chen JG, Ellis BE (2009) Arabidopsis mitogen-activated protein kinase MPK12 interacts with the MAPK phosphatase IBR5 and regulates auxin signaling. Plant J 57:975–985. https://doi.org/10.1111/j.1365-313X.2008.03741.x

Li B, Zhao Y, Liang L, Ren H, Xing Y, Chen L, Sun M, Wang Y, Han Y, Jia H, Huang C, Wu Z, Jia W (2012a) Purification and characterization of ZmRIP1, a novel reductant-inhibited protein tyrosine phosphatase from maize. Plant Physiol 159:671–681. https://doi.org/10.1104/pp.111.191510

Li L, Liu C, Lian X (2010) Gene expression profiles in rice roots under low phosphorus stress. Plant Mol Biol 72:423–432. https://doi.org/10.1007/s11103-009-9580-0

Li Z, Peng J, Wen X, Guo H (2012b) Gene network analysis and functional studies of senescence-associated genes reveal novel regulators of Arabidopsis leaf senescence. J Integr Plant Biol 54:526–539. https://doi.org/10.1111/j.1744-7909.2012.01136.x

Lian X, Wang S, Zhang J, Feng Q, Zhang L, Fan D, Li X, Yuan D, Han B, Zhang Q (2006) Expression profiles of 10,422 genes at early stage of low nitrogen stress in rice assayed using a cDNA microarray. Plant Mol Biol 60:617–631. https://doi.org/10.1007/s11103-005-5441-7

Liang F, Kumar S, Zhang Z-Y (2007) Proteomic approaches to studying protein tyrosine phosphatases. Mol BioSyst 3:308–316. https://doi.org/10.1039/b700704n

Lim WA, Pawson T (2010) Phosphotyrosine signaling: evolving a new cellular communication system. Cell 142:661–667. https://doi.org/10.1016/j.cell.2010.08.023

Lin W, Li B, Lu D, Chen S, Zhu N, He P, Shan L (2014) Tyrosine phosphorylation of protein kinase complex BAK1/BIK1 mediates Arabidopsis innate immunity. Proc Natl Acad Sci 111:3632–3637. https://doi.org/10.1073/pnas.1318817111

Lindberg RA, Quinn AM, Hunter T (1992) Dual-specificity protein kinases: will any hydroxyl do? Trends Biochem Sci 17:114–119. https://doi.org/10.1016/0968-0004(92)90248-8

Liu B, Fan J, Zhang Y, Mu P, Wang P, Su J, Lai H, Li S, Feng D, Wang J, Wang H (2012) OsPFA-DSP1, a rice protein tyrosine phosphatase, negatively regulates drought stress responses in transgenic tobacco and rice plants. Plant Cell Rep 31:1021–1032. https://doi.org/10.1007/s00299-011-1220-x

Liu R, Liu Y, Ye N, Zhu G, Chen M, Jia L, Xia Y, Shi L, Jia W, Zhang J (2015) AtDsPTP1 acts as a negative regulator in osmotic stress signaling during Arabidopsis seed germination and seedling establishment. J Exp Bot 66:1339–1353. https://doi.org/10.1093/jxb/eru484

Luan S (2000) Protein phosphatases: structure, regulation, and function. Adv Bot Res 32:67–107. https://doi.org/10.1016/S0065-2296(00)32022-5

Luan S (2002) Tyrosine phosphorylation in plant cell signaling. Proc Natl Acad Sci U S A 99:11567–11569. https://doi.org/10.1073/pnas.182417599

Luan S (2003) Protein phosphatases in plants. Annu Rev Plant Biol 54:63–92. https://doi.org/10.1146/annurev.arplant.54.031902.134743

Luan S, Ting J, Gupta R (2001) Protein tyrosine phosphatases in higher plants. New Phytol 151:155–164. https://doi.org/10.1046/j.1469-8137.2001.00168.x

Lumbreras V, Vilela B, Irar S, Solé M, Capellades M, Valls M, Coca M, Pagès M (2010) MAPK phosphatase MKP2 mediates disease responses in Arabidopsis and functionally interacts with MPK3 and MPK6. Plant J 63:1017–1030. https://doi.org/10.1111/j.1365-313X.2010.04297.x

Ma TL, Wu WH, Wang Y (2012) Transcriptome analysis of rice root responses to potassium deficiency. BMC Plant Biol 12:161. https://doi.org/10.1186/1471-2229-12-161

MacRobbie EAC (2002) Evidence for a role for protein tyrosine phosphatase in the control of ion release from the guard cell vacuole in stomatal closure. Proc Natl Acad Sci U S A 99:11963–11968. https://doi.org/10.1073/pnas.172360399

Manning G, Plowman GD, Hunter T, Sudarsanam S (2002) Evolution of protein kinase signaling from yeast to man. Trends Biochem Sci 27:514–520. https://doi.org/10.1016/S0968-0004(02)02179-5

Martin DMA, Miranda-Saavedra D, Barton GJ (2009) Kinomer v. 1.0: a database of systematically classified eukaryotic protein kinases. Nucleic Acids Res 37:D244–D250. https://doi.org/10.1093/nar/gkn834

Meekins DA, Guo H-F, Husodo S, Paasch BC, Bridges TM, Santelia D, Kotting O, Vander Kooi CW, Gentry MS (2013) Structure of the Arabidopsis glucan phosphatase LIKE SEX FOUR2 reveals a unique mechanism for starch dephosphorylation. Plant Cell 25:2302–2314. https://doi.org/10.1105/tpc.113.112706

Meekins DA, Raththagala M, Auger KD, Turner BD, Santelia D, Kötting O, Gentry MS, Vander Kooi CW (2015) Mechanistic insights into glucan phosphatase activity against polyglucan substrates. J Biol Chem 290:23361–23370. https://doi.org/10.1074/jbc.M115.658203

Meekins DA, Raththagala M, Husodo S, White CJ, Guo H-F, Kötting O, Vander Kooi CW, Gentry MS (2014) Phosphoglucan-bound structure of starch phosphatase Starch Excess4 reveals the mechanism for C6 specificity. Proc Natl Acad Sci U S A 111:7272–7277. https://doi.org/10.1073/pnas.1400757111

Meekins DA, Vander Kooi CW, Gentry MS (2016) Structural mechanisms of plant glucan phosphatases in starch metabolism. FEBS J 283:2427–2447. https://doi.org/10.1111/febs.13703

Meng TC, Fukada T, Tonks NK (2002) Reversible oxidation and inactivation of protein tyrosine phosphatases *in vivo*. Mol Cell 9:387–399. https://doi.org/10.1016/S1097-2765(02)00445-8

Miller WT (2012) Tyrosine kinase signaling and the emergence of multicellularity. Biochim Biophys Acta, Mol Cell Res 1823:1053–1057. https://doi.org/10.1016/j.bbamcr.2012.03.009

Miranda-Saavedra D, Barton GJ (2007) Classification and functional annotation of eukaryotic protein kinases. Proteins 68:893–914. https://doi.org/10.1002/prot.21444

Monroe-Augustus M, Zolman BK, Bartel B (2003) IBR5, a dual-specificity phosphatase-like protein modulating auxin and abscisic acid responsiveness in Arabidopsis. Plant Cell 15:2979–2991. https://doi.org/10.1105/tpc.017046

Moorhead GBG, De Wever V, Templeton G, Kerk D (2009) Evolution of protein phosphatases in plants and animals. Biochem J 417:401–409. https://doi.org/10.1042/BJ20081986

Mora-Garcia S, Vert G, Yin Y, Caño-Delgado A, Cheong H, Chory J (2004) Nuclear protein phosphatases with Kelch-repeat domains modulate the response to brassinosteroids in Arabidopsis. Genes Dev 18:448–460. https://doi.org/10.1101/gad.1174204

Mustelin T (2007) A brief introduction to the protein phosphatase families. In: Moorhead G (ed) Methods in molecular biology. Springer, Totowa, NJ, pp 9–22. https://doi.org/10.1385/1-59745-267-X:9

Naoi K, Hashimoto T (2004) A semidominant mutation in an Arabidopsis mitogen-activated protein kinase phosphatase-like gene compromises cortical microtubule organization. Plant Cell 16:1841–1853. https://doi.org/10.1105/tpc.021865

Ndimba BK, Chivasa S, Hamilton JM, Simon WJ, Slabas AR (2003) Proteomic analysis of changes in the extracellular matrix of Arabidopsis cell suspension cultures induced by fungal elicitors. Proteomics 3(6):1047–1059. https://doi.org/10.1002/pmic.200300413

Neel BG, Tonks NK (1997) Protein tyrosine phosphatases in signal transduction. Curr Opin Cell Biol 9:193–204. https://doi.org/10.1016/S0955-0674(97)80063-4

Nemoto K, Takemori N, Seki M, Shinozaki K, Sawasaki T (2015) Members of the plant CRK superfamily are capable of trans- and autophosphorylation of tyrosine residues. J Biol Chem 290:16665–16677. https://doi.org/10.1074/jbc.M114.617274

Niittylä T, Comparot-Moss S, Lue WL, Messerli G, Trevisan M, Seymour MDJ, Gatehouse JA, Villadsen D, Smith SM, Chen J, Zeeman SC, Smith AM (2006) Similar protein phosphatases control starch metabolism in plants and glycogen metabolism in mammals. J Biol Chem 281:11815–11818. https://doi.org/10.1074/jbc.M600519200

Oelze M, Kandlbinder A, Dietz K (2008) Redox regulation and overreduction control in the photosynthesizing cell: complexity in redox regulatory networks. Biochim Biophys Acta 1780:1261–1272. https://doi.org/10.1016/j.bbagen.2008.03.015

Östman A, Frijhoff J, Sandin Å, Böhmer F-D (2011) Regulation of protein tyrosine phosphatases by reversible oxidation. J Biochem 150:345–356. https://doi.org/10.1093/jb/mvr104

Park HC, Song EH, Nguyen XC, Lee K, Kim KE, Kim HS, Lee SM, Kim SH, Bae DW, Yun DJ, Chung WS (2011) Arabidopsis MAP kinase phosphatase 1 is phosphorylated and activated by its substrate AtMPK6. Plant Cell Rep 30:1523–1531. https://doi.org/10.1007/s00299-011-1064-4

Peters GH, Frimurer TM, Olsen OH (1998) Electrostatic evaluation of the signature motif (H/V) CX $_5$ R(S/T) in protein–tyrosine phosphatases. Biochemistry 37:5383–5393. https://doi.org/10.1021/bi971187i

Pincus D, Letunic I, Bork P, Lim WA (2008) Evolution of the phospho-tyrosine signaling machinery in premetazoan lineages. Proc Natl Acad Sci U S A 105:9680–9684. https://doi.org/10.1073/pnas.0803161105

Quettier AL, Bertrand C, Habricot Y, Miginiac E, Agnes C, Jeannette E, Maldiney R (2006) The *phs1-3* mutation in a putative dual-specificity protein tyrosine phosphatase gene provokes hypersensitive responses to abscisic acid in Arabidopsis *thaliana*. Plant J 47:711–719. https://doi.org/10.1111/j.1365-313X.2006.02823.x

Reyes D, Rodríguez D, Nicolás G, Nicolás C (2006) Evidence of a role for tyrosine dephosphorylation in the control of postgermination arrest of development by abscisic acid in Arabidopsis *thaliana* L. Planta 223:381–385. https://doi.org/10.1007/s00425-005-0135-6

Rodriguez M, Petersen M, Mundy J (2010) Mitogen-activated protein kinase signaling in plants. Annu Rev Plant Biol 61:621–649. https://doi.org/10.1146/annurev-arplant-042809-112252

Ross SH, Lindsay Y, Safrany ST, Lorenzo O, Villa F, Toth R, Clague MJ, Downes CP, Leslie NR (2007) Differential redox regulation within the PTP superfamily. Cell Signal 19:1521–1530. https://doi.org/10.1016/j.cellsig.2007.01.026

Ruddraraju KV, Zhang ZY (2017) Covalent inhibition of protein tyrosine phosphatases. Mol BioSyst 13:1257–1279. https://doi.org/10.1039/c7mb00151g

Rudrabhatla P, Rajasekharan R (2002) Developmentally regulated dual-specificity kinase from peanut that is induced by abiotic stresses. Plant Physiol 130:380–390. https://doi.org/10.1104/pp.005173

Rudrabhatla P, Reddy MM, Rajasekharan R (2006) Genome-wide analysis and experimentation of plant serine/threonine/tyrosine-specific protein kinases. Plant Mol Biol 60:293–319. https://doi.org/10.1007/s11103-005-4109-7

Santelia D, Kötting O, Seung D, Schubert M, Thalmann M, Bischof S, Meekins DA, Lutz A, Patron N, Gentry MS, Allain FH-T, Zeeman SC (2011) The Phosphoglucan phosphatase like sex Four2 dephosphorylates starch at the C3-position in Arabidopsis. Plant Cell 23:4096–4111. https://doi.org/10.1105/tpc.111.092155

Schreier TB, Umhang M, Lee S-K, Lue W-L, Shen Z, Silver D, Graf A, Müller A, Eicke S, Stadler-Waibel M, Seung D, Bischof S, Briggs SP, Kötting O, Moorhead GBG, Chen J, Zeeman SC (2019) LIKE SEX4 1 acts as a β-amylase-binding scaffold on starch granules during starch degradation. Plant Cell 31:2169–2186. https://doi.org/10.1105/tpc.19.00089

Seo S, Katou S, Seto H, Gomi K, Ohashi Y (2007) The mitogen-activated protein kinases WIPK and SIPK regulate the levels of jasmonic and salicylic acids in wounded tobacco plants. Plant J 49:899–909. https://doi.org/10.1111/j.1365-313X.2006.03003.x

Shankar A, Agrawal N, Sharma M, Pandey A, Pandey GK (2015) Role of protein tyrosine phosphatases in plants. Curr Genomics 16:224–236. https://doi.org/10.2174/1389202916666150424234300

Shankar A, Singh A, Kanwar P, Srivastava AK, Pandey A, Suprasanna P, Kapoor S, Pandey GK (2013) Gene expression analysis of rice seedling under potassium deprivation reveals major changes in metabolism and signaling components. PLoS One 8:e70321. https://doi.org/10.1371/journal.pone.0070321

Shankar A, Srivastava AK, Yadav AK, Sharma M, Pandey A, Raut VV, Das MK, Suprasanna P, Pandey GK (2014) Whole genome transcriptome analysis of rice seedling reveals alterations in Ca^{2+} ion signaling and homeostasis in response to Ca^{2+} deficiency. Cell Calcium 55:155–165. https://doi.org/10.1016/j.ceca.2014.02.011

Shiozaki K, Russell P (1995) Cell-cycle control linked to extracellular environment by MAP kinase pathway in fission yeast. Nature 378:739–743. https://doi.org/10.1038/378739a0

Silver DM, Kötting O, Moorhead GBG (2014) Phosphoglucan phosphatase function sheds light on starch degradation. Trends Plant Sci 19:471–478. https://doi.org/10.1016/j.tplants.2014.01.008

Silver DM, Silva LP, Issakidis-Bourguet E, Glaring MA, Schriemer DC, Moorhead GBG (2013) Insight into the redox regulation of the phosphoglucan phosphatase SEX4 involved in starch degradation. FEBS J 280:538–548. https://doi.org/10.1111/j.1742-4658.2012.08546.x

Singh A, Giri J, Kapoor S, Tyagi AK, Pandey GK (2010) Protein phosphatase complement in rice: genome-wide identification and transcriptional analysis under abiotic stress conditions and reproductive development. BMC Genomics 11:435. https://doi.org/10.1186/1471-2164-11-435

Skorey K, Ly HD, Kelly J, Hammond M, Ramachandran C, Huang Z, Gresser MJ, Wang Q (1997) How does alendronate inhibit protein-tyrosine phosphatases? J Biol Chem 272:22472–22480. https://doi.org/10.1074/jbc.272.36.22472

Smith RD, Walker JC (1996) Plant protein phosphatases. Plant Physiol Plant Mol Biol 47:101–125

Sokolov LN, Dominguez-Solis JR, Allary A-L, Buchanan BB, Luan S (2006) A redox-regulated chloroplast protein phosphatase binds to starch diurnally and functions in its accumulation. Proc Natl Acad Sci 103:9732–9737. https://doi.org/10.1073/pnas.0603329103

Stewart AE, Dowd S, Keyse SM, McDonald NQ (1999) Crystal structure of the MAPK phosphatase Pyst1 catalytic domain and implications for regulated activation. Nat Struct Biol 6:174–181. https://doi.org/10.1038/5861

Stoker AW (2005) Protein tyrosine phosphatases and signaling. J Endocrinol 185:19–33. https://doi.org/10.1677/joe.1.06069

Stone RL, Dixon JE (1994) Protein-tyrosine phosphatases. J Biol Chem 269:31323–31326. https://doi.org/10.1210/en.2011-1805

Strader LC, Monroe-Augustus M, Bartel B (2008) The IBR5 phosphatase promotes Arabidopsis auxin responses through a novel mechanism distinct from TIR1-mediated repressor degradation. BMC Plant Biol 8:41. https://doi.org/10.1186/1471-2229-8-41

Suzuki K, Shinshi H (1995) Transient activation and tyrosine phosphorylation of a protein kinase in tobacco cells treated with a fungal elicitor. Plant Cell 7:639–647. https://doi.org/10.1105/tpc.7.5.639

Tagliabracci VS, Turnbull J, Wang W, Girard JM, Zhao X, Skurat AV, Delgado-Escueta AV, Minassian BA, DePaoli-Roach AA, Roach PJ (2007) Laforin is a glycogen phosphatase, deficiency of which leads to elevated phosphorylation of glycogen *in vivo*. Proc Natl Acad Sci U S A 104:19262–19266. https://doi.org/10.1073/pnas.0707952104

Tanner JJ, Parsons ZD, Cummings AH, Zhou H, Gates KS (2011) Redox regulation of protein tyrosine phosphatases: structural and chemical aspects. Antioxid Redox Signal 15:77–97. https://doi.org/10.1089/ars.2010.3611

Tonks NK (2005) Redox redux: revisiting PTPs and the control of cell signaling. Cell 121:667–670. https://doi.org/10.1016/j.cell.2005.05.016

Tonks NK (2013) Protein tyrosine phosphatases—from housekeeping enzymes to master regulators of signal transduction. FEBS J 280:346–378. https://doi.org/10.1111/febs.12077

Tonks NK, Neel BG (1996) From form to function: Signaling by protein tyrosine phosphatases. Cell 87:365–368. https://doi.org/10.1016/s0092-8674(00)81357-4

Torruella M, Casano L, Vallejos R (1986) Evidence of the activity of tyrosine kinase(s) and of the presence of phosphotyrosine proteins in pea plantlets. J Biol Chem 261:6651–6653

Ulm R, Ichimura K, Mizoguchi T, Peck SC, Zhu T, Wang X, Shinozaki K, Paszkowski J (2002) Distinct regulation of salinity and genotoxic stress responses by Arabidopsis MAP kinase phosphatase 1. EMBO J 21:6483–6493. https://doi.org/10.1093/emboj/cdf646

Ulm R, Revenkova E, di Sansebastiano GP, Bechtold N, Paszkowski J (2001) Mitogen-activated protein kinase phosphatase is required for genotoxic stress relief in Arabidopsis. Genes Dev 15:699–709. https://doi.org/10.1101/gad.192601

Wang D, Pan Y, Zhao X, Zhu L, Fu B, Li Z (2011) Genome-wide temporal-spatial gene expression profiling of drought responsiveness in rice. BMC Genomics 12:149. https://doi.org/10.1186/1471-2164-12-149

Wasaki J, Shinano T, Onishi K, Yonetani R, Yazaki J, Fujii F, Shimbo K, Ishikawa M, Shimatani Z, Nagata Y, Hashimoto A, Ohta T, Sato Y, Miyamoto C, Honda S, Kojima K, Sasaki T, Kishimoto N, Kikuchi S, Osaki M (2006) Transcriptomic analysis indicates putative metabolic changes caused by manipulation of phosphorus availability in rice leaves. J Exp Bot 57:2049–2059. https://doi.org/10.1093/jxb/erj158

Wasaki J, Yonetani R, Kuroda S, Shinano T, Yazaki J, Fujii F, Shimbo K, Yamamoto K, Sakata K, Sasaki T, Kishimoto N, Kikuchi S, Yamagishi M, Osaki M (2003) Transcriptomic analysis of metabolic changes by phosphorus stress in rice plant roots. Plant Cell Environ 26:1515–1523. https://doi.org/10.1046/j.1365-3040.2003.01074.x

Wei K, Pan S (2014) Maize protein phosphatase gene family: identification and molecular characterization. BMC Genomics 15:773. https://doi.org/10.1186/1471-2164-15-773

Williams JG, Zvelebil M (2004) SH2 domains in plants imply new signaling scenarios. Trends Plant Sci 9:161–163. https://doi.org/10.1016/j.tplants.2004.02.001

Winterbourn CC, Hampton MB (2008) Thiol chemistry and specificity in redox signaling. Free Radic Biol Med 45:549–561. https://doi.org/10.1016/j.freeradbiomed.2008.05.004

Worby CA, Gentry MS, Dixon JE (2006) Laforin, a dual specificity phosphatase that dephosphorylates complex carbohydrates. J Biol Chem 281:30412–30418. https://doi.org/10.1074/jbc.M606117200

Xu Q, Fu HH, Gupta R, Luan S (1998) Molecular characterization of a tyrosine-specific protein phosphatase encoded by a stress-responsive gene in Arabidopsis. Plant Cell 10:849–857. https://doi.org/10.1105/tpc.10.5.849

Yoo S-D, Sheen J (2008) MAPK signaling in plant hormone ethylene signal transduction. Plant Signal Behav 3:848–849. https://doi.org/10.4161/psb.3.10.5995

Yoshimoto N, Kuroda S (2017) High-throughput analysis of mammalian receptor tyrosine kinase activation in yeast cells. In: Jimenez G (ed) ERK signaling, Methods in molecular biology. Humana Press, New York, pp 35–52. https://doi.org/10.1007/978-1-4939-6424-6_3

Yuvaniyama J, Denu JM, Dixon JE, Saper MA (1996) Crystal structure of the dual specificity protein phosphatase VHR. Science 272:1328–1331. https://doi.org/10.1126/science.272.5266.1328

Zhang Y, Li S, Zhou LZ, Fox E, Pao J, Sun W, Zhou C, McCormick S (2011) Overexpression of Arabidopsis *thaliana* PTEN caused accumulation of autophagic bodies in pollen tubes by disrupting phosphatidylinositol 3-phosphate dynamics. Plant J 68:1081–1092. https://doi.org/10.1111/j.1365-313X.2011.04761.x

Zhang Z-Y (2002) Protein tyrosine phosphatases: structure and function, substrate specificity, and inhibitor development. Annu Rev Pharmacol Toxicol 42:209–234. https://doi.org/10.1146/annurev.pharmtox.42.083001.144616

Zhao P, Sokolov LN, Ye J, Tang C-Y, Shi J, Zhen Y, Lan W, Hong Z, Qi J, Lu G-H, Pandey GK, Yang Y-H (2016) The LIKE SEX FOUR2 regulates root development by modulating reactive oxygen species homeostasis in Arabidopsis. Sci Rep 6:28683. https://doi.org/10.1038/srep28683

Zheng L, Huang F, Narsai R, Wu J, Giraud E, He F, Cheng L, Wang F, Wu P, Whelan J, Shou H (2009) Physiological and transcriptome analysis of iron and phosphorus interaction in rice seedlings. Plant Physiol 151:262–274. https://doi.org/10.1104/pp.109.141051

Zhou C, Cai Z, Guo Y, Gan S (2009) An Arabidopsis mitogen-activated protein kinase cascade, MKK9-MPK6, plays a role in leaf senescence. Plant Physiol 150:167–177. https://doi.org/10.1104/pp.108.133439

Zhou J, Wang X, Jiao Y, Qin Y, Liu X, He K, Chen C, Ma L, Wang J, Xiong L, Zhang Q, Fan L, Deng XW (2007) Global genome expression analysis of rice in response to drought and high-salinity stresses in shoot, flag leaf, and panicle. Plant Mol Biol 63:591–608. https://doi.org/10.1007/s11103-006-9111-1

Index

A
ABA-ABA receptor (ABAR), 73
ABA-binding proteins, 85, 91
ABA Hypersensitive Germination1 (AHG1), 209
ABA-Insensitive (ABI), 68
ABA Insensitive 1 (ABI1), 90–92, 129
ABA Insensitive 2 (ABI2), 91, 92
ABA Insensitive 5 (ABI5), 115
ABA receptors (ABARs), 68, 85, 127, 137, 138
ABA-responsive element (ABRE), 69
ABA signaling, 188, 190, 193
  ABA-binding proteins, 85, 91
  ABAR, 87
  ABI1, 90, 92
  ABI2, 92
  AHG3/AtPP2CA, 92, 93
  Arabidopsis, 84, 85
  bZIP, 84
  HAB1, 93
  phytohormone, 84
  in plants, 85
  PP2Cs, 84, 88, 90
  protein kinases, 87, 88
  PYR/PYL/RCAR proteins, 84, 86
  pyrabactin, 86
  RCARs, 91
  SnRK2s, 84
  START-domain proteins, 87
  stress tolerance, 84
ABA signaling module
  abiotic stress, 73
  abiotic stress conditions, 68
  activation loop, 68
  antagonistic correlation, 72
  BIN2 kinase, 73
  CDPKs/CPKs, 72
  clade A, 68
  CPK12, 72
  open ligand-binding pocket, 68
  SnRK-type protein kinases, 72
  TFs, 69
ABA signaling pathways, 110
ABCB transporters, 156
Abiotic stress, 149, 152, 157, 158, 163, 186, 188, 189, 191, 194, 195, 203–205, 208, 211, 217, 218, 220, 221, 224, 343
  AP2C1, 95
  Ca^{2+} signaling, 95
  CIPK23, 95
  plant health, 83
Abiotic stress signaling, 255
  ABA signal transduction pathway, 253
  drought and oxidative stress, 255
  MAPK signaling, 254
  Medicago truncatula, 255
  Nicotiana benthamiana, 254
  OsPP18, 255
  PIN1 apical-basal polarity, 254
  plant's life cycle, 252
  PP2B/calcineurin, 254
  SnRK1, 255
  stomatal opening, 253
ABRE-binding factor (ABF), 69, 84
Abscisic acid (ABA), 58, 76, 358
  abiotic stress, 83
  environmental stimuli, 83
  phytohormone, 83
  signaling pathway (*see* ABA signaling pathway)

G. K. Pandey (ed.), *Protein Phosphatases and Stress Management in Plants*,
https://doi.org/10.1007/978-3-030-48733-1

Abscisic acid (ABA) signaling, 253, 316, 361, 362
Acid phosphatases (APases), 11
 nonspecific, 31
 specific, 31
Actin filaments (AFs), 275
Adenosine diphosphate (ADP), 331
Adenosine monophosphate (AMP), 331
Adenosine triphosphate (ATP), 69, 331
AKT2, 93
AMP-activated kinase (AMPK), 113
Anti-nutrient, 45
Apoptosis signal-regulating kinase 1 (ASK1), 346
Arabidopsis, 84, 294
Arabidopsis K+ transporter 1 (AKT1), 204, 211, 212, 214–216, 219, 223
Arabidopsis PP2C family genes, 169
Arabidopsis Purple Acid Phosphatase 5 (AtPAP5), 42
Arabidopsis thaliana, 20, 87
Arabidopsis thaliana mutants, 138, 139
Arabidopsis thaliana SLP1 (AtSLP1), 4
Arabidopsis thaliana SLP2 (AtSLP2), 5
Arbuscular Mycorrhizal (AM) symbiosis, 43
Aspartate-based phosphatases, 334
Asp-based enzymes, 206
Aspergillus flavus (Afl), 315
Astragalus sinicus, 43
A. thaliana vacuolar H^+-pyrophosphatase 1 (AVP1), 60
AtPP2CA, 93
Auxin, 211
Auxin response factors (ARFs), 211
Auxin-responsive elements (AuxREs), 219
Auxin signaling, 362, 363

B
Bacterial leaf blight (BLB), 296
Bacterial pathogens
 Arabidopsis, 300, 304
 A. thaliana, 303
 AvrRpm1-induced defense responses, 303
 cytoplasmic kinase, 296
 genetic analyses, 301
 gram-negative bacterium, 296
 immune responses, 304
 immune signaling, 301
 molecular, 301
 O. sativa, 299
 P. syringae, 303
 pathogenesis-related genes, 304
 phytopathogenic bacteria, 304
 plant innate immune system, 296
 plant's innate immunity, 300
 Pseudomonas syringae, 300
 resistance genes, 296
Bare catalytic subunits, 7
Basic-domain leucine zipper (bZIP) transcription factors, 84
Bimolecular fluorescence complementation (BiFC), 70, 344
Biotic stress, 95, 344, 345
Blue light, 129, 131–134
BLUe light Signaling 1 (BLUS1), 133
Botrytis cinerea, 305, 344
Botrytis-induced kinase 1 (BIK1), 363
Brassinosteroid, 211
Brassinosteroid (BR) signaling pathway, 73
Brassinosteroid insensitive 1-associated kinase 1 (BAK1), 363
Brassinosteroid insensitive 2 (BIN2) kinase, 73
Brassinosteroid signaling kinase 5 (BSK5), 61
Brassinosteroid signaling, 363
Brassinosteroids (BRs), 61

C
Calcineurin A (CNA), 333
Calcineurin B-like proteins (CBLs), 72, 207, 215, 220, 254
Calcineurin B-like-interacting protein kinase (CIPK), 234
Calcium signaling, 214
Calcium-dependent protein kinases (CDPKs), 72, 87, 355
Calmodulin-dependent protein kinase (CCaMK), 255
Calmodulin-like protein 9 (CML9), 219
Canonical abiotic stress factors, 261
Carbohydrate-binding domain (CBD), 336
Cation/proton antiporter (CPA), 213
CBL interacting protein kinase (CIPK), 95, 156, 207, 208, 214, 216, 254
CBL proteins, 208
Cell division control (CDC), 206
Cellular processes, 295
Cereals and millets, PPase
 expression pattern, 251, 252
 genome-wide identification, 249, 250
Chitin elicitor receptor kinase 1 (CERK1), 307
Chloroplast sensor kinase (CSK), 3
Chloroplast transit peptide (cTP), 4, 5
Chromatin remodeling complex
 ABA signaling pathway, 70, 71
 Arabidopsis HAB1, 69

AT-hook domain, 70
ATP, 69
ATPase activities, 70
BiFC, 70
BRM/SWI3C complex, 71
dephosphorylation, BRM, 70
eukaryotes, 69
knock-out mutant, 70
mass spectrometry, 70
metazoan counterparts, 70
nuclear localization signals, 70
nuclear-localized interaction, 69
phosphorylation, 70
SWI/SNF-related ATP-dependent, 69
SWI3B, 70
ZZ zinc finger domains, 70
Chromodomain-helicase-DNA (CHD), 69
Clade A PP2Cs
ABA signaling pathway, 74
abiotic stress, 74
abiotic stress conditions, 76–77
characteristic, 74
homeodomain TF, 74
Clade B PP2Cs
AP2C1, 75
MPK3 and MPK6, 75
stomatal developmental pathway, 75
Cladosporium fulvum, 306
Crop cultivators, 53
Cyclin-dependent kinases (CDKs), 336
Cys residue, 357
Cysteine, 294
Cytoskeleton regulation
amino acid sequences, 277
ATP binding, 277
binding mechanism, 277
calyculin, 279
chromosome organization, 276
C-terminal region, 277
inhibitors, 278
integrated signaling networks, 275
ligand-protein interaction, 277
Magnoliophyta, 276
Medicago sativa, 276
microtubule functioning, 275
microtubules, 280
molecular docking, 279
MT orientation, 275
MT protein phosphatases, 276
Nicotiana tabacum, 277
NJ-clustering, 278
PDB complexes, 278
physiological processes, 275
post-translational modifications, 275
profile techniques, 279
protein kinases, 275
tautomycin, 279
tubulin code, 276

D
Damage-associated molecular patterns, 290
Dephosphorylation, 111, 216, 236
biological catalysis, 245
cellular processes, 245
tyrosine residue, 245
Developmental Signaling, 93–95
Dual-specificity kinase (DSKs), 354
Dual-specificity phosphatase 8 (DUSP8), 265
Dual specificity phosphatase 9 (DUSP9), 346
Dual-specificity phosphatases (DSP), 131, 165, 180, 189, 206, 246, 262, 353
abiotic stress, 343
amino acid, 333
animals, 345, 346
biotic stress, 344, 345
catalytic domain, 336
catalytic mechanism, 338, 339
cellular processes, 331, 332
cellular proteins, 331, 332
cellular signaling, 336
chemical reactions, 331
dephosphorylation, 331
discovery and structure, 337, 338
functional domain, 332
living organisms, 331
phosphorylation, 331
protein kinases, 337
protein phosphorylation, 331, 333, 335, 336
reactive oxygen species management, 342
starch metabolism, 340–342
tyrosine phosphatases, 337
Dual-specificity protein phosphatase (DSP), 90, 265
Dual-specificity PTPs (DsPTPs), 294

E
EF-Tu receptor (EFR), 313
E growth-regulating 2 (EGR2) phosphate, 274
Endoplasmic reticulum (ER), 18
Environmental stress, 83
Eukaryotes, 1
Eukaryotic protein phosphatase
animals, 22, 23
function, 22
plants

Eukaryotic protein phosphatase (*cont.*)
AtPAP25, 23
patatin, 23
Pi starvation signaling, 24
P-Tyr, 23
signaling pathway, 24
xyloglucan oligosaccharides, 23
reverse protein kinase action, 22
Extracellular matrix (ECM), 12
Extracellular signal-receptor kinase 2 (ERK-2), 345
Extracellular signal-regulated kinase (ERK), 345

F
Fungal pathogens
biotrophic bacterial pathogen, 306
defense-related proteins, 309
defense response, 308, 309
immune signaling, 305
isoflavonoid production, 309
phosphorylations, 305
physical association, 305
plant hosts, 305
plant productivity, 305
plant tissues, 305
programmed cell death, 305
protein phosphatase inhibitors, 308
PSP-type phosphatase, 308
soybean cotyledons, 309
tyrosine residue and forms, 307
Fusarium head blight (FHB), 315

G
Gene model, 166–183
Genes encoding phosphatases, 155
Genevestigator database, 366
Genome/proteome sequencing, 53
Genome-wide identification, 246
Glucose-6-phosphate (G6P), 113
Golovinomyces spp., 308
GORK-encoded channels, 212
G protein-coupled receptor (GPCR), 107
Guard cell Hydrogen peroxide-Resistant1 (GHR1), 135
Guard cells, 125, 126
ABA analogues, 129
ABA-receptor, 127
protein kinase, 135, 136
signal perception/transmission, 127

H
Haloacid dehalogenase (HAD)
cytosolic APase, 24
diverse assortment, 24
OsHAD1, 24
Haloacid dehydrogenase (HAD), 206
Heat shock protein 90 (Hsp90), 267
Heterodera schachtii, 310
High-affinity K+ transporter 5 (HAK5), 204, 211, 212, 214, 215
regulation, 218, 219
High Osmolarity Glycerol (HOG1), 58
Highly ABA-induced (HAI), 68
Highly ABA-Induced 1 (HAI1), 271
Holoenzymes
BSK5, 61
cortical cytoskeleton, 59
MAF1, 61
mevalonate pathway, 59
osmotic imbalance, 59
physiological functions, 59
phytohormone, 60
PP2A-C2 subunit, 60
RCN1p, 60
salinity stress, 59
salt stress, 61
seeds/guard cells, 61
Ser/Thr PP2A enzyme activation, 59
SOS, 60
steroidal hormones, 61
Tap46p expression, 61
TOR, 61
Homology to ABI1 (HAB1), 130
Hormone signaling, 108
Hormone signaling pathways
auxin, 362, 363
brassinosteroid, 363
Human ACP5 (HsACP5), 12
Human ACP7 (HsACP7), 12
Huntingtin-elongation-A subunit-TOR (HEAT), 247
Hydrogen peroxide (H_2O_2), 356
Hypersensitive to ABA (HAB1), 93
Hypersensitive to ABA 1 (HAB1), 235

I
Immune response, 332, 337, 340, 345
Inhibitor 2 (Inh2), 253
Initial syncytial cell (ISC), 310
Innate immune responses, 290
Inositol monophosphatase (IMP), 152
Inositol phosphate (IP), 152

Inositol signaling, 152
Interacting protein kinases (CIPKs), 72
Intermembrane space (IMS), 6

J
Jasmonic acid (JA) pathway, 316

K
Kinase-associated protein phosphatase (KAPP), 94, 151, 304
Kinase interaction motif (KIM), 72, 345

L
Last eukaryotic common ancestor (LECA), 112
Late-embryogenic abundant, 163
Leucine-rich repeat receptor-like kinases (LRR-RLK), 61, 135
Light harvesting complex (LHC), 236
Lipopolysaccharide (LPS), 354
Lys-ketoglutarate reductase (LKR), 114

M
Magnaporthe oryzae (*Mo*), 313
Maize genome encodes, 355
Mammalian signaling pathways, 367
MAP kinase phosphatases (MKPs), 268, 335
MAPK cascade
  activation loop, 71
  Clade B PP2Cs, 71
  divergent stimuli, 71
  KIM, 72
  Medicago sativa, 71
  MKKK18 functions, 72
  phosphorylation, 71
  regulatory functions, 73
  stress signals, 71
  T-glutamic acid (Glu/E)-Y phosphorylation motif, 72
  upstream signaling, 71
MAPK kinase kinases (MAPKKKs), 300
MAPK PHOSPHATASE 1 (MKP1), 272, 273, 316, 344
MAPK PHOSPHATASE 2 (MKP2), 273
MAPK signaling, 360
Mass spectrometry (MS), 20
Massively parallel signature sequence (MPSS), 251
Mechanosensitive (MS) channels, 58
Medicago sativa protein phosphatases 2C (MP2Cs), 71
Meloidogyne incognita, 311
Metal-dependent protein phosphatases (PPMs), 54, 165, 334
Methylation-sensitive amplified polymorphism (MSAP), 156
MFP1 associated factor 1 (MAF1), 61
Mia40, redox regulator
  Erv1, 6, 7
  IMS, 6, 7
  phosphatase activity, 7
  TOM, 6
Microtubule-associated stress protein 1 (MASP1), 270
Microtubules (MTs), 275
Mid1-complementing activity 1 (MCA1), 59
Mitogen-activated protein kinase (MAPK) cascade, 67, 73
Mitogen-activated protein kinase (MAPK) signaling, 254
Mitogen-activated protein kinase phosphatase (MAPKP), 335
Mitogen-activated protein kinases (MAPKs), 93, 95, 154, 209, 210, 268, 345
MscS-like (MSL), 59
Myotubularins (MTMs), 336

N
N. benthamiana, 256
Na+/H+ exchanger (NHX), 217
Nematode-associated molecular patterns (NAMPs), 309
Nicotiana benthamiana, 306
Nitrate reductase (NR), 237, 238
Nitrate transporters, 234
Nitrogen use efficiency (NUE), 233, 234, 236, 238–240
Non-alcoholic fatty liver disease (NAFLD), 346
Nonreceptor PTPs (NRPTPs), 335
N response, 234, 238–240
N-terminal amino acids, 3
Nucleotide-binding leucine-rich receptor (NB-LRR), 303

O
Oryza sativa, 355
Osmotic stress tolerance, 58
Osteosclerotic bone dysplasia/Raine syndrome, 20
Oxidative stress tolerance, 76

P

PAMP-induced genes, 210
Pathogen-associated molecular pattern (PAMP), 344, 365
Pattern recognition receptors (PRRs), 290
Phosphatases, 88, 150, 234, 236–238
Phosphatases of regenerating liver (PRLs), 335
Phosphatidic acid (PA), 134
Phosphoenolpyruvate carboxylase (PEPC), 240
Phospho-histidine (His) signaling, 353
Phosphoprotein phosphatases (PPPs), 2, 54, 129, 131, 204, 333
Phospho-pyruvate dehydrogenase phosphatase (PDP), 3
Phosphoregulatory mechanism, 204
Phosphorus use efficiency (PUE), 45
Phosphorylation, 1, 103, 234
Phosphorylations, 149
Phosphotyrosine phosphatases (PTPs), 129, 131, 150, 294
Phototropin2 (PHOT2), 106
Phytopathogenic microorganisms, 297, 298
Phytophthora infestans, 316
Pi-inducible phosphoenolpyruvate carboxylase (PEPC), 34
Pin-formed (PIN) proteins, 130
Pi-starvation-inducible (PSI), 14
Plant adaptation, 77
Plant homeo-domain (PHD), 69
Plant host-pathogen interactions
  abiotic and biotic agents, 289
  abscisic acid (ABA) signaling, 316
  amino acid(s), 291
  autotrophic green plants, 289
  cell death, 291
  cellular integrity, 291
  cellular responses, 291
  chemical inhibition, 316
  cytoplasmic domains, 290
  downstream signaling events, 291
  extracellular domain, 290
  factors, 289
  host biochemical pathways, 316
  immune regulatory network, 315
  immune signaling, 292
  immune system, 290
  inherent ability, 289
  innate immune responses, 290
  inner cytoplasmic domain, 290
  kinase domain, 290
  kinase-activated immune triggers, 291
  microbial pathogens, 289
  pathogenic microbes, 292
  phosphorylation events, 290
  PKs, 291
  plant growth, 291
  protein kinases, 292
  substrate recognition, 292
  virulence factors, 315
Plant immune signaling, 307
Plant PAPs
  abiotic/biotic stress, 42
  AM symbiosis, 43
  characteristic features, 45
  diverse physiological processes, 46
  effective breeding, 46
  eutrophication, 45
  functional details, 36–41
  humans/monogastric animals, 45
  multigene family, 30
  nitrogen fixation/nodule formation, 43
  nonspecific phosphatases, 31, 32
  organic P ulilization, 34, 35
  phosphate deficiency, 42
  phylogenetic classification, 31
  Pi deficiency, 45
  posttranscriptional/posttranslational modifications, 45
  PUE, 45
  regulation, 32–34
  regulation of flowering, 43, 44
  RSA, 42
  seed and pollen germination, 35
  seed development, 44
  senescing flag leaves, 45
  signal transduction pathways, 45
Plant protein phosphatases
  apicomplexan-induced inhibition, 267
  Arabidopsis thaliana, 263, 264
  Aspergillus fumigatus, 268
  BcPtpA and BcPtpB, 268
  biotic, 266–267
  biotic stress factor, 265
  Botrytis cinerea, 267
  calmodulin and calcineurin, 268
  Candida albicans, 268
  cellular protection/transcription factors, 262
  eukaryotic PP2Cs, 267
  heterogeneity, 262
  Leishmania, 265
  Msn2p localization, 268
  multicomponent system, 262
  phylogram, 264
  promoter level, 267

regulatory subunits, 263
signaling pathways, 262
signal perception, 262
stress factors, 263
Trypanosoma brucei, 267
Plant stress
biotic and abiotic, 261
cellular responses, 261
gene expression, 261
mammalian protein phosphatases, 262
PPP family, 262
protein phosphorylation state, 261
sequence homology, 262
serine and threonine residues (*see* Serine and threonine residues, dephosphorylation)
tyrosine phosphatases, 262
tyrosine residues (*see* Tyrosine residues, dephosphorylation)
Populus trichocarpa, 355
Posttranslational modification (PTM), 245
animals
classical secretion pathway, 18
ECM proteome, 17–18
Fam20C, 18
glycoprotein osteopontin, 20
Golgi network, 18, 19
phosphoproteomic screens, 17
secreted protein kinases, 18
VLK, 18
plants
AtGAL1, 21
ECM proteomics, 20
LC-MS/MS, 21
phosphorylation, 20
Pro-Q diamond phosphoprotein, 20
signal transduction cascade, 21
TMV, 21
Potassium deficiency, 191, 192, 211
Potassium homeostasis, 204, 216, 217, 220
Potassium uptake, 203
Potato virus X (PVX), 306
Pottassium transporters
non-voltage-gated class, 213
PP1 regulatory subunit2-like protein1 (PRSL1), 253
PP2As
CRISPR/Cas9, 62
omics approach, 62
phosphoprotemic approach, 62
serine/threonine phosphatases, 62
PP2B, 89
PP2C, 234, 238
PP2CA/AHG3, 92
PP2Cs
plant genomes code, 77
potential targets, 77
protein phosphatases, 77
PPP phosphatase
catalytic subunits, 246
regulatory subunits, 246
Propyzamide Hypersensitive 1 (PHS1), 90
Protein kinases (PKAs), 67, 87, 88, 90, 151, 234, 280, 290
γ-phosphate groups, 245
PPases, 245
proteome phosphorylation, 245
Protein Kinases With Unknown Function (PKUs), 3
Protein Phosphatase 1 (PP1), 184
dephosphorylation reaction, 247
holoenzyme, 247
regulatory subunits, 247
Ser/Thr phosphatase, 247
Protein Phosphatase 2A (PP2A), 104, 105, 112–113, 185, 186, 269
holoenzyme, 247
proteome fraction, 247
regulatory subunit, 247
stress signaling pathways, 247
stomata movements, 132
Protein phosphatase 2B (PP2B), 104, 186, 248
Protein phosphatase 2Cs (PP2C), 104, 106, 129, 193
vs. ABA receptors, 86
ABA signaling (*see* ABA signaling)
abiotic stress signaling, 95
biotic stress, 95
classification, 89, 90
developmental signaling, 93–95
evolution, 89, 90
phosphatases, 88
phosphorylation process, 88
protein kinases, 88
SnRK2-type kinases, 130
stomatal closure, 130
Protein phosphatase interaction (PPI), 157
Protein Phosphatase M (PPM), 184, 187
Protein phosphatases (PPs), 67, 129, 165, 184, 185, 189–195, 204–207, 222, 291
biotic stress factor, 280
catalytic site, 137
cellular signaling, 129
classification, 165
crop plants, 189
forms, 129

Protein phosphatases (PPs) (*cont.*)
functional analysis, 280
PYL proteins, 137
regulating subunits, 280
signal transduction, 208
stomatal opening, 132
structure, 205
types, 129
Protein phosphatases 1 (PP1), 104
Protein phosphatases in plant
agricultural pathogen, 309
ascaroside-induced immune triggers, 310
bacterial and fungal pathogens, 309
cellular signal transduction processes, 293
cellulase treatment, 310
downstream signaling pathways, 310
functional attributes, 295, 296
genome sequencing, 293
microbial pathogens, 309
nematode-induced resistance responses, 310
plant immune mechanisms, 309
PSP, 293, 294
PTPs, 294
regulatory and functional domains, 293
rhizospheric protozoan species, 311
Protein phosphatases *vs.* sugar signaling, 108
Protein phosphorylation, 87, 331, 333, 335, 336
chloroplasts, 1
eukaryotes, 2
mitochondria and chloroplasts, 2, 3, 7
MS-based substrate discovery, 7
phospho-proteomic data, 3
PTM (*see* Posttranslational modification (PTM))
quantitative mass spectometry, 7
regulatory subunits, 2
signaling pathway, 353
Protein Ser/Thr Phosphatase (PSP), 293, 294
Protein serine/threonine phosphatase 5 (PP5), 187
Protein tyrosine kinases (PTKs), 189, 332, 353
Protein Tyrosine Phosphatase 1 (PTP1), 301
Protein tyrosine phosphatases (PTPs), 54, 165, 188, 189, 332, 333, 353
abiotic stress, 366, 367
abiotic stress tolerance, 364, 365
biotic responses, 365, 366
catalytic domains, 356
Cys residue, 357
developmental pathways, 363, 364
drug development, 367
eukaryotic, 356
family, 166
glucan phosphatases, 368
hormone signaling (*see* Hormone signaling pathways)
hydrogen bonds, 357
hydrolytic mechanism, 356
intracellular redox state, 357, 358
kinase activity, 368
LMWPTPs, 356
MAPK signaling, 359–361
oxidation, 357
proximal amino acid, 357
PTP1B, 356
quantitative profiling, 368
redox signaling, 358–359 (*see also* Redox signaling)
variable inactivation, 368
whole-genome sequencing, 357
Pseudomonas syringae, 300
Purple acid phosphatases (PAPs)
abiotic and biotic stress, 25
adenylate-dependent reactions, 14, 16
Arabidopsis thaliana, 12, 13
charge-transfer transition, 12
Golgi and ECM phosphoproteins, 25
HsACP, 12
HsACP7, 12
isozymes, 25
mammalian, 29
metallohydrolases, 11
metallophosphoesterase family, 29
model of soil, 14, 15
Pi deficiency, 29
Pi starvation, 25
pink/purple color, 29
Pi-starvation response, 14
plants PAPs (*see* Plant PAPs)
proteomic and MS techniques, 25
PSI, 14
roles, 29
signal transduction cascades, 25
structure/classification, 30
tartrate-resistant, 12
PYLs and Kinases, 137
PYR/PYL/RCAR family proteins, 134
Pyrabactin, 86
Pyrabactin Resistance/Pyrabactin Resistance Like/Regulatory Component (PYR/PYL/RCAR), 84
Pyrabactin resistant-like/regulatory component of abscisic acid receptors (PYL/RCAR), 270
Pyruvate dehydrogenase complex (PDC), 3
Pyruvate dehydrogenase kinase (PDK), 3

Q

Quantitative mass spectrometry, 8
Quantitative real-time PCR (qRT-PCR), 251
Quick anion channel 1/aluminum-activated malate transporter 12 (QUAC1/ALMT12), 69

R

Ralstonia solanacearum, 344
Rapid alkalinization factor (RALF), 157
Reactive oxygen species (ROS), 76, 127, 157, 272, 301, 344
Receptor-like cytoplasmic kinase (RLCK), 301
Receptor-like enzymes (RPTPs), 335
Receptor-like kinases (RLKs), 57, 209, 290, 363
Receptor-like proteins (RLPs), 209, 290
Redox signaling
  Arabidopsis, 359
  LSF2, 359
  maize cells, 358
  MPK8, 359
  post-translational level, 358
  reductants/oxidants, 358
  SEX4 function, 359
  ZmRIP1/ZmPP1, 358
Reductant-inhibited PTPase1 (ZmRIP1), 358
Regulation of flowering
  direct mechanism, 44
  histochemical GUS assays, 44
  reproductive tissues, 43
Regulatory component of ABA response 1 (RCAR1), 86
Regulatory Components of ABA Receptor (RCARs), 68, 188
Reversible phosphorylation, 45, 103
Rhizobiales/Rhodobacterales/Rhodospirillaceae-like phosphatases (RLPHs), 54
Rhizoctonia cerealis, 307
Ribosomal protein, 112
Rice genome encodes, 355
RNA sequencing (RNA-seq), 251
Root system architecture (RSA), 256
  β-glucan synthesis, 42
  modulation, 42
  NtPAP12, 42
  Pi deficiency, 43

S

Saccharopine dehydrogenase (SDH), 114
Salicylic acid (SA), 73
Salinity stress sensor kinases
  cell wall connections, 57
  hormone abscisic acid, 59
  MAPKs, 58
  osmosensing pathways, 58
  RLKs, 57
  salt, 58
Salt overly sensitive (SOS), 60
Salt stress, 53, 185, 190–192
Ser/Thr phosphatase (STP), 333, 356
  PP1, 247
  PP2A, 247
  PP2B/calcineurin, 248
  PP2C, 248
Ser/Thr protein phosphatases
  ABA signaling, 222
  salt stress, 221
  stress conditions, 221
Serine and threonine residues, dephosphorylation
  ABA signaling, 270
  defense activation, 269
  HAB genes, 270
  HAI1, 271
  HDA14, 269
  II dephosphorylates α-tubulin, 270
  MAPK-mediated downstream defense pathway, 269
  MASP1, 270
  metabolic pathways and signaling components, 269
  phosphoproteomic analysis, 270
  physiological functions, 269
  PP2A, 269
  PP2A, 269
  PYL/RCAR, 270
Serine/threonine (Ser/Thr), 262
Serine/threonine (Ser/Thr) phosphatase, 245
Serine-threonine kinases (SnRKs), 109, 110, 135
Serine-threonine phosphatases
  okadaic acid, 54
  Oryza sativa genome, 56
  PPP family, 54
  protein kinases, 54
  protein phosphorylation, 54
  regulatory subunits, 55
  RLPHs, 54
Shewanella-like protein phosphatase 1 (SLP1)
  cTP, 4
  eukaryotic supergroups, 4
  sequence alignment, 4
  serine/threonine phosphatase, 3, 5

Shewanella-like protein phosphatase 2 (SLP2)
Arabidopsis thaliana, 4
serine/threonine phosphatase, 3
serine/threonine protein phosphatase, 5, 6
Signal transduction pathway, 54
Signaling molecule, sugar, 107
Slow anion channel-associated 1 (SLAC1), 69
Slowly Activating Anion Channel 1 (SLAC1), 216
SNF1/AMPK modules, 113
SNF1-related protein kinases 2 (SnRK2s), 84, 85, 87, 88, 156
Snf2-ATP coupling (SnAC), 70
SOS pathway, 157
Starch excess protein 4 (SEX4), 341
Starch metabolism, 340–342
Stomatal physiology, 223
Stress-activated protein kinase (SAPK), 265
Stress-responsive *NAC* transcription factor, 256
Stress signal, 53, 153
biochemical phosphorylation/dephosphorylation, 53
Stress tolerance, 164, 190, 194, 195
Suc nonfermenting-related kinase group 2 (SnRK2s), 209
Sucrose non-fermenting (SNF), 68
Sucrose non-fermenting-related kinase, 104, 113
Sugar hypersensitivity, 116
Sugar signaling, 108, 109, 111, 114, 116
Superoxide dismutase (SOD), 154
Suppressor of actin (SAC) phosphatases, 152
Synthetic lethal of N-end rule 1(SLN1), 58

T

Tandem affinity purification (TAP), 5
Target of rapamycin (TOR), 61, 104, 105, 107, 191, 193, 236, 267
Target-of-rapamycin-like signaling pathway (TOR), 191
Tetratricopeptide repeat (TPR), 187
Thylakoid-associated kinases (TAKs), 3
Tobacco mosaic virus (TMV), 21
Tobacco rattle virus (TRV)-based vectors, 311
TOR kinases, 112
TOR-AMPK module, 111
Transcription factors (TFs), 69
Transgenic plants
abiotic stress tolerance, 257
genetic improvement, 256, 257
stress-resistant crops, 257
Translocase of outer membrane (TOM), 6
Transmembrane, 335
Trehalose-6-phosphate (T6P), 113
Tryptophan (Trp), 68
Two-pore K^+ (TPK), 59
Type 2C protein phosphatases (PP2Cs), 73
abiotic stresses, 67
abscisic acid (ABA)-dependent signaling pathway, 67
MAPK cascade, 67
Mg^{2+}-/Mn^{2+}-dependent characteristics, 67
plant stress signaling pathways (*see* ABA signaling module)
Type III secretion system (TTSS), 305
Tyrosine phosphorylation
AtDsPTP1 protein sequence, 355
CDC25s, 355
cell cycle regulation, 354
cell cycle regulators, 355
cellular regulation, 354
Cys-based PTPs, 354
DsPTPs, 353
hormonal signaling pathways, 355
intra- and intercellular communication, 354
Laforin, 355
LPG, 354
mammalian systems, 354
MAPK kinases, 354
pathological conditions, 354
PTKs, 353
PTP classification, 356
PTPs, 353
SH2/PTP domains, 353
signal-to-noise ratio, 353
TK catalytic domains, 354
Tyrosine residues, dephosphorylation
ABA signal transduction pathway, 272
activation loops, 273
auxin responses, 273
biotic and abiotic stress, 273
cellular processes, 271
diverse environmental stimuli, 272
EGR2, 274
MAPK signaling cascades, #14, 273
microtubule depolymerisation and stress response, 271
MKP1, 273
mkp2 mutant plants, 272
myristoylation, 274
osmotic stress, 274
oxidative stress tolerance, 272
PHS1, 274
phs1-3 mutation, 272
proteomic approaches, 271
protoplast-based assays, 272
PTK gene expression, 271

PTP signaling, 274
root hair treatment, 271
salinity, 272
salt stress, 274
UV-C, 272
water stress, 272
Tyrosine-specific PTPs, 152
Tyr-specific phosphatases
DSP, 246
stress regulation, 246

U
Ubiquitin/26s proteasome system (UPS), 74
UDP-glucose pyrophosphorylase, 115
Ultraviolet (UV), 272

V
Vaccinia HI-related phosphatase (VHR), 337
Vertebrate lonesome kinase (VLK), 18
Virulence
animal pathogenic microbes, 312
bacterial PPs, 312–314
cellular processes, 312
fungal pathogen, 313, 315
microbial pathogens, 312
plant protein phosphatases, 312
protein phosphatases, 312

W
Wheat (*Triticum aestivum*), 307
WITH NO LYSINE KINASEs (WNKs), 108

X
Xanthomonas gardneri (*Xg*), 312

Y
Yeast 1-hybrid (Y1H), 84

The manufacturer's authorised representative in the EU is Springer Nature Customer Service Centre GmbH, Europaplatz 3, 69115 Heidelberg, Germany. If you have any concerns regarding our products, please contact ProductSafety@springernature.com

Printed and bound by CPI Group (UK) Ltd, Croydon, CR0 4YY
30/09/2026
02225503-0014